PHYSICAL CONSTANTS

Quantity	Symbol	Value
Universal gravitational constant	G	6.674×10^{-11} m³/(kg·s²)
Speed of light in vacuum	c	2.998×10^{8} m/s
Elementary charge	e	1.602×10^{-19} C
Planck's constant	h	6.626×10^{-34} J·s
		4.136×10^{-15} eV·s
	$\hbar = h/(2\pi)$	1.055×10^{-34} J·s
		6.582×10^{-16} eV·s
Universal gas constant	R	8.314 J/(mol·K)
Avogadro's number	N_{A}	6.022×10^{23} mol⁻¹
Boltzmann constant	k_{B}	1.381×10^{-23} J/K
		8.617×10^{-5} eV/K
Coulomb force constant	$k = 1/(4\pi\epsilon_0)$	8.988×10^{9} N·m²/C²
Permittivity of free space (electric constant)	ϵ_0	8.854×10^{-12} C²/(N·m²)
Permeability of free space (magnetic constant)	μ_0	$4\pi \times 10^{-7}$ T·m/A
Electron mass	m_{e}	9.109×10^{-31} kg
		0.000 548 580 u
Electron rest energy	$m_{\mathrm{e}}c^2$	0.5110 MeV
Proton mass	m_{p}	1.673×10^{-27} kg
		1.007 276 5 u
Proton rest energy	$m_{\mathrm{p}}c^2$	938.272 MeV
Neutron mass	m_{n}	1.675×10^{-27} kg
		1.008 664 9 u
Neutron rest energy	$m_{\mathrm{n}}c^2$	939.565 MeV
Compton wavelength of electron	λ_{C}	2.426×10^{-12} m
Stefan-Boltzmann constant	σ	5.670×10^{-8} W/(m²·K⁴)
Rydberg constant	R	1.097×10^{7} m⁻¹
Bohr radius of hydrogen atom	a_0	5.292×10^{-11} m
Ionization energy of hydrogen atom	$-E_1$	13.61 eV

Physics

Alan Giambattista

Cornell University

Betty McCarthy Richardson

Cornell University

Robert C. Richardson

Cornell University

Boston   Burr Ridge, IL   Dubuque, IA   New York   San Francisco   St. Louis
Bangkok   Bogotá   Caracas   Kuala Lumpur   Lisbon   London   Madrid   Mexico City
Milan   Montreal   New Delhi   Santiago   Seoul   Singapore   Sydney   Taipei   Toronto

 **Higher Education**

PHYSICS, SECOND EDITION

Published by McGraw-Hill, a business unit of The McGraw-Hill Companies, Inc., 1221 Avenue of the Americas, New York, NY 10020. Copyright © 2010 by The McGraw-Hill Companies, Inc. All rights reserved. Previous edition © 2008. No part of this publication may be reproduced or distributed in any form or by any means, or stored in a database or retrieval system, without the prior written consent of The McGraw-Hill Companies, Inc., including, but not limited to, in any network or other electronic storage or transmission, or broadcast for distance learning.

Some ancillaries, including electronic and print components, may not be available to customers outside the United States.

This book is printed on acid-free paper.

5 6 7 8 9 0 VNH/VNH 15 14 13 12

ISBN 978–0–07–733866–4
MHID 0–07–733866–9

Publisher: *Thomas D. Timp*
Sponsoring Editor: *Debra B. Hash*
Vice-President New Product Launches: *Michael Lange*
Senior Developmental Editor: *Mary E. Hurley*
Senior Marketing Manager: *Lisa Nicks*
Senior Project Manager: *Gloria G. Schiesl*
Senior Production Supervisor: *Laura Fuller*
Senior Media Project Manager: *Tammy Juran*
Senior Designer: *David W. Hash*
Cover/Interior Designer: *Rokusek Design*
(USE) Cover Image: *©Kazuya Shiota/Aflo/Jupiterimages*
Lead Photo Research Coordinator: *Carrie K. Burger*
Photo Research: *Danny Meldung/Photo Affairs, Inc*
Supplement Producer: *Mary Jane Lampe*
Compositor: *Laserwords Private Limited*
Typeface: *10/12 Times*
Printer: *R. R. Donnelley, Jefferson City, MO*

The credits section for this book begins on page C-1 and is considered an extension of the copyright page.

MCAT® is a registered trademark of the Association of American Medical Colleges. MCAT exam material included is printed with permission of the AAMC. The AAMC does not endorse this book.

Library of Congress Cataloging-in-Publication Data

Giambattista, Alan.
    Physics / Alan Giambattista, Betty McCarthy Richardson, Robert C. Richardson.—2nd ed.
        p. cm.
    Includes index.
    ISBN 978–0–07–340453–0 — ISBN 0–07–340453–5 (hard copy : alk. paper)
    1. Physics–Textbooks. I. Richardson, Betty McCarthy. II. Richardson, Robert C. III. Title.
QC21.3.G537 2010
530–dc22
                                                                                        2008034667

www.mhhe.com

About the Authors

Alan Giambattista grew up in Nutley, New Jersey. In his junior year at Brigham Young University he decided to pursue a physics major, after having explored math, music, and psychology. He did his graduate studies at Cornell University and has taught introductory college physics for over 20 years. When not found at the computer keyboard working on *Physics*, he can often be found at the keyboard of a harpsichord or piano. He is a member of the Cayuga Chamber Orchestra and has given performances of the Bach harpsichord concerti at several regional Bach festivals. He met his wife Marion in a singing group. They live in an 1824 parsonage built for an abolitionist minister, which is now surrounded by an organic dairy farm. Besides making music and taking care of the house, gardens, and fruit trees, they love to travel together.

Betty McCarthy Richardson was born and grew up in Marblehead, Massachusetts, and tried to avoid taking any science classes after eighth grade but managed to avoid only ninth grade science. After discovering that physics tells how things work, she decided to become a physicist. She attended Wellesley College and did graduate work at Duke University. While at Duke, Betty met and married fellow graduate student Bob Richardson and had two daughters, Jennifer and Pamela. Betty began teaching physics at Cornell in 1977. Many years later, she is still teaching the same course, Physics 101/102, an algebra-based course with all teaching done one-on-one in a Learning Center. From her own early experience of math and science avoidance, Betty has empathy with students who are apprehensive about learning physics. Betty's hobbies include collecting old children's books, reading, enjoying music, travel, and dining with royalty. A highlight for Betty during the Nobel Prize festivities in 1996 was being escorted to dinner on the arm of King Carl XVI Gustav of Sweden. Currently she is spending spare time enjoying grandsons Jasper (the 1-m child in Chapter 1), Dashiell and Oliver (the twins of Chapter 12), and Quintin, the newest arrival.

Robert C. Richardson was born in Washington, D.C., attended Virginia Polytechnic Institute, spent time in the United States Army, and then returned to graduate school in physics at Duke University where his thesis work involved NMR studies of solid helium-3. In the fall of 1966 Bob began work at Cornell University in the laboratory of David M. Lee. Their research goal was to observe the nuclear magnetic phase transition in solid ^{3}He that could be predicted from Richardson's thesis work with Professor Horst Meyer at Duke. In collaboration with graduate student Douglas D. Osheroff, they worked on cooling techniques and NMR instrumentation for studying low-temperature helium liquids and solids. In the fall of 1971, they made the accidental discovery that liquid ^{3}He undergoes a pairing transition similar to that of superconductors. The three were awarded the Nobel Prize for that work in 1996. Bob is currently the F. R. Newman Professor of Physics and the Senior Science Advisor at Cornell. In his spare time he enjoys gardening and photography.

In loving memory of Dad and of my niece, Natalie

Alan

*In memory of our daughter Pamela,
and for Quintin, Oliver, Dashiell, Jasper,
Jennifer, and Jim Merlis*

Bob and Betty

Brief Contents

Contents

PART THREE

Electromagnetism

PART FOUR

Electromagnetic Waves and Optics

List of Selected Applications

Note: Within the Problems, (#) = Chapter Number; CQ = Conceptual Question; MC = Multiple-Choice Question; P = Problem; R&S = Review & Synthesis.

Preface

Physics is intended for a two-semester college course in introductory physics using algebra and trigonometry. Our main goals in writing this book are

- to present the basic concepts of physics that students need to know for later courses and future careers,
- to emphasize that physics is a tool for understanding the real world, and
- to teach transferable problem-solving skills that students can use throughout their lives.

We have kept these goals in mind while developing the main themes of the book.

NEW TO THIS EDITION

Although the fundamental philosophy of the book has not changed, detailed feedback from almost 60 reviewers (many of whom used the first edition in the classroom) has enabled us to fine-tune our approach to make the text even more user-friendly, conceptually based, and relevant for students. The second edition also has some added features to further facilitate student learning.

A greater emphasis has been placed on fundamental physics concepts:

- **Connections identify areas in each chapter where important concepts are revisited.** A marginal Connections heading and summary adjacent to the coverage in the main text help students easily recognize that a previously introduced concept is being applied to the current discussion. Knowledge is being revisited and further developed—not newly introduced.
- **Checkpoint** questions have been added to applicable sections of the text to **allow students to pause and test their understanding of the concept** explored within the current section. The answers to the Checkpoints are found at the end of the chapter so that students can confirm their knowledge without jumping too quickly to the provided answer.
- The exercises in the **Review & Synthesis** sections have been revised to concentrate even more heavily on **helping students to realize through practice problems how the concepts in the previously covered group of chapters are interrelated**. The number of problems in the Review & Synthesis sections has also been increased in the new edition. (The MCAT review problems have been retained to also help pre-med students focus on the concepts covered in the upcoming exam.)
- **Nonessential coverage and derivations have been moved to the text's website.** This will help students not only to focus further on the fundamental, core concepts in their reading of the text but also allow them to go online for additional information or explanation on topics of interest. identifiers in the text direct students to additional information online.

In addition, the following general revisions occur in chapters of the text:

- The topical question from the chapter-opening vignette now appears in the margin (along with a reduced version of the chapter-opening image) to help students identify where in the main text the answer to the chapter-opening question is addressed.
- Applications have been clearly identified as such in the text with a complete listing in the front matter.
- Many helpful subheadings have been added to the text to help students quickly identify new subtopics.
- Portions of the text now caption images to establish a visual connection between the text's concepts and terms and the art and photos.

"G/R/R is as good as it gets as far as a college textbook in physics goes. One of the coauthors of this book has been teaching a course at this level for 30 years. This book is a direct result of her 30 years' worth of personal experience, and there is no better substitute for that. It is, without any doubt, one of the best of its kind."

Dr. Abu Fasihuddin, University of Connecticut

- Great care was taken by both the authors and the contributors to the second edition to revise the end-of-chapter and Review & Synthesis problems. Approximately 150 problems are new, and an emphasis has been placed on progressing difficulty level to help students gain confidence and reinforce new skills before tackling more challenging problems.

The following lists major chapter-specific revisions to the text:

Chapter 2:   Vector notation has been removed from Chapter 2. Discussion of vectors and components of vectors now begins in Chapter 3.

Chapter 3:   A discussion of Unit Vectors has been added to Section 3.2. A new example for finding average velocity has been added.

Chapter 4:   A more concise section on air resistance is provided with a more detailed discussion available online. A new Figure 4.20 emphasizes the normal and frictional forces as perpendicular components of a contact force.

Chapter 7:   Section 7.6 Motion of the Center of Mass has been simplified.

Chapter 8:   Example 8.1 has been replaced with a new problem on the rotational inertia of a barbell.

Chapter 10:   Section 10.8 The Pendulum has been made much more concise with a more detailed discussion of the physical pendulum available online.

Chapter 11:   A new "law box" highlights the physical properties that determine wave speed. The discussion on interference has been expanded for added clarity.

Chapter 12:   In Section 12.9, the discussion of shock waves has been shortened. A more detailed discussion is available online.

Chapter 14:   A detailed discussion of convection and Example 14.12 Roller Blading in Still Air have been moved online. Section 14.7 is now a brief, conceptual description of convection. Section 14.8 Thermal Radiation has been revised with a clearer description of solar radiation and global warming.

Chapter 15:   Section 15.5 Heat Engines has been revised to include a more accurate description of the development of the steam engine. The process of the internal combustion engine is now illustrated in Figure 15.12. Details of the Carnot cycle and discussion of the statistical interpretation of entropy are available online.

Chapter 16:   A new Example 16.7 Electric Field due to Three Point Charges has been added.

Chapter 22:   Section 22.1 has been simplified and is now titled Maxwell's Equations and Electromagnetic Waves. A more detailed discussion appears online. The material on antennas has been made more concise.

Chapter 27:   The derivation of the radii of the Bohr orbits has been moved online. The section on atomic energy levels has been revised and made more concise.

Chapter 28:   Section 28.8 Electron Energy Levels in a Solid has been made much more concise with a more detailed discussion available online.

Chapter 30:   The discussions of quarks and leptons have been expanded and clarified. The discussion of the standard model is significantly more concise. Twenty-first-century particle physics has been updated, and the most recent information will be provided online.

Please see your McGraw-Hill sales representative for a more detailed list of revisions.

ORGANIZATION OF CHAPTERS 2 THROUGH 4

In spite of the more traditional organization, Chapters 2–4 retain much of the flavor of the approach in *College Physics*. In particular, we use correct vector notation, diagrams, terminology, and methods from the very beginning. For example, we carefully distinguish components from magnitudes by writing "$v_x = -5$ m/s" and never "$v = -5$ m/s," even if the object moves only along the x-axis.

COMPREHENSIVE COVERAGE

Students should be able to get the whole story from the book. The text works well in our self-paced course, where students must rely on the textbook as their primary learning resource. Nonetheless, completeness and clarity are equally advantageous when the book is used in a more traditional classroom setting. *Physics* frees the instructor from having to try to "cover" everything. The instructor can then tailor class time to more important student needs—reinforcing difficult concepts, working through example problems, engaging the students in cooperative learning activities, describing applications, or presenting demonstrations.

INTEGRATING CONCEPTUAL PHYSICS INTO A QUANTITATIVE COURSE

Some students approach introductory physics with the idea that physics is just the memorization of a long list of equations and the ability to plug numbers into those equations. We want to help students see that a relatively small number of basic physics concepts are applied to a wide variety of situations. Physics education research has shown that students do not automatically acquire conceptual understanding; the concepts must be explained and the students given a chance to grapple with them. Our presentation, based on years of teaching this course, blends conceptual understanding with analytical skills. The **Conceptual Examples** and **Conceptual Practice Problems** in the text and a variety of Conceptual and Multiple-Choice Questions at the end of each chapter give students a chance to check and to enhance their conceptual understanding.

"Conceptual ideas are important, ideas must be motivated, physics should be integrated, a coherent problem-solving approach should be developed. I'm not sure other books are as explicit in these goals, or achieve them as well as Giambattista, Richardson, and Richardson."

Dr. Michael G. Strauss, University of Oklahoma

INTRODUCING CONCEPTS INTUITIVELY

We introduce key concepts and quantities in an informal way by establishing why the quantity is needed, why it is useful, and why it needs a precise definition. Then we make a transition from the informal, intuitive idea to a formal definition and name. Concepts motivated in this way are easier for students to grasp and remember than are concepts introduced by seemingly arbitrary, formal definitions.

For example, in Chapter 8, the idea of rotational inertia emerges in a natural way from the concept of rotational kinetic energy. Students can understand that a rotating rigid body has kinetic energy due to the motion of its particles. We discuss why it is useful to be able to write this kinetic energy in terms of a single quantity common to all the particles (the angular speed), rather than as a sum involving particles with many different speeds. When students understand why rotational inertia is defined the way it is, they are better prepared to move on to the concepts of torque and angular momentum.

We avoid presenting definitions or formulas without any motivation. When an equation is not derived in the text, we at least describe where the equation comes from or give a plausibility argument. For example, Section 9.9 introduces Poiseuille's law with two identical pipes in series to show why the volume flow rate must be proportional to the pressure drop per unit length. Then we discuss why $\Delta V/\Delta t$ is proportional to the fourth power of the radius (rather than to r^2, as it would be for an ideal fluid).

"The authors are clearly very able to communicate in written English. The text is well written, not concise to the point of density, but not discursive to the point of long-windedness. A real pleasure to read."

Dr. Galen T. Pickett, California State University, Long Beach

WRITTEN IN CLEAR AND FRIENDLY STYLE

We have kept the writing down-to-earth and conversational in tone—the kind of language an experienced teacher uses when sitting at a table working one-on-one with a student. We hope students will find the book pleasant to read, informative, and accurate without seeming threatening, and filled with analogies that make abstract concepts easier to grasp. We want students to feel confident that they can learn by studying the textbook.

While learning correct physics terminology is essential, we avoid all *unnecessary* jargon—terminology that just gets in the way of the student's understanding. For example, we never use the term *centripetal force,* since its use sometimes leads students to add a spurious "centripetal force" to their free-body diagrams. Likewise, we use *radial component of acceleration* because it is less likely to introduce or reinforce misconceptions than *centripetal acceleration.*

ACCURACY ASSURANCE

The authors and the publisher acknowledge the fact that inaccuracies can be a source of frustration for both the instructor and students. Therefore, throughout the writing and production of this edition, we have worked diligently to eliminate errors and inaccuracies. Bill Fellers of Fellers Math & Science conducted an independent accuracy check and worked all end-of-chapter questions and problems in the final draft of the manuscript. He then coordinated the resolution of discrepancies between accuracy checks, ensuring the accuracy of the text, the end-of-book answers, and the solutions manuals. Corrections were then made to the manuscript before it was typeset.

The page proofs of the text were double-proofread against the manuscript to ensure the correction of any errors introduced when the manuscript was typeset. The textual examples, practice problems and solutions, end-of-chapter questions and problems, and problem answers were accuracy checked by Fellers Math & Science again at the page proof stage after the manuscript was typeset. This last round of corrections was then cross-checked against the solutions manuals.

PROVIDING STUDENTS WITH THE TOOLS THEY NEED

Problem-Solving Approach

Problem-solving skills are central to an introductory physics course. We illustrate these skills in the example problems. Lists of problem-solving strategies are sometimes useful; we provide such strategies when appropriate. However, the most elusive skills—perhaps the most important ones—are subtle points that defy being put into a neat list. To develop real problem-solving expertise, students must learn how to think critically and analytically. Problem solving is a multidimensional, complex process; an algorithmic approach is not adequate to instill real problem-solving skills.

Strategy We begin each example with a discussion—in language that the students can understand—of the *strategy* to be used in solving the problem. The strategy illustrates the kind of analytical thinking students must do when attacking a problem: How do I decide what approach to use? What laws of physics apply to the problem and which of them are *useful* in this solution? What clues are given in the statement of the question? What information is implied rather than stated outright? If there are several valid approaches, how do I determine which is the most efficient? What assumptions can I make? What kind of sketch or graph might help me solve the problem? Is a simplification or approximation called for? If so, how can I tell if the simplification is valid? Can I make a preliminary estimate of the answer? Only after considering these questions can the student effectively solve the problem.

Solution Next comes the detailed *solution* to the problem. Explanations are intermingled with equations and step-by-step calculations to help the student understand the approach used to solve the problem. We want the student to be able to follow the mathematics without wondering, "Where did that come from?"

Discussion The numerical or algebraic answer is not the end of the problem; our examples end with a *discussion*. Students must learn how to determine whether their answer is consistent and reasonable by checking the order of magnitude of the answer,

"The major strength of this text is its approach, which makes students think out the problems, rather than always relying on a formula to get an answer. The way the authors encourage students to investigate whether the answer makes sense, and compare the magnitude of the answer with common sense is good also."

Dr. Jose D'Arruda,
University of North Carolina,
Pembroke

comparing the answer to a preliminary estimate, verifying the units, and doing an independent calculation when more than one approach is feasible. When there are several different approaches, the discussion looks at the advantages and disadvantages of each approach. We also discuss the implications of the answer—what can we learn from it? We look at special cases and look at "what if" scenarios. The discussion sometimes generalizes the problem-solving techniques used in the solution.

Practice Problem After each Example, a Practice Problem gives students a chance to gain experience using the same physics principles and problem-solving tools. By comparing their answers to those provided at the end of each chapter, they can gauge their understanding and decide whether to move on to the next section.

Our many years of experience in teaching the college physics course in a one-on-one setting has enabled us to anticipate where we can expect students to have difficulty. In addition to the consistent problem-solving approach, we offer several other means of assistance to the student throughout the text. A boxed problem-solving strategy gives detailed information on solving a particular type of problem, while an icon ⓘ for problem-solving tips draws attention to techniques that can be used in a variety of contexts. A hint in a worked example or end-of-chapter problem provides a clue on what approach to use or what simplification to make. A warning icon ⚠ emphasizes an explanation that clarifies a possible point of confusion or a common student misconception.

An important problem-solving skill that many students lack is the ability to extract information from a graph or to sketch a graph without plotting individual data points. Graphs often help students visualize physical relationships more clearly than they can do with algebra alone. We emphasize the use of graphs and sketches in the text, in worked examples, and in the problems.

Review & Synthesis with MCAT Review®

Eight **Review & Synthesis** sections appear throughout the text, following groups of related chapters. The *MCAT® Review* includes actual reading passages and questions written for the **Medical College Admission Test (MCAT)**. The *Review Exercises* are intended to serve as a bridge between textbook problems that are linked to a particular chapter and exam problems that are not. These exercises give students practice in formulating a problem-solving strategy without an external clue (section or chapter number) that indicates which concepts are involved. Many of the problems draw on material from more than one chapter to help the student integrate new concepts and skills with what has been learned previously.

Using Approximation, Estimation, and Proportional Reasoning

Physics is forthright about the constant use of simplified models and approximations in solving physics problems. One of the most difficult aspects of problem solving that students need to learn is that some kind of simplified model or approximation is usually required. We discuss how to know when it is reasonable to ignore friction, treat g as constant, ignore viscosity, treat a charged object as a point charge, or ignore diffraction.

Some Examples and Problems require the student to make an estimate—a useful skill both in physics problem solving and in many other fields. Similarly, we teach proportional reasoning as not only an elegant shortcut but also as a means to understanding patterns. We frequently use percentages and ratios to give students practice in using and understanding them.

Showcasing an Innovative Art Program

To help show that physics is more than a collection of principles that explain a set of contrived problems, in every chapter we have developed a system of illustration's, ranging from simpler diagrams to ellaborate and beautiful illustrations, that brings to life the

"I understood the math, mostly because it was worked out step-by-step, which I like."

Student, Bradley University

"The warning signs about many of the misconceptions, traps, and common mistakes is a very helpful and novel idea. Those of us who have taught undergraduate students in service courses have spent considerable time on these. It is good to see them in a book."

Dr. H.R. Chandrasekhar, University of Missouri, Columbia

"I have tried a number of texts in this course over the past 30 years that I have taught Physics 116–117, and I can assure you that G/R/R is the one I (and the students . . .) like the best. The explanations are clear, and the graphics are excellent—the best I have seen anywhere. And the structure of the question and problem sets is very good. G/R/R is the best standard algebra-based text I have ever seen."

Dr. Carey E. Stronach, Virginia State University

connections between physics concepts and the complex ways in which they are applied. We believe these illustrations, with subjects ranging from three-dimensional views of electric field lines to the biomechanics of the human body and from representations of waves to the distribution of electricity in the home, will help students see the power and beauty of physics.

Helping Students See the Relevance of Physics in Their Lives

Students in an introductory college physics course have a wide range of backgrounds and interests. We stimulate interest in physics by relating the principles to applications relevant to students' lives and in line with their interests. The text, examples, and end-of-chapter problems draw from the everyday world; from familiar technological applications; and from other fields such as biology, medicine, archaeology, astronomy, sports, environmental science, and geophysics. (Applications in the text are identified with a text heading or marginal note. An icon ( ) identifies applications in the biological or medical sciences.)

The **Physics at Home** experiments give students an opportunity to explore and see physics principles operate in their everyday lives. These activities are chosen for their simplicity and for the effective demonstration of physics principles.

Each **Chapter Opener** includes a photo and vignette, designed to capture student interest and maintain it throughout the chapter. The vignette describes the situation shown in the photo and asks the student to consider the relevant physics. A reduced version of the chapter opener photo and question marks where the topic from the vignette is addressed within the chapter.

Focusing on the Concepts

To focus on the basic, core concepts of physics and reinforce for students that all of physics is based on a few, fundamental ideas, within chapters we have developed **Connections** to identify areas where important concepts are revisited. A marginal Connections heading and summary adjacent to the coverage in the main text help students easily recognize that a previously introduced concept is being applied to the current discussion. Knowledge is being built-up—not newly introduced.

The exercises in the **Review & Synthesis sections** have been revised to increase the number of available exercises and to also concentrate even more heavily on helping students to realize through practice problems how the concepts in the previously covered group of chapters are interrelated.

Checkpoint questions have been added to applicable sections of the text to allow students to pause and test their understanding of the concept explored within the current section. The answers to the Checkpoints are found at the end of the chapter so that students can confirm their knowledge without jumping too quickly to the provided answer.

Applications are clearly identified as such in the text with a complete listing in the front matter. With Applications, students have the opportunity to see how physics concepts are experienced through their everyday lives.

 icons identify opportunities for students to access additional information or explanation of topics of interest online. This will help students to focus even further on just the very fundamental, core concepts in their reading of the text.

ADDITIONAL RESOURCES FOR INSTRUCTORS AND STUDENTS

Online Homework and Resources

McGraw-Hill's *Physics* website offers online electronic homework along with a myriad of resources for both instructors and students. Instructors can create homework with easy-to-assign algorithmically generated problems from the text and the simplicity of automatic grading and reporting:

- The end-of-chapter problems and Review & Synthesis exercises appear in the online homework system in diverse formats and with various tools.
- The online homework system incorporates new and exciting interactive tools and problem types: ranking problems, a graphing tool, a free-body diagram drawing tool, symbolic entry, a math palette, and multi-part problems.

Instructors also have access to PowerPoint lecture outlines, an Instructor's Resource Guide with solutions, suggested demonstrations, electronic images from the text, clicker questions, quizzes, tutorials, interactive simulations, and many other resources directly tied to text-specific materials in *Physics*. Students have access to self-quizzing, interactive simulations, tutorials, selected solutions for the text's problems, and more.

See www.mhhe.com/grr to learn more and to register.

Electronic Media Integrated with the Text

McGraw-Hill is proud to bring you an assortment of outstanding interactives and tutorials like no other. These activities offer a fresh and dynamic method to teach the physics basics by providing students with activities that work with real data. W icons identify areas in the text where additional understanding can be gained through work with an interactive or tutorial on the text website.

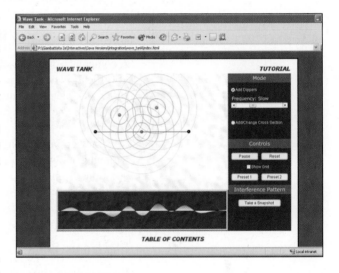

The interactives allow students to manipulate parameters and gain a better understanding of the more difficult physics concepts by watching the effect of these manipulations. Each interactive includes:

- Analysis tool (interactive model)
- Tutorial describing its function
- Content describing its principle themes

The text website contains accompanying interactive quizzes. An instructor's guide for each interactive with a complete overview of the content and navigational tools, a quick demonstration description, further study with the textbook, and suggested end-of-chapter follow-up questions is also provided as an online instructor's resource.

The tutorials, developed and integrated by Raphael Littauer of Cornell University, provide the opportunity for students to approach a concept in steps. Detailed feedback is provided when students enter an incorrect response, which encourages students to further evaluate their responses and helps them progress through the problem.

Electronic Book Images and Assets for Instructors

Build instructional materials wherever, whenever, and however you want!

Accessed from the *Physics* website, an online digital library containing photos, artwork, interactives, and other media types can be used to create customized lectures, visually enhanced tests and quizzes, compelling course websites, or attractive printed support materials. Assets are copyrighted by McGraw-Hill Higher Education, but can be used by instructors for classroom purposes. The visual resources in this collection include

- **Art** Full-color digital files of all illustrations in the book can be readily incorporated into lecture presentations, exams, or custom-made classroom materials. In addition, all files are preinserted into PowerPoint slides for ease of lecture preparation.

- **Active Art Library** These key art pieces—formatted as PowerPoint slides—allow you to illustrate difficult concepts in a step-by-step manner. The artwork is broken into small, incremental pieces, so you can incorporate the illustrations into your lecture in whatever sequence or format you desire.
- **Photos** The photos collection contains digital files of photographs from the text, which can be reproduced for multiple classroom uses.
- **Worked Example Library, Table Library, and Numbered Equations Library** Access the worked examples, tables, and equations from the text in electronic format for inclusion in your classroom resources.
- **Interactives** Flash files of the physics interactives described earlier are included so that you can easily make use of the interactives in a lecture or classroom setting.

Also residing on your textbook's website are

- **PowerPoint Lecture Outlines** Ready-made presentations that combine art and lecture notes are provided for each chapter of the text.
- **PowerPoint Slides** For instructors who prefer to create their lectures from scratch, all illustrations and photos are preinserted by chapter into blank PowerPoint slides.

Computerized Test Bank Online

A comprehensive bank of over 2000 test questions in multiple-choice format at a variety of difficulty levels is provided within a computerized test bank powered by McGraw-Hill's flexible electronic testing program—EZ Test Online (www.eztestonline.com). EZ Test Online allows you to create paper and online tests or quizzes in this easy-to-use program!

Imagine being able to create and access your test or quiz anywhere, at any time without installing the testing software. Now, with EZ Test Online, instructors can select questions from multiple McGraw-Hill test banks or create their own, and then either print the test for paper distribution or give it online. See www.mhhe.com/grr for more information.

Electronic Books

If you or your students are ready for an alternative version of the traditional textbook, McGraw-Hill brings you innovative and inexpensive electronic textbooks. By purchasing E-books from McGraw-Hill, students can save as much as 50% on selected titles delivered on the most advanced E-book platforms available.

E-books from McGraw-Hill are smart, interactive, searchable, and portable, with such powerful built-in tools as detailed searching, highlighting, note taking, and student-to-student or instructor-to-student note sharing. E-books from McGraw-Hill will help students to study smarter and quickly find the information they need. E-books also saves students money. Contact your McGraw-Hill sales representative to discuss E-book packaging options.

Personal Response Systems

Personal response systems, or "clickers," bring interactivity into the classroom or lecture hall. Wireless response systems give the instructor and students immediate feedback from the entire class. The wireless response pads are essentially remotes that are easy to use and engage students, allowing instructors to motivate student preparation, interactivity, and active learning. Instructors receive immediate feedback to gauge which concepts students understand. Questions covering the content of the *Physics* text (formatted in PowerPoint) are available on the website for *Physics*.

Instructor's Resource Guide

The *Instructor's Resource Guide* includes many unique assets for instructors, such as demonstrations, suggested reform ideas from physics education research, and ideas for incorporating just-in-time teaching techniques. It also includes answers to the end-of-chapter conceptual questions and complete, worked-out solutions for all the end-of-chapter problems from the text. The Instructors Resource Guide is available in the Instructor Resources on the text's website.

ALEKS®

Help students master the math skills needed to understand difficult physics problems. ALEKS® [Assessment and LEarning in Knowledge Spaces] is an artificial intelligence–based system for individualized math learning available via the World Wide Web.

ALEKS® is

- A robust course management system. It tells you exactly what your students know and don't know.
- Focused and efficient. It enables students to quickly master the math needed for college physics.
- Artificial intelligence. It totally individualizes assessment and learning.
- Customizable. Click on or off each course topic.
- Web based. Use a standard browser for easy Internet access.
- Inexpensive. There are no setup fees or site license fees.

ALEKS® is a registered trademark of ALEKS Corporation.

Student Solutions Manual

The *Student Solutions Manual* contains complete worked-out solutions to selected end-of-chapter problems and questions, selected Review & Synthesis problems, and the MCAT Review Exercises from the text. The solutions in this manual follow the problem-solving strategy outlined in the text's examples and also guide students in creating diagrams for their own solutions.

For more information, contact a McGraw-Hill customer service representative at (800) 338–3987, or by email at www.mhhe.com. To locate your sales representative, go to www.mhhe.com for Find My Sales Rep.

To the Student

HOW TO SUCCEED IN YOUR PHYSICS CLASS

It's true—how much you get out of your studies depends on how much you put in. Success in a physics class requires:

- Commitment of time and perseverance
- Knowing and motivating yourself
- Getting organized
- Managing your time

This section will help you learn how to be effective in these areas, as well as offer guidance in:

- Getting the most out of your lecture
- Finding extra help when you need it
- Getting the most out of your textbook
- How to study for an exam

Commitment of Time and Perseverance

Learning and mastering takes time and patience. Nothing worthwhile comes easily. Be committed to your studies and you will reap the benefits in the long run. A regular, sustained effort is much more effective than sporadic bouts of cramming.

Knowing and Motivating Yourself

What kind of learner are you? When are you most productive? Know yourself and your limits, and work within them. Know how to motivate yourself to give your all to your studies and achieve your goals.

There are many types of learners, and no right or wrong way of learning. Which category do you fall into?

- **Visual learner** You respond best to "seeing" processes and information. Focus on text illustrations and graphs. Use course handouts and the animations on the course and text websites to help you. Draw diagrams in your notes to illustrate concepts.
- **Auditory learner** You work best by listening to—and possibly recording—the lecture and by talking information through with a study partner.
- **Tactile/Kinesthetic Learner** You learn best by being "hands on." You'll benefit by applying what you've learned during lab time. Writing and drawing are physical activities, so don't neglect taking notes on your reading and the lecture to explain the content in your own words. Try pacing while you read the text. Stand up and write on a chalkboard during discussions in your study group.

Identify your own personal preferences for learning and seek out the resources that will best help you with your studies. Also remember, even though you have a preferred style of learning, most learners benefit when they engage in all styles of learning.

Getting Organized

It's simple, yet it's fundamental. It seems the more organized you are, the easier things come. Take the time before your course begins to analyze your life and your study habits. Get organized now and you'll find you have a little more time—and a lot less stress.

- **Find a calendar system that works for you.** The best kind is one that you can take with you everywhere. To be truly organized, you should integrate all aspects of your life into this one calendar—school, work, and leisure. Some people also find it helpful to have an additional monthly calendar posted by their desk for "at a

A good rule of thumb is to allow 2 hours of study time for every hour you spend in lecture. For instance, a 3-hour lecture deserves 6 hours of study time per week. If you commit to studying for this course daily, you're investing a little less than one hour per day, including the weekend.

Begin each of the tasks assigned in your course with the goal of understanding the material. Simply completing the assignment does not mean that learning has taken place. Your fellow students, your instructor, and this textbook can all be important resources in broadening your knowledge.

glance" dates and to have a visual planner. If you do this, be sure you are consistently synchronizing both calendars so as not to miss anything. *More tips for organizing your calendar can be found in the time management discussion below.*

- By the same token, **keep everything for your course or courses in one place**—and at your fingertips. A three-ring binder works well because it allows you to add or organize handouts and notes from class in any order you prefer. Incorporating your own custom tabs helps you flip to exactly what you need at a moment's notice.
- **Find your space.** Find a place that helps you be organized and focused. If it's your desk in your dorm room or in your home, keep it clean. Clutter adds confusion and stress and wastes time. Perhaps your "space" is at the library. If that's the case, keep a backpack or bag that's fully stocked with what you might need—your text, binder or notes, pens, highlighters, Post-its, phone numbers of study partners. [*Hint:* A good place to keep phone numbers is in your "one place for everything calendar."]

Managing Your Time

Managing your time is the single most important thing you can do to help yourself, but it's probably one of the most difficult tasks to successfully master.

In college, you are expected to work much harder and to learn much more than you ever have before. To be successful you need to invest in your education with a commitment of time. We all lead busy lives, but we all make choices as to how we spend our time. Choose wisely.

- **Know yourself and when you'll be able to study most efficiently.** When are you most productive? Are you a night owl? Or an early bird? Plan to study when you are most alert and can have uninterrupted segments. This could include a quick 5-minute review before class or a one-hour problem-solving study session with a friend.
- **Create a set daily study time for yourself.** Having a set schedule helps you commit to studying and helps you plan instead of cram. Find—and use—a planner that is small enough that you can take it with you everywhere. This may be a simple paper calendar or an electronic version. They all work on the same premise: **organize *all* of your activities in one place**.
- **Schedule study time using shorter, focused blocks with small breaks.** Doing this offers two benefits: (1) You will be less fatigued and gain more from your effort and (2) Studying will seem less overwhelming, and you will be less likely to procrastinate.
- **Plan time for leisure, friends, exercise, and sleep.** Studying should be your main focus, but you need to balance your time—and your life.
- **Log your homework deadlines and exam dates** in your personal calendar.
- Try to **complete tasks ahead of schedule**. This will give you a chance to carefully review your work before it is due. You'll feel less stressed in the end.
- **Know where help can be found.** At the beginning of the semester, find your instructor's office hours, your lab partner's contact information, and the "Help Desk" or Learning Resource Center if your course offers one. Make use of all of the support systems that your college or university has to offer. Ask questions both in class and during your instructor's office hours. Don't be shy—your instructor is there to help you learn.
- **Prioritize!** In your calendar or planner, highlight or number key projects; do them first, and then cross them off when you've completed them. Give yourself a pat on the back for getting them done!
- **Review your calendar and reprioritize** *daily.*
- **Resist distractions by setting and sticking to a designated study time.**
- **Multitask when possible.** You may find a lot of extra time you didn't think you had. Review material in your head or think about how to tackle a tough problem while walking to class or doing laundry.

Add extra "padding" into your personal deadlines. If you have a report due on Friday, set a goal for yourself to have it done on Wednesday. Then, take time on Thursday to look over your project with a fresh eye. Make any corrections or enhancements and have it ready to turn in on Friday.

Plan to study and plan for leisure. Being well balanced will help you focus when it is time to study.

Try combining social time with studying in a group, or social time with mealtime or exercise. Being a good student doesn't mean you have to be a hermit. It does mean you need to know how to smartly budget your time.

Getting the Most Out of Lectures

Your instructors want you to succeed. They put a lot of effort into preparing their lectures and other materials designed to help you learn. Attending class is one of the simplest, most valuable things you can do to help yourself. But it doesn't end there—getting the most out of your lectures means being organized. Here's how:

Prepare Before You Go to Class Study the text on the lecture topic *before* attending class. Familiarizing yourself with the material gives you the ability to take notes selectively rather than scrambling to write everything down. You'll be able to absorb more of the subtleties and difficult points from the lecture. You may also develop some good questions to ask your instructor.

Don't feel overwhelmed by this task. Spend time the night before class gaining a general overview of the topics for the next lecture using your syllabus. If your schedule does not allow this, plan to arrive at class 5–15 minutes before lecture. Bring your text with you and skim the chapter before lecture begins.

Don't try to read an entire chapter in one sitting; study one or two sections at a time. It's difficult to maintain your concentration in a long session with so many new concepts and skills to learn.

Be a Good Listener Most people think they are good listeners, but few really are. Are you?

Important points to remember:

- You can't listen if you are talking.
- You aren't listening if you are daydreaming or constantly distracted by other concerns.
- Listening and comprehending are two different things. Listen carefully in class. The language of science is precise; be sure you understand your instructor. If you don't understand something your instructor is saying, ask a question or jot a note and visit the instructor during office hours. You are likely doing others a favor when you ask questions because there are probably others in the class who have the same questions.

Take Good Notes

- Use a standard size notebook, or better yet, a three-ring binder with loose leaf notepaper. The binder will allow you to organize and integrate your notes and handouts, integrate easy-to-reference tabs, and the like.
- Color-code your notes. Use one color of ink pen to take your initial notes. You can annotate later using a pencil, which can be erased if need be.
- Start a new page with each lecture or note-taking session.
- Label each page with the date and a heading for each day.

- Focus on main points and try to use an outline format to take notes to capture key ideas and organize sub-points.
- Take your text to lecture, and keep it open to the topics being discussed. You can also take brief notes in your textbook margin or reference textbook pages in your notebook to help you study later.
- Review and edit your notes shortly after class—within 24 hours—to make sure they make sense and that you've recorded core thoughts. You may also want to compare your notes with a study partner later to make sure neither of you have missed anything.
- This is a very IMPORTANT point: *You can and should also add notes from your reading of the textbook.*

Get a Study Partner Find a few study partners and get together regularly. Four or five study partners to a group is a good number. Too many students make the group unwieldy, but you want enough students to ensure the group can meet even if one or two people can't make it. Having study partners has many benefits. First, they can help you keep your commitment to this class. By having set study dates, you can combine study and social time, and maybe even make it fun! In addition, you now have several minds to help digest the information from the lecture and the text:

- Talk through concepts and go over the difficulties you may be having. Take turns explaining things to each other. You learn a tremendous amount when you teach someone else.
- Compare your notes and solutions with the Practice Problems.
- Try a new approach to a problem or look at the problem from the perspective of your partner. There are often many ways to do the same problem. You can benefit from the insights of others—and they from you—but resist the temptation to simply copy solutions. You need to learn how to solve the problem yourself.
- Quiz each other and discuss some of the Conceptual Questions from the end of the chapter.
- Don't take advantage of your study partner by skipping class or skipping study dates. You obviously won't have a study partner—or a friend—much longer if it's not a mutually beneficial arrangement!

Getting the Most Out of Your Textbook

We hope that you enjoy your physics course using this text. While studying physics does require hard work, we have tried to remove the obstacles that sometimes make introductory physics unnecessarily difficult. We have also tried to reveal the beauty inherent in the principles of physics and how these principles are manifest all around you.

In our years of teaching experience, we have found that studying physics is a skill that must be learned. It's much more effective to *study* a physics textbook, which involves active participation on your part, than to read through passively. Even though active study takes more time initially, in the long run it will save you time; you learn more in one active study session than in three or four superficial readings.

As you study, take particular note of the following elements:

Consider the **chapter opener**. It will help you make the connection between the physics you are about to study and how it affects the world around you. Each chapter opener includes a photo and vignette designed to pique your interest in the chapter. The vignette describes the situation shown in the photo and asks you to consider the relevant physics. The question is then answered within the chapter. Look for the reduced opener photo and question on the referenced page.

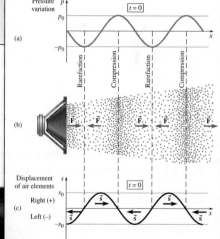

CHAPTER

24

Optical Instruments

The Hubble Space Telescope, orbiting Earth at an altitude of about 600 km, was launched in 1990 by the crew of the Space Shuttle *Discovery*. What is the advantage of having a telescope in space when there are telescopes on Earth with larger light-gathering capabilities? What justifies the cost of $2 billion to place this 12.5-ton instrument into orbit? (See p. 910 for the answer.)

Evaluate the **Concepts & Skills to Review** on the first page of each chapter. It lists important material from previous chapters that you should understand before you start reading. If you have problems recalling any of the concepts, you can revisit the sections referenced in the list.

Concepts & Skills to Review

- distinction between real and virtual images (Section 23.6)
- magnification (Section 23.8)
- refraction (Section 23.3)
- thin lenses (Section 23.9)
- finding images with ray diagrams (Section 23.6)
- small-angle approximations (Appendix A.7)

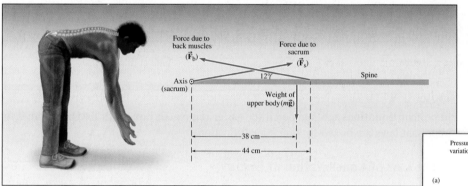

Study the figures and graphs carefully. **Some elaborate illustrations** and more straightforward **diagrammatic illustrations** are used in combination throughout the text to help you grasp concepts. Complex illustrations help you visualize the most difficult concepts. When looking at graphs, try to see the wealth of information displayed. Ask yourself about the physical meaning of the slope, the area under the curve, the overall shape of the graph, the vertical and horizontal intercepts, and any maxima and minima.

Marginal **Connections** headings and summaries adjacent to the coverage in the main text identify areas where important concepts are revisited. Consider the notes carefully to help you recognize how a previously introduced concept is being applied to the current discussion.

Checkpoint questions appear in applicable sections of the text to allow you to test your understanding of the concept explored within the current section. The answers to the Checkpoints are found at the end of the chapter so that you can confirm your knowledge without jumping too quickly to the provided answer.

✓ CHECKPOINT 8.2

You are trying to loosen a nut, without success. Why might it help to switch to a wrench with a longer handle?

 icons identify opportunities for you to access additional information or explanation of topics of interest online.

Various **Reinforcement Notes** appear in the margin to emphasize the important points in the text.

$$\vec{v} = \lim_{\Delta t \to 0} \frac{\Delta \vec{r}}{\Delta t} \qquad (3\text{-}12)$$

($\Delta\vec{r}$ is the displacement during a *very short* time interval Δt)

If an object moves along a curved path, the direction of the velocity vector at any point is tangent to the path at that point.

Important **Equations** are numbered for easier reference. Equations that correspond to important laws are boxed for quick identification.

Statements of important physics **Rules and Laws** are boxed to highlight the most important and central concepts.

The Law of Conservation of Energy

The total energy in the universe is unchanged by any physical process:

total energy before = total energy after.

Problem-Solving Strategy for Newton's Second Law

- Decide what objects will have Newton's second law applied to them.
- Identify all the *external* forces acting on that object.
- Draw an FBD to show all the forces acting on the object.
- Choose a coordinate system. If the direction of the net force is known, choose axes so that the net force (and the acceleration) are along one of the axes.
- Find the net force by adding the forces as vectors.
- Use Newton's second law to relate the net force to the acceleration.
- Relate the acceleration to the change in the velocity vector during a time interval of interest.

Boxed **Problem-Solving Strategies** give detailed information on solving a particular type of problem. These are supplied for the most fundamental physical rules and laws.

⚠ A **warning note** describes possible points of confusion or any common misconceptions that may apply to a particular concept.

tor; the length of the arrow is proportional to the magnitude of the vector. By contrast, a **scalar** quantity can have magnitude, algebraic sign, and units, but not a direction in space. It wouldn't make sense to draw an arrow to represent a scalar such as mass!

In this book, an arrow over a boldface symbol indicates a vector quantity ($\vec{r}$). (Some books use boldface without the arrow or the arrow without boldface.) When writing by hand, always draw an arrow over a vector symbol to distinguish it from a scalar. When the symbol for a vector is written without the arrow and in italics rather than boldface (r), it stands for the *magnitude* of the vector (which is a scalar). Absolute value bars are also used to stand for the magnitude of a vector, so $r = |\vec{r}|$. The magnitude of a vector may have units and is never negative; it can be positive or zero.

 A **problem-solving tip** will guide you in applying problem-solving techniques.

When scalars are added or subtracted, they do so in the usual way: 3 kg of water plus 2 kg of water is equal to 5 kg of water. Adding or subtracting vectors is different. Vectors follow rules of addition and subtraction that take into account the *directions* of the vectors as well as their magnitudes. Whenever you need to add or subtract quantities, check whether they are vectors. If so, be sure to add or subtract them correctly *as vectors. Do not just add or subtract their magnitudes.*

Example 6.4

Bungee Jumping

A bungee jumper makes a jump in the Gorge du Verdon in southern France. The jumping platform is 182 m above the bottom of the gorge. The jumper weighs 780 N. If the jumper falls to within 68 m of the bottom of the gorge, how much work is done by the bungee cord on the jumper during his descent? Ignore air resistance.

Strategy Ignoring air resistance, only two forces act on the jumper during the descent: gravity and the tension in the cord. Since the jumper has zero kinetic energy at both the highest and lowest points of the jump, the change in kinetic energy for the descent is zero. Therefore, the total work done by the two forces on the jumper must equal zero.

Solution Let W_g and W_c represent the work done on the jumper by gravity and by the cord. Then

$$W_{total} = W_g + W_c = \Delta K = 0$$

The work done by gravity is

$$W_g = F_y \, \Delta y = -mg \, \Delta y$$

where the weight of the jumper is $mg = 780$ N. With $y = 0$ at the bottom of the gorge, the vertical component of the displacement is

$$\Delta y = y_f - y_i = 68 \text{ m} - 182 \text{ m} = -114 \text{ m}$$

Then the work done by gravity is

$$W_g = -(780 \text{ N}) \times (-114 \text{ m}) = +89 \text{ kJ}$$

The work done by the cord is $W_c = W_{total} - W_g = -89$ kJ.

Discussion The work done by gravity is positive, since the force and the displacement are in the same direction (downward). If not for the negative work done by the cord, the jumper would have a kinetic energy of 89 kJ after falling 114 m.

   The length of the bungee cord is not given, but it does not affect the answer. At first the jumper is in free fall as the cord plays out to its full length; only then does the cord begin to stretch and exert a force on the jumper, ultimately bringing him to rest again. Regardless of the length of the cord, the total work done by gravity and by the cord must be zero since the change in the jumper's kinetic energy is zero.

Practice Problem 6.4   **The Bungee Jumper's Speed**

Suppose that during the jumper's descent, at a height of 111 m above the bottom of the gorge, the cord has done −21.7 kJ of work on the jumper. What is the jumper's speed at that point?

√ **CHECKPOINT 6.3**

Kinetic energy and work are related. Can kinetic energy ever be negative? Can work ever be negative?

6.4   GRAVITATIONAL POTENTIAL ENERGY (1)

Gravitational Potential Energy When Gravitational Force Is Constant

Toss a stone up with initial speed v_i. Ignoring air resistance, how high does the stone go? We can solve this problem with Newton's second law, but let's use work and energy instead. The stone's initial kinetic energy is $K_i = \frac{1}{2}mv_i^2$. For an upward displacement Δy, gravity does negative work $W_{grav} = -mg \, \Delta y$. No other forces act, so this is the total work done on the stone. The stone is momentarily at rest at the top, so $K_f = 0$. Then

---

When you come to an **Example**, pause after you've read the problem. Think about the strategy you would use to solve the problem. See if you can work through the problem on your own. Now study the *Strategy*, *Solution*, and *Discussion* in the textbook. Sometimes you will find that your own solution is right on the mark; if not, you can focus your attention on the areas of misunderstanding or any mistakes you may have made.

   Work the *Practice Problem* after each Example to practice applying the physics concepts and problem-solving skills you've just learned. Check your answer with the one given at the end of the chapter. If your answer isn't correct, review the previous section in the textbook to try to find your mistake.

Application headings identify places in the text where physics can be applied to other areas of your life. Familiar topics and interests are discussed in the accompanying text, including examples from biology, archaeology, astronomy, sports, and the everyday world. The biology/life science examples have a special icon. ⟶

Banked Curves   To help prevent cars from going into a skid or losing control, the roadway is often banked (tilted at a slight angle) around curves so that the outer portion of the road—the part farthest from the center of curvature—is higher than the inner portion. Banking changes the angle and magnitude of the normal force, $\vec{N}$, so that it has a horizontal component N_x directed toward the center of curvature (in the

Application of radial acceleration and contact forces: banked roadways

 Application of the manometer: measuring blood pressure

Try the *Physics at Home* experiments in your dorm room or at home. They reinforce key physics concepts and help you see how these concepts operate in the world around you.

PHYSICS AT HOME

Drop a very tiny speck of dust or lint into a container of water and push the speck below the surface. The motion of the speck—called *Brownian motion*—is easily observed as it is pushed and bumped about randomly by collisions with water molecules. The water molecules themselves move about randomly, but at much higher speeds than the speck of dust due to their much smaller mass.

5.1 Description of Uniform Circular Motion

1. A carnival swing is fixed on the end of an 8.0-m-long beam. If the swing and beam sweep through an angle of 120°, what is the distance through which the riders move?

2. A soccer ball of diameter 31 cm rolls without slipping at a linear speed of 2.8 m/s. Through how many revolutions has the soccer ball turned as it moves a linear distance of 18 m?

3. Find the average angular speed of the second hand of a

Problems

©	Combination conceptual/quantitative problem
▼	Biological or medical application
✦	Challenging problem
Blue #	Detailed solution in the Student Solutions Manual
① ②	Problems paired by concept
🌐	Text website interactive or tutorial

5.1 Description of Uniform Circular Motion

1. A carnival swing is fixed on the end of an 8.0-m-l beam. If the swing and beam sweep through an angl 120 mov

✦114. A student's head is bent over her physics book. The head weighs 50.0 N and is supported by the muscle force $\vec{\mathbf{F}}_m$ exerted by the neck extensor muscles and by the contact force $\vec{\mathbf{F}}_c$ exerted at the atlantooccipital joint. Given that the magnitude of $\vec{\mathbf{F}}_m$ is 60.0 N and is directed 35° below the horizontal, find (a) the magnitude and (b) the direction of $\vec{\mathbf{F}}_c$.

Write your *own* chapter summary or outline, adding notes from class where appropriate, and then compare it with the **Master the Concepts** provided at the end of the chapter. This will help you identify the most important and fundamental concepts in each chapter.

Along with working the problems assigned by your instructor, try quizzing yourself on the **Multiple-Choice Questions.** Check your answers against the answers at the end of the book. Consider the **Conceptual Questions** to check your qualitative understanding of the key ideas from the chapter. Try writing some responses to practice your writing skills and to help prepare for any essay problems on the exam.

When working the **Problems** and **Comprehensive Problems** assigned by your instructor, pay special attention to the explanatory paragraph below the Problem heading and the keys accompanying each problem.

- *Paired Problems* are connected with a bracket. Your instructor may assign the even-numbered problem, which has no answer at the end of the book. However, working the connected odd-numbered problem will allow you to check your answer at the back of the book and apply what you have learned to working the even-numbered problem.
- Problem numbers highlighted in blue have a solution available in the *Student Solutions Manual* if you need additional help or would like to double-check your work.
- The *difficulty level* for each problem is indicated. The least difficult problems and problems of intermediate difficulty have no diamond. The more challenging problems have one diamond ✦.

Read through all of the assigned problems and budget your time accordingly.
- © indicates a combination **Conceptual and Quantitative** problem.
- ▼ indicates a problem with a biological or medical application.
- 🌐 indicates a problem that has an accompanying interactive or tutorial online.

While working your solutions to problems, try to **keep your work in symbolic form** until the very end. Symbolic solutions will allow you to view which factors affect the results and how the answer would change should any one of the variables in the problem change their value. In this fashion, your solution to any one problem becomes a solution to a whole series of similar problems.

Substituting values into your final symbolic solution will then enable you to judge if your answer is reasonable and provide greater ease in troubleshooting your error if it is not. Always perform a "reality check" at the end of each problem. Did you obtain a reasonable answer given the question being asked?

Review & Synthesis: Chapters 1–5

Review Exercises

1. From your knowledge of Newton's second law and dimensional analysis, find the units (in SI base units) of the spring constant k in the equation $F = kx$, where F is a force and x is a distance.

2. Harrison traveled 2.00 km west, then 5.00 km in a direction 53.0° south of west, then 1.00 km in a direction 60.0° north of west. (a) In what direction, and for how far, should Harrison travel to return to his starting point? (b) If Harrison returns directly to his starting point with a speed of 5.00 m/s, how long will the return trip take?

3. (a) How many center-stripe road reflectors, separated by 17.6 yd, are required along a 2.20-mile section of curving mountain roadway? (b) Solve the same problem for a road

his rapid descent and lost control? (It turns out that aircraft altitudes are given in feet throughout the world except in China, Mongolia, and the former Soviet states where meters are used.)

8. Paula swims across a river that is 10.2 m wide. She can swim at 0.833 m/s in still water, but the river flows with a speed of 1.43 m/s. If Paula swims in such a way that she crosses the river in as short a time as possible, how far downstream is she when she gets to the opposite shore?

9. Peter is collecting paving stones from a quarry. He harnesses two dogs, Sandy and Rufus, in tandem to the loaded cart. Sandy pulls with force $\vec{F}$ at a 15° angle to the north of east; Rufus pulls with 1.5 times the force of Sandy and at an angle of 30.0° south of east. Use a ruler

After a group of related chapters, you will find a **Review & Synthesis** section. This section will provide *Review Exercises* that require you to combine two or more concepts learned in the previous chapters. Working these problems will help you to prepare for cumulative exams. This section also contains *MCAT Review* exercises. These problems were written for the actual MCAT exam and will provide additional practice if this exam is part of your future plans.

How to Study for an Exam

- Be an active learner:
  - read
  - be an active participant in class; ask questions
  - apply what you've learned; think through scenarios rather than memorizing your notes
- Finish reading all material—text, notes, handouts—at least three days prior to the exam.
- Three days prior to the exam, set aside time each day to do self-testing, work practice problems, and review your notes. Useful tools to help:
  - end-of-chapter summaries
  - questions and practice problems
  - text website
  - your professor's course website
  - the Student Solutions Manual
  - your study partner
- Analyze your weaknesses, and create an "I don't know this yet" list. Focus on strengthening these areas and narrow your list as you study.
- If you find that you were unable to allow the full three days to study for the exam, the most important thing you can do is try some practice problems that are similar to those your instructor assigned for homework. Choose odd-numbered problems so that you can check your answer. The Review & Synthesis problems are designed to help you prepare for exams. Try to solve each problem under exam conditions—use a formula sheet, if your instructor provides one with the exam, but don't look at the book or your notes. If you can't solve the problem, then you have found an area of weakness. Study the material needed to solve that problem and closely related material. Then try another similar problem.
- VERY IMPORTANT—Be sure to sleep and eat well before the exam. Staying up late and memorizing the night before an exam doesn't help much in physics. On a physics exam, you will be asked to demonstrate reasoning and analytical skills acquired by much practice. If you are fatigued or hungry, you won't perform at your highest level.

We hope that these suggestions will help you get the most out of your physics course. After many years working with students, both in the classroom and one-on-one in a self-paced course, we wrote this book so you could benefit from our experience. In *Physics*, we have tried to address the points that have caused difficulties for our students in the past. We also wish to share with you some of the pleasure and excitement we have found in learning about the physical laws that govern our world.

Alan Giambattista

Betty Richardson

Bob Richardson

Acknowledgments

We are grateful to the faculty, staff, and students at Cornell University, who helped us in a myriad of ways. We especially thank our friend and colleague Bob Lieberman who shepherded us through the process as our literary agent and who inspired us as an exemplary physics teacher. Donald F. Holcomb, Persis Drell, Peter Lepage, and Phil Krasicky read portions of the manuscript and provided us with many helpful suggestions. Raphael Littauer contributed many innovative ideas and served as a model of a highly creative, energetic teacher.

We are indebted to Jeevak Parpia, David G. Cassel, Edith Cassel, Richard Galik, Lou Hand, Chris Henley, and Tomás Arias for many helpful discussions while they taught Physics 101–102 using the second edition. We also appreciate the assistance of Leonard J. Freelove and Rosemary French. We thank our enthusiastic and capable graduate teaching assistants and, above all, the students in Physics 101–102, who patiently taught us how to teach physics.

We are grateful for the guidance and enthusiasm of Debra Hash and Mary Hurley at McGraw-Hill, whose tireless efforts were invaluable in bringing this project to fruition. We would like to thank the entire team of talented professionals assembled by McGraw-Hill to publish this book, including Traci Andre, Tammy Ben, Carrie Burger, Linda Davoli, Laura Fuller, David Hash, Tammy Juran, Mary Jane Lampe, Lisa Nicks, Mary Reeg, Gloria Schiesl, Thomas Timp, Dan Wallace, and many others whose hard work has contributed to making the book a reality.

We are grateful to Bill Fellers for accuracy-checking the manuscript and for many helpful suggestions.

Our thanks to Janet Scheel, Warren Zipfel, Rebecca Williams, and Mike Nichols for contributing some of the medical and biological applications; to Nick Taylor and Mike Strauss for contributing to the end-of-chapter and Review & Synthesis problems; and to Nick Taylor for writing answers to the Conceptual Questions.

From Alan: Above all, I am deeply grateful to my family. Marion, Katie, Charlotte, Julia, and Denisha, without your love, support, encouragement, and patience, this book could never have been written.

From Bob and Betty: We thank our daughter Pamela's classmates and friends at Cornell and in the Vanderbilt Master's in Nursing program who were an early inspiration for the book, and we thank Dr. Philip Massey who was very special to Pamela and is dear to us. We thank our friends at *blur,* Alex, Damon, Dave, and Graham, who love physics and are inspiring young people of Europe to explore the wonders of physics through their work with the European Space Agency's Mars mission. Finally we thank our daughter Jennifer, our grandsons Jasper, Dashiell, Oliver, and Quintin, and son-in-law Jim who endured our protracted hours of distraction while this book was being written.

REVIEWERS, CLASS TESTERS, AND ADVISORS

This text reflects an extensive effort to evaluate the needs of college physics instructors and students, to learn how well we met those needs, and to make improvements where we fell short. We gathered information from numerous reviews, class tests, and focus groups.

The primary stage of our research began with commissioning reviews from instructors across the United States and Canada. We asked them to submit suggestions for improvement on areas such as content, organization, illustrations, and ancillaries. The detailed comments of these reviewers constituted the basis for the revision plan.

We organized focus groups across the United States from 2006 through 2008. Participants reviewed our text in comparison to other books and suggested improvements to *Physics* and ways in which we as publishers could help to improve the content of the college physics course.

Finally, we received extremely useful advice on the instructional design, quality, and content of the print and media ancillary packages from Pete Anderson, Gerry Feldman, Ajawad Haija, Hong Luo, David Mast, John Prineas, Michael Pravica, and Craig Wiegert.

Considering the sum of these opinions, the Giambattista/Richardson/Richardson texts now embody the collective knowledge, insight, and experience of hundreds of college physics instructors. Their influence can be seen in everything from the content, accuracy, and organization of the text to the quality of the illustrations.

We are grateful to the following instructors for their thoughtful comments and advice:

REVIEWERS AND FOCUS GROUP ATTENDEES

David Aaron *South Dakota State University*

Rhett Allain *Southeastern Louisiana University*

Peter Anderson *Oakland Community College*

Natalie Batalha *San Jose State*

Thomas K. Bolland *The Ohio State University*

Juan Burciaga *Whitman College*

Peng Chen Dai *University of Tennessee—Knoxville*

Carl Covatto *Arizona State University*

Michael Crescimanno *Youngstown State*

Steven Ellis *University of Kentucky—Lexington*

Abbas Faridi *Orange Coast College*

Gerald Feldman *George Washington University*

David Gerdes *University of Michigan*

Robert Hagood *Washtenaw Community College*

Ajawad Haija *Indiana University of Pennsylvania*

Grady Hendricks *Blinn Community College*

Klaus Honschied *The Ohio State University*

John Hopkins *The Pennsylvania State University*

Brad Johnson *Western Washington University*

Kyungseon Joo *University of Connecticut*

Linda Jones *College of Charleston*

Arya Karamjeet *San Jose State*

Daniel Kennefick *University of Arkansas*

Yuri Kholodenko *Albany College of Pharmacy*

Dana Klinck *Hillsborough Community College, Tampa*

Allen Landers *Auburn University*

Paul Lee *California State University Northridge*

Hong Luo *University at Buffalo*

Stephanie Magelby *Brigham Young University*

George Marion *Texas State University, San Marcos*

David Mast *University of Cincinnati*

Dan Mazilu *Virginia Polytechnic Institute & State University*

Rahul Mehta *University of Central Arkansas*

Meredith Newby *Clemson University*

Miroslav Peric *California State University Northridge*

Amy Pope *Clemson University*

Michael Pravica *University of Nevada, Las Vegas*

Kent Price *Morehead State*

John Prineas *University of Iowa*

Oren Quist *South Dakota State University*

Larry Rowan *University of North Carolina—Chapel Hill*

Ajit Rupaal *Western Washington University*

Douglas Sherman *San Jose State University*

Bjoern Siepel *Portland State University*

Michael Sobel *Brooklyn College*

Xiang-Ning Song *Richland College*

Tim Stelzer *University of Illinois*

James Taylor *University of Central Missouri*

Marshall Thomsen *Eastern Michigan University*

Ralf Widenhorn *Portland State University*

Craig Wiegert *University of Georgia*

Karen Williams *East Central University*

Scott Wissink *Indiana University*

Pei Xiong-Skiba *Austin Peay State University*

Capp Yess *Morehead State*

David Young *Louisiana State University—Baton Rouge*

Michael Ziegler *The Ohio State University*

We are also grateful to our international reviewers for their comments and suggestions:

Goh Hock Leong   *National Junior College—Singapore*

Mohammed Saber Musazay   *King Fahd University of Petroleum and Minerals*

Contributors

We are deeply indebted to:

Professor Suzanne Willis of Northern Illinois University and Professor Susanne M. Lee, Visiting Scientist Rensselaer Polytechnic Institute, for creating the instructor resources and demonstrations in the *Physics Instructors' Resource Guide.*

Professor Jack Cuthbert of Holmes Community College, Ridgeland for the Test Bank to accompany *Physics.*

Professor Lorin Swint Matthews of Baylor University for the clicker questions to accompany *Physics.*

Professor Carl Covatto of Arizona State University for the PowerPoint Lectures to accompany *Physics.*

Professor Allen Landers of Auburn University for his work on the *Physics* collection of Active Art on the text's website.

Introduction

In 2004, the exploration rovers *Spirit* and *Opportunity* landed on sites on opposite sides of Mars. The primary goal of the mission was to examine a wide variety of rocks and soils that might provide evidence of the past presence of water on Mars and clues to where the water went. The mission sent back tens of thousands of photographs and a wealth of geologic data. By contrast, in a previous mission to Mars, a simple mistake caused the loss of the Mars Climate Orbiter as it entered orbit around Mars. In this chapter, you will learn how to avoid making this same mistake. (See p. 9.)

The Mars Exploration Rover *Opportunity* looks back toward its lander in "Eagle Crater" on the surface of Mars.

Concepts & Skills to Review

- algebra, geometry, and trigonometry (Appendix A)
- To the Student: How to Succeed in Your Physics Class (p. xxii)

1.1   WHY STUDY PHYSICS?

Physics is the branch of science that describes matter, energy, space, and time at the most fundamental level. Whether you are planning to study biology, architecture, medicine, music, chemistry, or art, some principles of physics are relevant to your field.

Physicists look for patterns in the physical phenomena that occur in the universe. They try to explain what is happening, and they perform experiments to see if the proposed explanation is valid. The goal is to find the most basic laws that govern the universe and to formulate those laws in the most precise way possible.

The study of physics is valuable for several reasons:

- Since physics describes matter and its basic interactions, all natural sciences are built on a foundation of the laws of physics. A full understanding of chemistry requires a knowledge of the physics of atoms. A full understanding of biological processes in turn is based on the underlying principles of physics and chemistry. Centuries ago, the study of *natural philosophy* encompassed what later became the separate fields of biology, chemistry, geology, astronomy, and physics. Today there are scientists who call themselves biophysicists, chemical physicists, astrophysicists, and geophysicists, demonstrating how thoroughly the sciences are intertwined.

- In today's technological world, many important devices can be understood correctly only with a knowledge of the underlying physics. Just in the medical world, think of laser surgery, magnetic resonance imaging, instant-read thermometers, x-ray imaging, radioactive tracers, heart catheterizations, sonograms, pacemakers, microsurgery guided by optical fibers, ultrasonic dental drills, and radiation therapy.

- By studying physics, you acquire skills that are useful in other disciplines. These include thinking logically and analytically; solving problems; making simplifying assumptions; constructing mathematical models; using valid approximations; and making precise definitions.

- Society's resources are limited, so it is important to use them in beneficial ways and not squander them on scientifically impossible projects. Political leaders and the voting public are too often led astray by a lack of understanding of scientific principles. Can a nuclear power plant supply energy safely to a community? What is the truth about the greenhouse effect, the ozone hole, and the danger of radon in the home? By studying physics, you learn some of the basic scientific principles and acquire some of the intellectual skills necessary to ask probing questions and to formulate informed opinions on these important matters.

- Finally, by studying physics, we hope that you develop a sense of the beauty of the fundamental laws governing the universe.

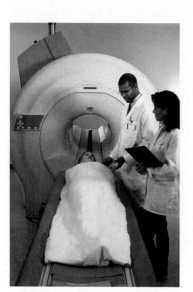

A patient being prepared for magnetic resonance imaging (MRI). MRI provides a detailed image of the internal structures of the patient's body.

1.2   TALKING PHYSICS

Some of the words used in physics are familiar from everyday speech. This familiarity can be misleading, since the scientific definition of a word may differ considerably from its common meaning. In physics, words must be precisely defined so that anyone reading a scientific paper or listening to a science lecture understands exactly what is meant. Some of the basic defined quantities, whose names are also words used in everyday speech, include time, length, force, velocity, acceleration, mass, energy, momentum, and temperature.

In everyday language, *speed* and *velocity* are synonyms. In physics, there is an important distinction between the two. In physics, *velocity* includes the *direction* of motion as well as the distance traveled per unit time. When a moving object changes direction, its velocity changes even though its speed may not have changed. Confusion of the scientific definition of *velocity* with its everyday meaning will prevent a correct understanding of some of the basic laws of physics and will lead to incorrect answers.

Mass, as used in everyday language, has several different meanings. Sometimes *mass* and *weight* are used interchangeably. In physics, mass and weight are *not* interchangeable. Mass is a measure of inertia—the tendency of an object at rest to remain at rest or, if moving, to continue moving with the same velocity. Weight, on the other hand, is a measure of the gravitational pull on an object. (Mass and weight are discussed in more detail in Chapter 4.)

There are two important reasons for the way in which we define physical quantities. First, physics is an experimental science. The results of an experiment must be stated unambiguously so that other scientists can perform similar experiments and compare their results. Quantities must be defined precisely to enable experimental measurements to be uniform no matter where they are made. Second, physics is a mathematical science. We use mathematics to quantify the relationships among physical quantities. These relationships can be expressed mathematically only if the quantities being investigated have precise definitions.

1.3 THE USE OF MATHEMATICS

A working knowledge of algebra, trigonometry, and geometry is essential to the study of introductory physics. Some of the more important mathematical tools are reviewed in Appendix A. If you know that your mathematics background is shaky, you might want to test your mastery by doing some problems from a math textbook. You may find it useful to visit www.mhhe.com to explore the Schaum's Outline series, especially the Schaum's Outlines of *Precalculus, College Physics,* or *Physics for Pre-Med, Biology, and Allied Health Students.*

Mathematical equations are shortcuts for expressing concisely in symbols relationships that are cumbersome to describe in words. Algebraic symbols in the equations stand for quantities that consist of numbers *and units.* The number represents a measurement and the measurement is made in terms of some standard; the unit indicates what standard is used. In physics, a number to specify a quantity is useless unless we know the unit attached to the number. When buying silk to make a sari, do we need a length of 5 millimeters, 5 meters, or 5 kilometers? Is the term paper due in 3 minutes, 3 days, or 3 weeks? Systems of units are discussed in Section 1.5.

There are not enough letters in the alphabet to assign a unique letter to each quantity. The same letter V can represent volume in one context and voltage in another. Avoid attempting to solve problems by picking equations that seem to have the correct letters. A skilled problem-solver understands *specifically* what quantity each symbol in a particular equation represents, can specify correct units for each quantity, and understands the situations to which the equation applies.

Ratios and Proportions In the language of physics, the word **factor** is used frequently, often in a rather idiosyncratic way. If the power emitted by a radio transmitter has doubled, we might say that the power has "increased by a factor of two." If the concentration of sodium ions in the bloodstream is half of what it was previously, we might say that the concentration has "decreased by a factor of two," or, in a blatantly inconsistent way, someone else might say that it has "decreased by a factor of one-half." The *factor* is the number by which a quantity is multiplied or divided when it is changed from one value to another. In other words, the factor is really a ratio. In the case of the radio transmitter, if P_0 represents the initial power and P represents the power after new equipment is installed, we write

$$\frac{P}{P_0} = 2$$

It is also common to talk about "increasing 5%" or "decreasing 20%." If a quantity increases $n\%$, that is the same as saying that it is multiplied by a factor of $1 + (n/100)$. If a quantity decreases $n\%$, then it is multiplied by a factor of $1 - (n/100)$. For example, an increase of 5% means something is 1.05 times its original value and a decrease of 4% means it is 0.96 times the original value.

Physicists talk about increasing "by some factor" because it often simplifies a problem to think in terms of **proportions**. When we say that A is proportional to B (written $A \propto B$), we mean that if B increases by some factor, then A must increase by the same factor. For instance, the circumference of a circle equals 2π times the radius: $C = 2\pi r$. Therefore $C \propto r$. If the radius doubles, the circumference also doubles. The area of a circle is proportional to the *square* of the radius ($A = \pi r^2$, so $A \propto r^2$). The area must increase by the same factor as the radius *squared*, so if the radius doubles, the area increases by a factor of $2^2 = 4$.

$A \propto B$ means $A_1/A_2 = B_1/B_2$

Example 1.1

Effect of Increasing Radius on the Volume of a Sphere

The volume of a sphere is given by the equation

$$V = \tfrac{4}{3}\pi r^3$$

where V is the volume and r is the radius of the sphere. If a basketball has a radius of 12.4 cm and a tennis ball has a radius of 3.20 cm, by what factor is the volume of the basketball larger than the volume of the tennis ball?

Strategy   The problem gives the values of the radii for the two balls. To keep track of which ball's radius and volume we mean, we use subscripts "b" for basketball and "t" for tennis ball. The radius of the basketball is r_b and the radius of the tennis ball is r_t. Since $\tfrac{4}{3}$ and π are constants, we can work in terms of proportions.

Solution   The ratio of the basketball radius to that of the tennis ball is

$$\frac{r_b}{r_t} = \frac{12.4\ [\text{cm}]}{3.20\ [\text{cm}]} = 3.875$$

The volume of a sphere is proportional to the cube of its radius:

$$V \propto r^3$$

Since the basketball radius is larger by a factor of 3.875, and volume is proportional to the cube of the radius, the new volume should be bigger by a factor of $3.875^3 \approx 58.2$.

Discussion   A slight variation on the solution is to write out the proportionality in terms of ratios of the corresponding sides of the two equations:

$$\frac{V_b}{V_t} = \frac{\tfrac{4}{3}\pi r_b^3}{\tfrac{4}{3}\pi r_t^3} = \left(\frac{r_b}{r_t}\right)^3$$

Substituting the ratio of r_b to r_t yields

$$\frac{V_b}{V_b} = 3.875^3 \approx 58.2$$

which says that V_b is approximately 58.2 times V_t.

Practice Problem 1.1   Power Dissipated by a Lightbulb

The electric power P dissipated by a lightbulb of resistance R is $P = V^2/R$, where V represents the line voltage. During a brownout, the line voltage is 10.0% less than its normal value. How much power is drawn by a lightbulb during the brownout if it normally draws 100.0 W (watts)? Assume that the resistance does not change.

✓ CHECKPOINT 1.3

If the radius of the sphere is increased by a factor of 3, by what factor does the volume of the sphere change?

1.4   SCIENTIFIC NOTATION AND SIGNIFICANT FIGURES

In physics, we deal with some numbers that are very small and others that are very large. It can get cumbersome to write numbers in conventional decimal notation. In **scientific notation**, any number is written as a number between 1 and 10 times an integer power of ten. Thus the radius of Earth, approximately 6 380 000 m at the equator, can be written 6.38×10^6 m; the radius of a hydrogen atom, 0.000 000 000 053 m, can be written 5.3×10^{-11} m. Scientific notation eliminates the need to write zeros to locate the decimal point correctly.

Learn how to use the button on your calculator (usually labeled EE) to enter a number in scientific notation. To enter 1.2×10^8, press 1.2, EE, 8.

In science, a measurement or the result of a calculation must indicate the **precision** to which the number is known. The precision of a device used to make a measurement is limited by the finest division on the scale. Using a meterstick with millimeter divisions as the smallest separations, we can measure a length to a precise number of millimeters and we can estimate a fraction of a millimeter between two divisions. If the meterstick has centimeter divisions as the smallest separations, we measure a precise number of centimeters and estimate the fraction of a centimeter that remains.

Significant Figures   The most basic way to indicate the precision of a quantity is to write it with the correct number of **significant figures**. The significant figures are all the digits that are known accurately plus the one estimated digit. If we say that the distance from here to the state line is 12 km, that does not mean we know the distance to be *exactly* 12 kilometers. Rather, the distance is 12 km *to the nearest kilometer.* If instead we said that the distance is 12.0 km, that would indicate that we know the distance to the nearest *tenth* of a kilometer. More significant figures indicate a greater degree of precision.

Rules for Identifying Significant Figures

1. Nonzero digits are always significant.
2. Final or ending zeros written to the right of the decimal point are significant.
3. Zeros written to the right of the decimal point for the purpose of spacing the decimal point are not significant.
4. Zeros written to the left of the decimal point may be significant, or they may only be there to space the decimal point. For example, 200 cm could have one, two, or three significant figures; it's not clear whether the distance was measured to the nearest 1 cm, to the nearest 10 cm, or to the nearest 100 cm. On the other hand, 200.0 cm has four significant figures (see rule 5). Rewriting the number in scientific notation is one way to remove the ambiguity. In this book, when a number has zeros to the left of the decimal point, you may *assume a minimum of two significant figures.*
5. Zeros written between significant figures are significant.

Example 1.2

Identifying the Number of Significant Figures

For each of these values, identify the number of significant figures and rewrite it in standard scientific notation.

(a) 409.8 s
(b) 0.058700 cm
(c) 9500 g
(d) 950.0×10^1 mL

Strategy   We follow the rules for identifying significant figures as given. To rewrite a number in scientific notation, we move the decimal point so that the number to the left of the decimal point is between 1 and 10 and compensate by multiplying by the appropriate power of ten.

continued on next page

Example 1.2 continued

Solution (a) All four digits in 409.8 s are significant. The zero is between two significant figures, so it is significant. To write the number in scientific notation, we move the decimal point two places to the left and compensate by multiplying by 10^2: 4.098×10^2 s.

(b) The first two zeros in 0.058700 cm are not significant; they are used to place the decimal point. The digits 5, 8, and 7 are significant, as are the two final zeros. The answer has five significant figures: 5.8700×10^{-2} cm.

(c) The 9 and 5 in 9500 g are significant, but the zeros are ambiguous. This number could have two, three, or four significant figures. If we take the most cautious approach and assume the zeros are not significant, then the number in scientific notation is 9.5×10^3 g.

(d) The final zero in 950.0×10^1 mL is significant since it comes after the decimal point. The zero to its left is also significant since it comes between two other significant digits.

The result has four significant figures. The number is not in *standard* scientific notation since 950.0 is not between 1 and 10; in scientific notation we write 9.500×10^3 mL.

Discussion Scientific notation clearly indicates the number of significant figures since all zeros are significant; none are used only to place the decimal point. In (c), if we want to show that the zeros were significant, we would write 9.500×10^3 g.

Practice Problem 1.2 Identifying Significant Figures

State the number of significant figures in each of these measurements and rewrite them in standard scientific notation.

(a) 0.000 105 44 kg   (b) 0.005 800 cm   (c) 602 000 s

Significant Figures in Calculations

1. When two or more quantities are added or subtracted, the result is as precise as the *least precise* of the quantities (Example 1.3). If the quantities are written in scientific notation with different powers of ten, first rewrite them with the same power of ten. After adding or subtracting, round the result, keeping only as many decimal places as are significant in *all* of the quantities that were added or subtracted.

2. When quantities are multiplied or divided, the result has the same number of significant figures as the quantity with the *smallest number of significant figures* (Example 1.4).

3. In a series of calculations, rounding to the correct number of significant figures should be done only at the end, *not at each step*. Rounding at each step would increase the chance that roundoff error could snowball and have an adverse effect on the accuracy of the final answer. It's a good idea to keep *at least two* extra significant figures in calculations, then round at the end.

Example 1.3

Significant Figures in Addition

Calculate the sum 44.56005 s + 0.0698 s + 1103.2 s.

Strategy The sum cannot be more precise than the least precise of the three quantities. The quantity 44.56005 s is known to the nearest 0.00001 s, 0.0698 s is known to the nearest 0.0001 s, and 1103.2 s is known to the nearest 0.1 s.

Therefore the least precise is 1103.2 s. The sum has the same precision; it is known to the nearest tenth of a second.

Solution According to the calculator,

$$44.56005 + 0.0698 + 1103.2 = 1147.82985$$

continued on next page

Example 1.3 continued

We do *not* want to write all of those digits in the answer. That would imply greater precision than we actually have. Rounding to the nearest tenth of a second, the sum is written

$$= 1147.8 \text{ s}$$

and there are five significant figures in the result.

⚠️ **Discussion** Note that the least precise measurement is not necessarily the one with the fewest number of significant figures. The least precise is the one whose rightmost significant figure represents the largest unit: the "2" in 1103.2 s represents 2 tenths of a second. In addition or subtraction, we are concerned with the precision rather than the number of significant figures. The three quantities to be added have seven, three, and five significant figures, respectively, while the sum has five significant figures.

Practice Problem 1.3   Significant Figures in Subtraction

Calculate the difference 568.42 m − 3.924 m and write the result in scientific notation. How many significant figures are in the result?

Example 1.4

Significant Figures in Multiplication

Find the product of 45.26 m/s and 2.41 s. How many significant figures does the product have?

Strategy The product should have the same number of significant figures as the factor with the least number of significant figures.

Solution A calculator gives

$$45.26 \times 2.41 = 109.0766$$

Since the answer should have only three significant figures, we round the answer to

$$45.26 \text{ m/s} \times 2.41 \text{ s} = 109 \text{ m}$$

Discussion Writing the answer as 109.0766 m would give the false impression that we know the answer to a precision of about 0.0001 m, whereas we actually have a precision of only about 1 m.

Note that although both factors were known to two decimal places, our solution is properly given with no decimal places. It is the number of significant figures that matters in multiplication or division. In scientific notation, we write 1.09×10^2 m.

Practice Problem 1.4   Significant Figures in Division

Write the solution to 28.84 m divided by 6.2 s with the correct number of significant figures.

When an integer, or a fraction of integers, is used in an equation, the precision of the result is not affected by the integer or the fraction; the number of significant figures is limited only by the measured values in the problem. The fraction $\frac{1}{2}$ in an equation is *exact*; it does not reduce the number of significant figures to one. In an equation such as $C = 2\pi r$ for the circumference of a circle of radius r, the factors 2 and π are exact. We use as many digits for π as we need to maintain the precision of the other quantities.

Order-of-Magnitude Estimates Sometimes a problem may be too complicated to solve precisely, or information may be missing that would be necessary for a precise calculation. In such a case, an **order-of-magnitude** solution is the best we can do. By *order of magnitude*, we mean "roughly what power of ten?" An order of magnitude calculation is done to at most one significant figure. Even when a more precise solution is feasible, it is often a good idea to start with a quick, "back-of-the-envelope estimate." Why? Because we can often make a good guess about the correct order of magnitude of the answer to a problem, even before we start solving the problem. If the answer comes out with a different order of magnitude, we go back and search out an error. Suppose a problem concerns a vase that is knocked off a fourth-story window ledge. We can guess by experience the order of magnitude of the time it takes the vase to hit the ground. It might be 1 s, or 2 s, but we are certain that it is *not* 1000 s or 0.00001 s.

Back-of-the-envelope estimate: a calculation so short that it could easily fit on the back of an envelope

CHECK POINT 1.4

What are some of the reasons for making order-of-magnitude estimates?

1.5   UNITS

A **metric system** of units has been used for many years in scientific work and in European countries. The metric system is based on powers of ten (Fig. 1.1). In 1960, the General Conference of Weights and Measures, an international authority on units, proposed a revised metric system called the *Système International d'Unités* in French (abbreviated **SI**), which uses the meter (m) for length, the kilogram (kg) for mass, the second (s) for time, and four more base units (Table 1.1). **Derived units** are constructed from combinations of the base units. For example, the SI unit of force is $kg \cdot m/s^2$; the combination of $kg \cdot m/s^2$ is given a special name, the newton (N), in honor of Isaac Newton. The newton is a derived unit because it is composed of a combination of base units. When units are named after famous scientists, the name of the unit is written with a lowercase letter, even though it is based on a proper name; the *abbreviation* for the unit is written with an uppercase letter. The inside front cover of the book has a complete listing of the derived SI units used in this book.

$kg \cdot m/s^2$ can also be written $kg \cdot m \cdot s^{-2}$

As an alternative to explicitly writing powers of ten, SI uses prefixes for units to indicate power of ten factors. Table 1.2 shows some of the powers of ten and the SI prefixes used for them. These are also listed on the inside front cover of the book. Note that when an SI unit with a prefix is raised to a power, the prefix is *also* raised to that power. For example, $8 \ cm^3 = 2 \ cm \times 2 \ cm \times 2 \ cm$.

SI units are preferred in physics and are emphasized in this book. Since other units are sometimes used, we must know how to convert units. Various scientific fields, even in physics, do use units other than SI units, whether for historical or practical reasons.

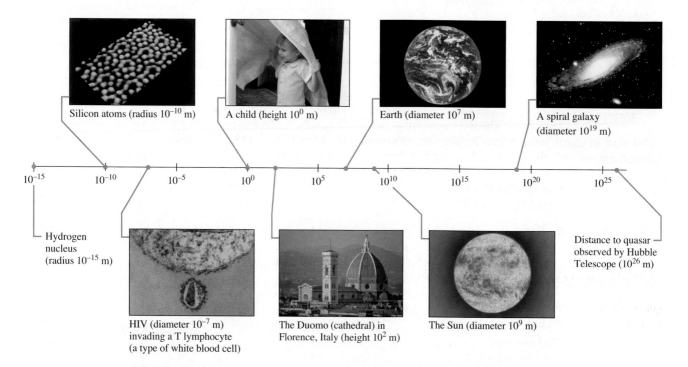

Silicon atoms (radius 10^{-10} m)   A child (height 10^0 m)   Earth (diameter 10^7 m)   A spiral galaxy (diameter 10^{19} m)

10^{-15} 10^{-10} 10^{-5} 10^0 10^5 10^{10} 10^{15} 10^{20} 10^{25}

Hydrogen nucleus (radius 10^{-15} m)   HIV (diameter 10^{-7} m) invading a T lymphocyte (a type of white blood cell)   The Duomo (cathedral) in Florence, Italy (height 10^2 m)   The Sun (diameter 10^9 m)   Distance to quasar observed by Hubble Telescope (10^{26} m)

Figure 1.1 Scientific notation uses powers of ten to express quantities that have a wide range of values.

Table 1.1	SI Base Units		
Quantity	**Unit Name**	**Symbol**	**Definition**
Length	meter	m	The distance traveled by light in vacuum during a time interval of 1/299 792 458 s.
Mass	kilogram	kg	The mass of the international prototype of the kilogram.
Time	second	s	The duration of 9 192 631 770 periods of the radiation corresponding to the transition between the two hyperfine levels of the ground state of the cesium-133 atom.
Electric current	ampere	A	The constant current in two long, thin, straight, parallel conductors placed 1 m apart in vacuum that would produce a force on the conductors of 2×10^{-7} N per meter of length.
Temperature	kelvin	K	The fraction 1/273.16 of the thermodynamic temperature of the triple point of water.
Amount of substance	mole	mol	The amount of substance that contains as many elementary entities as there are atoms in 0.012 kg of carbon-12.
Luminous intensity	candela*	cd	The luminous intensity, in a given direction, of a source that emits radiation of frequency 540×10^{12} Hz and that has a radiant intensity in that direction of 1/683 watts per steradian.

*Not used in this book

For example, in atomic and nuclear physics, the SI unit of energy (the joule, J) is rarely used; instead the energy unit used is usually the electron-volt (eV). Biologists and chemists use units that are not familiar to physicists. One reason that SI is preferred is that it provides a common denominator—all scientists are familiar with the SI units.

In most of the world, SI units are used in everyday life and in industry. In the United States, the U.S. customary units—sometimes called English units—are still used. The base units for this system are the foot, the second, and the pound. The pound is legally defined in the United States as a unit of mass, but it is also commonly used as a unit of force (in which case it is sometimes called *pound-force*). Since mass and force are entirely different concepts in physics, this inconsistency is one good reason to use SI units.

In the autumn of 1999, to the chagrin of NASA, a $125 million spacecraft was destroyed as it was being maneuvered into orbit around Mars. The company building the booster rocket provided information about the rocket's thrust in U.S. customary units, but the NASA scientists who were controlling the rocket thought the figures provided were in metric units. Arthur Stephenson, chairman of the Mars Climate Orbiter Mission Failure Investigation Board, stated that, "The 'root cause' of the loss of the spacecraft was the failed translation of English units into metric units in a segment of ground-based, navigation-related mission software." After a journey of 122 million miles, the Climate Orbiter dipped about 15 miles too deep into the Martian atmosphere, causing the propulsion system to overheat. The discrepancy in units unfortunately caused a dramatic failure of the mission.

Converting Units   If the statement of a problem includes a mixture of different units, the units must be converted to a single, consistent set before the problem is solved. Quantities to be added or subtracted *must be expressed in the same units.* Usually the best way is to convert everything to SI units. Common conversion factors are listed on the inside front cover of this book.

Examples 1.5 and 1.6 illustrate the technique for converting units. The quantity to be converted is multiplied by one or more conversion factors written as a fraction equal to 1. The units are multiplied or divided as algebraic quantities.

What happened to the Mars Climate Orbiter?

Table 1.2	SI Prefixes
Prefix (abbreviation)	**Power of Ten**
peta- (P)	10^{15}
tera- (T)	10^{12}
giga- (G)	10^{9}
mega- (M)	10^{6}
kilo- (k)	10^{3}
deci- (d)	10^{-1}
centi- (c)	10^{-2}
milli- (m)	10^{-3}
micro- (μ)	10^{-6}
nano- (n)	10^{-9}
pico- (p)	10^{-12}
femto- (f)	10^{-15}

Some conversions are exact by definition. One meter is defined to be *exactly* equal to 100 cm; all SI prefixes are exactly a power of ten. The use of an exact conversion factor such as 1 m = 100 cm, or 1 foot = 12 inches, does not affect the precision of the result; the number of significant figures is limited only by the other quantities in the problem.

Example 1.5

Buying Clothes in a Foreign Country

Michel, an exchange student from France, is studying in the United States. He wishes to buy a new pair of jeans, but the sizes are all in *inches*. He does remember that 1 m = 3.28 ft and that 1 ft = 12 in. If his waist size is 82 cm, what is his waist size in inches?

Strategy   Each conversion factor can be written as a fraction. If 1 m = 3.28 ft, then

$$\frac{3.28 \text{ ft}}{1 \text{ m}} = 1$$

We can multiply any quantity by 1 without changing its value. We arrange each conversion factor in a fraction and multiply one at a time to get from centimeters to inches.

Solution   We first convert cm to meters.

$$82 \text{ cm} \times \frac{1 \text{ m}}{100 \text{ cm}}$$

Now, we convert meters to feet.

$$82 \text{ cm} \times \frac{1 \text{ m}}{100 \text{ cm}} \times \frac{3.28 \text{ ft}}{1 \text{ m}}$$

Finally, we convert feet to inches.

$$82 \text{ cm} \times \frac{1 \text{ m}}{100 \text{ cm}} \times \frac{3.28 \text{ ft}}{1 \text{ m}} \times \frac{12 \text{ in}}{1 \text{ ft}} = 32 \text{ in}$$

In each case, the fraction is written so that the unit we are converting *from* cancels out.

As a check:

$$\text{cm} \times \frac{\text{m}}{\text{cm}} \times \frac{\text{ft}}{\text{m}} \times \frac{\text{in}}{\text{ft}} = \text{in}$$

Discussion   This problem could have been done in one step using a direct conversion factor from inches to cm (1 in = 2.54 cm). One of the great advantages of SI units is that all the conversion factors are powers of ten (see Table 1.2); there is no need to remember that there are 12 inches in a foot, 4 quarts in a gallon, 16 ounces in a pound, 5280 feet in a mile, and so on.

Practice Problem 1.5   Driving on the Autobahn

A BMW convertible travels on the German autobahn at a speed of 128 km/h. What is the speed of the car (a) in meters per second? (b) in miles per hour?

Example 1.6

Conversion of Volume

A beaker of water contains 255 mL of water. (1 mL = 1 milliliter; 1 L = 1000 cm³.) What is the volume of the water in (a) cubic centimeters? (b) cubic meters?

Strategy   First convert milliliters to liters; then convert liters to cubic centimeters. To convert cubic centimeters to cubic meters, use 100 cm = 1 m. Since there are *three* factors of centimeters to convert, we have to multiply by $\left(\frac{1 \text{ m}}{100 \text{ cm}}\right)$ *three times.*

Solution   (a) The prefix milli- means 10^{-3}, so 1 mL = 10^{-3} L. Then

$$255 \text{ mL} \times \frac{10^{-3} \text{ L}}{1 \text{ mL}} \times \frac{1000 \text{ cm}^3}{1 \text{ L}} = 255 \text{ cm}^3$$

(b) 1 m = 100 cm. Since we need to convert *cubic* centimeters to *cubic* meters, we must raise the conversion factor to the third power:

$$255 \text{ cm}^3 \times \left(\frac{1 \text{ m}}{100 \text{ cm}}\right)^3 = 255 \text{ cm}^3 \times \frac{(1 \text{ m})^3}{(100 \text{ cm})^3}$$

$$255 \text{ cm}^3 \times \frac{1 \text{ m}^3}{100^3 \text{ cm}^3} = 2.55 \times 10^{-4} \text{ m}^3$$

Discussion   Be careful when a unit is raised to a power other than one; the conversion factor must be raised to the same power. Writing out the units to make sure they cancel prevents mistakes. When a quantity is raised to a power, both the number and the unit must be raised to the same power. $(100 \text{ cm})^3$ is equal to $100^3 \text{ cm}^3 = 10^6 \text{ cm}^3$; it is *not* equal to 100 cm^3, nor is it equal to 10^6 cm.

Practice Problem 1.6   Surface Area of Earth

The radius of Earth is 6.4×10^3 km. Find the surface area of Earth in square meters and in square miles. (Surface area of a sphere = $4\pi r^2$.)

Whenever a calculation is performed, always write out the units with each quantity. Combine the units algebraically to find the units of the result. This small effort has three important benefits:

1. It shows what the units of the result are. A common mistake is to get the correct numerical result of a calculation but to write it with the wrong units, making the answer wrong.
2. It shows where unit conversions must be done. If units that should have canceled do not, we go back and perform the necessary conversion. When a distance is calculated and the result comes out with units of meter-seconds per hour (m·s/h), we should convert hours to seconds.
3. It helps locate mistakes. If a distance is calculated and the units come out as m/s, we know to look for an error.

✓ CHECKPOINT 1.5

If 1 fluid ounce (fl oz) is approximately 30 mL, how many liters are in a half gallon (64 fl oz) of milk?

1.6 DIMENSIONAL ANALYSIS

Dimensions are basic *types* of units, such as time, length, and mass. (Warning: The word *dimension* has several other meanings, such as in "three-dimensional space" or "the dimensions of a soccer field.") Many different units of length exist: meters, inches, miles, nautical miles, fathoms, leagues, astronomical units, angstroms, and cubits, just to name a few. All have dimensions of length; each can be converted into any other.

We can add, subtract, or equate quantities only if they have the same dimensions (although they may not necessarily be given in the same units). It is possible to add 3 meters to 2 inches (after converting units), but it is not possible to add 3 meters to 2 kilograms. To analyze dimensions, treat them as algebraic quantities, just as we did

Example 1.7

Dimensional Analysis for a Distance Equation

Analyze the dimensions of the equation $d = vt$, where d is distance traveled, v is speed, and t is elapsed time.

Strategy Replace each quantity with its dimensions. Distance has dimensions [L]. Speed has dimensions of length per unit time [L/T]. The equation is dimensionally consistent if the dimensions are the same on both sides.

Solution The right side has dimensions
$$\frac{[L]}{[T]} \times [T] = [L]$$
Since both sides of the equation have dimensions of length, the equation is dimensionally consistent.

Discussion If, by mistake, we wrote $d = v/t$ for the relation between distance traveled and elapsed time, we could quickly catch the mistake by looking at the dimensions. On the right side, v/t would have dimensions $[L/T^2]$, which is not the same as the dimensions of d on the left side.

A quick dimensional analysis of this sort is a good way to catch algebraic errors. Whenever we are unsure whether an equation is correct, we can check the dimensions.

Practice Problem 1.7 Testing Dimensions of Another Equation

Test the dimensions of the following equation:
$$d = \frac{1}{2}at$$
where d is distance traveled, a is acceleration (which has SI units m/s^2), and t is the elapsed time. If incorrect, can you suggest what might have been omitted?

with units in Section 1.5. Usually [M], [L], and [T] are used to stand for mass, length, and time dimensions, respectively. Equivalently, we can use the SI base units: kg for mass, m for length, and s for time.

Applying Dimensional Analysis   Dimensional analysis is good for more than just checking equations. In some cases, we can completely solve a problem—up to a dimensionless factor like $1/(2\pi)$ or $\sqrt{3}$—using dimensional analysis. To do this, first list all the relevant quantities on which the answer might depend. Then determine what combinations of them have the same dimensions as the answer for which we are looking. If only one such combination exists, then we have the answer, except for a possible dimensionless multiplicative constant.

Example 1.8

Violin String Frequency

While it is being played, a violin string produces a tone with frequency f in s^{-1}; the frequency is the number of vibrations *per second* of the string. The string has mass m, length L, and tension T. If the tension is increased 5.0%, how does the frequency change? Tension has SI unit $kg \cdot m/s^2$.

Strategy   We could make a study of violin strings, but let us see what we can find out by dimensional analysis. We want to find out how the frequency f can depend on m, L, and T. We won't know if there is a dimensionless constant involved, but we can work by proportions so any such constant will divide out.

Solution   The unit of tension T is $kg \cdot m/s^2$. The units of f do not contain kg or m; we can get rid of them from T by dividing the tension by the length and the mass:

$$\frac{T}{mL} \text{ has SI unit } s^{-2}$$

That is almost what we want; all we have to do is take the square root:

$$\sqrt{\frac{T}{mL}} \text{ has SI unit } s^{-1}$$

Therefore,

$$f = C\sqrt{\frac{T}{mL}}$$

where C is some dimensionless constant. To answer the question, let the original frequency and tension be f and T and the new frequency and tension be f' and T', where $T' = 1.050T$. Frequency is proportional to the square root of tension, so

$$\frac{f'}{f} = \sqrt{\frac{T'}{T}} = \sqrt{1.050} = 1.025$$

The frequency increases 2.5%.

Discussion   We'll learn in Chapter 11 how to calculate the value of C, which is 1/2. That is the *only* thing we cannot get by dimensional analysis. There is *no* other way to combine T, m, and L to come up with a quantity that has the units of frequency.

Practice Problem 1.8   Increase in Kinetic Energy

When a body of mass m is moving with a speed v, it has kinetic energy associated with its motion. Energy is measured in $kg \cdot m^2 \cdot s^{-2}$. If the speed of a moving body is increased by 25% while its mass remains constant, by what percentage does the kinetic energy increase?

✓ CHECKPOINT 1.6

If two quantities have different dimensions, is it possible to (a) multiply; (b) divide; (c) add; (d) subtract them?

1.7 PROBLEM-SOLVING TECHNIQUES

No single method can be used to solve every physics problem. We demonstrate useful problem-solving techniques in the examples in every chapter of this text. Even for a particular problem, there may be more than one correct way to approach the solution. Problem-solving techniques are *skills* that must be *practiced* to be learned.

Think of the problem as a puzzle to be solved. Only in the easiest problems is the solution method immediately apparent. When you do not know the entire path to a solution, see where you can get by using the given information—find whatever you can. Exploration of this sort may lead to a solution by suggesting a path that had not been considered. Be willing to take chances. You may even find the challenge enjoyable!

When having some difficulty, it helps to work with a classmate or two. One way to clarify your thoughts is to put them into words. After you have solved a problem, try to explain it to a friend. If you can explain the problem's solution, you really do understand it. Both of you will benefit. But do not rely too much on help from others; the goal is for each of you to develop your own problem-solving skills.

General Guidelines for Problem Solving

1. Read the problem *carefully* and *all the way through.*

2. Reread the problem one sentence at a time and draw a sketch or diagram to help you visualize what is happening.

3. Write down and organize the given information. Some of the information can be written in labels on the diagram. Be sure that the labels are unambiguous. Identify in the diagram the object, the position, the instant of time, or the time interval to which the quantity applies. Sometimes information might be usefully written in a table beside the diagram. Look at the wording of the problem again for information that is implied or stated indirectly.

4. Identify the goal of the problem. What quantities need to be found?

5. If possible, make an estimate to determine the order of magnitude of the answer. This estimate is useful as a check on the final result to see if it is reasonable.

6. Think about how to get from the given information to the final desired information. Do not rush this step. Which principles of physics can be applied to the problem? Which will help get to the solution? How are the known and unknown quantities related? Are all of the known quantities relevant, or might some of them not affect the answer? Which equations are relevant and may lead to the solution to the problem? This step requires skills developed only with much practice in problem solving.

7. Frequently, the solution involves more than one step. Intermediate quantities might have to be found first and then used to find the final answer. Try to map out a path from the given information to the solution. Whenever possible, a good strategy is to divide a complex problem into several simpler subproblems.

8. Perform algebraic manipulations with algebraic symbols (letters) as far as possible. Substituting the numbers in too early has a way of hiding mistakes.

9. Finally, if the problem requires a numerical answer, substitute the known numerical quantities, *with their units*, into the appropriate equation. Leaving out the units is a common source of error. Writing the units shows when a unit conversion needs to be done—and also may help identify an algebra mistake.

10. Once the solution is found, don't be in a hurry to move on. Check the answer—is it reasonable? Try to think of other ways to solve the same problem. Many problems can be solved in several different ways. Besides providing a check on the answer, finding more than one method of solution deepens our understanding of the principles of physics and develops problem-solving skills that will help solve other problems.

1.8   APPROXIMATION

Physics is about building conceptual and mathematical models and comparing observations of the real world with the model. Simplified models help us to analyze complex situations. In various contexts we assume there is no friction, or no air resistance, no heat loss, or no wind blowing, and so forth. If we tried to take all these things into consideration with every problem, the problems would become vastly more complicated to solve. We never can take account of *every* possible influence. We freely make approximations whenever possible to turn a complex problem into an easier one, as long as the answer will be accurate enough for our purposes.

A valuable skill to develop is the ability to know when an assumption or approximation is reasonable. It might be permissible to ignore air resistance when dropping a stone, but not when dropping a beach ball. Why? We must always be prepared to justify any approximation we make by showing the answer is not changed very much by its use.

As well as making simplifying approximations in models, we also recognize that measurements are approximate. Every measured quantity has some uncertainty; it is impossible for a measurement to be exact to an arbitrarily large number of significant figures. Every measuring device has limits on the precision and accuracy of its measurements.

Approximating the Surface Area of the Human Body   Sometimes it is difficult or impossible to measure precisely a quantity that is needed for a problem. Then we have to make a reasonable estimate. Suppose we need to know the surface area of a human being to determine the heat loss by radiation in a cold room. We can estimate the height of an average person. We can also estimate the average distance around the waist or hips. Approximating the shape of a human body as a cylinder, we can estimate the surface area by calculating the surface area of a cylinder with the same height and circumference (Fig. 1.2a).

If we need a better estimate, we use a slightly more refined model. For instance, we might approximate the arms, legs, trunk, and head and neck as cylinders of various sizes (Fig. 1.2b). How different is the sum of these areas from the original estimate? That gives an idea of how close the first estimate is.

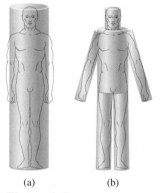

(a)   (b)

Figure 1.2 Approximation of human body by one or more cylinders.

Example 1.9

Number of Cells in the Human Body

Average-sized cells in the human body are about 10 μm in length (Fig. 1.3). How many cells are in the human body? Make an order-of-magnitude estimate.

Strategy   We divide this problem into three subproblems: estimating the volume of a human, estimating the volume of the average cell, and finally estimating the number of cells.

To find the volume of a human body, we approximate the body as a cylinder, as previously discussed. Next we assume the cells are cubical to find the volume of a cell. Third, the ratio of the two volumes (volume of the body to volume of the cell) shows how many cells are in the body.

Solution   Model the body as a cylinder. A typical height is about 2 m. A typical *maximum* circumference (think hip size) is about 1 m. The corresponding radius is $1/(2\pi)$ m, or about 1/6 m. The *average* radius is somewhat smaller; say

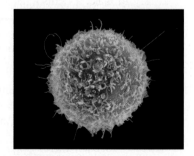

Figure 1.3

Scanning electron micrograph of a precursor T lymphocyte (a type of white blood cell in the human body). The cell is approximately 12 μm in diameter.

0.1 m. The volume of a cylinder is the height times the cross-sectional area:

$$V = Ah = \pi r^2 h \approx 3 \times (0.1 \text{ m})^2 \times (2 \text{ m}) = 0.06 \text{ m}^3$$

continued on next page

Example 1.9 continued

The volume of a cube is $V = s^3$. Then the volume of an average cell is about

$$V_{cell} \approx (1 \times 10^{-5} \text{ m})^3 = 1 \times 10^{-15} \text{ m}^3$$

The number of cells is the ratio of the two volumes:

$$N = \frac{\text{volume of body}}{\text{average volume of cell}} \approx \frac{6 \times 10^{-2} \text{ m}^3}{1 \times 10^{-15} \text{ m}^3} \approx 6 \times 10^{13}$$

Discussion Based on this rough estimate, we cannot rule out the possibility that a better estimate might be 3×10^{13}. On the other hand, we *can* rule out the possibility that the number of cells is, say, 100 million ($= 10^8$).

Practice Problem 1.9 Drinking Water Consumed in the United States

How many liters of water are swallowed by the people living in the United States in one year? This is a type of problem made famous by the physicist Enrico Fermi (1901–1954), who was a master at this sort of back-of-the-envelope calculation. Such problems are often called *Fermi problems* in his honor. (1 liter $= 10^{-3}$ m$^3 \approx 1$ quart.)

1.9 GRAPHS

Graphs are used to help us see a pattern in the relationship between two quantities. It is much easier to see a pattern on a graph than to see it in a table of numerical values. When we do experiments in physics, we change one quantity (the **independent variable**) and see what happens to another (the **dependent variable**). We want to see how one variable *depends on* another. The value of the independent variable is usually plotted along the horizontal axis of the graph. In a plot of p versus q, which means p is plotted on the vertical axis and q on the horizontal axis, normally p is the dependent variable and q is the independent variable.

Some general guidelines for recording data and making graphs are given next.

Recording Data and Making Data Tables

1. Label columns with the names of the data being measured and be sure to include the units for the measurements. Do not erase any data, but just draw a line through data that you think are erroneous. Sometimes you may decide later that the data were correct after all.

2. Try to make a realistic estimate of the precision of the data being taken when recording numbers. For example, if the timer says 2.3673 s, but you know your reaction time can vary by as much as 0.1 s, the time should be recorded as 2.4 s. When doing calculations using measured values, remember to round the final answer to the correct number of significant figures.

3. Do not wait until you have collected all of your data to start a graph. It is much better to graph each data point as it is measured. By doing so, you can often identify equipment malfunction or measurement errors that make your data unreliable. You can also spot where something interesting happens and take data points closer together there. Graphing as you go means that you need to find out the range of values for both the independent and dependent variables.

Graphing Data

1. Make *large*, *neat* graphs. A tiny graph is not very illuminating. Use at least half a page. A graph made carelessly obscures the pattern between the two variables.

2. Label axes with the name of the quantities graphed and their units. Write a meaningful title.

The equation of a straight line on a graph of y versus x can be written $y = mx + b$, where m is the slope and b is the y-intercept (the value of y corresponding to $x = 0$).

The symbol Δ, the Greek uppercase letter delta, stands for the difference between two measurements. The notation Δy is read aloud as "delta y" and represents a change in the value of y.

3. When a linear relation is expected, use a ruler or straightedge to draw the best-fit straight line. Do not *assume* that the line must go through the origin—make a measurement to find out, if possible. Some of the data points will probably fall above the line and some will fall below the line.

4. Determine the slope of a best-fit line by measuring the ratio $\Delta y / \Delta x$ using as large a range of the graph as possible. Do not choose two data points to calculate the slope; instead, read values from two points on the best-fit line. Show the calculations. Do not forget to write the units; slopes of graphs in physics have units, since the quantities graphed have units.

5. When a nonlinear relationship is expected between the two variables, the best way to test that relationship is to manipulate the data algebraically so that a linear graph is expected. The human eye is a good judge of whether a straight line fits a set of data points. It is not so good at deciding whether a curve is parabolic, cubic, or exponential. To test the relationship $x = \frac{1}{2}at^2$, where x and t are the quantities measured, graph x versus t^2 instead of x versus t.

6. If one data point does not lie near the line or smooth curve connecting the other data points, that data point should be investigated to see whether an error was made in the measurement or whether some interesting event is occurring at that point. If something unusual is happening there, obtain additional data points in the vicinity.

7. When the slope of a graph is used to calculate some quantity, pay attention to the equation of the line and the units along the axes. The quantity to be found may be the inverse of the slope or twice the slope or one half the slope. For example, if you wish to find the value of a in the relationship $x = \frac{1}{2}at^2$, and you make a graph of x versus t^2, then the slope of the line is $\frac{1}{2}a$. The value of a you seek is twice the slope.

Example 1.10

Length of a Spring

In an introductory physics laboratory experiment, students are investigating how the length of a spring varies with the weight hanging from it. Various weights (accurately calibrated to 0.01 N) ranging up to 6.00 N can be hung from the spring; then the length of the spring is measured with a meterstick (Fig. 1.4). The goal is to see if the weight F and length L are related by

$$F = kx$$

where $x = (L - L_0)$, L_0 is the length of the spring when no weight is hanging from it, and k is called the *spring constant* of the spring. Graph the data in the table and calculate k for this spring.

F (N):	0	0.50	1.00	2.50	3.00	3.50	4.00	5.00	6.00
L (cm):	9.4	10.2	12.5	17.9	19.7	22.5	23.0	28.8	29.5

Strategy   Weight is the independent variable, so it is plotted on the horizontal axis. After plotting the data points, we draw the best-fit straight line. Then we calculate the slope of the line, using two points on the line that are widely separated

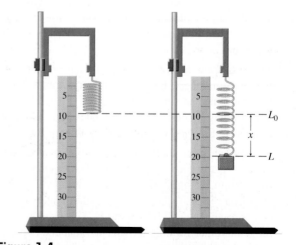

Figure 1.4

A weight causes an extension in the length of a spring.

and that cross gridlines of the graph (so the values are easy to read). The slope of the graph is not k; we must solve the equation for L, since length is plotted on the vertical axis.

continued on next page

Example 1.10 continued

Solution   Figure 1.5 shows a graph with data points and a best-fit straight line. There is some scatter in the data, but a linear relationship is plausible.

Two points where the line crosses gridlines of the graph are (0.80 N, 12.0 cm) and (4.40 N, 25.0 cm). From these, we calculate the slope:

$$\text{slope} = \frac{\Delta L}{\Delta F} = \frac{25.0\ \text{cm} - 12.0\ \text{cm}}{4.40\ \text{N} - 0.80\ \text{N}} = 3.61\ \frac{\text{cm}}{\text{N}}$$

By analyzing the units of the equation $F = k(L - L_0)$, it is clear that the slope cannot be the spring constant; k has the same units as weight divided by length (N/cm). Is the slope equal to $1/k$? The units would be correct for that case. To be sure, we solve the equation of the line for L:

$$L = \frac{F}{k} + L_0$$

We recognize the equation of a line with a slope of $1/k$. Therefore,

$$k = \frac{1}{3.61\ \text{cm/N}} = 0.277\ \text{N/cm}$$

Discussion   As discussed in the graphing guidelines, the slope of the straight-line graph is calculated from two widely spaced values *along the best-fit line*. We do not subtract values of actual data points. We are looking for an average value from the data; using two data points to find the slope would defeat the purpose of plotting a graph or of taking more than two data measurements. The values read from the graph, including the units, are indicated in Fig. 1.5. The units for the slope are cm/N, since we plotted centimeters versus newtons. For this particular problem the *inverse* of the slope is the quantity we seek, the spring constant in N/cm.

Practice Problem 1.10   Another Weight on Spring

What is the length of the spring of Example 1.10 when a weight of 8.00 N is suspended? Assume that the relationship found in Example 1.10 still holds for this weight.

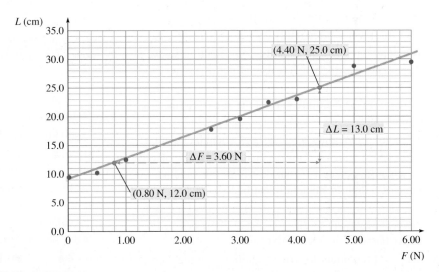

Figure 1.5

Spring length versus weight hanging.

Master the Concepts

- Terms used in physics must be precisely defined. A term may have a different meaning in physics from the meaning of the same word in other contexts.

- A working knowledge of algebra, geometry, and trigonometry is essential in the study of physics.

- The *factor* by which a quantity is increased or decreased is the ratio of the new value to the original value.

- When we say that A is *proportional* to B (written $A \propto B$), we mean that if B increases by some factor, then A must increase by the same factor.

- In *scientific notation*, a number is written as the product of a number between 1 and 10 and a whole-number power of ten.

- *Significant figures* are the basic *grammar* of precision. They enable us to communicate quantitative information and indicate the precision to which that information is known.

- When two or more quantities are added or subtracted, the result is as precise as the *least precise* of the quantities. When quantities are multiplied or divided, the result has

continued on next page

Master the Concepts continued

the same number of significant figures as the quantity with the *smallest number of significant figures.*

- Order-of-magnitude estimates and calculations are made to be sure that the more precise calculations are realistic.

- The units used for scientific work are those from the *Système International* (*SI*). SI uses seven *base units*, which include the meter (m), the kilogram (kg), and the second (s) for length, mass, and time, respectively. Using combinations of the base units, we can construct other *derived units.*

- If the statement of a problem includes a mixture of different units, the units should be converted to a single, consistent set before the problem is solved. Usually the best way is to convert everything to SI units.

- Dimensional analysis is used as a quick check on the validity of equations. Whenever quantities are added, subtracted, or equated, they must have the same dimensions (although they may not necessarily be given in the same units).

- Mathematical approximations aid in simplifying complicated problems.

- Problem-solving techniques are *skills* that must be *practiced* to be learned.

- A graph is plotted to give a picture of the data and to show how one variable changes with respect to another. Graphs are used to help us see a pattern in the relationship between two variables.

- Whenever possible, make a careful choice of the variables plotted so that the graph displays a linear relationship.

Conceptual Questions

1. Give a few reasons for studying physics.
2. Why must words be carefully defined for scientific use?
3. Why are simplified models used in scientific study if they do not exactly match real conditions?
4. By what factor does tripling the radius of a circle increase (a) the circumference of the circle? (b) the area of the circle?
5. What are some of the advantages of scientific notation?
6. After which numeral is the decimal point usually placed in scientific notation? What determines the number of numerical digits written in scientific notation?
7. Are all the digits listed as "significant figures" precisely known? Might any of the significant digits be less precisely known than others? Explain.
8. Why is it important to write quantities with the correct number of significant figures?
9. List three of the base units used in SI.
10. What are some of the differences between the SI and the customary U.S. system of units? Why is SI preferred for scientific work?
11. Sort the following units into three groups of dimensions and identify the dimensions: fathoms, grams, years, kilometers, miles, months, kilograms, inches, seconds.
12. What are the first two steps to be followed in solving almost any physics problem?
13. Why do scientists plot graphs of their data instead of just listing values?
14. A student's lab report concludes, "The speed of sound in air is 327." What is wrong with that statement?
15. Once the solution of a problem has been found, what should be done before moving on to solve another problem?

Multiple-Choice Questions

1. One kilometer is approximately
   (a) 2 miles (b) 1/2 mile (c) 1/10 mile (d) 1/4 mile
2. 55 mi/h is approximately
   (a) 90 km/h (b) 30 km/h (c) 10 km/h (d) 2 km/h
3. By what factor does the volume of a cube increase if the length of the edges are doubled?
   (a) 16   (b) 8   (c) 4   (d) 2   (e) $\sqrt{2}$
4. If the length of a box is reduced to one third of its original value and the width and height are doubled, by what factor has the volume changed?
   (a) 2/3   (b) 1   (c) 4/3   (d) 3/2
   (e) depends on relative proportion of length to height and width
5. If the area of a circle is found to be half of its original value after the radius is multiplied by a certain factor, what was the factor used?
   (a) $1/(2\pi)$ (b) 1/2   (c) $\sqrt{2}$ (d) $1/\sqrt{2}$ (e) 1/4
6. In terms of the original diameter d, what new diameter will result in a new spherical volume that is a factor of eight times the original volume?
   (a) $8d$ (b) $2d$ (c) $d/2$ (d) $d \times \sqrt[3]{2}$ (e) $d/8$
7. An equation for potential energy states $U = mgh$. If U is in $\text{kg·m}^2\text{·s}^{-2}$, m is in kg, and g is in m·s^{-2}, what are the units of h?
   (a) s   (b) s^2 (c) m^{-1} (d) m   (e) g^{-1}

8. The equation for the speed of sound in a gas states that $v = \sqrt{\gamma k_B T / m}$. Speed v is measured in m/s, γ is a dimensionless constant, T is temperature in kelvins (K), and m is mass in kg. What are the units for the Boltzmann constant, k_B?
   (a) $kg \cdot m^2 \cdot s^2 \cdot K$ (b) $kg \cdot m^2 \cdot s^{-2} \cdot K^{-1}$ (c) $kg^{-1} \cdot m^{-2} \cdot s^2 \cdot K$
   (d) $kg \cdot m/s$ (e) $kg \cdot m^2 \cdot s^{-2}$

9. How many significant figures should be written in the sum 4.56 g $+ 9.032$ g $+ 580.0078$ g $+ 540.439$ g?
   (a) 3   (b) 4   (c) 5   (d) 6   (e) 7

10. How many significant figures should be written in the product 0.0078406 m $\times 9.45020$ m?
    (a) 3   (b) 4   (c) 5   (d) 6   (e) 7

Problems

Ⓒ	Combination conceptual/quantitative problem
✚	Biological or medical application
✦	Challenging problem
Blue #	Detailed solution in the Student Solutions Manual
①②	Problems paired by concept
🌐	Text website interactive or tutorial

1.3 The Use of Mathematics

1. The gardener is told that he must increase the height of his fences 37% if he wants to keep the deer from jumping in to eat the foliage and blossoms. If the current fence is 1.8 m high, how high will the new fence be?

2. What is the ratio of the number of seconds in a day to the number of hours in a day?

3. A spherical balloon expands when it is taken from the cold outdoors to the inside of a warm house. If its surface area increases 16.0%, by what percentage does the radius of the balloon change?

4. A spherical balloon is partially blown up and its surface area is measured. More air is then added, increasing the volume of the balloon. If the surface area of the balloon expands by a factor of 2.0 during this procedure, by what factor does the radius of the balloon change? (🌐 tutorial: car on curve)

5. For any cube with edges of length s, what is the ratio of the surface area to the volume?

6. Samantha is 1.50 m tall on her eleventh birthday and 1.65 m tall on her twelfth birthday. By what factor has her height increased? By what percentage?

7. The "scale" of a certain map is 1/10000. This means the length of, say, a road as represented on the map is 1/10000 the actual length of the road. What is the ratio of the *area* of a park as represented on the map to the actual area of the park? (🌐 tutorial: scaling)

8. On Monday, a stock market index goes up 5.00%. On Tuesday, the index goes down 5.00%. What is the net percentage change in the index for the two days?

9. According to Kepler's third law, the orbital period T of a planet is related to the radius R of its orbit by $T^2 \propto R^3$. Jupiter's orbit is larger than Earth's by a factor of 5.19. What is Jupiter's orbital period? (Earth's orbital period is 1 yr.)

10. If the radius of a circular garden plot is increased by 25%, by what percentage does the area of the garden increase?

11. A poster advertising a student election candidate is too large according to the election rules. The candidate is told she must reduce the length and width of the poster by 20.0%. By what percentage will the area of the poster be reduced?

12. An architect is redesigning a rectangular room on the blueprints of the house. He decides to double the width of the room, increase the length by 50%, and increase the height by 20%. By what factor has the volume of the room increased?

1.4 Scientific Notation and Significant Figures

13. Perform these operations with the appropriate number of significant figures.
    (a) 3.783×10^6 kg $+ 1.25 \times 10^8$ kg
    (b) $(3.783 \times 10^6$ m$) \div (3.0 \times 10^{-2}$ s$)$

14. Write these numbers in scientific notation: (a) the U.S. population, 290000000; (b) the diameter of a helium nucleus, 0.0000000000000038 m.

15. In the following calculations, be sure to use an appropriate number of significant figures.
    (a) 3.68×10^7 g $- 4.759 \times 10^5$ g
    (b) $\dfrac{6.497 \times 10^4 \text{ m}^2}{5.1037 \times 10^2 \text{ m}}$

16. Write your answer to the following problems with the appropriate number of significant figures.
    (a) 6.85×10^{-5} m $+ 2.7 \times 10^{-7}$ m
    (b) 702.35 km $+ 1897.648$ km
    (c) 5.0 m $\times 4.3$ m
    (d) $(0.04/\pi)$ cm
    (e) $(0.040/\pi)$ m

17. Solve the following problem and express the answer in scientific notation with the appropriate number of significant figures: $(3.2$ m$) \times (4.0 \times 10^{-3}$ m$) \times (1.3 \times 10^{-8}$ m$)$.

18. How many significant figures are in each of these measurements?
    (a) 7.68 g   (b) 0.420 kg
    (c) 0.073 m   (d) 7.68×10^5 g
    (e) 4.20×10^3 kg   (f) 7.3×10^{-2} m
    (g) 2.300×10^4 s

19. Solve the following problem and express the answer in meters per second (m/s) with the appropriate number of significant figures. $(3.21 \text{ m})/(7.00 \text{ ms}) = ?$ [*Hint:* Note that ms stands for milliseconds.]

20. Solve the following problem and express the answer in meters with the appropriate number of significant figures and in scientific notation:

$$3.08 \times 10^{-1} \text{ km} + 2.00 \times 10^3 \text{ cm}$$

1.5 Units

21. A cell membrane is 7.0 nm thick. How thick is it in inches?

22. The label on a small soda bottle lists the volume of the drink as 355 mL. (a) How many fluid ounces are in the bottle? A competitor's drink is labeled 16.0 fl oz. (b) How many milliliters are in that drink?

23. The length of the river span of the Brooklyn Bridge is 1595.5 ft. The total length of the bridge is 6016 ft. Find the length and the order of magnitude in meters of (a) the river span and (b) the total bridge length?

24. Convert 1.00 km/h to meters per second (m/s).

25. A sprinter can run at a top speed of 0.32 miles per minute. Express her speed in (a) m/s and (b) mi/h.

26. The first modern Olympics in 1896 had a marathon distance of 40 km. In 1908, for the Olympic marathon in London, the length was changed to 42.195 km to provide the British royal family with a better view of the race. This distance was adopted as the official marathon length in 1921 by the International Amateur Athletic Federation. What is the official length of the marathon in miles?

27. At the end of 2006 an expert economist from the Global Economic Institute in Kiel, Germany, predicted a drop in the value of the dollar against the euro of 10% over the next 5 years. If the exchange rate was $1.27 to 1 euro on November 5, 2006, and was $1.45 to 1 euro on November 5, 2007, what was the actual drop in the value of the dollar over the first year?

28. The intensity of the Sun's radiation that reaches Earth's atmosphere is 1.4 kW/m^2 (kW = kilowatt; W = watt). Convert this to W/cm^2.

29. Density is the ratio of mass to volume. Mercury has a density of 1.36×10^4 kg/m^3. What is the density of mercury in units of g/cm^3?

30. A molecule in air is moving at a speed of 459 m/s. How many meters would the molecule move during 7.00 ms (milliseconds) if it didn't collide with any other molecules?

31. Express this product in units of km^3 with the appropriate number of significant figures: $(3.2 \text{ km}) \times (4.0 \text{ m}) \times (13 \times 10^{-3} \text{ mm})$.

32. (a) How many square centimeters are in 1 square foot? (1 in. = 2.54 cm.) (b) How many square centimeters are in 1 square meter? (c) Using your answers to parts (a) and (b), but without using your calculator, roughly how many square feet are in one square meter?

33. A snail crawls at a pace of 5.0 cm/min. Express the snail's speed in (a) ft/s and (b) mi/h.

34. An average-sized capillary in the human body has a cross-sectional area of about 150 μm^2. What is this area in square millimeters (mm^2)?

1.6 Dimensional Analysis

35. An equation for potential energy states $U = mgh$. If U is in joules, with m in kg, h in m, and g in m/s^2, find the combination of SI base units that are equivalent to joules.

36. One equation involving force states that $F_{net} = ma$, where F_{net} is in newtons, m is in kg, and a is in m·s^{-2}. Another equation states that $F = -kx$, where F is in newtons, k is in kg·s^{-2}, and x is in m. (a) Analyze the dimensions of ma and kx to show they are equivalent. (b) What are the dimensions of the force unit newton?

37. An equation for the period T of a planet (the time to make one orbit about the Sun) is $4\pi^2 r^3/(GM)$, where T is in s, r is in m, G is in m^3/(kg·s^2), and M is in kg. Show that the equation is dimensionally correct.

38. The relationship between kinetic energy K (SI unit kg·m^2·s^{-2}) and momentum p is $K = p^2/(2m)$, where m stands for mass. What is the SI unit of momentum?

39. An expression for buoyant force is $F_B = \rho g V$, where F_B has dimensions [MLT^{-2}], ρ (density) has dimensions [ML^{-3}], and g (gravitational field strength) has dimensions [LT^{-2}]. (a) What must be the dimensions of V? (b) Which could be the correct interpretation of V: velocity or volume?

40. Use dimensional analysis to determine how the linear speed (v in m/s) of a particle traveling in a circle depends on some, or all, of the following properties: r is the radius of the circle; ω is an angular frequency in s^{-1} with which the particle orbits about the circle, and m is the mass of the particle. There is no dimensionless constant involved in the relation.

1.8 Approximation

41. What is the approximate distance from your eyes to a book you are reading?

42. What is the approximate volume of your physics textbook in cubic centimeters (cm^3)?

43. (a) Estimate the average mass of a person's leg.
    (b) Estimate the length of a full-size school bus.

44. Estimate the number of times a human heart beats during its lifetime.

45. Estimate the number of automobile repair shops in the city you live in by considering its population, how often an automobile needs repairs, and how many cars each shop can service per day. Then look in the yellow pages of your phone directory to see how accurate your estimate is. By what percentage was your estimate off?

46. What is the order of magnitude of the number of seconds in one year?

47. What is the order of magnitude of the height (in meters) of a 40-story building?

1.9 Graphs

48. You have just performed an experiment in which you measured many values of two quantities, A and B. According to theory, $A = cB^3 + A_0$. You want to verify that the values of c and A_0 are correct by making a graph of your data that enables you to determine their values from a slope and a vertical axis intercept. What quantities do you put on the vertical and horizontal axes of the plot?

49. A nurse recorded the values shown in the temperature chart for a patient's temperature. Plot a graph of temperature versus elapsed time and from the graph find (a) an estimate of the temperature at noon and (b) the slope of the graph. (c) Would you expect the graph to follow the same trend over the next 12 hours? Explain.

Time	Temp (°F)
10:00 A.M.	100.00
10:30 A.M.	100.45
11:00 A.M.	100.90
11:30 A.M.	101.35
12:45 P.M.	102.48

50. A graph of x versus t^4, with x on the vertical axis and t^4 on the horizontal axis, is linear. Its slope is 25 m/s^4 and its vertical axis intercept is 3 m. Write an equation for x as a function of t.

51. A patient's temperature was 97.0°F at 8:05 A.M. and 101.0°F at 12:05 P.M. If the temperature change with respect to elapsed time was linear throughout the day, what would the patient's temperature be at 3:35 P.M.?

52. The weight of a baby measured over an 11-mon period is given in the weight chart for this problem. (a) Plot the baby's weight versus age over the 11 mon. (b) What was the average monthly weight gain for this baby over the period from birth to 5 mon? How do you find this value from the graph? (c) What was the average monthly weight gain for the baby over the period from 5 mon to 10 mon? (d) If a baby continued to grow at the same rate as in the first five months of life, what would the child weigh at age 12 yr?

Weight (lb)	Age (mon)
6.6	0 (birth)
7.4	1.0
9.6	2.0
11.2	3.0
12.0	4.0
13.6	5.0
13.8	6.0
14.8	7.0
15.0	8.0
16.6	9.0
17.5	10.0
18.4	11.0

53. A physics student plots results of an experiment as v versus t. The equation that describes the line is given by $at = v - v_0$. (a) What is the slope of this line? (b) What is the vertical axis intercept of this line?

54. A linear plot of speed versus elapsed time has a slope of 6.0 m/s^2 and a vertical intercept of 3.0 m/s. (a) What is the change in speed in the time interval between 4.0 s and 6.0 s? (b) What is the speed when the elapsed time is equal to 5.0 s?

55. In a laboratory you measure the decay rate of a sample of radioactive carbon. You write down the following measurements:

Time (min)	0	15	30	45	60	75	90
Decays/s	405	237	140	90	55	32	19

(a) Plot the decays per second versus time. (b) Plot the natural logarithm of the decays per second versus the time. Why might the presentation of the data in this form be useful?

56. An object is moving in the x-direction. A graph of the distance it has moved as a function of time is shown. (a) What are the slope and vertical axis intercept? (Be sure to include units.) (b) What physical significance do the slope and intercept on the vertical axis have for this graph?

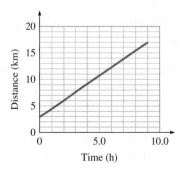

Comprehensive Problems

57. It is useful to know when a small number is negligible. Perform the following computations. (a) 186.300 + 0.0030 (b) 186.300 − 0.0030 (c) 186.300 × 0.0030 (d) 186.300/0.0030 (e) For cases (a) and (b), what percent error will result if you ignore the 0.0030? Explain why you can never ignore the smaller number, 0.0030, for case (c) and case (d)? (f) What rule can you make about ignoring small values?

58. The weight of an object at the surface of a planet is proportional to the planet's mass and inversely proportional to the square of the radius of the planet. Jupiter's radius is 11 times Earth's and its mass is 320 times Earth's. An apple weighs 1.0 N on Earth. How much would it weigh on Jupiter?

59. In cleaning out the artery of a patient, a doctor increases the radius of the opening by a factor of 2.0. By what factor does the cross-sectional area of the artery change?

60. A scanning electron micrograph of xylem vessels in a corn root shows the vessels magnified by a factor of 600. In the micrograph the xylem vessel is 3.0 cm in diameter. (a) What is the diameter of the vessel itself? (b) By what factor has the cross-sectional area of the vessel been increased in the micrograph?

61. The average speed of a nitrogen molecule in air is proportional to the square root of the temperature in kelvins (K). If the average speed is 475 m/s on a warm summer day (temperature = 300.0 K), what is the average speed on a cold winter day (250.0 K)?

62. A furlong is 220 yd; a fortnight is 14 d. How fast is 1 furlong per fortnight (a) in μm/s? (b) in km/day?

63. Given these measurements, identify the number of significant figures and rewrite in scientific notation.
    (a) 0.00574 kg   (b) 2 m   (c) 0.450×10^{-2} m
    (d) 45.0 kg   (e) 10.09×10^4 s   (f) 0.09500×10^5 mL

64. A car has a gas tank that holds 12.5 U.S. gal. Using the conversion factors from the inside front cover, (a) determine the size of the gas tank in cubic inches. (b) A cubit is an ancient measurement of length that was defined as the distance from the elbow to the tip of the finger, about 18 in. long. What is the size of the gas tank in cubic cubits?

65. You are given these approximate measurements: (a) the radius of Earth is 6×10^6 m, (b) the length of a human body is 6 ft, (c) a cell's diameter is 10^{-6} m, (d) the width of the hemoglobin molecule is 3×10^{-9} m, and (e) the distance between two atoms (carbon and nitrogen) is 3×10^{-10} m. Write these measurements in the simplest possible metric prefix forms (in either nm, Mm, μm, or whatever works best).

66. A typical virus is a packet of protein and DNA (or RNA) and can be spherical in shape. The influenza A virus is a spherical virus that has a diameter of 85 nm.

If the volume of saliva coughed onto you by your friend with the flu is 0.010 cm³ and 10^{-9} of that volume consists of viral particles, how many influenza viruses have just landed on you?

67. The smallest "living" thing is probably a type of infectious agent known as a viroid. Viroids are plant pathogens that consist of a circular loop of single-stranded RNA, containing about 300 bases. (Think of the bases as beads strung on a circular RNA string.) The distance from one base to the next (measured along the circumference of the circular loop) is about 0.35 nm. What is the diameter of a viroid in (a) m, (b) μm, and (c) in.?

68. The largest living creature on Earth is the blue whale, which has an average length of 70 ft. The largest blue whale on record (and therefore the largest animal ever found) was 1.10×10^2 ft long. (a) Convert this length to meters. (b) If a double-decker London bus is 8.0 m long, how many double-decker-bus lengths is the record whale?

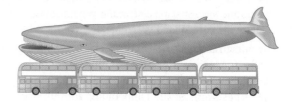

69. The record blue whale in Problem 68 had a mass of 1.9×10^5 kg. Assuming that its average density was 0.85 g/cm³, as has been measured for other blue whales, what was the volume of the whale in cubic meters (m³)? (Average density is the ratio of mass to volume.)

70. A sheet of paper has length 27.95 cm, width 8.5 in., and thickness 0.10 mm. What is the volume of a sheet of paper in m³? (Volume = length × width × thickness.)

71. An object moving at constant speed v around a circle of radius r has an acceleration a directed toward the center of the circle. The SI unit of acceleration is m/s². (a) Use dimensional analysis to find a as a function of v and r. (b) If the speed is increased 10.0%, by what percentage does the radial acceleration increase?

72. The speed of ocean waves depends on their wavelength λ (measured in meters) and the gravitational field strength g (measured in m/s²) in this way:

$$v = K\lambda^p g^q$$

where K is a dimensionless constant. Find the values of the exponents p and q.

73. In the United States, we often use miles per hour (mi/h) when discussing speed, but the SI unit of speed is m/s. What is the conversion factor for changing m/s to mi/h? If you want to make a quick approximation of the speed in mi/h given the speed in m/s, what might be the easiest conversion factor to use?

74. How many cups of water are required to fill a bathtub?

75. Without looking up any data, make an order-of-magnitude estimate of the annual consumption of gasoline (in gallons) by passenger cars in the United States. Make reasonable order-of-magnitude estimates for any quantities you need. Think in terms of average quantities. (1 gal ≈ 4 L.)

76. Some thieves, escaping after a bank robbery, drop a sack of money on the sidewalk. (a) Estimate the mass of the sack if it contains $5000 in half-dollar coins. (b) Estimate the mass if the sack contains $1 000 000 in $20 bills.

77. The weight W of an object is given by $W = mg$, where m is the object's mass and g is the gravitational field strength. The SI unit of field strength g, expressed in SI base units, is m/s^2. What is the SI unit for weight, expressed in base units?

78. Kepler's law of planetary motion says that the square of the period of a planet (T^2) is proportional to the cube of the distance of the planet from the Sun (r^3). Mars is about twice as far from the Sun as Venus. How does the period of Mars compare with the period of Venus?

79. One morning you read in the *New York Times* that the net worth of the richest man in the world, Carlos Slim Helu of Mexico, is $59 000 000 000. Later that day you see him on the street, and he gives you a $100 bill. What is his net worth now? (Think of significant figures.)

80. Estimate the number of hairs on the average human head. [*Hint:* Consider the number of hairs in an area of 1 $in.^2$ and then consider the area covered by hair on the head.]

81. Suppose you have a pair of Seven League Boots. These are magic boots that enable you to stride along a distance of 7.0 leagues with each step. (a) If you march along at a military march pace of 120 paces/min, what will be your speed in km/h? (b) Assuming you could march on top of the oceans when you step off the continents, how long (in minutes) will it take you to march around the Earth at the equator? (1 league = 3 mi = 4.8 km.)

82. The electrical power P drawn from a generator by a lightbulb of resistance R is $P = V^2/R$, where V is the line voltage. The resistance of bulb B is 42% greater than the resistance of bulb A. What is the ratio P_B/P_A of the power drawn by bulb B to the power drawn by bulb A if the line voltages are the same?

83. Three of the fundamental constants of physics are the speed of light, $c = 3.0 \times 10^8$ m/s, the universal gravitational constant, $G = 6.7 \times 10^{-11}$ $m^3 \cdot kg^{-1} \cdot s^{-2}$, and Planck's constant, $h = 6.6 \times 10^{-34}$ $kg \cdot m^2 \cdot s^{-1}$.

    (a) Find a combination of these three constants that has the dimensions of time. This time is called the *Planck time* and represents the age of the universe before which the laws of physics as presently understood cannot be applied. (b) Using the formula for the Planck time derived in part (a), what is the time in seconds?

84. Use dimensional analysis to determine how the period T of a swinging pendulum (the elapsed time for a complete cycle of motion) depends on some, or all, of these properties: the length L of the pendulum, the mass m of the pendulum bob, and the gravitational field strength g (in m/s^2). Assume that the amplitude of the swing (the maximum angle that the string makes with the vertical) has no effect on the period.

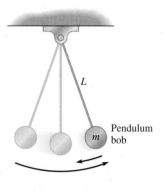

85. The Space Shuttle astronauts use a *massing chair* to measure their mass. The chair is attached to a spring and is free to oscillate back and forth. The frequency of the oscillation is measured and that is used to calculate the total mass m attached to the spring. If the spring constant of the spring k is measured in kg/s^2 and the chair's frequency f is 0.50 s^{-1} for a 62-kg astronaut, what is the chair's frequency for a 75-kg astronaut? The chair itself has a mass of 10.0 kg. [*Hint:* Use dimensional analysis to find out how f depends on m and k.]

86. The average depth of the oceans is about 4 km, and oceans cover about 70% of Earth's surface. Make an order-of-magnitude estimate of the volume of water in the oceans. Do not look up any data in books. (Use your ingenuity to estimate the radius or circumference of Earth.)

87. The population of a culture of yeast cells is studied in the laboratory to see the effects of limited resources (food, space) on population growth. At 2-h intervals, the size of the population (measured as total mass of yeast cells) is recorded (see table on p. 24). (a) Make a graph of the yeast population as a function of elapsed time. Draw a best-fit smooth curve. (b) Notice from the graph of part (a) that after a long time, the population asymptotically approaches a maximum known as the *carrying capacity*. From the graph, estimate the carrying capacity for this population. (c) When the population is much smaller than the carrying capacity, the growth is expected to be exponential: $m(t) = m_0 e^{rt}$, where m is the population at any time t, m_0 is the initial population, r is the *intrinsic growth rate* (i.e., the growth rate in the absence of limits), and e is the base of natural logarithms (see Appendix A.3). To obtain a straight line graph from this exponential relationship, we can plot the natural logarithm of m/m_0:

$$\ln \frac{m}{m_0} = \ln e^{rt} = rt$$

Make a graph of $\ln (m/m_0)$ versus t from $t = 0$ to $t = 6.0$ h, and use it to estimate the intrinsic growth rate r for the yeast population. (The term ln stands for the natural logarithm; see Appendix A.3 if you need help with natural logs.)

Time (h)	Mass (g)
0.0	3.2
2.0	5.9
4.0	10.8
6.0	19.1
8.0	31.2
10.0	46.5
12.0	62.0
14.0	74.9
16.0	83.7
18.0	89.3
20.0	92.5
22.0	94.0
24.0	95.1

Answers to Practice Problems

1.1 81.0 W

1.2 (a) five; 1.0544×10^{-4} kg; (b) four; 5.800×10^{-3} cm; (c) ambiguous, three to six; if three, 6.02×10^5 s

1.3 The least precise value is to the nearest hundredth of a meter, so we round the result to the nearest hundredth of a meter: 564.50 m or, in scientific notation, 5.6450×10^2 m; five significant figures.

1.4 4.7 m/s

1.5 (a) 35.6 m/s; (b) 79.5 mi/h

1.6 5.1×10^{14} m^2; 2.0×10^8 mi^2

1.7 The equation is dimensionally inconsistent; the right side has dimensions [L/T]. To have matching dimensions we must multiply the right side by [T]; the equation must involve time squared: $d = \frac{1}{2}at^2$.

1.8 kinetic energy = (constant) $\times mv^2$; kinetic energy increases by 56%.

1.9 10^{11} L (Make a rough estimate of the population to be about 3×10^8 people, each drinking about 1.5 L/day.)

1.10 38.3 cm

Answers to Checkpoints

1.3 The volume increases by a factor of 27.

1.4 Order-of-magnitude estimates provide a quick method for obtaining limited precision solutions to problems. Even if greater accuracy is required, order-of-magnitude calculations are still useful as they provide a check as to the accuracy of the higher precision calculation.

1.5 1.9 L

1.6 (a) and (b) It is possible to multiply or divide quantities with different dimensions. (c) and (d) To be added or subtracted, quantities *must* have the same dimensions.

CHAPTER

2

Motion Along a Line

Despite its enormous mass (425 to 900 kg), the Cape buffalo is capable of running at a top speed of about 55 km/h (34 mi/h). Since the top speed of the African lion is about the same, how is it ever possible for a lion to catch the buffalo, especially since the lion typically makes its move from a distance of 20 to 30 m from the buffalo? (See p. 34 for the answer.)

Concepts & Skills to Review

- scientific notation and significant figures (Section 1.4)
- converting units (Section 1.5)
- problem-solving techniques (Section 1.7)
- meaning of *velocity* in physics (Section 1.2)

2.1   POSITION AND DISPLACEMENT

Position

CONNECTION:

The topic of Chapters 2 and 3 is **kinematics**: the mathematical description of motion. Beginning in Chapter 4, we will learn the principles of physics that predict and explain *why* objects move the way they do.

To describe motion unambiguously, we need a way to say *where* an object is located. Suppose that at 3:00 P.M. a train stops on an east-west track as a result of an engine problem. The engineer wants to call the railroad office to report the problem. How can he tell them where to find the train? He might say something like "three kilometers east of the old trestle bridge." Notice that he uses a point of reference: the old trestle bridge. Then he states how far the train is from that point and in what direction. If he omits any of the three pieces (the reference point, the distance, and the direction), then his description of the train's whereabouts is ambiguous.

The same thing is done in physics. First, we choose a reference point, called the **origin**. Then, to describe the location of something, we give its distance from the origin and the direction. For motion along a line, we can choose the line of motion to be the *x*-axis of a coordinate system. The origin is the point $x = 0$. The position of an object can be described by its *x*-coordinate, which tells us both how far the object is from the origin and on which side. For the train in Fig. 2.1, we choose the origin at the center of the bridge and the +*x*-direction to the east. Then $x = +3$ km means the train is 3 km east of the bridge and $x = -26$ km means the train is 26 km west of the bridge.

Displacement

Once the train's engine is repaired and it goes on its way, we might want to describe its motion. At 3:14 P.M., it leaves its initial position, 3 km east of the origin (see Fig. 2.1). At 3:56 P.M., the train is 26 km west of the origin, which is 29 km to the west of its initial position. **Displacement** is defined as the change of the position—the final position minus the initial position. The displacement is written Δx where the symbol Δ (the uppercase Greek letter delta) means *the change in* the quantity that follows.

Displacement: final position minus initial position

Displacement:

$$\Delta x = x_f - x_i \tag{2-1}$$

Figure 2.1 Initial (x_i) and final (x_f) positions of a train. (Train not to scale.)

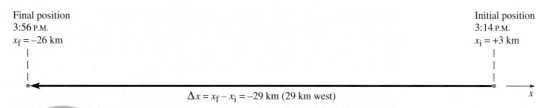

Figure 2.2 With the x-axis pointing east, $\Delta x = x_f - x_i = -26$ km $-(+3$ km$) = -29$ km. The train's displacement is 29 km west.

We can subtract x-coordinates to find the displacement of the train. If we choose the x-axis to the east, then $x_i = +3$ km and $x_f = -26$ km. The displacement is

$$\Delta x = x_f - x_i = (-26 \text{ km}) - (+3 \text{ km}) = -29 \text{ km}$$

The displacement is 29 km in the $-x$-direction (west) (Fig. 2.2).

Displacement Versus Distance   Notice that the magnitude of the displacement is not necessarily equal to the *distance traveled*. Suppose the train first travels 7 km to the east, putting it 10 km east of the origin, and then reverses direction and travels 36 km to the west. The total distance traveled in that case is $(7 \text{ km} + 36 \text{ km}) = 43$ km, but the magnitude of the displacement—which is the distance between the initial and final positions—is 29 km. The displacement depends only on the starting and ending positions, not on the path taken.

Example 2.1

A Mule Hauling Corn to Market

A mule hauls the farmer's wagon along a straight road for 4.3 km directly east to the neighboring farm where a few bushels of corn are loaded onto the wagon. Then the farmer drives the mule back along the same straight road, heading west for 7.2 km to the market. Find the displacement of the mule from the starting point to the market. (The train first travels 7 km to the east, then reverses direction and travels 36 km to the west.)

Strategy   The problem gives us two successive displacements along a straight line. Let's choose the $+x$-axis to point east and an arbitrary point along the road to be the origin. Suppose the mule starts at position x_1 (Fig. 2.3). It goes east until it reaches the neighbor's farm at position x_2. The displacement to the neighbor's farm is $x_2 - x_1 = 4.3$ km east.

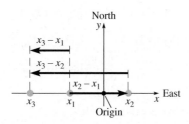

Figure 2.3

The total displacement is the sum of two successive displacements.

Then the mule goes 7.2 km west to reach the market at position x_3. The displacement from the neighbor's farm to the market is $x_3 - x_2 = -7.2$ km (negative because the displacement is in the $-x$-direction). The problem asks for the displacement of the mule from x_1 to x_3.

Solution   We can eliminate x_2, the intermediate position, by adding the two displacements:

$$(x_3 - x_2) + (x_2 - x_1) = -7.2 \text{ km} + 4.3 \text{ km}$$

$$x_3 - x_1 = -2.9 \text{ km}$$

The displacement is 2.9 km west.

Discussion   When we added the two displacements, the intermediate position x_2 dropped out, as it must since the displacement is independent of the path taken from the initial position to the final position. The result does not depend on the choice of origin.

Practice Problem 2.1   A Nervous Squirrel

A nervous squirrel, trying to cross a road, first moves 3.0 m east, then 4.0 m west, then 1.2 m west, then 6.0 m east. What is the squirrel's total displacement?

Adding Displacements   Generalizing the result of Example 2.1, the total displacement for a trip with several parts is the sum of the displacements for each part of the trip. Although x-coordinates depend on the choice of origin, displacements (*changes* in x-coordinates) do *not* depend on the choice of origin.

✓ CHECKPOINT 2.1

In Example 2.1, is the magnitude of the displacement equal to the distance traveled? Explain.

2.2   VELOCITY: RATE OF CHANGE OF POSITION

We introduced *velocity* as a quantity with magnitude and direction in Section 1.2. The magnitude is the speed with which the object moves and the direction is the direction of motion. Now we develop a mathematical definition of velocity that fits that description. Note that displacement indicates by how much and in what direction the position has changed, but implies nothing about *how long* it took to move from one point to the other. Velocity depends on both the displacement and the time interval.

Average Velocity

Reminder: the symbol Δ stands for *the change in.* If the initial value of a quantity Q is Q_i and the final value is Q_f, then $\Delta Q = Q_f - Q_i$. ΔQ is read "delta Q."

When a displacement Δx occurs during a time interval Δt, the **average velocity** during that time interval is

Average velocity:

$$v_{av,x} = \frac{\Delta x}{\Delta t} \tag{2-2}$$

Since Δt is always positive, the direction of the average velocity is the same as the direction of the displacement.

⚠ The symbol Δ does not stand alone and cannot be canceled in equations because it *modifies* the quantity that follows it; $\frac{\Delta x}{\Delta t}$ means $\frac{x_f - x_i}{t_f - t_i}$, which is *not* the same as x/t.

Example 2.2

Average Velocity of a Train

Find the average velocity in kilometers per hour of the train shown in Fig. 2.1 during the time interval between 3:14 P.M., when the train is 3 km east of the origin, and 3:56 P.M., when it is 26 km west of the origin.

Strategy   We choose the +x-axis to the east, as before. Then the displacement is $\Delta x = -29$ km, which means 29 km to the west. The average velocity is also to the west, so $v_{av,x}$ is negative. We convert Δt to hours to find the average velocity in kilometers per hour.

Solution   The time interval is $\Delta t = 56$ min -14 min $= 42$ min. Converting to hours,

$$\Delta t = 42 \text{ min} \times \frac{1 \text{ h}}{60 \text{ min}} = 0.70 \text{ h}$$

The average velocity is

$$v_{av,x} = \frac{\Delta x}{\Delta t} = \frac{-29 \text{ km}}{0.70 \text{ h}} = -41 \text{ km/h}$$

The negative sign means that the average velocity is directed along the negative x-axis, or to the west.

continued on next page

Example 2.2 continued

Discussion   If the train had started at the same instant of time, 3:14 P.M., and had traveled directly west at a constant 41 km/h, it would have ended up in exactly the same place—26 km west of the trestle bridge—at 3:56 P.M.

Had we started measuring time from when we first spotted the motionless train at 3:00 P.M., instead of 3:14 P.M., we would have found the average velocity over a different time interval, changing the average velocity. The average velocity depends on the time interval considered.

The magnitude of the train's average velocity is *not* equal to the total distance traveled divided by the time interval for the complete trip. The latter quantity is called the average *speed*:

$$\text{average speed} = \frac{\text{distance traveled}}{\text{total time}} = \frac{43 \text{ km}}{0.70 \text{ h}} = 61 \text{ km/h}$$

The distinction arises because the average velocity is the constant velocity that would result in the same *displacement* (during the given time interval), while the average speed is the constant speed that would result in the same *distance traveled* (during the same time interval).

Practice Problem 2.2   Average Velocity for a Different Time Interval

What is the average velocity of the same train during the time interval from 3:28 P.M., when it is at $x = 10$ km, to 3:56 P.M., when it is at $x = -26$ km?

Average Speed Versus Average Velocity   The *average* velocity does not convey detailed information about the motion during the corresponding time interval Δt. The average velocity would be the same for any other motion that takes the object through the same displacement in the same amount of time. However, the average *speed*, defined as the total *distance* traveled divided by the time interval, depends on the path traveled.

√ CHECKPOINT 2.2

Can average speed ever be greater than the magnitude of the average velocity? Explain.

Instantaneous Velocity

The speedometer of a car does not indicate the average speed for an entire trip. When a speedometer reads 55 mi/h, it does *not* necessarily mean that the car travels 55 miles in the next hour; the car could change its speed or direction or stop during that hour. The speedometer reading can be used to calculate how far the car travels during a *very short time interval*—short enough that the speed does not change appreciably. For instance, at 55 mi/h (= 25 m/s), we can calculate that in 0.010 s the car moves 25 m/s × 0.010 s = 0.25 m—as long as the speed does not change significantly during that 0.010-s interval.

Similarly, the **instantaneous velocity** is a quantity whose magnitude is the speed and whose direction is the direction of motion. The instantaneous velocity can be used to calculate the *displacement* of the object *during a very short time interval*, as long as *neither the speed nor the direction of motion* change significantly during that time interval. Repeating the word *instantaneous* can get cumbersome. When we refer simply to *the velocity*, we always mean the *instantaneous* velocity.

CONNECTION:

Couldn't we omit "*x*" sub-scripts in average ($v_{av,x}$) and instantaneous (v_x) velocity? If we wanted to understand only motion along a line, then we certainly would. However, in Chapter 3 we generalize the definitions of position, displacement, velocity, and acceleration as vector quantities in three dimensions. Using the "*x*" subscripts now lets us carry forward everything in Chapter 2 *without requiring a change in notation.* Then, when you look back to review Chapter 2, you won't have to remember different definitions for the same symbol. For example, in Chapter 3 we'll learn that v (without the subscript) stands for the *magnitude* of the velocity (the speed), which can never be negative.

Thus, the velocity at some instant of time t is the average velocity during a *very short* time interval:

Instantaneous velocity:

$$v_x = \lim_{\Delta t \to 0} \frac{\Delta x}{\Delta t} \tag{2-3}$$

(Δx is the displacement during a *very short* time interval Δt)

The notation lim is read "the limit, as Δt approaches zero, of . . . ." In other words, let the time interval get smaller and smaller, *approaching*—but never reaching—zero. This notation in Eq. (2-3) reminds you that Δt must be a *very short* time interval. How short a time interval is short enough? If you use a shorter time interval and the calculation of v_x always gives the same value (to within the precision of your measurements), then Δt is short enough. In other words, Δt must be short enough that we can treat the velocity as constant during that time interval. When v_x is constant, cutting Δt in half also cuts the displacement in half, giving the same value for $\Delta x / \Delta t$.

Graphical Relationships Between Position and Velocity

For motion along the x-axis, the displacement is Δx. The average velocity can be represented on the graph of $x(t)$ as the slope of a line connecting two points (called a *chord*). In Fig. 2.4a, the displacement $\Delta x = x_3 - x_1$ is the *rise* of the graph (the change along the vertical axis) and the time interval $\Delta t = t_3 - t_1$ is the *run* of the graph (the change along the horizontal axis). The slope of the chord is the rise over the run:

$$\text{slope of chord} = \frac{\text{rise}}{\text{run}} = \frac{\Delta x}{\Delta t} = v_{av,x} \tag{2-4}$$

The slope of the chord is the average velocity for that time interval.

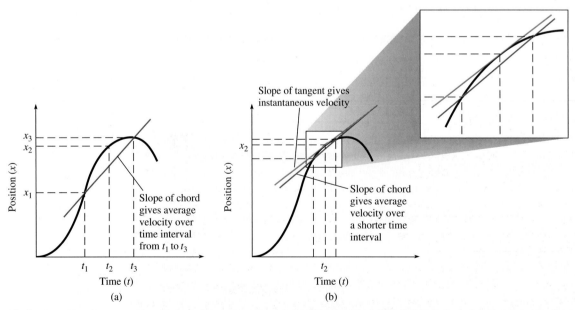

Figure 2.4 A graph of $x(t)$ for an object moving along the x-axis. (a) The average velocity $v_{x,av}$ for the time interval t_1 to t_3 is the slope of the chord connecting those two points on the graph. (b) The average velocity measured over a shorter time interval. As the time interval gets shorter and shorter, the average velocity approaches the *instantaneous* velocity v_x at the instant t_2. The slope of the *tangent line* to the graph is v_x at that instant.

Finding v_x on a Graph of $x(t)$ To find the *instantaneous* velocity at some time $t = t_2$, we draw lines showing the average velocity for shorter and shorter time intervals. As the time interval is reduced (Fig. 2.4b), the average velocity changes. As Δt gets shorter and shorter, the chord approaches a tangent line to the graph at t_2. Thus, v_x is the *slope of the line tangent to the graph of $x(t)$* at the chosen time.

The slope of the tangent line on a graph of $x(t)$ is v_x.

In Fig. 2.5, the position of the train considered in Example 2.2 is graphed as a function of time, where 3:00 P.M. is chosen as $t = 0$.

The graph of position versus time shows a curving line, but that does not mean the train travels along a curved path. The motion of the train is along a straight line since the track runs in an east-west direction. The graph shows the train's position as a function of time.

A horizontal portion of the graph (as from $t = 0$ to $t = 14$ min and from $t = 23$ min to $t = 28$ min) indicates that the position is not changing during that time interval and, therefore, it is at rest (its velocity is zero). Sloping portions of the graph indicate that the train is moving. The steeper the graph, the larger the speed of the train. The sign of the slope indicates the direction of motion. A positive slope ($t = 14$ min to $t = 23$ min) indicates motion in the $+x$-direction, and a negative slope ($t = 28$ min to $t = 56$ min) indicates motion in the $-x$-direction.

Example 2.3

Velocity of the Train

Use Fig. 2.5 to estimate the velocity of the train in kilometers per hour at $t = 40$ min.

Strategy   Figure 2.5 is a graph of $x(t)$. The slope of a line tangent to the graph at $t = 40$ min is v_x at that instant. After sketching a tangent line on the graph, we find its slope from the rise divided by the run.

Solution   Figure 2.6 shows a tangent line drawn on the graph. Using the endpoints of the tangent line, the rise is $(-25 \text{ km}) - (15 \text{ km}) = -40 \text{ km}$. The run is approximately $(57 \text{ min}) - (30 \text{ min}) = 27 \text{ min} = 0.45 \text{ h}$. Then

$$v_x \approx -40 \text{ km}/(0.45 \text{ h}) \approx -89 \text{ km/h}$$

The velocity is approximately 89 km/h in the $-x$-direction (west).

Discussion   Since the slope of a line is constant, any two points *on the tangent line* would give the same value for the slope. Using widely spaced points tends to give a more accurate estimate for the slope.

Practice Problem 2.3   Maximum Eastward Velocity

Estimate the maximum velocity of the train in kilometers per hour during the time it moves east ($t = 14$ min to $t = 23$ min).

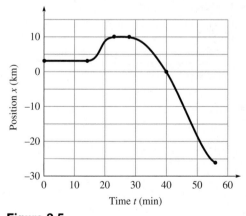

x (km)	t (min)
+3	0
+3	14
+10	23
+10	28
0	40
−26	56

Figure 2.5

Graph of position x versus time t for the train. The positions of the train at various times are marked with a dot. The position of the train would have to be measured at more frequent time intervals to accurately trace out the shape of the graph.

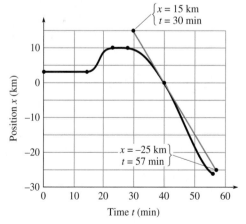

Figure 2.6

On the graph of $x(t)$, the slope of a line tangent to the graph at $t = 40$ min is v_x at $t = 40$ min.

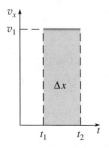

Figure 2.7 Displacement Δx between t_1 and t_2 is represented by the shaded area under the red $v_x(t)$ graph.

Finding Δx with Constant Velocity   What about the other way around? Given a graph of $v_x(t)$, how can we determine the displacement (change in position)? If v_x is constant during a time interval, then the average velocity is equal to the instantaneous velocity:

$$v_x = v_{av,x} = \frac{\Delta x}{\Delta t} \quad \text{(for constant } v_x) \tag{2-5}$$

and therefore

$$\Delta x = v_x \, \Delta t \quad \text{(for constant } v_x) \tag{2-6}$$

The graph of Fig. 2.7 shows v_x versus t for an object moving along the x-axis with constant velocity v_1 from time t_1 to t_2. The displacement Δx during the time interval $\Delta t = t_2 - t_1$ is $v_1 \Delta t$. The shaded rectangle has "height" v_1 and "width" Δt. Since the area of a rectangle is the product of the height and width, the displacement Δx is represented by the area of the rectangle between the graph of $v_x(t)$ and the time axis for the time interval considered.

When we speak of the area under a graph, we are not talking about the literal number of square centimeters of paper or computer screen. The figurative area under a graph usually does not have dimensions of an ordinary area [L^2]. In a graph of $v_x(t)$, v_x has dimensions [L/T] and time has dimensions [T]; areas on such a graph have dimensions [L/T] $\times$ [T] = [L], which is correct for a displacement. The *units* of Δx are determined by the units used on the axes of the graph. If v_x is in meters per second and t is in seconds, then the displacement is in meters.

Finding Δx with Changing Velocity   What if the velocity is not constant? The displacement Δx during a *very small* time interval Δt can be found in the same way as for constant velocity since, during a short enough time interval, the velocity does not change appreciably. Then v_x and Δt are the height and width of a narrow rectangle (Fig. 2.8a) and the displacement during that short time interval is the area of the rectangle. To find the total displacement during any time interval, the areas of all the narrow rectangles are added together (Fig. 2.8b). To improve the approximation, we let the time interval Δt approach zero and find that the displacement Δx during any time interval equals the area under the graph of $v_x(t)$ (Fig. 2.8c). When v_x is negative, x is decreasing and the displacement is in the $-x$-direction, so we must count the area as negative when it is below the time axis.

Δx is the area under the graph of $v_x(t)$. The area is negative when the graph is beneath the time axis ($v_x < 0$).

The magnitude of the train's displacement is represented as the shaded areas in Fig. 2.9. The train's displacement from $t = 14$ min to $t = 23$ min is +7 km (area *above* the t-axis means displacement in the +x-direction) and from $t = 28$ min to $t = 56$ min it is -36 km (area *below* the t-axis means displacement in the $-x$-direction). The total displacement from $t = 0$ to $t = 56$ min is $\Delta x = (+7 \text{ km}) + (-36 \text{ km}) = -29$ km.

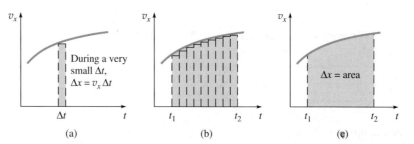

Figure 2.8 (a) Displacement Δx during a short time interval is approximately the area of a rectangle of height v_x and width Δt. (b) During a longer time interval, the displacement is approximately the sum of the areas of the rectangles. (c) The area under the v_x versus t graph for any time interval represents the displacement during that interval.

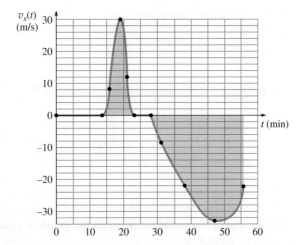

Figure 2.9 A graph of train velocity versus time. The train's displacement from $t = 14$ min to $t = 23$ min is the shaded area under the graph during that time interval. To estimate the area, count the number of grid boxes under the curve, estimating the fraction of the boxes that are only partly below the curve. Each box is 2 m/s in height and 5 min ($= 300$ s) in width, so each box represents an "area" (displacement) of 2 m/s $\times$ 300 s $= 600$ m $= 0.60$ km. The total number of shaded boxes for this time interval is about 12, so the displacement is about $\Delta x \approx 12 \times 0.60$ km $= +7.2$ km, which is close to the actual value of 7 km (during this time interval the train went from +3 km to +10 km). The shaded area for the time interval $t = 28$ min to $t = 56$ min is below the time axis; this negative area represents displacement in the $-x$-direction (west). The number of shaded grid boxes in this interval is about 60, so the displacement during this time interval is $\Delta x \approx -(60) \times 0.60$ km $= -36$ km.

2.3   ACCELERATION: RATE OF CHANGE OF VELOCITY

The rate of change of the velocity is called the **acceleration**. The use of the word *acceleration* in everyday language is often imprecise and not in accord with its scientific definition. In everyday language, it usually means "an increase in speed" but sometimes is used almost as a synonym for speed itself. In physics, acceleration does not necessarily indicate an increase in speed. Acceleration can indicate any kind of change in velocity.

The concept of acceleration is much less intuitive for most people than the concept of velocity. Keep reminding yourself that the acceleration tells you how the velocity *is changing*. The direction of the *change* in velocity is not necessarily the same as the direction of either the initial or final velocities.

Average Acceleration

The **average acceleration** during a time interval Δt is:

$$a_{av,x} = \frac{\Delta v_x}{\Delta t} \tag{2-7}$$

Since average acceleration is the change in velocity divided by the corresponding time interval, the SI units of acceleration are (m/s)/s $=$ m/s^2, read as "meters per second squared." Thinking of m/s^2 as (m/s)/s can help you develop an understanding of what acceleration is. Suppose an object has a constant acceleration $a_x = +3.0$ m/s^2. Then v_x increases 3.0 m/s during every second of elapsed time (the change in v_x is +3.0 m/s per second). If $a_x = -2.0$ m/s^2, then v_x would decrease 2.0 m/s during every second (the change in v_x is -2.0 m/s per second).

For example, suppose it takes 30 s for a truck to slow down from 25 m/s to 10 m/s while traveling east. With the x-axis pointing east, the truck's average acceleration during that time interval is

$$a_{av,x} = \frac{\Delta v_x}{\Delta t} = \frac{-15 \text{ m/s}}{30 \text{ s}} = -0.50 \text{ m/s}^2$$

or 0.50 m/s^2 to the west.

CONNECTION:

Compare average acceleration [Eq. (2-7)] and average velocity [Eq. (2-2)]. Each is the change in a quantity divided by the time interval during which the change occurs. Each can have different values for different time intervals.

Instantaneous Acceleration

To find the **instantaneous acceleration**, we calculate the average acceleration during a *very short time interval*:

Definition of instantaneous acceleration:

$$a_x = \lim_{\Delta t \to 0} \frac{\Delta v_x}{\Delta t}$$

(2-8)

(Δv_x is the change in velocity during a *very short* time interval Δt)

The time interval Δt must be short enough that we can treat the acceleration as constant during that time interval. Just as with instantaneous velocity, the word *instantaneous* is not always repeated. *Acceleration* without the adjective means *instantaneous* acceleration.

Can the lion catch the buffalo?

The chapter opener asked how an African lion can ever catch a Cape buffalo. Although Cape buffaloes and African lions have about the same top *speed*, lions are capable of much larger *accelerations* than are buffaloes. Starting from rest, it takes a buffalo much longer to get to its top speed. On the other hand, lions have much less stamina. Once the buffalo reaches its top speed, it can maintain that speed much longer than can the lion. Thus, a Cape buffalo is capable of outrunning a lion unless the stalking lion can get fairly close before charging.

Conceptual Example 2.4

Direction of Acceleration While Slowing Down

Damon moves in the $-x$-direction on his motor scooter. He "decelerates" as he approaches a stop sign. While slowing down, is the scooter's acceleration a_x positive or negative? What is the direction of the acceleration?

Strategy   The acceleration has the same direction as the *change* in the velocity.

Solution and Discussion   The term *decelerate* is not a scientific term. In common usage it means the scooter is slowing: the scooter's velocity is decreasing in magnitude. Damon is moving in the $-x$-direction, so v_x is negative. He is slowing down, so the *absolute value* of v_x, $|v_x|$, is getting smaller. To reduce the magnitude of a negative number, we have to add a positive number. Therefore, the change in v_x is positive ($\Delta v_x > 0$). In other words, v_x is increasing. Since Δv_x is positive, a_x is positive. The acceleration is in the $+x$-direction.

Conceptual Practice Problem 2.4   Continuing on His Way

As Damon pulls away from the stop sign, continuing in the $-x$-direction, his speed gradually increases. What is the sign of a_x? What is the direction of the acceleration?

The Direction of the Acceleration

Generalizing Example 2.4, suppose an object moves along the x-axis. When the acceleration is in the same direction as the velocity, the object is speeding up. If v_x and a_x are both positive, the object is moving in the $+x$-direction and is speeding up. If they are both negative, the object is moving in the $-x$-direction and is speeding up.

When the acceleration and velocity are in opposite directions, the object is slowing down. When v_x is positive and a_x is negative, the object is moving in the positive x-direction and is slowing down. When v_x is negative and a_x is positive, the object is moving in the negative x-direction and is slowing down.

In straight-line motion, the acceleration is always in the same direction as the velocity, in the direction opposite to the velocity, or zero.

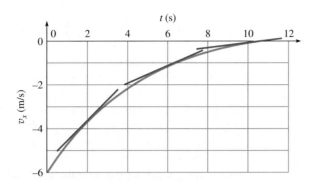

Figure 2.10 In this graph of v_x versus t, as Damon is stopping, v_x is negative, but a_x (the slope) is positive. The value of v_x is increasing, but—since it is less than zero to begin with and is getting closer to zero as time goes on—the speed is *decreasing*. The slopes of the three tangent lines shown represent the instantaneous accelerations (a_x) at three different times.

Graphical Relationships Between Velocity and Acceleration

Both velocity and acceleration measure rates of change: velocity is the rate of change of position and acceleration is the rate of change of velocity. Therefore, the graphical relationship of acceleration to velocity is the same as the graphical relationship of velocity to position: a_x is the slope on a graph of $v_x(t)$ and Δv_x is the area under a graph of $a_x(t)$.

Figure 2.10 shows a graph of v_x versus t for Damon slowing down on his scooter. He is moving in the $-x$-direction, so $v_x < 0$, and his speed is decreasing, so $|v_x|$ is decreasing. The slope of a tangent line to the graph is a_x at that instant. Three tangent lines are drawn, showing that a_x is positive (the slopes are positive) and is not constant (the slopes are not all the same).

CONNECTION:

On a graph of *any* quantity Q as a function of time, the slope of the graph represents the instantaneous rate of change of Q. On a graph of the *rate of change of Q* as a function of time, the area under the graph represents ΔQ.

Example 2.5

Acceleration of a Sports Car

A sports car starting at rest can achieve 30.0 m/s in 4.7 s according to the advertisements. Figure 2.11 shows data for v_x as a function of time as the sports car starts from rest and travels in a straight line in the $+x$-direction. (a) What is the average acceleration of the sports car from 0 to 30.0 m/s? (b) What is the maximum acceleration of the car? (c) What is the car's displacement from $t = 0$ to $t = 19.1$ s (when it reaches 60.0 m/s)? (d) What is the car's average velocity during the entire 19.1 s interval?

Strategy (a) To find the average acceleration, the change in velocity for the time interval is divided by the time interval. (b) The instantaneous acceleration is the slope of the velocity graph, so it is maximum where the graph is steepest. At that point, the velocity is changing at a high rate. We expect the maximum acceleration to take place early on; the magnitude of acceleration must decrease as the velocity gets higher and higher—there is a maximum velocity for the car, after all. (c) The displacement Δx is the area under the $v_x(t)$ graph. The graph is not a simple shape such as a triangle or rectangle, so an estimate of the area is made. (d) Once we have a value for the displacement, we can apply the definition of average velocity.

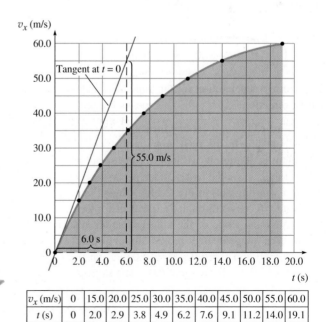

v_x (m/s)	0	15.0	20.0	25.0	30.0	35.0	40.0	45.0	50.0	55.0	60.0
t (s)	0	2.0	2.9	3.8	4.9	6.2	7.6	9.1	11.2	14.0	19.1

Figure 2.11

Data table and graph of $v_x(t)$ for a sports car.

continued on next page

Example 2.5 continued

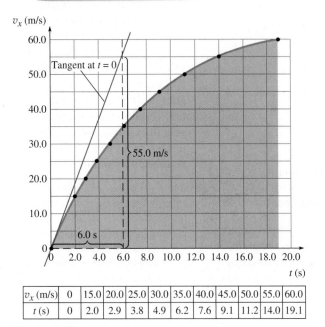

v_x (m/s)	0	15.0	20.0	25.0	30.0	35.0	40.0	45.0	50.0	55.0	60.0
t (s)	0	2.0	2.9	3.8	4.9	6.2	7.6	9.1	11.2	14.0	19.1

Figure 2.11

Data table and graph of $v_x(t)$ for a sports car.

Given: Graph of $v_x(t)$ in Fig. 2.11.

To find: (a) $a_{av,x}$ for $v_x = 0$ to 30.0 m/s; (b) maximum value of a_x; (c) Δx from $v_x = 0$ to 60.0 m/s; (d) $v_{av,x}$ from $t = 0$ to 19.1 s

Solution   (a) The car starts from rest, so $v_{xi} = 0$. It reaches $v_x = 30.0$ m/s at $t = 4.9$ s, according to the data table. Then for this time interval,

$$a_{av,x} = \frac{\Delta v_x}{\Delta t} = \frac{30.0 \text{ m/s} - 0 \text{ m/s}}{4.9 \text{ s} - 0 \text{ s}} = 6.1 \text{ m/s}^2$$

The average acceleration for this time interval is 6.1 m/s² in the +x-direction.

(b) The acceleration a_x, at any instant of time, is the slope of the tangent line to the $v_x(t)$ graph at that time. To find the maximum acceleration, we look for the steepest part of the graph. In this case, the largest slope occurs near $t = 0$, just as the car is starting out. In Fig. 2.11, a tangent line to the $v_x(t)$ graph at $t = 0$ passes through $t = 0$. Values for the rise and run to calculate the slope of the tangent line are read from the graph. The tangent line passes through the two points $(t = 0, v_x = 0)$ and $(t = 6.0 \text{ s}, v_x = 55.0 \text{ m/s})$ on the graph, so the rise is 55.0 m/s for a run of 6.0 s. The slope of this line is

$$a_x = \frac{\text{rise}}{\text{run}} = \frac{55.0 \text{ m/s} - 0 \text{ m/s}}{6.0 \text{ s} - 0 \text{ s}} = +9.2 \text{ m/s}^2$$

The maximum acceleration is 9.2 m/s² in the +x-direction.

(c) Δx is the area under the $v_x(t)$ graph shown shaded in Fig. 2.11. The area can be estimated by counting the number of grid boxes under the curve. Each box is 5.0 m/s in height and 2.0 s in width, so each represents an "area" (displacement) of 10 m. When counting the number of boxes under the curve, a best estimate is made for the fraction of the boxes that are only partly below the curve. Approximately 75 boxes lie below the curve, so the displacement is $\Delta x = 75 \times 10$ m = 750 m. Since the car travels along a straight line and does not change direction, 750 m is also the distance traveled. (d) The average velocity during the 19.1-s interval is

$$v_{av,x} = \frac{\Delta x}{\Delta t} = \frac{750 \text{ m}}{19.1 \text{ s}} = 39 \text{ m/s}$$

Discussion   The graph of velocity as a function of time is often the most helpful graph to have when solving a problem. If that graph is not given in the problem, it is useful to sketch one. The $v_x(t)$ graph shows displacement, velocity, and acceleration at once: the velocity v_x is given by the points or the curve graphed, the displacement Δx is the area under the curve, and the acceleration a_x is the slope of the curve.

Why is the average velocity 39 m/s? Why is it not halfway between the initial velocity (0 m/s) and the final velocity (60 m/s)? If the acceleration were constant, the average velocity would indeed be $\frac{1}{2}(0 + 60$ m/s$) = 30$ m/s. The actual average velocity is somewhat higher than that—the acceleration is greater at the start, so less of the time interval is spent going (relatively) slow and more is spent going fast. The speed is less than 30 m/s for only 4.9 s, but is greater than 30 m/s for 14.2 s.

Practice Problem 2.5   Braking a Car

An automobile is traveling along a straight road heading to the southeast at 24 m/s when the driver sees a deer begin to cross the road ahead of her. She steps on the brake and brings the car to a complete stop in an elapsed time of 8.0 s. A data recording device, triggered by the sudden braking action, records the following velocities and times as the car slows. Let the positive x-axis be directed to the southeast. Plot a graph of v_x versus t and find (a) the average acceleration as the car comes to a stop and (b) the instantaneous acceleration at $t = 2.0$ s.

v_x (m/s)	24	17.3	12.0	8.7	6.0	3.5	2.0	0.75	0
t (s)	0	1.0	2.0	3.0	4.0	5.0	6.0	7.0	8.0

✓ CHECKPOINT 2.3

What physical quantity does the slope of the tangent to a graph of v_x versus time represent?

2.4   MOTION ALONG A LINE WITH CONSTANT ACCELERATION

The graphical and mathematical relationships between position, velocity, and acceleration presented so far apply regardless of whether the acceleration is changing or is constant. In the important special case of an object whose acceleration is *constant* (both in magnitude and direction), we can write these relationships as algebraic equations. First, let us agree on a consistent notation:

- Choose an origin and a direction for the positive axis. (For vertical motion, it is conventional to use the y-axis instead of the x-axis, where the $+y$-direction is up.)
- At an initial time t_i, the initial position and velocity are x_i and v_{ix}.
- At a later time $t_f = t_i + \Delta t$, the final position and velocity are x_f and v_{fx}.

From the following two essential relationships the others can be derived:

1. Since the acceleration a_x is constant, the change in velocity over a given time interval $\Delta t = t_f - t_i$ is the acceleration—the rate of change of velocity—times the elapsed time:

$$\Delta v_x = v_{fx} - v_{ix} = a_x \Delta t \qquad (2\text{-}9)$$

(if a_x is constant during the entire time interval)

Equation (2-9) is the definition of a_x [Eq. (2-8)] with the assumption that a_x is constant.

2. Since the velocity changes linearly with time, the average velocity is given by:

$$v_{av,x} = \tfrac{1}{2}(v_{fx} + v_{ix}) \quad (\text{constant } a_x) \qquad (2\text{-}10)$$

Equation (2-10) is *not* true *in general*, but it is true for constant acceleration. To see why, refer to the $v_x(t)$ graph in Fig. 2.12a. The graph is linear because the acceleration—the slope of the graph—is constant. The displacement during any time interval is represented by the area under the graph. The average velocity is found by forming a rectangle with an area equal to the area under the curve in Fig. 2.12a, because the average velocity should give the same displacement in the same time interval. Figure 2.12b shows that, to make the excluded area above $v_{av,x}$ (triangle 1) equal to the extra area under $v_{av,x}$ (triangle 2), the average velocity must be exactly halfway between the initial and final velocities. Combining Eq. (2-10) with the definition of average velocity,

$$\Delta x = x_f - x_i = v_{av,x} \Delta t \qquad (2\text{-}2)$$

gives our second essential relationship for constant acceleration:

$$\Delta x = \tfrac{1}{2}(v_{fx} + v_{ix}) \Delta t \qquad (2\text{-}11)$$

(if a_x is constant during the entire time interval)

If the acceleration is *not* constant, there is no reason why the average velocity has to be exactly halfway between the initial and the final velocity. As an illustration, imagine a trip where you drive along a straight highway at 80 km/h for 50 min and

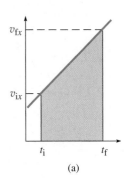

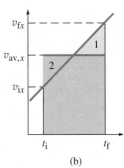

Figure 2.12 Finding the average velocity when the acceleration is constant.

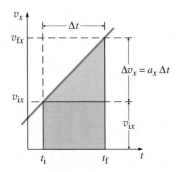

Figure 2.13 Graphical interpretation of Eq. (2-12).

then at 60 km/h for 30 min. Your acceleration is zero for the entire trip *except* during the few seconds while you slowed from 80 km/h to 60 km/h. The magnitude of your average velocity is *not* 70 km/h. You spent more time going 80 km/h than you did going 60 km/h, so the magnitude of your average velocity would be greater than 70 km/h.

Other Useful Relationships for Constant Acceleration   Two more useful relationships can be formed between the various quantities (displacement, initial and final velocities, acceleration, and time interval) by eliminating some quantity from Eqs. (2-9) and (2-11). For example, suppose we don't know the final velocity v_{fx}. Then we can solve Eq. (2-9) for v_{fx}, substitute into Eq. (2-11), and simplify:

$$\Delta x = \tfrac{1}{2}(v_{fx} + v_{ix})\,\Delta t = \tfrac{1}{2}[(v_{ix} + a_x\Delta t) + v_{ix}]\,\Delta t$$

$$\Delta x = v_{ix}\Delta t + \tfrac{1}{2}a_x(\Delta t)^2 \quad \text{(constant } a_x) \tag{2-12}$$

We can interpret Eq. (2-12) graphically. Figure 2.13 shows a $v_x(t)$ graph for motion with constant acceleration. The displacement that occurs between t_i and a later time t_f is the area under the graph for that time interval. Partition this area into a rectangle plus a triangle. The area of the rectangle is

$$\text{base} \times \text{height} = v_{ix}\,\Delta t$$

The height of the triangle is the change in velocity, which is equal to $a_x\,\Delta t$. The area of the triangle is

$$\tfrac{1}{2}\text{base} \times \text{height} = \tfrac{1}{2}\Delta t \times a_x\,\Delta t = \tfrac{1}{2}a_x(\Delta t)^2$$

Adding these areas gives Eq. (2-12).

Another useful relationship comes from eliminating the time interval Δt:

$$\Delta x = \tfrac{1}{2}(v_{fx} + v_{ix})\,\Delta t = \tfrac{1}{2}(v_{fx} + v_{ix})\left(\frac{v_{fx} - v_{ix}}{a_x}\right) = \frac{v_{fx}^2 - v_{ix}^2}{2a_x}$$

Rearranging terms,

$$v_{fx}^2 - v_{ix}^2 = 2a_x\,\Delta x \quad \text{(constant } a_x) \tag{2-13}$$

✓ CHECKPOINT 2.4

At 3:00 P.M., an airplane is moving due west at 460 km/h. At 3:05 P.M., it is moving due west at 480 km/h. Is its average velocity during the time interval necessarily 470 km/h west? Explain.

Example 2.6

A Sliding Brick

Starting from rest, a brick slides along a straight line down an icy roof with a constant acceleration of magnitude 4.9 m/s² (Fig. 2.14). How fast is the brick moving when it reaches the edge of the roof 0.90 s later?

Strategy   What is the direction of the acceleration? It has to be downward along the roof, in the same direction as the brick's velocity. An acceleration opposite the velocity would make the brick slow down, but since it starts from rest, a

continued on next page

Example 2.6 continued

constant acceleration can only make it speed up. We choose the +x-axis in the direction of the acceleration. Then we use the acceleration to find how the velocity changes during the time interval.

Solution   With the x-axis in the direction of the acceleration, $a_x = +4.9$ m/s². The brick is initially at rest so $v_{ix} = 0$.

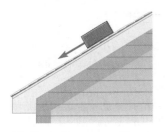

Figure 2.14

A brick sliding down an icy roof.

We want to know v_{fx} at the end of the time interval $\Delta t = 0.90$ s. Since a_x is constant, v_x changes at a constant rate:

$$\Delta v_x = v_{fx} - v_{ix} = a_x \, \Delta t = (+4.9 \text{ m/s}^2) \times (0.90 \text{ s}) = 4.4 \text{ m/s}$$

At the edge of the roof, the brick is moving at 4.4 m/s parallel to the roof.

Discussion   Conceptual check: $a_x = +4.9$ m/s² means that v_x increases 4.9 m/s every second. The brick slides for a bit less than 1 s, so the increase in v_x is a bit less than 4.9 m/s.

Practice Problem 2.6   Displacement of the Brick

How far from the edge of the roof was the brick when it started sliding?

Example 2.7

Displacement of a Motorboat

A motorboat starts from rest at a dock and heads due east with a constant acceleration of magnitude 2.8 m/s². After traveling for 140 m, the motor is throttled down to slow down the boat at 1.2 m/s² (while still moving east) until its speed is 16 m/s. Just as the boat attains the speed of 16 m/s, it passes a buoy due east of the dock. (a) Sketch a qualitative graph of $v_x(t)$ for the motorboat from the dock to the buoy. Let the +x-axis point east. (b) What is the distance between the dock and the buoy?

Strategy   This problem involves two different values of acceleration, so it must be divided into two subproblems. The equations for constant acceleration cannot be applied to a time interval during which the acceleration changes. But for each of two time intervals, the acceleration of the boat is constant: from t_1 to t_2, $a_{1x} = +2.8$ m/s²; from t_2 to t_3, $a_{2x} = -1.2$ m/s². The two subproblems are connected by the position and velocity of the boat at the instant the acceleration changes. This is reflected in the graph of $v_x(t)$: It consists of two different straightline segments with different slopes that connect with the same value of v_x at time t_2.

For subproblem 1, the boat speeds up with a constant acceleration of 2.8 m/s² to the east. We know the acceleration, the displacement (140 m east), and the initial velocity: the boat starts from rest, so the initial velocity v_{1x} is zero. We need to calculate the final velocity v_{2x}, which then becomes the initial velocity for the second subproblem.

The boat is always headed to the east, so we choose east as the positive x-direction.

> **Subproblem 1:**
>
> Known: $v_{1x} = 0$; $a_{1x} = +2.8$ m/s²;
> $\Delta x_{21} = x_2 - x_1 = 140$ m.
>
> To find: v_{2x}.

For subproblem 2, we know acceleration, final velocity v_{3x}, and we have just found the initial velocity v_{2x} from subproblem 1. Because the boat is slowing down, its acceleration is in the direction opposite its velocity; therefore, $a_{2x} < 0$. From these three quantities we can find the displacement of the boat during the second time interval.

> **Subproblem 2:**
>
> Known: v_{2x} from subproblem 1;
> $a_{2x} = -1.2$ m/s²; $v_{3x} = +16$ m/s.
>
> To find: $\Delta x_{32} = x_3 - x_2$.

Adding the displacements for the two time intervals gives the total displacement. The magnitude of the total displacement is the distance between the dock and the buoy.

continued on next page

Example 2.7 continued

Solution   (a) The graph starts with $v_x = 0$ at $t = t_1$. We choose $t_1 = 0$ for simplicity. The graph is a straight line with slope $+2.8$ m/s^2 until $t = t_2$. Then, starting from where the graph left off, the graph continues as a straight line with slope -1.2 m/s^2 until the graph reaches $v_x = 16$ m/s at $t = t_3$. Figure 2.15 shows the $v_x(t)$ graph. It is not quantitatively accurate because we have not calculated the values of t_2 and t_3.

(b1) To find v_{2x} without knowing the time interval, we eliminate Δt from Eqs. (2-9) and (2-11) for constant acceleration:

$$\Delta x_{21} = \frac{1}{2}(v_{2x} + v_{1x})\,\Delta t = \frac{1}{2}(v_{2x} + v_{1x})\left(\frac{v_{2x} - v_{1x}}{a_{1x}}\right) = \frac{v_{2x}^2 - v_{1x}^2}{2a_{1x}}$$

Solving for v_{2x},

$$v_{2x} = \pm\sqrt{v_{1x}^2 + 2a_{1x}\,\Delta x} = \pm\sqrt{0 + 2 \times 2.8 \text{ m/s}^2 \times 140 \text{ m}}$$
$$= \pm 28 \text{ m/s}$$

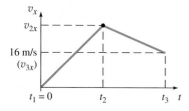

Figure 2.15

Graph of v_x versus t for the motorboat.

The boat is moving east, in the $+x$-direction, so the correct sign here is positive: $v_{2x} = +28$ m/s.

(b2) The final velocity for the first interval (v_{2x}) is the *initial* velocity for the second interval. The final velocity is v_{3x}. Using the same equation just derived for this time interval,

$$\Delta x_{32} = \frac{v_{3x}^2 - v_{2x}^2}{2a_{2x}} = \frac{(16 \text{ m/s})^2 - (28 \text{ m/s})^2}{2 \times (-1.2 \text{ m/s}^2)} = +220 \text{ m}$$

The total displacement is

$$x_3 - x_1 = (x_3 - x_2) + (x_2 - x_1) = 220 \text{ m} + 140 \text{ m} = +360 \text{ m}$$

The buoy is 360 m from the dock.

Discussion   The natural division of the problem into two parts occurs because the boat has two different constant accelerations during two different time periods. In problems that can be subdivided in this way, the final velocity and position found in the first part becomes the initial velocity and position for the second part.

Practice Problem 2.7   Time to Reach the Buoy

What is the time required by the boat in Example 2.7 to reach the buoy?

2.5   VISUALIZING MOTION ALONG A LINE WITH CONSTANT ACCELERATION

Motion Diagrams   In Fig. 2.16, three carts move in the same direction with three different values of constant acceleration. The position of each cart is depicted in a **motion diagram** as it would appear in a stroboscopic photograph with pictures taken at equal time intervals (here, the time interval is 1.0 s).

The yellow cart has zero acceleration and, therefore, constant velocity. During each 1.0-s time interval its displacement is the same: 1.0 m/s × 1.0 s = 1.0 m to the right.

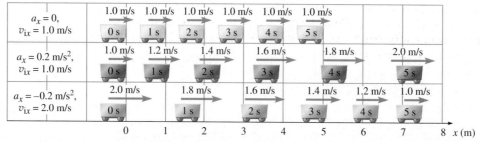

Figure 2.16   Each cart is shown as if photographs were taken at 1.0-s time intervals of 1.0 s. The arrows above each cart indicate the instantaneous velocities.

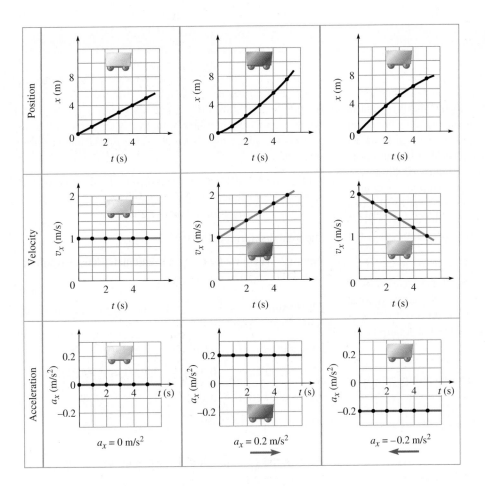

Figure 2.17 Graphs of position, velocity, and acceleration for the carts of Fig. 2.16.

The red cart has a constant acceleration of 0.2 m/s^2 to the right. Although m/s^2 is normally read "meters per second squared," it can be useful to think of it as "m/s per second": the cart's velocity changes by 0.2 m/s during each 1.0-s time interval. In this case, acceleration is in the same direction as the velocity, so the velocity increases. The displacement of the cart during successive 1.0-s time intervals gets larger and larger.

The blue cart experiences a constant acceleration of 0.2 m/s^2 in the $-x$-direction—the direction *opposite* to the velocity. The magnitude of the velocity then decreases; during each 1.0-s interval, the speed decreases by 0.2 m/s. Now the displacements during 1.0-s intervals get smaller and smaller.

Graphs   Figure 2.17 shows graphs of $x(t)$, $v_x(t)$, and $a_x(t)$ for each of the carts. The acceleration graphs are horizontal since each of the carts has a constant acceleration. All three v_x graphs are straight lines. Since a_x is the rate of change of v_x, the slope of the v_x graph at any value of t is a_x at that value of t. With constant acceleration, the slope is the same everywhere and the graph is linear. Remember that a positive a_x does mean that v_x is increasing, but not necessarily that the *speed* is increasing. If v_x is negative, then a positive a_x indicates a *decreasing* speed. (See Conceptual Example 2.4.) Speed is increasing when the acceleration and velocity are in the same direction (a_x and v_x both positive *or* both negative). Speed is decreasing when acceleration and velocity are in opposite directions—when a_x and v_x have opposite signs.

The position graph is linear for the yellow cart because it has constant velocity. For the red cart, the $x(t)$ graph curves with increasing slope, showing that v_x is increasing. For the blue cart, the $x(t)$ graph curves with decreasing slope, showing that v_x is decreasing.

Example 2.8

Two Spaceships

Two spaceships are moving from the same starting point in the $+x$-direction with constant accelerations. The silver spaceship has an initial velocity of $+2.00$ km/s and an acceleration of $+0.400$ km/s^2. The black spaceship has an initial velocity of $+6.00$ km/s and an acceleration of -0.400 km/s^2. (a) Find the time at which the silver spaceship just overtakes the black spaceship. (b) Sketch graphs of $v_x(t)$ for the two spaceships. (c) Sketch a motion diagram (similar to Fig. 2.16) showing the positions of the two spaceships at 1.0-s intervals.

Strategy We can find the positions of the spaceships at later times from the initial velocities and the accelerations. At first, the black spaceship is moving faster, so it pulls out ahead. Later, the silver ship overtakes the black ship at the instant their *positions are equal.*

Solution (a) The position of either spaceship at a later time is given by Eq. (2-12):

$$x_f = x_i + \Delta x = x_i + v_{ix}\,\Delta t + \tfrac{1}{2}a_x(\Delta t)^2$$

We set the final position of the silver spaceship equal to that of the black spaceship ($x_{fs} = x_{fb}$):

$$x_{is} + v_{isx}\,\Delta t + \tfrac{1}{2}a_{sx}\,(\Delta t)^2 = x_{ib} + v_{ibx}\,\Delta t + \tfrac{1}{2}a_{bx}(\Delta t)^2$$

Subscripts are useful for preventing you from mixing up similar quantities. The subscripts s and b stand for silver and black, respectively. The subscripts i and f stand for initial and final, respectively. A skilled problem-solver must be able to come up with algebraic symbols that are explicit and unambiguous.

The initial positions are the same: $x_{is} = x_{ib}$. Subtracting the initial positions from each side, moving all terms to one side, and factoring out one power of Δt yields

$$\Delta t(v_{isx} + \tfrac{1}{2}a_{sx}\,\Delta t - v_{ibx} - \tfrac{1}{2}a_{bx}\,\Delta t) = 0$$

This equation has two solutions—there are two times at which the spaceships are at the same position. One solution is $\Delta t = 0$. We already knew that the two spaceships started at the same *initial* position. The other solution, which gives the time at which one spaceship overtakes the other, is found by setting the expression in parentheses equal to zero. Solving for Δt,

$$\Delta t = \frac{2(v_{isx} - v_{ibx})}{a_{bx} - a_{sx}} = \frac{2 \times (2.00 \text{ km/s} - 6.00 \text{ km/s})}{-0.400 \text{ km/s}^2 - 0.400 \text{ km/s}^2} = 10.0 \text{ s}$$

The silver spaceship overtakes the black spaceship 10.0 s after they leave the starting point.

(b) Figure 2.18 shows the $v_x(t)$ graphs with $t_i = 0$. Note that the area under the graphs from t_i to t_f is the same in the two graphs: the spaceships have the same displacement during that interval.

(c) Equation (2-12) can be used to find the position of each spaceship as a function of time. Choosing $x_i = 0$, $t_i = 0$, and $t = t_f$, the position at time t is

$$x(t) = 0 + v_{ix}t + \tfrac{1}{2}a_x t^2$$

Figure 2.19 shows the data table calculated this way and the corresponding motion diagram.

Discussion Quick check: the two ships must have the same displacement at $\Delta t = 10.0$ s.

$$\Delta x_s = v_{isx}\,\Delta t + \tfrac{1}{2}a_{sx}(\Delta t)^2$$
$$= 2.00 \text{ km/s} \times 10.0 \text{ s} + \tfrac{1}{2} \times 0.400 \text{ km/s}^2 \times (10.0 \text{ s})^2$$
$$= 40.0 \text{ km}$$

$$\Delta x_b = v_{ibx}\,\Delta t + \tfrac{1}{2}a_{bx}(\Delta t)^2$$
$$= 6.00 \text{ km/s} \times 10.0 \text{ s} + \tfrac{1}{2} \times (-0.400 \text{ km/s}^2) \times (10.0 \text{ s})^2$$
$$= 40.0 \text{ km}$$

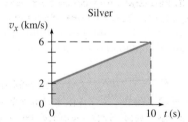

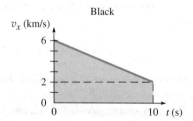

Figure 2.18

Graphs of v_x versus t for the silver and black spaceships. The shaded area under each graph represents the displacement Δx during the time interval.

continued on next page

Example 2.8 continued

Practice Problem 2.8   Time to Reach the Same Velocity

When do the two spaceships have the same *velocity*? What is the value of the velocity then?

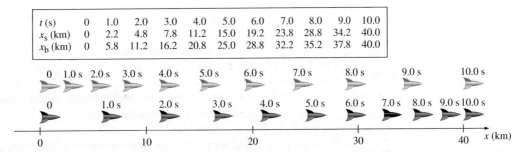

Figure 2.19
Calculated positions of the spaceships at 1.0-s time intervals and a motion diagram.

2.6   FREE FALL

Suppose you are standing on a bridge over a deep gorge. If you drop a stone into the gorge, how fast does it fall? You know from experience that it does not fall at a constant velocity; the longer it falls, the faster it goes. A better question is: What is the stone's acceleration?

First, let us simplify the problem. If the stone were moving very fast, air resistance would oppose its motion. When it is not falling so fast, the effect of air resistance is negligibly small. In **free fall**, no forces act on an object other than the gravitational force that makes the object fall. On Earth, free fall is an idealization since there is always *some* air resistance. We also assume that the stone's change in altitude is small enough that Earth's gravitational pull on it is constant.

CONNECTION:

Free fall is an example of motion with constant acceleration.

Free-fall Acceleration   An object in free fall has a constant downward acceleration, called the *free-fall acceleration.* The magnitude of this acceleration varies a little from one place to another near Earth's surface, but at any given place, it has the same value for every object, regardless of the mass of the object. Unless another value is given in a particular problem, please assume that the magnitude of the free-fall acceleration near Earth's surface is

$$a_{\text{free fall}} = g = 9.80 \text{ m/s}^2 \qquad (2\text{-}14)$$

The symbol g represents the magnitude of the free-fall acceleration.

When dealing with vertical motion, the y-axis is usually chosen to be positive pointing upward. The direction of the free-fall acceleration is down, so $a_y = -g$. The same techniques and equations used for other constant acceleration situations are used with free fall.

Earth's gravity always pulls downward, so the acceleration of an object in free fall is always downward and constant in magnitude, *regardless of whether the object is moving up, down, or is at rest, and independent of its speed.* If the object is moving downward, the downward acceleration makes it speed up; if it is moving upward, the downward acceleration makes it slow down.

In free fall, $a_y = -g$ (if the y-axis points up).

Figure 2.20 Graph of v_y versus t for an object thrown upward.

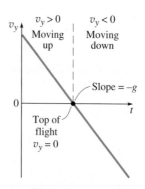

Acceleration at Highest Point   If an object is thrown straight up, its velocity is zero at the highest point of its flight. Why? On the way up, its velocity v_y is positive (if the positive y-axis is pointing up). On the way down, v_y is negative. Since v_y changes continuously, it must pass through zero to change sign (Fig. 2.20). At the highest point, the velocity is zero but the *acceleration* is *not* zero. If the acceleration were to suddenly become zero at the top of flight, the velocity would no longer change; the object would get *stuck at the top* rather than fall back down. The velocity is zero at the top but it does not *stay* zero; it keeps changing at the same rate.

✓ CHECKPOINT 2.6

Is it possible for an object in free fall to be moving upward? Explain.

Example 2.9

Throwing Stones

Standing on a bridge, you throw a stone straight upward. The stone hits a stream, 44.1 m below the point at which you release it, 4.00 s later. (a) What is the velocity of the stone just after it leaves your hand? (b) What is the velocity of the stone just before it hits the water? (c) Draw a motion diagram for the stone, showing its position at 0.1-s intervals during the first 0.9 s of its motion. (d) Sketch graphs of $y(t)$ and $v_y(t)$. The positive y-axis points up.

Strategy   Ignoring air resistance, the stone is in free fall once your hand releases it and until it hits the water. For the time interval during which the stone is in free fall, the initial velocity is the velocity of the stone *just after* it leaves your hand and the final velocity is the velocity *just before* it hits the water. During free fall, the stone's acceleration is constant and assumed to be 9.80 m/s² downward. Known: $a_y = -9.80$ m/s²; $\Delta y = -44.1$ m at $\Delta t = 4.00$ s. To find: v_{iy} and v_{fy}.

Solution   (a) Equation (2-12) can be used to solve for v_{iy} since all the other quantities in it (Δy, Δt, and a_y) are known and the acceleration is constant.

$$\Delta y = v_{iy}\,\Delta t + \tfrac{1}{2}a_y(\Delta t)^2$$

Solving for v_{iy},

$$v_{iy} = \frac{\Delta y}{\Delta t} - \frac{1}{2}a_y\,\Delta t \qquad (1)$$

$$= \frac{-44.1 \text{ m}}{4.00 \text{ s}} - \frac{1}{2}(-9.80 \text{ m/s}^2 \times 4.00 \text{ s})$$

$$= -11.0 \text{ m/s} + 19.6 \text{ m/s} = 8.6 \text{ m/s}$$

The initial velocity is 8.6 m/s upward.

(b) The change in v_y is $a_y\,\Delta t$ from Eq. (2-9):

$$v_{fy} = v_{iy} + a_y\,\Delta t$$

Substituting the expression for v_{iy} in the preceding equation,

$$v_{fy} = \left(\frac{\Delta y}{\Delta t} - \frac{1}{2}a_y\,\Delta t\right) + a_y\,\Delta t = \frac{\Delta y}{\Delta t} + \frac{1}{2}a_y\,\Delta t \qquad (2)$$

$$= \frac{-44.1 \text{ m}}{4.00 \text{ s}} + \frac{1}{2}(-9.80 \text{ m/s}^2 \times 4.00 \text{ s})$$

$$= -11.0 \text{ m/s} - 19.6 \text{ m/s} = -30.6 \text{ m/s}$$

The final velocity is 30.6 m/s downward.

continued on next page

Example 2.9 continued

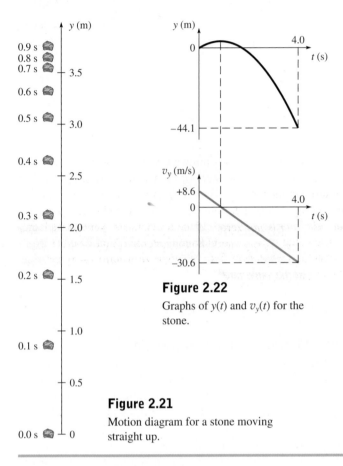

Figure 2.22

Graphs of $y(t)$ and $v_y(t)$ for the stone.

Figure 2.21

Motion diagram for a stone moving straight up.

(c) Choosing $y_i = 0$ and $t_i = 0$, the position of the stone as a function of time is

$$y(t) = v_{iy}t + \tfrac{1}{2}a_y t^2$$

The motion diagram is shown in Fig. 2.21.

(d) The graphs are shown in Fig. 2.22.

Discussion The final speed is greater than the initial speed, as expected. Equations (1) and (2) have a direct interpretation, which is a good check on their validity. The first term, $\Delta y/\Delta t$, is the average velocity of the stone during the 4.00 s of free fall. The second term, $\tfrac{1}{2}a_y\Delta t$, is *half* the change in v_y since $\Delta v_y = a_y \Delta t$. Because the acceleration is constant, the average velocity is halfway between the initial and final velocities. Therefore, the initial velocity is the average velocity minus half of the change, while the final velocity is the average velocity plus half of the change.

Practice Problem 2.9   Height Attained by Stone

(a) How high above the bridge does the stone go? [*Hint:* What is v_y at the highest point?] (b) If you dropped the stone instead of throwing it, how long would it take to hit the water?

Master the Concepts

- Displacement is the change in position: $\Delta x = x_f - x_i$. The displacement depends only on the starting and ending positions, not on details of the motion. The magnitude of the displacement is not necessarily equal to the total distance traveled; it is the straight-line distance from the initial position to the final position.
- Average velocity is the constant velocity that would cause the same displacement in the same amount of time.

$$v_{av,x} = \frac{\Delta x}{\Delta t} \text{ (for any time interval } \Delta t) \quad (2\text{-}2)$$

- Velocity is a measure of how fast and in what direction something moves. Its direction is the direction of the object's motion and its magnitude is the instantaneous speed. It is the instantaneous rate of change of the position.

$$v_x = \lim_{\Delta t \to 0} \frac{\Delta x}{\Delta t} \text{ (for a } \textit{very short} \text{ time interval } \Delta t) \quad (2\text{-}3)$$

- Average acceleration is the constant acceleration that would cause the same velocity change in the same amount of time.

$$a_{av,x} = \frac{\Delta v_x}{\Delta t}\text{(for any time interval } \Delta t) \quad (2\text{-}7)$$

- Acceleration is the instantaneous rate of change of the velocity.

$$a_x = \lim_{\Delta t \to 0} \frac{\Delta v_x}{\Delta t}\text{(for a } \textit{very short} \text{ time interval } \Delta t) \quad (2\text{-}8)$$

Acceleration does not necessarily mean speeding up. A velocity can change by decreasing speed or by changing direction.

- Interpreting graphs: On a graph of $x(t)$, the slope at any point is v_x. On a graph of $v_x(t)$, the slope at any point is a_x, and the area under the graph during any time interval is the displacement Δx during that time interval. If v_x is negative, the displacement is also

continued on next page

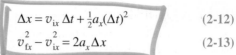

$$\Delta x = v_{ix}\,\Delta t + \tfrac{1}{2}a_x(\Delta t)^2 \qquad (2\text{-}12)$$

$$v_{fx}^2 - v_{ix}^2 = 2a_x\,\Delta x \qquad (2\text{-}13)$$

Master the Concepts continued

negative, so we must count the area as negative when it is below the time axis.

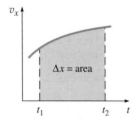

On a graph of $a_x(t)$, the area under the curve is Δv_x, the change in v_x during that time interval.

These same relationships hold for position, velocity, and acceleration along the y-axis if a_y is constant.

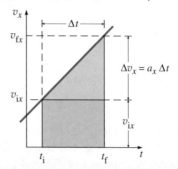

- Essential relationships for constant acceleration problems: if a_x is constant during the entire time interval Δt from t_i until a later time $t_f = t_i + \Delta t$,

$$\Delta v_x = v_{fx} - v_{ix} = a_x\,\Delta t \qquad (2\text{-}9)$$

$$\Delta x = \tfrac{1}{2}(v_{fx} + v_{ix})\,\Delta t \qquad (2\text{-}11)$$

- An object in free fall has a constant downward acceleration. The magnitude of the acceleration g varies a little from place to place near Earth's surface. A typical value is $g = 9.80\ \text{m/s}^2$.

Conceptual Questions

1. Explain the difference between distance traveled, displacement, and displacement magnitude.

2. Explain the difference between speed and velocity.

3. On a graph of v_x versus time, what quantity does the area under the graph represent?

4. On a graph of v_x versus time, what quantity does the slope of the graph represent?

5. On a graph of a_x versus time, what quantity does the area under the graph represent?

6. On a graph of x versus time, what quantity does the slope of the graph represent?

7. What is the relationship between average velocity and instantaneous velocity? An object can have different instantaneous velocities at different times. Can the same object have different average velocities? Explain.

8. Can the velocity of an object be zero and the acceleration be nonzero at the same time? Explain.

9. You are bicycling along a straight north-south road. Let the x-axis point north. Describe your motion in each of the following cases. Example: $a_x > 0$ and $v_x > 0$ means you are moving north and speeding up. (a) $a_x > 0$ and $v_x < 0$. (b) $a_x = 0$ and $v_x < 0$. (c) $a_x < 0$ and $v_x = 0$. (d) $a_x < 0$ and $v_x < 0$. (e) Based on your answers, explain why it is not a good idea to use the expression "negative acceleration" to mean slowing down.

10. When a coin is tossed straight up, what can you say about its velocity and acceleration at the highest point of its motion?

Multiple-Choice Questions

1. A ball is thrown straight up into the air. Ignore air resistance. While the ball is in the air its acceleration
   (a) increases.
   (b) is zero.
   (c) remains constant.
   (d) decreases on the way up and increases on the way down.
   (e) changes direction.

2. Which car has a westward acceleration?
   (a) a car traveling westward at constant speed
   (b) a car traveling eastward and speeding up
   (c) a car traveling westward and slowing down
   (d) a car traveling eastward and slowing down
   (e) a car starting from rest and moving toward the east

Questions 3 and 4. A toy rocket is propelled straight upward from the ground and reaches a height Δy. After an elapsed time Δt, measured from the time the rocket was first fired off, the rocket has fallen back down to the ground, landing at the same spot from which it was launched.

Answer choices:

(a) zero

(b) $2\dfrac{\Delta y}{\Delta t}$

(c) $\dfrac{\Delta y}{\Delta t}$

(d) $\dfrac{1}{2}\dfrac{\Delta y}{\Delta t}$

3. What is the magnitude of the average velocity of the rocket during this time?

4. What is the average speed of the rocket during this time?

5. A leopard starts from rest at $t = 0$ and runs in a straight line with a constant acceleration until $t = 3.0$ s. The distance covered by the leopard between $t = 1.0$ s and $t = 2.0$ s is
   (a) the same as the distance covered during the first second.
   (b) twice the distance covered during the first second.
   (c) three times the distance covered during the first second.
   (d) four times the distance covered during the first second.

Multiple-Choice Questions 6–15. A jogger is exercising along a long, straight road that runs north-south. She starts out heading north. A graph of $v_x(t)$ follows Question 10.

6. What is the displacement of the jogger from $t = 18.0$ min to $t = 24.0$ min?
   (a) 720 m, south
   (b) 720 m, north
   (c) 2160 m, south
   (d) 3600 m, north

7. What is the displacement of the jogger for the entire 30.0 min?
   (a) 3120 m, south
   (b) 2400 m, north
   (c) 2400 m, south
   (d) 3840 m, north

8. What is the total distance traveled by the jogger in 30.0 min?
   (a) 3840 m   (b) 2340 m   (c) 2400 m   (d) 3600 m

9. What is the average velocity of the jogger during the 30.0 min?
   (a) 1.3 m/s, north
   (b) 1.7 m/s, north
   (c) 2.1 m/s, north
   (d) 2.9 m/s, north

10. What is the average speed of the jogger for the 30 min?
    (a) 1.4 m/s
    (b) 1.7 m/s
    (c) 2.1 m/s
    (d) 2.9 m/s

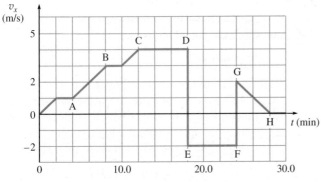

Multiple-Choice Questions 6–15

11. In what direction is she running at time $t = 20$ min?
    (a) south
    (b) north
    (c) not enough information

12. In which region of the graph is a_x positive?
    (a) A to B
    (b) C to D
    (c) E to F
    (d) G to H

13. In which region is a_x negative?
    (a) A to B
    (b) C to D
    (c) E to F
    (d) G to H

14. In which region is the velocity directed to the south?
    (a) A to B
    (b) C to D
    (c) E to F
    (d) G to H

15. What distance does the jogger travel during the first 10.0 min ($t = 0$ to 10.0 min)?
    (a) 8.5 m
    (b) 510 m
    (c) 900 m
    (d) 1020 m

16. The figure shown here has four graphs of x versus time. Which graph shows a constant, positive, nonzero velocity?

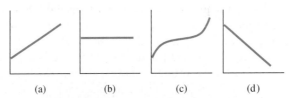

(a)   (b)   (c)   (d)

Multiple-Choice Questions 16 and 17

17. The four graphs show v_x versus time. (a) Which graph shows a constant velocity? (b) Which graph shows a_x constant and positive? (c) Which graph shows a_x constant and negative? (d) Which graph shows a changing a_x that is always positive?

Problems

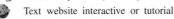

ⓒ   Combination conceptual/quantitative problem
♥   Biological or medical application
✦   Challenging problem
Blue #   Detailed solution in the Student Solutions Manual
① ②   Problems paired by concept
〰   Text website interactive or tutorial

2.1 Position and Displacement

1. A displacement of magnitude 32 cm toward the east is followed by displacements of magnitude 48 cm to the east and then 64 cm to the west. What is the total displacement?

2. A squirrel is trying to locate some nuts he buried for the winter. He moves 4.0 m to the right of a stone and digs unsuccessfully. Then he moves 1.0 m to the left of his hole, changes his mind, and moves 6.5 m to the right of that position and digs a second hole. No luck. Then he moves 8.3 m to the left and digs again. He finds a nut at last. What is the squirrel's total displacement from its starting point?

3. A runner, jogging along a straight line path, starts at a position 60 m east of a milestone marker and heads west. After a short time interval he is 20 m west of the mile marker. Choose east to be the positive x-direction. (a) What is the runner's displacement from his starting point? (b) What is his displacement from the milestone? (c) The runner then turns around and heads east. If at a later time the runner is 140 m east of the milestone, what is his displacement from the starting point at this time? (d) What is the total distance traveled from the starting point if the runner stops at the final position listed in part (c)?

4. Johannes bicycles from his dorm to the pizza shop that is 3.00 mi east. Darren's apartment is located 1.50 mi west of Johannes's dorm. If Darren is able to meet Johannes at the pizza shop by bicycling in a straight line, what is the distance and direction he must travel?

5. At 3 P.M. a car is located 20 km south of its starting point. One hour later it is 96 km farther south. After two more hours, it is 12 km south of the original starting point. (a) What is the displacement of the car between 3 P.M. and 6 P.M.? (b) What is the displacement of the car from the starting point to the location at 4 P.M.? (c) What is the displacement of the car between 4 P.M. and 6 P.M.?

2.2 Velocity: Rate of Change of Position

6. For the train of Example 2.2, find the average velocity between 3:14 P.M. when the train is at 3 km east of the origin and 3:28 P.M. when it is 10 km east of the origin.

7. A cyclist travels 10.0 km east in a time of 11 min 40 s. What is his average velocity in meters per second?

8. In a game against the White Sox, baseball pitcher Nolan Ryan threw a pitch measured at 45.1 m/s. If it was 18.4 m from Nolan's position on the pitcher's mound to home plate, how long did it take the ball to get to the batter waiting at home plate? Treat the ball's velocity as constant and ignore any gravitational effects.

9. Jason drives due west with a speed of 35.0 mi/h for 30.0 min, then continues in the same direction with a speed of 60.0 mi/h for 2.00 h, then drives farther west at 25.0 mi/h for 10.0 min. What is Jason's average velocity for the entire trip?

10. Two cars, a Toyota Yaris and a Jeep, are traveling in the same direction, although the Yaris is 186 m behind the Jeep. The speed of the Yaris is 24.4 m/s and the speed of the Jeep is 18.6 m/s. How much time does it take for the Yaris to catch the Jeep? [*Hint:* What must be true about the displacement of the two cars when they meet?] (tutorial: catchup)

11. Speedometer readings are obtained and graphed as a car comes to a stop along a straight-line path. How far does the car move between $t = 0$ and $t = 16$ s? (tutorial: start/stop traffic)

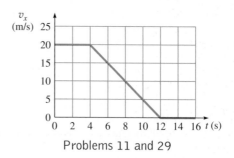

Problems 11 and 29

12. A graph is plotted of the vertical velocity v_y of an elevator versus time. The y-axis points up. (a) How high is the elevator above the starting point ($t = 0$) after 20 s has elapsed? (b) When is the elevator at its highest location above the starting point?

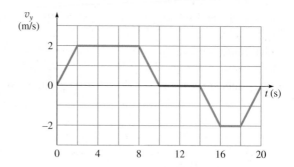

13. A bicycle is moving along a straight line. The graph in the figure shows its position from the starting point as a function of time. (a) In which section(s) of the graph does the object have the highest speed? (b) At which time(s) does the object reverse its direction of

motion? (c) How far does the object move from $t = 0$ to $t = 3$ s?

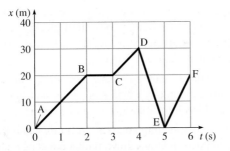

14. A ball thrown by a pitcher on a women's softball team is timed at 65.0 mph. The distance from the pitching rubber to home plate is 43.0 ft. In major league baseball the corresponding distance is 60.5 ft. If the batter in the softball game and the batter in the baseball game are to have equal times to react to the pitch, with what speed must the baseball be thrown? Assume the ball travels with a constant velocity. [*Hint:* There is no need to convert units; set up a ratio.]

15. A motor scooter travels east at a speed of 12 m/s. The driver then reverses direction and heads west at 15 m/s. What is the change in velocity of the scooter? Give magnitude and direction.

16. To pass a physical fitness test, Massimo must run 1000 m at an average rate of 4.0 m/s. He runs the first 900 m in 250 s. Is it possible for Massimo to pass the test? If so, how fast must he run the last 100 m to pass the test? Explain.

17. The graph shows speedometer readings, in meters per second (on the vertical axis), obtained as a skateboard travels along a straight-line path. How far does the board move between $t = 3.00$ s and $t = 8.00$ s?

18. The graph shows values of $x(t)$ in meters, on the vertical axis, for a skater traveling in a straight line. (a) What is $v_{av,x}$ for the interval from $t = 0$ to $t = 4.0$ s? (b) from $t = 0$ to $t = 5.0$ s?

19. The graph shows values of $x(t)$ in meters for a skater traveling in a straight line. What is v_x at $t = 2.0$ s?

20. The graph shows values of $x(t)$ in meters for an object traveling in a straight line. Plot v_x as a function of time for this object from $t = 0$ to $t = 8$ s.

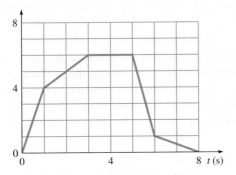

Problems 17, 18, 19, and 20

21. A chipmunk, trying to cross a road, first moves 80 cm to the right, then 30 cm to the left, then 90 cm to the right, and finally 310 cm to the left. (a) What is the chipmunk's total displacement? (b) If the elapsed time was 18 s, what was the chipmunk's average speed? (c) What was its average velocity?

22. Rita Jeptoo of Kenya was the first female finisher in the 110th Boston Marathon. She ran the first 10.0 km in a time of 0.5689 h. Assume the race course to be along a straight line. (a) What was her average speed during the first 10.0 km segment of the race? (b) She completed the entire race, a distance of 42.195 km, in a time of 2.3939 h. What was her average speed for the race?

23. A relay race is run along a straight-line track of length 300.0 m running south to north. The first runner starts at the south end of the track and passes the baton to a teammate at the north end of the track. The second runner races back to the start line and passes the baton to a third runner who races 100.0 m northward to the finish line. The magnitudes of the average velocities of the first, second, and third runners during their parts of the race are 7.30 m/s, 7.20 m/s, and 7.80 m/s, respectively. What is the average velocity of the baton for the entire race? [*Hint:* You will need to find the time spent by each runner in completing her portion of the race.]

2.3 Acceleration: Rate of Change of Velocity

24. If a pronghorn antelope accelerates from rest in a straight line with a constant acceleration of 1.7 m/s², how long does it take for the antelope to reach a speed of 22 m/s?

25. If a car traveling at 28 m/s is brought to a full stop in 4.0 s after the brakes are applied, find the average acceleration during braking.

26. An 1100-kg airplane starts from rest; 8.0 s later it reaches its takeoff speed of 35 m/s. What is the average acceleration of the airplane during this time?

27. A rubber ball is attached to a paddle by a rubber band. The ball is initially moving away from the paddle with a speed of 4.0 m/s. After 0.25 s, the ball is moving toward the paddle with a speed of 3.0 m/s. What is the average acceleration of the ball during that 0.25 s? Give magnitude and direction.

28. (a) In Fig. 2.11, what is the instantaneous acceleration of the sports car of Example 2.5 at the time of 14 s from the start? (b) What is the displacement of the car from $t = 12.0$ s to $t = 16.0$ s? (c) What is the average velocity of the car in the 4.0-s time interval from 12.0 s to 16.0 s?

29. The graph with Problem 11 shows speedometer readings as a car comes to a stop. What is the magnitude of the acceleration at $t = 7.0$ s?

◆30. The figure shows a plot of $v_x(t)$ for a car traveling in a straight line. (a) What is $a_{av,x}$ between $t = 6$ s and $t = 11$ s? (b) What is $v_{av,x}$ for the same time interval? (c) What is $v_{av,x}$ for the interval $t = 0$ to $t = 20$ s? (d) What is the increase in the car's speed between 10 s and 15 s? (e) How far does the car travel from time $t = 10$ s to time $t = 15$ s?

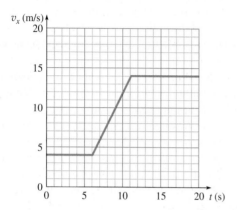

31. The graph shows v_x versus t for a body moving along a straight line. (a) What is a_x at $t = 11$ s? (b) What is a_x at $t = 3$ s? (c) How far does the body travel from $t = 12$ s to $t = 14$ s? ([tutorial]: x, v, a)

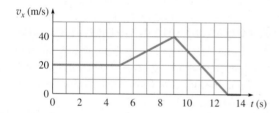

2.4 Motion Along a Line with Constant Acceleration; 2.5 Visualizing Motion Along a Line with Constant Acceleration

32. A toboggan is sliding in a straight line down a snowy slope. The table shows the speed of the toboggan at various times during its trip. (a) Make a graph of the speed as a function of time. (b) Judging by the graph, is it plausible that the toboggan's acceleration is constant? If so, what is the acceleration?

Time Elapsed, t (s)	Speed of Toboggan, v (m/s)
0	0
1.14	2.8
1.62	3.9
2.29	5.6
2.80	6.8

33. The St. Charles streetcar in New Orleans starts from rest and has a constant acceleration of 1.20 m/s² for 12.0 s. (a) Draw a graph of v_x versus t. (b) How far has the train traveled at the end of the 12.0 s? (c) What is the speed of the train at the end of the 12.0 s? (d) Draw a motion diagram, showing the streetcar's position at 2.0-s intervals.

34. An airplane lands and starts down the runway with a southwest velocity of 55 m/s. What constant acceleration allows it to come to a stop in 1.0 km?

35. A train is traveling south at 24.0 m/s when the brakes are applied. It slows down with a constant acceleration to a speed of 6.00 m/s in a time of 9.00 s. (a) Draw a graph of v_x versus t for a 12-s interval (starting 2 s before the brakes are applied and ending 1 s after the brakes are released). Let the x-axis point to the north. (b) What is the acceleration of the train during the 9.00-s interval? (c) How far does the train travel during the 9.00 s?

◆36. A 1200-kg airplane starts from rest and moves forward with a constant acceleration of magnitude 5.00 m/s² along a runway that is 250 m long. (a) How long does it take the plane to reach a speed of 46.0 m/s? (b) How far along the runway has the plane moved when it reaches 46.0 m/s?

37. A car is speeding up and has an instantaneous velocity of 1.0 m/s in the +x-direction when a stopwatch reads 10.0 s. It has a constant acceleration of 2.0 m/s² in the +x-direction. (a) What change in speed occurs between $t = 10.0$ s and $t = 12.0$ s? (b) What is the speed when the stopwatch reads 12.0 s?

38. You are driving your car along a country road at a speed of 27.0 m/s. As you come over the crest of a hill, you notice a farm tractor 25.0 m ahead of you on the road, moving in the same direction as you at a speed of 10.0 m/s. You immediately slam on your brakes and slow down with a constant acceleration of magnitude 7.00 m/s². Will you hit the tractor before you stop? How far will you travel before you stop or collide with the tractor? If you stop, how far is the tractor in front of you when you finally stop?

39. A train is traveling along a straight, level track at 26.8 m/s (60.0 mi/h). Suddenly the engineer sees a truck stalled on the tracks 184 m ahead. If the maximum possible braking acceleration has magnitude 1.52 m/s², can the train be stopped in time?

40. In a cathode ray tube in an old TV, electrons are accelerated from rest with a constant acceleration of magnitude 7.03×10^{13} m/s² during the first 2.0 cm of the tube's length; then they move at essentially constant velocity another 45 cm before hitting the screen. (a) Find the speed of the electrons when they hit the

screen. (b) How long does it take them to travel the length of the tube?

41. The graph is of v_x versus t for an object moving along the x-axis. How far does the object move between $t = 9.0$ s and $t = 13.0$ s? Solve using two methods: a graphical analysis and an algebraic solution.

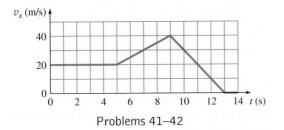

Problems 41–42

42. The graph is of v_x versus t for an object moving along the x-axis. What is the average acceleration between $t = 5.0$ s and $t = 9.0$ s? Solve using two methods: a graphical analysis and an algebraic solution.

43. A train, traveling at a constant speed of 22 m/s, comes to an incline with a constant slope. While going up the incline, the train slows down with a constant acceleration of magnitude 1.4 m/s². (a) Draw a graph of v_x versus t where the x-axis points up the incline. (b) What is the speed of the train after 8.0 s on the incline? (c) How far has the train traveled up the incline after 8.0 s? (d) Draw a motion diagram, showing the trains position at 2.0-s intervals.

2.6 Free Fall

In the problems, please assume the free-fall acceleration $g = 9.80$ m/s² unless a more precise value is given in the problem statement. Ignore air resistance.

44. A brick is thrown vertically upward with an initial speed of 3.00 m/s from the roof of a building. If the building is 78.4 m tall, how much time passes before the brick lands on the ground?

45. A penny is dropped from the observation deck of the Empire State building (369 m above ground). With what velocity does it strike the ground?

46. (a) How long does it take for a golf ball to fall from rest for a distance of 12.0 m? (b) How far would the ball fall in twice that time?

47. Grant Hill jumps 1.3 m straight up into the air to slam-dunk a basketball into the net. With what speed did he leave the floor?

48. During a walk on the Moon, an astronaut accidentally drops his camera over a 20.0-m cliff. It leaves his hands with zero speed, and after 2.0 s it has attained a velocity of 3.3 m/s downward. How far has the camera fallen after 4.0 s?

49. Glenda drops a coin from ear level down a wishing well. The coin falls a distance of 7.00 m before it strikes the water. If the speed of sound is 343 m/s, how long after Glenda releases the coin will she hear a splash?

50. A stone is launched straight up by a slingshot. Its initial speed is 19.6 m/s and the stone is 1.50 m above the ground when launched. (a) How high above the ground does the stone rise? (b) How much time elapses before the stone hits the ground?

51. A 55-kg lead ball is dropped from the leaning tower of Pisa. The tower is 55 m high. (a) How far does the ball fall in the first 3.0 s of flight? (b) What is the speed of the ball after it has traveled 2.5 m downward? (c) What is the speed of the ball 3.0 s after it is released? (d) If the ball is thrown vertically upward from the top of the tower with an initial speed of 4.80 m/s, where will it be after 2.42 s?

52. A balloonist, riding in the basket of a hot air balloon that is rising vertically with a constant velocity of 10.0 m/s, releases a sandbag when the balloon is 40.8 m above the ground. What is the bag's speed when it hits the ground?

53. Superman is standing 120 m horizontally away from Lois Lane. A villain throws a rock vertically downward with a speed of 2.8 m/s from 14.0 m directly above Lois. (a) If Superman is to intervene and catch the rock just before it hits Lois, what should be his minimum constant acceleration? (b) How fast will Superman be traveling when he reaches Lois?

54. A student, looking toward his fourth-floor dormitory window, sees a flowerpot with nasturtiums (originally on a window sill above) pass his 2.0-m high window in 0.093 s. The distance between floors in the dormitory is 4.0 m. From a window on which floor did the flowerpot fall?

55. You drop a stone into a deep well and hear it hit the bottom 3.20 s later. This is the time it takes for the stone to fall to the bottom of the well, plus the time it takes for the sound of the stone hitting the bottom to reach you. Sound travels about 343 m/s in air. How deep is the well?

Comprehensive Problems

In the problems, please assume the free-fall acceleration $g = 9.80$ m/s² unless a more precise value is given in the problem statement. Ignore air resistance.

56. (a) If a freestyle swimmer traveled 1500 m in a time of 14 min 53 s, how fast was his average speed? (b) If the pool was rectangular and 50 m in length, how does the

speed you found compare with his sustained swimming speed of 1.54 m/s during one length of the pool after he had been swimming for 10 min? What might account for the difference?

57. While passing a slower car on the highway, you accelerate uniformly from 17.4 m/s to 27.3 m/s in a time of 10.0 s. (a) How far do you travel during this time? (b) What is your acceleration magnitude?

58. A cheetah can accelerate from rest to 24 m/s in 2.0 s. Assuming the acceleration is constant over the time interval, (a) what is the magnitude of the acceleration of the cheetah? (b) What is the distance traveled by the cheetah in these 2.0 s? (c) A runner can accelerate from rest to 6.0 m/s in the same time, 2.0 s. What is the magnitude of the acceleration of the runner? By what factor is the cheetah's average acceleration magnitude greater than that of the runner?

59. A rocket is launched from rest. After 8.0 min, it is 160 km above the Earth's surface and is moving at a speed of 7.6 km/s. Assuming the rocket moves up in a straight line, what are its (a) average velocity and (b) average acceleration?

60. A streetcar named Desire travels between two stations 0.60 km apart. Leaving the first station, it accelerates for 10.0 s at 1.0 m/s^2 and then travels at a constant speed until it is near the second station, when it brakes at 2.0 m/s^2 in order to stop at the station. How long did this trip take? [*Hint:* What's the average velocity?]

61. An unmarked police car starts from rest just as a speeding car passes at a speed of v. If the police car speeds up with a constant acceleration of magnitude a, what is the speed of the police car when it catches up to the speeder, who does not realize she is being pursued and does not vary her speed?

62. A stone is thrown vertically downward from the roof of a building. It passes a window 16.0 m below the roof with a speed of 25.0 m/s. It lands on the ground 3.00 s after it was thrown. What was (a) the initial velocity of the stone and (b) how tall is the building?

63. A car traveling at 29 m/s (65 mi/h) runs into a bridge abutment after the driver falls asleep at the wheel. (a) If the driver is wearing a seat belt and comes to rest within a 1.0-m distance, what is his acceleration (assumed constant)? (b) A passenger who isn't wearing a seat belt is thrown into the windshield and comes to a stop in a distance of 10.0 cm. What is the acceleration of the passenger?

64. To pass a physical fitness test, Marcella must run 1000 m at an average speed of 4.00 m/s. She runs the first 500 m at an average of 4.20 m/s. (a) How much time does she have to run the last 500 m? (b) What

should be her average speed over the last 500 m in order to finish with an overall average speed of 4.00 m/s?

65. At 3:00 P.M., a bank robber is spotted driving north on I-15 at milepost 126. His speed is 112.0 mi/h. At 3:37 P.M., he is spotted at milepost 185 doing 105.0 mi/h. During this time interval, what are the bank robber's displacement, average velocity, and average acceleration? (Assume a straight highway.)

66. Based on the information given in Problem 59, is it possible that the rocket moves with constant acceleration? Explain.

67. An elevator starts at rest on the ninth floor. At $t = 0$, a passenger pushes a button to go to another floor. The graph for this problem shows the acceleration a_y of the elevator as a function of time. Let the y-axis point upward. (a) Has the passenger gone to a higher or lower floor? (b) Sketch a graph of the velocity v_y of the elevator versus time. (c) Sketch a graph of the position y of the elevator versus time.

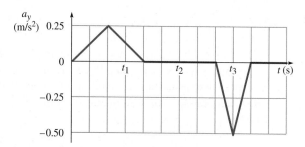

68. The graph for this problem shows the vertical velocity v_y of a bouncing ball as a function of time. The y-axis points up. Answer these questions based on the data in the graph. (a) At what time does the ball reach its maximum height? (b) For how long is the ball in contact with the floor? (c) What is the maximum height of the ball? (d) What is the acceleration of the ball while in the air? (e) What is the average acceleration of the ball while in contact with the floor?

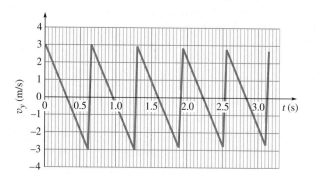

+69. A rocket engine can accelerate a rocket launched from rest vertically up with an acceleration of 20.0 m/s^2. However, after 50.0 s of flight the engine fails. (a) What is the rocket's altitude when the engine fails? (b) When does it reach its maximum height? (c) What is the maximum height reached? [*Hint:* A graphical solution may be easiest.] (d) What is the velocity of the rocket just before it hits the ground?

+70. The graph shows the position x of a switch engine in a rail yard as a function of time t. At which of the labeled times t_0 to t_7 is (a) $a_x < 0$, (b) $a_x = 0$, (c) $a_x > 0$, (d) $v_x = 0$, (e) the speed decreasing?

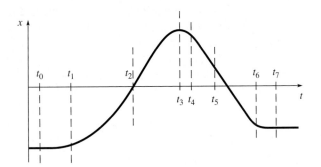

+71. An airtrack glider, 8.0 cm long, blocks light as it goes through a photocell gate. The glider is released from rest on a frictionless inclined track and the gate is positioned so that the glider has traveled 96 cm when it is in the middle of the gate. The timer gives a reading of 333 ms for the glider to pass through this gate. Friction is negligible. What is the acceleration (assumed constant) of the glider along the track?

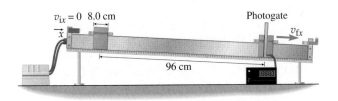

+72. Find the point of no return for an airport runway of 1.50 mi in length if a jet plane can accelerate at 10.0 ft/s^2 and decelerate at 7.00 ft/s^2. The point of no return occurs when the pilot can no longer abort the takeoff without running out of runway. What length of time is available from the start of the motion in which to decide on a course of action?

+73. In the human nervous system, signals are transmitted along neurons as *action potentials* that travel at speeds of up to 100 m/s. (An action potential is a traveling influx of sodium ions through the membrane of a neuron.) The signal is passed from one neuron to another by the release of neurotransmitters in the synapse. Suppose someone steps on your toe. The pain signal travels along a 1.0-m-long sensory neuron to the spinal column, across a synapse to a second 1.0-m-long neuron, and across a second synapse to the brain. Suppose that the synapses are each 100 nm wide, that it takes 0.10 ms for the signal to cross each synapse, and that the action potentials travel at 100 m/s. (a) At what average speed does the signal cross a synapse? (b) How long does it take the signal to reach the brain? (c) What is the average speed of propagation of the signal?

Answers to Practice Problems

2.1 3.8 m east

2.2 77 km/h in the $-x$-direction (west)

2.3 About 100 to 110 km/h in the $+x$-direction (east)

2.4 The velocity is increasing in magnitude, so the acceleration is in the same direction as the velocity (the $-x$-direction). Thus, a_x is negative; the acceleration is in the $-x$-direction.

2.5

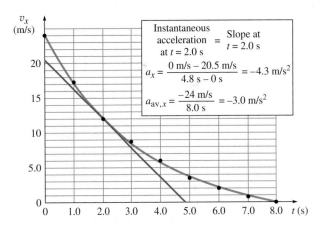

(a) $a_{\text{av},x} = -3.0 \text{ m/s}^2$ where the negative sign means the average acceleration is directed to the northwest;

(b) $a_x = -4.3 \text{ m/s}^2$ **(northwest)**

2.6 2.0 m

2.7 20 s

2.8 5.00 s after they leave the starting point; 4.00 km/s in the $+x$-direction

2.9 (a) 3.8 m; (b) 3.00 s

Answers to Checkpoints

2.1 No. The magnitude of the displacement is the shortest distance between two points. The distance traveled can be greater than or equal to the displacement, depending on the path taken. In Example 2.1 the displacement is 2.9 km to the west, and the distance traveled is 11.5 km.

2.2 Yes. Average speed is the distance traveled divided by the time interval in moving from point A to point B. Average velocity is the displacement from point A to point B divided by the same time interval. The magnitude of the displacement is the shortest possible distance from A to B. Thus the average velocity magnitude is less than or equal to the average speed.

2.3 The slope of the tangent to a graph of v_x versus time is the instantaneous acceleration a_x at the time.

2.4 Only if the plane's acceleration is constant must its average velocity be 470 km/h west. If its acceleration is not constant, the average velocity is not necessarily 470 km/h west. To find the average velocity, we would divide the plane's displacement by the time interval.

2.6 Yes. If you throw a ball upward, it is in free fall as soon as it loses contact with your hand.

Motion in a Plane

A gull scoops up a clam and takes it high above the ground. While flying parallel to the ground, the gull lets go of the clam. The clam lands on a rock below and cracks open. Then the gull alights and enjoys lunch. A beachcomber on the beach sees the clam fall along a parabolic path, just as a projectile would. Why does the clam not drop straight down? What does the path of the falling clam look like to the gull? (See pp. 73 and 76–77 for the answers.)

- trigonometric functions: sine, cosine, and tangent (Appendix A.7)
- Pythagorean theorem (Appendix A.6)
- position, displacement, velocity, and acceleration (Sections 2.1–2.3)
- average and instantaneous quantities (Sections 2.2–2.3)
- motion along a line with constant acceleration (Sections 2.4–2.6)

3.1   GRAPHICAL ADDITION AND SUBTRACTION OF VECTORS

Chapter 2 introduced the quantities position, displacement, velocity, and acceleration to describe motion along a line—that is, motion in one dimension of space. To describe motion in more than one dimension, we need a full treatment of vector addition and subtraction because position, displacement, velocity, and acceleration are vectors. (Other vectors you will study in this book include force, momentum, angular momentum, torque, and the electric and magnetic fields.)

Vectors and Scalars   All **vectors** have a direction as well as a magnitude. The direction of any vector is always a *physical direction in space* such as up, down, north, or 35° south of west.

Vector quantities are usually drawn as arrows pointing in the direction of the vector; the length of the arrow is proportional to the magnitude of the vector. By contrast, a **scalar** quantity can have magnitude, algebraic sign, and units, but not a direction in space. It wouldn't make sense to draw an arrow to represent a scalar such as mass!

In this book, an arrow over a boldface symbol indicates a vector quantity ($\vec{\mathbf{r}}$). (Some books use boldface without the arrow or the arrow without boldface.) When writing by hand, always draw an arrow over a vector symbol to distinguish it from a scalar. When the symbol for a vector is written without the arrow and in italics rather than boldface (r), it stands for the *magnitude* of the vector (which is a scalar). Absolute value bars are also used to stand for the magnitude of a vector, so $r = |\vec{\mathbf{r}}|$. The magnitude of a vector may have units and is never negative; it can be positive or zero.

Vector quantities have both magnitude and direction.

Conceptual Example 3.1

Vector or Scalar?

Is temperature a vector quantity?

Strategy   If a quantity is a vector, it must have both a magnitude and a physical direction in space.

Solution and Discussion   Does temperature have a direction? A temperature in Fahrenheit or Celsius can be above or below zero—is that a direction? No. A vector must have a *physical direction* in space. It does not make sense to say that the temperature of your coffee is "85 degrees Celsius in the southwest direction." "The temperature is up 5 degrees today," means that it has increased, not that it is pointing vertically upward. Temperature is a scalar, not a vector.

Conceptual Practice Problem 3.1   Bank Balance

When you deposit a paycheck, the balance of your checking account "goes up." When you pay a bill, it "goes down." Is the balance of your account a vector quantity?

When scalars are added or subtracted, they do so in the usual way: 3 kg of water plus 2 kg of water is equal to 5 kg of water. Adding or subtracting vectors is different. Vectors follow rules of addition and subtraction that take into account the *directions* of the vectors as well as their magnitudes. Whenever you need to add or subtract quantities, check whether they are vectors. If so, be sure to add or subtract them correctly *as vectors. Do not just add or subtract their magnitudes.*

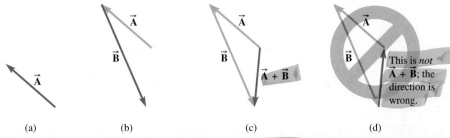

(a)   (b)   (c)   (d)

Figure 3.1 Adding two vectors graphically. (a) Draw one vector arrow. (b) Draw the second, starting where the first arrow ended. (c) The sum of the two. (d) A common mistake.

Graphical Vector Addition

We start with a graphical method to help develop your intuition. To add two vectors graphically, first draw an arrow to represent one of them (Fig. 3.1a). (It does not matter in what order vectors are added; $\vec{\mathbf{A}} + \vec{\mathbf{B}} = \vec{\mathbf{B}} + \vec{\mathbf{A}}$.) The arrow points in the direction of the vector and its length is proportional to the magnitude of the vector. It doesn't matter where you start drawing the arrow. The value of a vector is not changed by moving it as long as its direction and magnitude are not changed.

Now draw the second vector arrow starting where the first ends. In other words, place the "tail" of the second arrow at the "tip" of the first (Fig 3.1b). Finally, draw an arrow starting from the *tail* of the first and ending at the *tip* of the second. This arrow represents the sum of the two vectors (Fig. 3.1c). A common error is to draw the sum from the tip of the second to the tail of the first (Fig. 3.1d). If the lengths and directions of the vectors are drawn accurately to scale, using a ruler and a protractor, then the length and direction of the sum can be determined with the ruler and protractor. To add more than two vectors, continue drawing them tip to tip.

Vector Subtraction

To subtract a vector is to add its opposite (that is, a vector with the same magnitude but opposite direction): $\vec{\mathbf{r}}_f - \vec{\mathbf{r}}_i = \vec{\mathbf{r}}_f + (-\vec{\mathbf{r}}_i)$. Multiplying a vector by the scalar -1 reverses the vector's direction while leaving its magnitude unchanged, so $-\vec{\mathbf{r}}_i = -1 \times \vec{\mathbf{r}}_i$ is a vector equal in magnitude and opposite in direction to $\vec{\mathbf{r}}_i$.

Using Compass Headings

It is common to use compass headings to specify vector directions in a horizontal plane. For example, the direction of the vector in Fig. 3.2 is "20° north of east," which means that the vector makes a 20° angle with the east direction and is on the north (rather than the south) side of east. The same direction could be described as "70° east of north," although it is customary to use the smaller angle. Northeast means "45° north of east" or, equivalently, "45° east of north."

Position and Displacement

The position $\vec{\mathbf{r}}$ of an object can be represented as a vector arrow drawn from the origin to the location of the object (Fig. 3.3). Its magnitude is the distance from the origin. The displacement is literally the *change in position* (the final position vector minus the initial position vector):

$$\Delta\vec{\mathbf{r}} = \vec{\mathbf{r}}_f - \vec{\mathbf{r}}_i \qquad (3\text{-}1)$$

Figure 3.4 shows the graphical subtraction of two position vectors to illustrate the displacement for a trip from Killarney to Kenmare. This same procedure is used to subtract any kind of vector quantity (velocity, acceleration, etc.).

Addition of Displacement Vectors

As in Example 2.1, the total displacement for a trip with several parts is the vector sum of the displacements for each part of the trip because

$$\vec{\mathbf{r}}_3 - \vec{\mathbf{r}}_1 = (\vec{\mathbf{r}}_3 - \vec{\mathbf{r}}_2) + (\vec{\mathbf{r}}_2 - \vec{\mathbf{r}}_1) \qquad (3\text{-}2)$$

Example 3.2 explores this idea further.

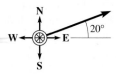

Figure 3.2 Measuring angles with respect to compass headings. The direction of this vector is 20° north of east (20° N of E).

A plus sign (+) between vector quantities indicates *vector addition*, not ordinary addition. An equals sign (=) between vector quantities means that the vectors are identical in magnitude *and* direction, *not* simply that their magnitudes are equal.

Vector Subtraction: $\vec{\mathbf{A}} - \vec{\mathbf{B}} = \vec{\mathbf{A}} + (-\vec{\mathbf{B}})$, where $-\vec{\mathbf{B}}$ has the same magnitude as $\vec{\mathbf{B}}$ but is opposite in direction. Note that the order matters: $\vec{\mathbf{B}} - \vec{\mathbf{A}} = -(\vec{\mathbf{A}} - \vec{\mathbf{B}})$.

Figure 3.3 A position vector $\vec{\mathbf{r}}$.

Figure 3.4 (a) Two position vectors, $\vec{r}_i$ and $\vec{r}_f$, drawn from an *arbitrary origin* to the starting point (Killarney) and to the ending point (Cork) of a trip. (b) The final position vector minus the initial position vector is the displacement $\Delta\vec{r}$, found by adding $-\vec{r}_i + \vec{r}_f$.

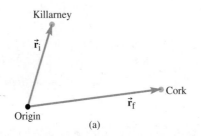
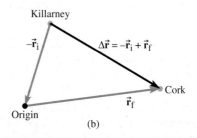
(a)   (b)

Example 3.2

An Irish Adventure (1)

In a trip from Killarney to Cork, Charlotte and Shona drive at a compass heading of 27° west of south for 18 km to Kenmare, then directly south for 17 km to Glengariff, then at a compass heading of 13° north of east for 48 km to Cork. Find the displacement vector for the entire trip by adding the three displacements graphically.

Blarney castle.

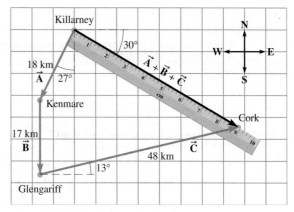

Figure 3.5
Graphical addition of the displacement vectors for the trip from Killarney to Cork via Kenmare and Glengariff.

$$18 \text{ km} \times \frac{0.2 \text{ cm}}{1 \text{ km}} = 3.6 \text{ cm}$$

Similarly, the arrows for $\vec{B}$ and $\vec{C}$ should be 3.4 cm and 9.6 cm long, respectively.

After drawing the three vector arrows tip to tail, the arrow from the tail of the first vector to the tip of the last vector represents the sum (Fig. 3.5). This arrow is measured to have length 8.9 cm and its direction is 30° south of east. The total displacement has magnitude

$$8.9 \text{ cm} \times \frac{1 \text{ km}}{0.2 \text{ cm}} = 44.5 \text{ km}$$

Rounding to two significant figures, the total displacement $\vec{A} + \vec{B} + \vec{C}$ has magnitude 45 km and is directed 30° south of east.

Strategy To add the displacement vectors, place the tail of each successive vector at the tip of the preceding vector. The value of a vector is not changed by moving it as long as its direction and magnitude are not changed, so a vector can be drawn starting at any point. The sum of the three displacements is then drawn from the tail of the first vector to the tip of the last vector. To add vectors graphically and get an accurate result, we use a ruler and a protractor. The protractor is used to draw the vector arrows in the correct directions and the ruler is used to draw them with the correct lengths. Then the length and direction of the sum can be determined with the ruler and protractor.

Solution Let's call the four positions $\vec{r}_1$ (Killarney), $\vec{r}_2$ (Kenmare), $\vec{r}_3$ (Glengariff), and $\vec{r}_4$ (Cork). The displacement for the whole trip is $\vec{r}_4 - \vec{r}_1$. The problem gives the displacements for the three parts of the trip; let's call them $\vec{A} = \vec{r}_2 - \vec{r}_1 = 18$ km, 27° west of south; $\vec{B} = \vec{r}_3 - \vec{r}_2 = 17$ km, south; and $\vec{C} = \vec{r}_4 - \vec{r}_3 = 48$ km, 13° north of east. The sum of these three displacements is the total displacement because

$$\vec{A} + \vec{B} + \vec{C} = (\vec{r}_2 - \vec{r}_1) + (\vec{r}_3 - \vec{r}_2) + (\vec{r}_4 - \vec{r}_3) = \vec{r}_4 - \vec{r}_1$$

Next we choose a convenient scale for the lengths of the vector arrows. Here we choose to represent 1 km as an arrow length of 0.2 cm, so the length of the vector arrow for $\vec{A}$ should be

Discussion Note that the answer includes both the magnitude and direction of the displacement. If a homework or exam question has you calculate a vector quantity such as position or velocity, don't forget to specify the direction as well as the magnitude in your answer. One without the other is incomplete.

Although the magnitude and direction of a position vector depends on the choice of origin, the magnitude and

continued on next page

Example 3.2 continued

~~direction of a displacement (*change* of position) does *not* depend on the choice of origin.~~
    ~~The total *distance* traveled by Charlotte and Shona is 18 km + 17 km + 48 km = 83 km, which is *not* equal to the magnitude of the total displacement. Finding the total distance involves adding three *scalars*, while finding the total displacement involves adding three *vectors*. The magnitude of the total displacement is the *straight-line* distance from Killarney to Cork.~~

Practice Problem 3.2 A Traveling Executive

An executive flies from Kansas City to Chicago (displacement = 400 mi in the direction 30° north of east) and then from Chicago to Tulsa (600 mi, 45° south of west). Add the two displacements graphically to find the total displacement from Kansas City to Tulsa.

3.2   VECTOR ADDITION AND SUBTRACTION USING COMPONENTS

Components of a Vector

Any vector can be expressed as the sum of vectors parallel to the *x*-, *y*-, and (if needed) *z*-axes. The *x*-, *y*-, and *z*-components of a vector indicate the magnitude and direction of the three vectors along the three perpendicular axes. The sign of a component indicates the direction along that axis. The *x*-, *y*-, and *z*-components of vector $\vec{A}$ are written with subscripts as follows: A_x, A_y, and A_z. One exception to this otherwise consistent notation is that the *x*-, *y*-, and *z*-components of a position vector $\vec{r}$ are usually written x, y, and z (instead of r_x, r_y, and r_z). For now we will deal only with vectors in the *xy*-plane.

The *x*-component of a position vector $\vec{r}$ is x, the *x*-coordinate. For all other vectors, the *x*-component is designated by a subscript *x*. For example, the *x*-component of a velocity vector $\vec{v}$ is written v_x. Components of vectors have magnitude, units, and an algebraic sign. The sign indicates the direction: a positive *x*-component indicates the direction of the positive *x*-axis, while a negative *x*-component indicates the opposite direction (the negative *x*-axis).

Finding Components   The process of finding the components of a vector is called **resolving** the vector into its components. Consider the velocity vector $\vec{v}$ in Fig. 3.6. We can think of $\vec{v}$ as the sum of two vectors, one parallel to the *x*-axis and the other parallel to the *y*-axis. The magnitudes of these two vectors are the *magnitudes* (absolute values) of the *x*- and *y*-components of $\vec{v}$. We can find the magnitudes of the components using the right triangle in Fig. 3.6 and the trigonometric functions in Fig. 3.7. The length of the arrow represents the magnitude of the vector ($v = 9.4$ m/s), so

$$\cos 58° = \frac{\text{adjacent}}{\text{hypotenuse}} = \frac{|v_x|}{v} \quad \text{and} \quad \sin 58° = \frac{\text{opposite}}{\text{hypotenuse}} = \frac{|v_y|}{v} \quad (3\text{-}3)$$

Now we must determine the correct algebraic sign for each of the components. From Fig. 3.6, the vector along the *x*-axis points in the *positive x*-direction and the vector along the *y*-axis points in the *negative y*-direction, so in this case,

$$v_x = +v \cos 58° = 5.0 \text{ m/s} \quad \text{and} \quad v_y = -v \sin 58° = -8.0 \text{ m/s} \quad (3\text{-}4)$$

Using the right triangle in Fig. 3.8 gives the same values for the *x*- and *y*-components of $\vec{v}$ since $\cos 32° = \sin 58°$ and $\sin 32° = \cos 58°$.

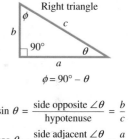

Figure 3.6 Resolving a velocity vector $\vec{v}$ into *x*- and *y*-components.

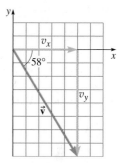

$$\phi = 90° - \theta$$

$$\sin \theta = \frac{\text{side opposite } \angle\theta}{\text{hypotenuse}} = \frac{b}{c}$$

$$\cos \theta = \frac{\text{side adjacent } \angle\theta}{\text{hypotenuse}} = \frac{a}{c}$$

$$\tan \theta = \frac{\text{side opposite } \angle\theta}{\text{side adjacent } \angle\theta} = \frac{b}{a}$$

Figure 3.7 Trigonometric functions (see Appendix A.7 for more information).

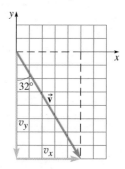

Figure 3.8 Resolving the velocity vector into components using a different right triangle.

Problem-Solving Strategy: Finding the x- and y-Components of a Vector from Its Magnitude and Direction

1. Draw a right triangle with the vector as the hypotenuse and the other two sides parallel to the x- and y-axes.
2. Determine one of the unknown angles in the triangle.
3. Use trigonometric functions to find the magnitudes of the components. Make sure your calculator is in "degree mode" to evaluate trigonometric functions of angles in degrees and "radian mode" for angles in radians.
4. Determine the correct algebraic sign for each component.

Finding Magnitude and Direction   We must also know how to reverse the process to find a vector's magnitude and direction from its component.

Problem-Solving Strategy: Finding the Magnitude and Direction of a Vector Ā from Its x- and y-Components

1. Sketch the vector on a set of x- and y-axes in the correct quadrant, according to the signs of the components.
2. Draw a right triangle with the vector as the hypotenuse and the other two sides parallel to the x- and y-axes.
3. In the right triangle, choose which of the unknown angles you want to determine.
4. Use the inverse tangent function to find the angle. The lengths of the sides of the triangle represent $|A_x|$ and $|A_y|$. If θ is opposite the side parallel to the x-axis, then tan θ = opposite/adjacent = $|A_x/A_y|$. If θ is opposite the side parallel to the y-axis, then tan θ = opposite/adjacent = $|A_y/A_x|$. If your calculator is in "degree mode," then the result of the inverse tangent operation will be in degrees. [In general, the inverse tangent has two possible values between 0 and 360° because tan α = tan (α + 180°). However, when the inverse tangent is used to find one of the angles in a right triangle, the result can never be greater than 90°, so the value the calculator returns is the one you want.]
5. Interpret the angle: specify whether it is the angle below the horizontal, or the angle west of south, or the angle clockwise from the negative y-axis, etc.
6. Use the Pythagorean theorem to find the magnitude of the vector.

$$A = \sqrt{A_x^2 + A_y^2} \qquad (3\text{-}5)$$

Suppose we knew the components of the velocity vector in Fig. 3.6, but not the magnitude and direction. Let us find the angle θ between $\vec{v}$ and the +x-axis:

$$\theta = \tan^{-1}\frac{\text{opposite}}{\text{adjacent}} = \tan^{-1}\frac{|v_y|}{|v_x|} = \tan^{-1}\frac{8.0 \text{ m/s}}{5.0 \text{ m/s}} = 58° \qquad (3\text{-}6)$$

The magnitude of $\vec{v}$ is

$$v = \sqrt{v_x^2 + v_y^2} = \sqrt{(+5.0 \text{ m/s})^2 + (-8.0 \text{ m/s})^2} = 9.4 \text{ m/s}$$

Adding Vectors Using Components

It is generally easier and more accurate to add vectors algebraically rather than graphically. The algebraic method relies on adding the components of the vectors. Remember that each vector is thought of as the sum of vectors parallel to the axes (Fig. 3.9a). When

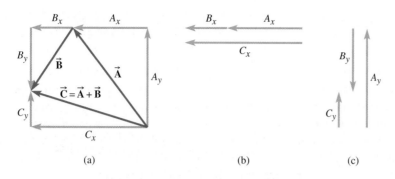

Figure 3.9 (a) $\vec{C} = \vec{A} + \vec{B}$, shown graphically with the x- and y-components of each vector illustrated. (b) $C_x = A_x + B_x$; (c) $C_y = A_y + B_y$.

(a)   (b)   (c)

adding vectors, we can add them in any order and group them as we please. So we can sum the x-components to find the x-component of the sum (Fig. 3.9b) and then do the same with the y-components (Fig. 3.9c):

$$\vec{C} = \vec{A} + \vec{B} \quad \text{if and only if} \quad C_x = A_x + B_x \quad \text{and} \quad C_y = A_y + B_y \quad (3\text{-}7)$$

In Eq. (3-7), remember that $A_x + B_x$ represents ordinary addition since the signs of the components carry the direction information.

Problem-Solving Strategy: Adding Vectors Using Components

1. Find the x- and y-components of each vector to be added.
2. Add the x-components (*with their algebraic signs*) of the vectors to find the x-component of the sum. (If the signs are not correct, the sum will not be correct.)
3. Add the y-components (with their algebraic signs) of the vectors to find the y-component of the sum.
4. If necessary, use the x- and y-components of the sum to find the magnitude and direction of the sum.

Estimation Using Graphical Addition Even when using the component method to add vectors, the graphical method is an important first step. A rough sketch of vector addition, even one made without carefully measuring the lengths or the angles, has important benefits. Sketching the vectors makes it much easier to get the signs of the components correct. The graphical addition also serves as a check on the answer—it provides an estimate of the magnitude and direction of the sum, which can be used to check the algebraic answer. Graphical addition gives you a mental picture of what is going on and an intuitive feel for the algebraic calculations.

✓ **CHECKPOINT 3.2**

Two displacements $\vec{A}$ and $\vec{B}$ have x- and y-components as follows: $A_x = +3.0$ km, $A_y = -6.0$ km, $B_x = -8.5$ km, $B_y = -1.2$ km. The total displacement is $\vec{C} = \vec{A} + \vec{B}$. What are the x- and y-components of $\vec{C}$?

Choosing x- and y-Axes

A problem can be made easier to solve with a good choice of axes. We can choose any direction we want for the x- and y-axes, as long as they are perpendicular to one another. Three common choices are

- x-axis horizontal and y-axis vertical, when the vectors all lie in a vertical plane;
- x-axis east and y-axis north, when the vectors all lie in a horizontal plane; and
- x-axis parallel to an inclined surface and y-axis perpendicular to it.

Example 3.3

An Irish Adventure (2)

In the trip of Example 3.2, Charlotte and Shona drive at a compass heading of 27° west of south for 18 km to Kenmare, then directly south for 17 km to Glengariff, then at a compass heading of 13° north of east for 48 km to Cork. Use the component method to find the magnitude and direction of the displacement vector for the entire trip.

Strategy   As before, let's call the three successive displacements $\vec{\mathbf{A}}$, $\vec{\mathbf{B}}$, and $\vec{\mathbf{C}}$, respectively. To add the vectors using components, we first choose directions for the x- and y-axes. Then we find the x- and y-components of the three displacements. Adding the x- or y-components of the three displacements gives the x- or y-component of the total displacement. Finally, from the components we find the magnitude and direction of the total displacement.

Solution   A good choice is the conventional one: x-axis to the east and the y-axis to the north. The first displacement ($\vec{\mathbf{A}}$) is directed 27° west of south. Both of its components are negative since west is the $-x$-direction and south is the $-y$-direction. Using the right triangle in Fig. 3.10, the side of the triangle opposite the 27° angle is parallel to the x-axis. The sine function relates the opposite side to the hypotenuse:

$$A_x = -A \sin 27° = -18 \text{ km} \times 0.454 = -8.17 \text{ km}$$

where A is the magnitude of $\vec{\mathbf{A}}$. The cosine relates the adjacent side to the hypotenuse:

$$A_y = -A \cos 27° = -18 \text{ km} \times 0.891 = -16.0 \text{ km}$$

Displacement $\vec{\mathbf{B}}$ has no x-component since its direction is south. Therefore,

$$B_x = 0 \quad \text{and} \quad B_y = -17 \text{ km}$$

The direction of $\vec{\mathbf{C}}$ is 13° north of east. Both its components are positive. From Fig. 3.10, the side of the right triangle opposite the 13° angle is parallel to the y-axis, so

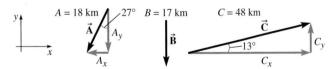

Figure 3.10
Resolving $\vec{\mathbf{A}}$, $\vec{\mathbf{B}}$, and $\vec{\mathbf{C}}$ into x- and y-components.

$$C_x = +C \cos 13° = +48 \text{ km} \times 0.974 = +46.8 \text{ km}$$
$$C_y = +C \sin 13° = +48 \text{ km} \times 0.225 = +10.8 \text{ km}$$

Now we sum the x- and y-components separately to find the x- and y-components of the total displacement:

$$\Delta x = A_x + B_x + C_x$$
$$= (-8.17 \text{ km}) + 0 + 46.8 \text{ km} = +38.63 \text{ km}$$
$$\Delta y = A_y + B_y + C_y$$
$$= (-16.0 \text{ km}) + (-17 \text{ km}) + 10.8 \text{ km} = -22.2 \text{ km}$$

The magnitude and direction of $\Delta\vec{\mathbf{r}}$ can be found from the right triangle in Fig. 3.11. The magnitude is represented by the hypotenuse:

$$\Delta r = \sqrt{(\Delta x)^2 + (\Delta y)^2} = \sqrt{(38.63 \text{ km})^2 + (-22.2 \text{ km})^2}$$
$$= 45 \text{ km}$$

The angle θ is

$$\theta = \tan^{-1} \frac{\text{opposite}}{\text{adjacent}} = \tan^{-1} \frac{22.2 \text{ km}}{38.63 \text{ km}} = 30°$$

Since $+x$ is east and $-y$ is south, the direction of the displacement is 30° south of east. The magnitude and direction of the displacement found using components agree with the displacement found graphically in Fig. 3.5.

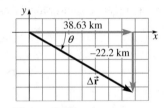

Figure 3.11
Finding the magnitude and direction of $\Delta\vec{\mathbf{r}}$.

Discussion   Note that the x-component of one displacement was found using the sine function while another was found using the cosine. The x-component (or the y-component) of the vector can be related to *either* the sine or the cosine, depending on which angle in the right triangle is used.

Practice Problem 3.3   Changing the Coordinate Axes

Find the x- and y-components of the displacements for the three legs of the trip if the x-axis points south and the y-axis points east.

Unit Vectors

The connection between a vector and its components may be expressed using the **unit vectors** $\hat{\mathbf{x}}$ (read aloud as "x hat"), $\hat{\mathbf{y}}$, and $\hat{\mathbf{z}}$, which are defined as vectors of magnitude 1 that point in the $+x$-, $+y$-, and $+z$-directions, respectively. (In some books, you may see them written as $\hat{\mathbf{i}}, \hat{\mathbf{j}},$ and $\hat{\mathbf{k}}.$) They are called *unit* vectors because the magnitude of each is the pure number 1—they do *not* have physical units such as kilograms or meters. Any vector $\vec{\mathbf{A}}$ can be written as the sum of three vectors along the coordinate axes:

$$\vec{\mathbf{A}} = A_x\hat{\mathbf{x}} + A_y\hat{\mathbf{y}} + A_z\hat{\mathbf{z}} \tag{3-8}$$

Here A_x is the x-component of $\vec{\mathbf{A}}$, which has physical units and can be positive or negative. $A_x\hat{\mathbf{x}}$ is a vector of magnitude $|A_x|$ directed in the $+x$-direction if $A_x > 0$ and in the $-x$-direction if $A_x < 0$. For example, consider the velocity vector $\vec{\mathbf{v}}$ of Fig. 3.8. $\vec{\mathbf{v}}$ has x-component $v_x = +5.0$ m/s and y-component $v_y = -8.0$ m/s, so $\vec{\mathbf{v}} = (+5.0 \text{ m/s})\hat{\mathbf{x}} + (-8.0 \text{ m/s})\hat{\mathbf{y}}$.

Using unit vector notation is one way to keep track of vector components in vector addition and subtraction without writing separate equations for each component. Adding two vectors in the xy-plane looks like this:

$$\vec{\mathbf{A}}_1 + \vec{\mathbf{A}}_2 = \left(A_{1x}\hat{\mathbf{x}} + A_{1y}\hat{\mathbf{y}}\right) + \left(A_{2x}\hat{\mathbf{x}} + A_{2y}\hat{\mathbf{y}}\right) \tag{3-9}$$

Regrouping the terms shows that the x-component of the sum is the sum of the x-components and likewise for the y-components:

$$\vec{\mathbf{A}}_1 + \vec{\mathbf{A}}_2 = \left(A_{1x} + A_{2x}\right)\hat{\mathbf{x}} + \left(A_{1y} + A_{2y}\right)\hat{\mathbf{y}} \tag{3-10}$$

3.3   VELOCITY

The definitions of average velocity, instantaneous velocity, average acceleration, and instantaneous acceleration from Chapter 2 still apply when the motion is not in a straight line as long as we add and subtract them as vectors. Suppose we want to know the instantaneous velocity of a race car at point P as it goes around a curved section of a racetrack (Fig. 3.12a). At a slightly later time the race car is at point Q. Let $\vec{\mathbf{r}}_i$ be the position of the car at P and $\vec{\mathbf{r}}_f$ be the position at point Q.

Average Velocity   The displacement $\Delta\vec{\mathbf{r}} = \vec{\mathbf{r}}_f - \vec{\mathbf{r}}_i$ is represented as an arrow from P to Q. Alternatively, to subtract $\vec{\mathbf{r}}_i$ from $\vec{\mathbf{r}}_f$, the two vectors can be drawn with their tails at the same point. After reversing the direction of $\vec{\mathbf{r}}_i$ to represent $-\vec{\mathbf{r}}_i$, the arrows are tip to tail and ready to add $\vec{\mathbf{r}}_f + (-\vec{\mathbf{r}}_i)$—see Fig. 3.12b. The average velocity during this time interval is the displacement $\Delta\vec{\mathbf{r}}$ divided by the time interval:

$$\vec{\mathbf{v}}_{av} = \frac{\vec{\mathbf{r}}_f - \vec{\mathbf{r}}_i}{t_f - t_i} = \frac{\Delta\vec{\mathbf{r}}}{\Delta t} \tag{3-11}$$

The direction of the average velocity is the direction of the displacement $\Delta\vec{\mathbf{r}}$.

Instantaneous Velocity   The instantaneous velocity at P is the limit of the average velocity as Δt approaches zero. As we shorten the time interval between the initial and final positions by moving point Q closer and closer to P, the direction of the displacement

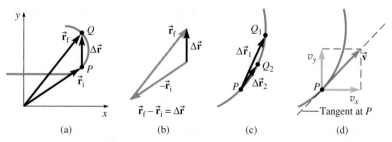

Figure 3.12 (a) Position vectors for two points on the curve. (b) The displacement $\Delta\vec{\mathbf{r}}$ from point P to point Q. (c) As the time interval is decreased, the final point moves closer and closer to P; the direction of the displacement $\Delta\vec{\mathbf{r}}$ approaches the tangent to the curve at P. (d) Instantaneous velocity can be resolved into components along perpendicular axes.

vector $\Delta \vec{\mathbf{r}}$ gradually changes, approaching the tangent to the curved path at P (Fig. 3.12c). Expressed in mathematical terminology, the instantaneous velocity is the limit of $\Delta \vec{\mathbf{r}}/\Delta t$ as the time interval approaches zero:

$$\vec{\mathbf{v}} = \lim_{\Delta t \to 0} \frac{\Delta \vec{\mathbf{r}}}{\Delta t} \qquad (3\text{-}12)$$

($\Delta \vec{\mathbf{r}}$ is the displacement during a *very short* time interval Δt)

If an object moves along a curved path, the direction of the velocity vector at any point is tangent to the path at that point.

With this definition, the instantaneous velocity at P becomes tangent to the curve at P (Fig. 3.12d). Here we are talking about a tangent to the actual path through space, *not* a tangent line on a graph of position versus time.* The magnitude of the velocity vector is the speed at which the object moves and the direction of the velocity vector is the direction of motion.

Component Equations   A vector equation is always equivalent to a set of equations, one for each component. The x- and y-components of the average velocity are

$$v_{av,x} = \frac{\Delta x}{\Delta t} \quad \text{and} \quad v_{av,y} = \frac{\Delta y}{\Delta t} \qquad (3\text{-}13)$$

The x- and y-components of the instantaneous velocity are

$$v_x = \lim_{\Delta t \to 0} \frac{\Delta x}{\Delta t} \quad \text{and} \quad v_y = \lim_{\Delta t \to 0} \frac{\Delta y}{\Delta t} \qquad (3\text{-}14)$$

To put Eq. (3-14) into words, the x-component of an object's velocity is the rate of change of its x-coordinate and the y-component of its velocity is the rate of change of its y-coordinate.

Example 3.4

An Irish Adventure (3)

In their trip from Kenmare to Cork via Glengariff, Charlotte and Shona travel a total distance of 83 km in 1.4 h. The total displacement for the trip is 45 km, 30° south of east. What is their average velocity? Contrast it with their average speed, defined as the total distance divided by the time interval.

Strategy   The average velocity is calculated from the displacement—not from the distance traveled.

Solution   The magnitude of the average velocity is

$$|\vec{\mathbf{v}}_{av}| = \frac{|\Delta \vec{\mathbf{r}}|}{\Delta t} = \frac{45 \text{ km}}{1.4 \text{ h}} = 32 \text{ km/h}$$

The average velocity has the same direction as the displacement, so $\vec{\mathbf{v}}_{av} = 32$ km/h, 30° south of east. The average speed is

$$\text{average speed} = \frac{83 \text{ km}}{1.4 \text{ h}} = 59 \text{ km/h}$$

Therefore, $|\vec{\mathbf{v}}_{av}|$ is not equal to the average speed. Furthermore, average velocity is a vector quantity with a direction in space, and average speed is a scalar.

Practice Problem 3.4   Average Velocity Versus Average Speed

In Example 3.4, $|\vec{\mathbf{v}}_{av}|$ was less than the average speed. Can $|\vec{\mathbf{v}}_{av}|$ ever be greater than the average speed? Can $|\vec{\mathbf{v}}_{av}|$ ever be equal to the average speed? Explain.

3.4   ACCELERATION

The average acceleration $\vec{\mathbf{a}}_{av}$ is the change in velocity divided by the elapsed time:

$$\vec{\mathbf{a}}_{av} = \frac{\vec{\mathbf{v}}_f - \vec{\mathbf{v}}_i}{t_f - t_i} = \frac{\Delta \vec{\mathbf{v}}}{\Delta t} \qquad (3\text{-}15)$$

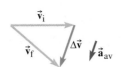

Turning while keeping speed constant

$\vec{v}_i$

$\vec{v}_f$ $\Delta\vec{v}$ $\vec{a}_{av}$

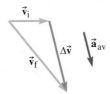

Turning while increasing speed

$\vec{v}_i$

$\Delta\vec{v}$ $\vec{a}_{av}$

$\vec{v}_f$

Figure 3.13 Two examples to illustrate that the average acceleration is always in the same direction as the change in velocity $\Delta\vec{v}$ during the same time interval.

For motion in a plane, this vector equation is equivalent to two component equations:

$$a_{av,x} = \frac{\Delta v_x}{\Delta t} \quad \text{and} \quad a_{av,y} = \frac{\Delta v_y}{\Delta t} \qquad (3\text{-}16)$$

The direction of $\vec{a}_{av}$ is the same as the direction of $\Delta\vec{v}$ (Fig. 3.13).

Instantaneous acceleration is the limit of the average acceleration as the time interval approaches zero:

$$\vec{a} = \lim_{\Delta t \to 0} \frac{\Delta\vec{v}}{\Delta t} \qquad (3\text{-}17)$$

($\Delta\vec{v}$ is the change in velocity during a *very short* time interval Δt)

In component form,

$$a_x = \lim_{\Delta t \to 0} \frac{\Delta v_x}{\Delta t} \quad \text{and} \quad a_y = \lim_{\Delta t \to 0} \frac{\Delta v_y}{\Delta t} \qquad (3\text{-}18)$$

In straight-line motion the acceleration is always along the same line as the velocity. For motion in two dimensions, the acceleration vector can make any angle with the velocity vector because the velocity vector can change in magnitude, in direction, or both. The direction of the acceleration is the direction of the *change* in velocity $\Delta\vec{v}$ during a *very short* time interval.

✓ CHECKPOINT 3.4

An airplane is initially moving due north at 400 km/h. After making a slight course correction, it is moving at the same speed but in a direction 2.0° east of north. Is the plane's average acceleration during this time interval zero? Explain.

Example 3.5

Skating Uphill

An inline skater is traveling on a level road with a speed of 8.94 m/s; 120.0 s later she is climbing a hill with a 15.0° angle of incline at a speed of 7.15 m/s. (a) What is the change in her velocity? (b) What is her average acceleration during the 120.0-s time interval?

Strategy The change in velocity is *not* 1.79 m/s (= 8.94 m/s −7.15 m/s). That is the change in *speed.* The change in velocity is found by subtracting the initial velocity *vector* from the final velocity *vector.* After first making a graphical sketch, we use the component method. The average acceleration is the change in velocity divided by the elapsed time.

continued on next page

Example 3.5 continued

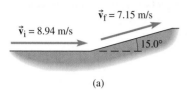

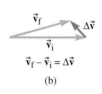

(a)

(b)

Figure 3.14

(a) Change in velocity as the skater slows going uphill and (b) graphical subtraction of velocity vectors.

Solution   (a) Figure 3.14a shows the initial and final velocity vectors and the slope of the hill. The initial velocity is horizontal as the skater skates on level ground. The final velocity is 15.0° above the horizontal. To subtract the two velocity vectors graphically, we place the tails of the vectors together. The change in velocity $\Delta\vec{v}$ is found by drawing a vector arrow from the tip of $\vec{v}_i$ to the tip of $\vec{v}_f$. Judging by the graphical subtraction in Fig. 3.14b, the change in velocity is roughly at a 45° angle above the $-x$-axis. Its magnitude is smaller than the magnitudes of the initial and final velocity vectors—something like 2 to 3 m/s.

The components v_{fx} and v_{fy} can be found from a right triangle (Fig. 3.15):

$$v_{fx} = v_f \cos \theta = 7.15 \text{ m/s} \times 0.9659 = 6.91 \text{ m/s}$$

$$v_{fy} = v_f \sin \theta = 7.15 \text{ m/s} \times 0.2588 = 1.85 \text{ m/s}$$

Since v_i has only an x-component,

$$v_{iy} = 0 \quad \text{and} \quad v_{ix} = v_i = 8.94 \text{ m/s}$$

Now we subtract the components to find the components of $\Delta\vec{v}$:

$$\Delta v_x = v_{fx} - v_{ix} = (6.91 - 8.94) \text{ m/s} = -2.03 \text{ m/s}$$

and

$$\Delta v_y = v_{fy} - v_{iy} = (1.85 - 0) \text{ m/s} = +1.85 \text{ m/s}$$

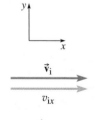

Figure 3.15

Initial and final velocity vectors resolved into components.

To find the magnitude of $\Delta\vec{v}$, we apply the Pythagorean theorem (Fig. 3.16):

$$|\Delta\vec{v}|^2 = (\Delta v_x)^2 + (\Delta v_y)^2 = (-2.03 \text{ m/s})^2 + (1.85 \text{ m/s})^2$$

$$= 7.54 \text{ (m/s)}^2$$

$$|\Delta\vec{v}| = 2.75 \text{ m/s}$$

The angle is found from

$$\tan \phi = \frac{\text{opposite}}{\text{adjacent}} = \left|\frac{\Delta v_y}{\Delta v_x}\right| = \frac{1.85 \text{ m/s}}{2.03 \text{ m/s}} = 0.9113$$

$$\phi = \tan^{-1} 0.9113 = 42.3°$$

The direction of the change in velocity $\Delta\vec{v}$ is 42.3° above the negative x-axis.

(b) The magnitude of the average acceleration is

$$|\vec{a}_{av}| = \frac{|\Delta\vec{v}|}{\Delta t} = \frac{2.75 \text{ m/s}}{120.0 \text{ s}} = 0.0229 \text{ m/s}^2$$

The direction of the average acceleration is the same as the direction of $\Delta\vec{v}$: 42.3° above the negative x-axis.

Discussion   Checking back with the graphical subtraction in Fig. 3.14b, the magnitude of $\Delta\vec{v}$ appears to be roughly $\frac{1}{4}$ to $\frac{1}{3}$ the magnitude of $\vec{v}_i$. Since $\frac{1}{4} \times 8.94$ m/s = 2.24 m/s and $\frac{1}{3} \times 8.94$ m/s = 2.98 m/s, the answer of 2.75 m/s is reasonable.

Figure 3.14b also shows the direction of $\Delta\vec{v}$ to be roughly midway between the $+y$- and $-x$-axes. We found the direction of $\Delta\vec{v}$ to be 42.3° above the $-x$-axis and, therefore, 47.7° from the $+y$-axis. So the direction we calculated is also reasonable based on the graphical subtraction.

Figure 3.16

Reconstruction of $\Delta\vec{v}$ from its components (not to scale).

Practice Problem 3.5   Change in Sailboat Velocity

A C&C 30 sailboat is sailing at 12.0 knots (6.17 m/s) heading directly east across the harbor. When a gust of wind comes up, the boat changes its heading to 11.0° north of east and its speed increases to 14.0 knots (7.20 m/s). [A boat's speed is customarily expressed in knots, which means nautical miles per hour. A nautical mile (6076 ft) is a little longer than a statute mile (5280 ft).] (a) What is the magnitude and direction of the change in velocity of the sailboat in m/s? (b) If this velocity change occurs during a 2.0-s time interval, what is the average acceleration of the sailboat during that interval?

3.5 MOTION IN A PLANE WITH CONSTANT ACCELERATION

If an object moves in the xy-plane with constant acceleration, then both a_x and a_y are constant. By looking separately at the motion along two perpendicular axes, the y-direction and the x-direction, each component becomes a one-dimensional problem, which we studied in Chapter 2. We can apply any of the constant acceleration relationships from Section 2.4 separately to the x-components and to the y-components.

It is generally easiest to choose the axes so that the acceleration has only one non-zero component. Suppose we choose the axes so that the acceleration is in the positive or negative y-direction. Then $a_x = 0$ and v_x is constant. With this choice, the constant acceleration relationships [Eqs. (2-9) through (2-13)] become

x-axis: $a_x = 0$	**y-axis: constant a_y**
$\Delta v_x = 0$ (v_x is constant)	$\Delta v_y = a_y \Delta t$ (3-19)
$\Delta x = v_x \Delta t$	$\Delta y = \frac{1}{2}(v_{fy} + v_{iy})\Delta t$ (3-20)
	$\Delta y = v_{iy}\Delta t + \frac{1}{2}a_y(\Delta t)^2$ (3-21)
	$v_{fy}^2 - v_{iy}^2 = 2a_y\Delta y$ (3-22)

Why are only two equations shown in the column for the x-axis? The other two are redundant when $a_x = 0$.

Note that there is no mixing of components in Eqs. (3-19) through (3-22). Each equation pertains either to the x-components or to the y-components; none contains the x-component of one vector quantity and the y-component of another. The only quantity that appears in both x- and y-component equations is the time interval—a scalar.

Motion of Projectiles

An object in free fall near the Earth's surface has a constant acceleration. As long as air resistance is negligible, the constant downward pull of gravity gives the object a constant downward acceleration with magnitude g. In Section 2.6 we considered objects in free fall, but only when they had no horizontal velocity component, so they moved straight up or straight down. Now we consider objects (called **projectiles**) in free fall that have a *nonzero* horizontal velocity component. The motion of a projectile takes place in a vertical plane.

Suppose some medieval marauders are attacking a castle. They have a catapult that propels large stones into the air to bombard the walls of the castle (Fig. 3.17). Picture a stone leaving the catapult with initial velocity $\vec{v}_i$. ($\vec{v}_i$ is the *initial* velocity for the time interval *during which it moves as a projectile.* It is also the *final* velocity for the time interval during which it is in contact with the catapult.) The **angle of elevation** is the angle of the initial velocity above the horizontal. Once the stone is in the air, the only force acting on it is the downward gravitational force, provided that the air resistance has a negligible effect on the motion. The **trajectory** (path) of the stone is shown in Fig. 3.18. The positive x-axis is chosen in the horizontal direction (to the right) and the positive y-axis is upward.

If the initial velocity $\vec{v}_i$ is at an angle θ above the horizontal, then resolving it into components gives

$$v_{ix} = v_i \cos\theta \quad \text{and} \quad v_{iy} = v_i \sin\theta \qquad (3\text{-}23)$$

(+y-axis up, θ measured from the horizontal x-axis)

With the y-axis pointing up, $a_y = -g$ because the acceleration is downward (in the $-y$-direction). The acceleration has no x-component ($a_x = 0$), so the stone's horizontal

Figure 3.17 A medieval catapult.

Figure 3.18 Motion diagram showing the trajectory of a projectile. The position is drawn at equal time intervals. Superimposed are the velocity vectors along with their x- and y-components.

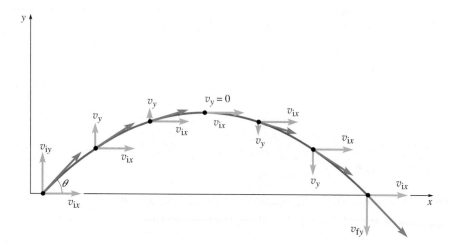

velocity component v_x is *constant.* The vertical velocity component v_y changes at a constant rate, exactly as if the stone were propelled straight up with an initial speed of v_{iy}. The initially positive v_y decreases until, at the top of flight, $v_y = 0$. Then the pull of gravity makes the projectile fall back downward. During the downward trip, v_y is still changing at the same constant rate with which it changed on the way up and at the top of the path. The acceleration has the same constant value—magnitude and direction—for the entire path.

The motion of a projectile when air resistance is negligible is the superposition of horizontal motion with constant velocity and vertical motion with constant acceleration. The vertical and horizontal motions each proceed independently, as if the other motion were not present. In the experiment of Fig. 3.19, one ball was dropped and, at the same instant, another was projected horizontally. The strobe photo shows snapshots of the two balls at equally spaced time intervals. The *vertical* motion of the two is identical; at every instant, the two are at the same height. The fact that they have different horizontal motion does not affect their vertical motion. (This statement would *not* be true if air resistance were significant.)

The horizontal and vertical motions of a projectile can be treated separately; they are independent of each other.

PHYSICS AT HOME

Take a nickel and a penny to a room with a high table or countertop. Place the penny at the edge of the table and then slide the nickel so it collides with the penny. Listen for the sound of the two coins hitting the floor. The two coins will slide off the table with different horizontal velocities but will land at the same time.

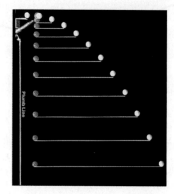

Figure 3.19 Independence of horizontal and vertical motion of a projectile in the absence of air resistance. The vertical motion of the projectile (white) is the same as that of an object (red) that falls straight down.

Trajectory of a Projectile

The graph of an equation of the form

$$y = kx^2, \; k = \text{a nonzero constant}$$

is a parabola. Show that the trajectory of a projectile is a parabola. [*Hint:* Choose the origin at the highest point of the trajectory and let $t_i = 0$ at that instant.]

Strategy and Solution We start at the high point of the path and look at displacements from there. The horizontal displacement is proportional to the elapsed time t since the horizontal velocity is constant. The vertical displacement is the average vertical velocity component times the elapsed time t. The average vertical velocity component is itself proportional to t since it changes at a constant rate. Therefore, the vertical displacement is proportional to t^2. Thus, the vertical displacement y is proportional to the square of the horizontal displacement x and $y = kx^2$, where k is a constant of proportionality. The path followed by a projectile in free fall is a parabola.

Discussion The same conclusion can be drawn algebraically. With the $+y$-axis upward and the origin and $t = 0$ at the top of flight, x_i, y_i, and v_{iy} are all zero. Then $x = v_{ix}t$ and

$$y = v_{iy}t + \frac{1}{2}a_y t^2 = -\frac{1}{2}gt^2 = -\frac{1}{2}g\left(\frac{x}{v_{ix}}\right)^2 = -\left(\frac{g}{2v_{ix}^2}\right)x^2$$

So y is proportional to x^2 and the constant of proportionality is $-g/(2v_{ix}^2)$.

Conceptual Practice Problem 3.6   Throwing Stones

You stand at the edge of a cliff and throw stones horizontally into the river below. To double the horizontal displacement of a stone from the cliff to where it lands, by what factor must you increase the stone's initial speed? Ignore air resistance.

Graphing Projectile Motion Figure 3.20 shows graphs of the x- and y-components of the velocity and position of a projectile as functions of time. In this case, the projectile is launched above flat ground at $t = 0$ and returns to the same elevation at a later time t_f. Note that the y-component graphs are *symmetrical* about the vertical line through the highest point in the trajectory. The y-component of velocity decreases linearly from its initial value; the slope of the line is $a_y = -g$. When $v_y = 0$, the projectile is at the apex of its trajectory. Then v_y continues to decrease at the same rate and is now negative with its magnitude getting larger and larger. At t_f, when the projectile has returned to its original altitude, the y-component of the velocity has the same magnitude as at $t = 0$ but with the opposite sign ($v_y = -v_{iy}$).

The graph of $y(t)$ indicates that the projectile moves upward, quickly at first and then gradually slowing, until it reaches the maximum height. The slope of the tangent to the $y(t)$ graph at any particular moment of time is v_y at that instant. At the highest point

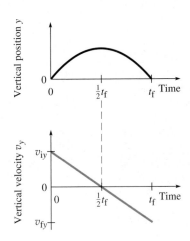
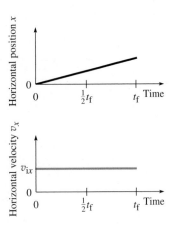

Figure 3.20 Projectile motion: separate vertical and horizontal quantities versus time.

of the $y(t)$ graph, the tangent is horizontal and $v_y = 0$. After that, gravity makes the projectile start to fall downward.

The horizontal velocity is constant, so the graph of $v_x(t)$ is a horizontal line. The horizontal position x increases uniformly in time because the object is moving with a constant v_x.

CHECKPOINT 3.5

When a basketball is thrown in an arc toward the net, what can you say about its velocity and acceleration at the highest point of the arc?

Example 3.7

Attacking the Castle Walls

The catapult used by the marauders hurls a stone with a velocity of 50.0 m/s at a 30.0° angle of elevation (Fig. 3.21). (a) What is the maximum height reached by the stone? (b) What is its *range* (defined as the horizontal distance traveled when the stone returns to its original height)? (c) How long has the stone been in the air when it returns to its original height?

Strategy The problem gives both the magnitude and direction of the initial velocity of the stone. Ignoring air resistance, the stone has a constant downward acceleration once it has been launched—until it hits the ground or some obstacle. We choose the positive y-axis upward and the positive x-axis in the direction of horizontal motion of the stone (toward the castle). When the stone reaches its maximum height, the velocity component in the y-direction is zero since the stone goes no higher. When the stone returns to its original height, $\Delta y = 0$ and $v_y = -v_{iy}$. The range can be found once the time of flight t_f is known—time is the quantity that connects the x-component equations to the y-component equations. Therefore, we solve (c) before (b). One way to find t_f is to find the time to reach maximum height and then double it (see Fig. 3.20). (Other methods include setting $\Delta y = 0$ or setting $v_y = -v_{iy}$.)

Solution (a) First we find the x- and y-components of the initial velocity for an angle of elevation $\theta = 30.0°$.

$$v_{iy} = v_i \sin \theta \quad \text{and} \quad v_{ix} = v_i \cos \theta$$

The maximum height is the vertical displacement Δy when $v_{fy} = 0$.

$$\Delta y = \tfrac{1}{2}(v_{fy} + v_{iy})\,\Delta t = \tfrac{1}{2}(0 + v_i \sin \theta)\,\Delta t$$

Eliminating the time interval using $v_{fy} - v_{iy} = a_y \Delta t$ yields

$$\Delta y = \frac{1}{2}(v_i \sin \theta)\left(\frac{0 - v_i \sin \theta}{a_y}\right) = -\frac{(v_i \sin \theta)^2}{2a_y}$$

$$= \frac{-(50.0 \text{ m/s} \times \sin 30.0°)^2}{2 \times (-9.80 \text{ m/s}^2)} = 31.9 \text{ m}$$

The maximum height of the projectile is 31.9 m above its launch height.

(c) The initial and final heights are the same. Due to this symmetry, the time of flight (t_f) is *twice* the time it takes the projectile to reach its maximum height. The time to reach the maximum height can be found from

$$v_{fy} = 0 = v_{iy} + a_y \Delta t$$

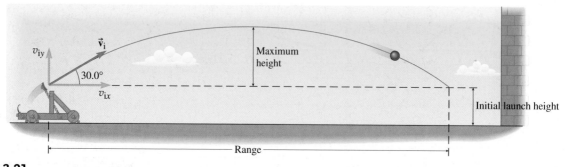

Figure 3.21

A catapult projects a stone into the air in an attack on a castle wall.

continued on next page

Example 3.7 continued

Solving for Δt,

$$\Delta t = \frac{-v_{iy}}{a_y}$$

The time of flight is

$$t_f = 2\,\Delta t = 2 \times \frac{-50.0 \text{ m/s} \times \sin 30.0°}{-9.80 \text{ m/s}^2} = 5.10 \text{ s}$$

(b) The range is

$$\Delta x = v_{ix}\, t_f = (50.0 \text{ m/s} \times \cos 30.0°) \times 5.10 \text{ s} = 221 \text{ m}$$

Discussion   Quick check: using

$$y_f - y_i = v_{iy}\,\Delta t + \tfrac{1}{2} a_y\,(\Delta t)^2$$

we can check that $\Delta y = 31.9$ m when $\Delta t = \tfrac{1}{2} \times 5.10$ s and that $\Delta y = 0$ when $\Delta t = 5.10$ s. Here we check the first of these:

$$\Delta y = (50.0 \text{ m/s} \times \sin 30.0°) \times 2.55 \text{ s} + \tfrac{1}{2} \times (-9.80 \text{ m/s}^2) \times (2.55 \text{ s})^2$$

$$= 63.8 \text{ m} + (-31.9 \text{ m}) = 31.9 \text{ m}$$

which is correct. This is not an *independent* check, since this equation can be derived from the others, but it can reveal algebra or calculation errors.

Since we analyze the horizontal motion independently from the vertical motion, we start by resolving the given initial velocity into *x*- and *y*-components. Time is what connects the horizontal and vertical motions.

Practice Problem 3.7   Maximum Height for Arrows

Archers have joined in the attack on the castle and are shooting arrows over the walls. If the angle of elevation for an arrow is 45°, find an expression for the maximum height of the arrow in terms of v_i and g. [*Hint:* Simplify the expression using $\sin 45° = \cos 45° = 1/\sqrt{2}$.]

PHYSICS AT HOME

On a warm day, take a garden hose and aim the nozzle so that the water streams upward at an angle above the horizontal. Set the nozzle for a fast, narrow stream for best effect. Once the water leaves the nozzle, it becomes a projectile with a constant downward acceleration (ignoring the small effect of air resistance). The continuous stream of water lets us see the parabolic path easily. Stand in one place and try aiming the nozzle at different angles of elevation to find an angle that gives the maximum range. Aim for a particular spot on the ground (at a distance less than the maximum range) and see if you can find two different angles of elevated nozzle position that allow the stream to hit the target spot (see Fig. 3.22).

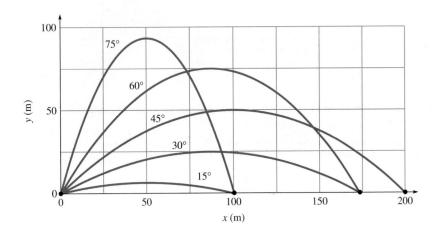

Figure 3.22 Parabolic trajectories of projectiles launched with the same initial speed ($v_i = 44.3$ m/s) at five different angles. The ranges of projectiles launched at angles θ and $90° - \theta$ are the same. The maximum range occurs for $\theta = 45°$.

Monkey and Hunter

An inexperienced hunter aims and shoots an arrow straight at a coconut that is being held by a monkey in a tree (Fig. 3.23). At the same instant that the arrow leaves the bow, the monkey drops the coconut. Ignoring air resistance, does the arrow hit the coconut, the monkey, or neither?

Strategy and Solution If there were no gravity, the arrow would fly straight to the coconut (along the dashed blue line in Fig. 3.23). Since gravity gives the dropped coconut and the released arrow the same constant acceleration downward, they each fall the same vertical distance below the positions they would have had with no gravity. The coconut falls along the dashed red line; the distance fallen at 0.25-s intervals is marked. The arrow falls below the blue dashed line by the same distances, marked along its trajectory at 0.25-s intervals.

The arrow ends up hitting the coconut no matter what the initial speed of the arrow (as long as the arrow's range is at least as large as the horizontal distance to the coconut). The

higher the speed of the arrow, the sooner they meet and the shorter the vertical distance that the coconut falls before being hit.

Discussion An experienced hunter would have aimed *above* the initial position of the coconut to compensate for gravity; he would have missed the coconut but might have hit the monkey unless the monkey jumped down to retrieve the coconut.

Conceptual Practice Problem 3.8   Changes in Position and Velocity for Consecutive Arrows

An arrow is shot into the air. One second later, a second arrow is shot with the same initial velocity. While the two are both in the air, does the difference in their positions ($\vec{r}_2 - \vec{r}_1$) stay constant or does it change with time? Does the difference in their velocities ($\vec{v}_2 - \vec{v}_1$) stay constant or does it change with time?

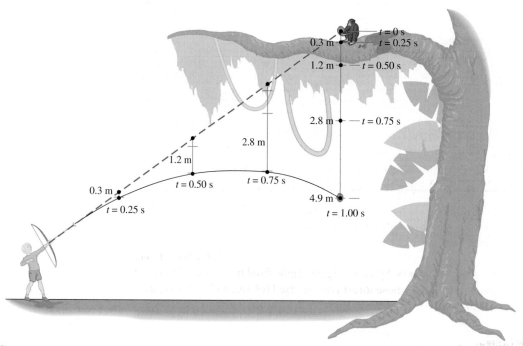

Figure 3.23

A monkey drops a coconut at the very instant an arrow is shot toward the coconut. In each quarter second, the coconut and arrow have fallen the same distance below where their positions would be if there were no gravity.

Example 3.9

A Bullet Fired Horizontally

A bullet is fired horizontally from the top of a cliff that is 20.0 m above a long lake. If the muzzle speed of the bullet is 500.0 m/s, how far from the bottom of the cliff does the bullet strike the surface of the lake? Ignore air resistance.

Strategy We need to find the total time of flight so that we can find the horizontal displacement. The bullet is starting from the high point of the parabolic path because $v_{iy} = 0$. As usual in projectile problems, we choose the y-axis to be the positive vertical direction.

Known: $\Delta y = -20.0$ m; $v_{iy} = 0$; $v_{ix} = 500.0$ m/s. To find: Δx.

Solution The vertical displacement through which the bullet falls is 20.0 m. The relationship between Δy and Δt is

$$\Delta y = \tfrac{1}{2}(v_{fy} + v_{iy})\, \Delta t$$

Substituting $v_{iy} = 0$ and $v_{fy} = v_{iy} + a_y \Delta t = a_y \Delta t$ yields

$$\Delta y = \tfrac{1}{2} a_y (\Delta t)^2 \Rightarrow \Delta t = \sqrt{\frac{2\,\Delta y}{a_y}}$$

The horizontal displacement of the bullet is

$$\Delta x = v_{ix} \Delta t = v_{ix} \sqrt{\frac{2\,\Delta y}{a_y}}$$

$$= 500.0 \text{ m/s} \times \sqrt{\frac{2 \times (-20.0 \text{ m})}{-9.8 \text{ m/s}^2}} = 1.01 \text{ km}$$

Discussion How did we know to start with the y-component equation when the question asks about the *horizontal* displacement? The question gives v_{ix} and asks for Δx. The missing information needed is the time during which the bullet is in the air; the time can be found from analysis of the *vertical* motion.

We ignored air resistance in this problem, which is not very realistic. The actual distance would be less than 1.01 km.

Practice Problem 3.9   Bullet Velocity

Find the horizontal and vertical components of the bullet's velocity just before it hits the surface of the lake. At what angle does it strike the surface?

At the beginning of the chapter, we asked why the clam does not fall straight down when the gull lets go. The gull is flying horizontally with the clam, so the clam has the same horizontal velocity as the gull. When the gull lets go, the clam falls toward Earth, but since $a_x = 0$ the clam retains the same horizontal component of velocity as the gull. Therefore, the clam is a projectile starting at the top of its parabolic trajectory.

Why does the clam not drop straight down?

3.6   VELOCITY IS RELATIVE; REFERENCE FRAMES

The idea of *relativity* arose in physics centuries before Einstein's theory. Nicole Oresme (1323–1382) wrote that motion of one object can only be perceived relative to some other object. Until now, we have tacitly assumed in most situations that displacements, velocities, and accelerations should be measured in a **reference frame** attached to Earth's surface—that is, by choosing an origin fixed in position relative to Earth's surface and a set of axes whose directions are fixed relative to Earth's surface. After learning about relative velocities, we will take another look at this assumption.

Relative Velocity

Suppose Wanda is walking down the aisle of a train moving along the track at a constant velocity (Fig. 3.24). Imagine asking, "How fast is Wanda moving?" This question is not well defined. Do we mean her speed as measured by Tim, a passenger on the train, or her speed as measured by Greg, who is standing on the ground and looking into the train as it passes by? The answer to the question "How fast?" depends on the observer.

Figure 3.25 shows Wanda walking from one end of the car to the other during a time interval Δt. The displacement of Wanda as measured by Tim—her displacement *relative*

Figure 3.24 Tim and Greg watch Wanda walk down the aisle of a train. Wanda's velocity with respect to Tim (or with respect to the train) is $\vec{v}_{WT}$; Tim's velocity with respect to Greg (or with respect to the ground) is $\vec{v}_{TG}$.

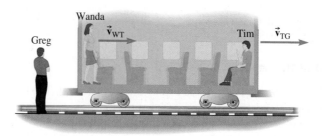

to the train—is $\Delta\vec{r}_{WT} = \vec{v}_{WT}\Delta t$. During the same time interval, the *train's* displacement *relative to Greg* is $\Delta\vec{r}_{TG} = \vec{v}_{TG}\Delta t$. As measured by Greg, Wanda's displacement is partly due to her motion relative to the train and partly due to the motion of the train relative to the ground. Figure 3.25 shows that $\Delta\vec{r}_{WT} + \Delta\vec{r}_{TG} = \Delta\vec{r}_{WG}$. Dividing by the time interval Δt gives the relationship between the three velocities:

$$\vec{v}_{WT} + \vec{v}_{TG} = \vec{v}_{WG} \qquad (3\text{-}24)$$

 To be sure that you are adding the velocity vectors correctly, think of the subscripts as if they were fractions that get multiplied when the velocity vectors are added. In Eq. (3-24), $\dfrac{W}{T} \times \dfrac{T}{G} = \dfrac{W}{G}$ so the equation is correct.

Applications of Relative Velocities for Pilots and Sailors   Relative velocities are of enormous practical interest to pilots of aircraft, sailors, and captains of ocean freighters. The pilot of an airplane is ultimately concerned with the motion of the plane with respect to the ground—the takeoff and landing points are fixed points on the ground. However, the controls of the plane (engines, rudder, ailerons, and spoilers) affect the motion of the plane *with respect to the air.* A sailor has to consider three different velocities of the boat: with respect to shore (for launching and landing), with respect to the air (for the behavior of the sails), and with respect to the water (for the behavior of the rudder).

✓ CHECKPOINT 3.6

In Fig. 3.24, if the train is moving at 18.0 m/s with respect to the ground and Wanda walks at 1.5 m/s with respect to the train, how fast is Wanda moving (a) with respect to Greg and (b) with respect to Tim?

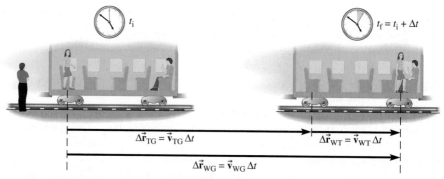

Figure 3.25 Wanda's displacement relative to the ground is the sum of her displacement relative to the train and the displacement of the train relative to the ground.

Example 3.10

Flight from Denver to Chicago

An airplane flies from Denver to Chicago (1770 km) in 4.4 h when no wind blows. On a day with a tailwind, the plane makes the trip in 4.0 h. (a) What is the wind speed? (b) If a headwind blows with the same speed, how long does the trip take?

Strategy We assume the plane has the same *airspeed*—the same speed relative to the air—in both cases. Once the plane is up in the air, the behavior of the wings, control surfaces, etc., depends on how fast the air is rushing by; the ground speed is irrelevant. But it is not irrelevant for the passengers, who are interested in a displacement relative to the ground.

Solution Let $\vec{v}_{PG}$ and $\vec{v}_{PA}$ represent the velocity of the plane relative to the ground and the velocity of the plane relative to the air, respectively. The wind velocity—the velocity of the air relative to the ground—can be written $\vec{v}_{AG}$. Then $\vec{v}_{PA} + \vec{v}_{AG} = \vec{v}_{PG}$. The equation is correct since $\frac{P}{A} \times \frac{A}{G} = \frac{P}{G}$. With no wind,

$$v_{PA} = v_{PG} = \frac{1770 \text{ km}}{4.4 \text{ h}} = 400 \text{ km/h}$$

(a) On the day with the tailwind,

$$v_{PG} = \frac{1770 \text{ km}}{4.0 \text{ h}} = 440 \text{ km/h}$$

We expect v_{PA} to be the same regardless of whether there is a wind or not. Since we are dealing with a tailwind, $\vec{v}_{PA}$ and

$\vec{v}_{AG}$ are in the same direction, which we label as the +x-direction in Fig. 3.26. Then,

$$v_{PAx} + v_{AGx} = v_{PGx}$$

$$v_{AGx} = v_{PGx} - v_{PAx} = 440 \text{ km/h} - 400 \text{ km/h} = 40 \text{ km/h}$$

$v_{AGy} = 0$, so the wind speed is $v_{AG} = 40$ km/h.

(b) With a 40 km/h headwind, $\vec{v}_{PA}$ and $\vec{v}_{AG}$ are in opposite directions (Fig. 3.27). The velocity of the plane with respect to the ground is

$$v_{PGx} = v_{PAx} + v_{AGx} = 400 \text{ km/h} + (-40 \text{ km/h}) = 360 \text{ km/h}$$

The ground speed of the plane is 360 km/h and the trip takes

$$\frac{1770 \text{ km}}{360 \text{ km/h}} = 4.9 \text{ h}$$

Discussion Quick check: the trip takes longer with a headwind (4.9 h) than with no wind (4.4 h), as we expect.

Practice Problem 3.10 Rowing Across the Bay

Jamil, practicing to get on the crew team at school, rows a one-person racing shell to the north shore of the bay for a distance of 3.6 km to his friend's dock. On a day when the water is still (no current flowing), it takes him 20 min (1200 s) to reach his friend. On another day when a current flows southward, it takes him 30 min (1800 s) to row the same course. Ignore air resistance. (a) What is the speed of the current in m/s? (b) How long does it take Jamil to return home with that same current flowing?

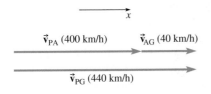

$\vec{v}_{PA}$ (400 km/h) $\vec{v}_{AG}$ (40 km/h)

$\vec{v}_{PG}$ (440 km/h)

Figure 3.26
Addition of velocity vectors in the case of a tailwind. Lengths of vectors are not to scale.

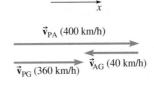

$\vec{v}_{PA}$ (400 km/h)

$\vec{v}_{PG}$ (360 km/h) $\vec{v}_{AG}$ (40 km/h)

Figure 3.27
Addition of velocity vectors in the case of a headwind. Lengths of vectors are not to scale.

The vector equation (3-24) applies to situations where the velocities are not all along the same line, as illustrated in Example 3.11.

Example 3.11

Rowing Across a River

Jack wants to row directly across a river from the east shore to a point on the west shore. The width of the river is 250 m and the current flows from north to south at 0.61 m/s. The

trip takes Jack 4.2 min. In what direction did he head his rowboat to follow a course due west across the river? At what speed with respect to still water is Jack able to row?

continued on next page

Example 3.11 continued

Strategy   We start with a sketch of the situation (Fig. 3.28). To keep the various velocities straight, we choose subscripts as follows: R = rowboat; W = water; S = shore. The velocity of the current given is the velocity of the water relative to the shore: $\vec{v}_{WS}$ = 0.61 m/s, south. The velocity of the rowboat relative to shore ($\vec{v}_{RS}$) is due west. The magnitude of $\vec{v}_{RS}$ can be found from the displacement relative to shore and the time interval, both of which are given. The question asks for the magnitude and direction of the velocity of the rowboat relative to the water ($\vec{v}_{RW}$). The three velocities are related by

$$\vec{v}_{RW} + \vec{v}_{WS} = \vec{v}_{RS}$$

To compensate for the current carrying the rowboat south with respect to shore, Jack heads (points) the rowboat upstream (against the current) at some angle to the north of west.

Solution   In a sketch of the vector addition (Fig. 3.29), the velocity of the rowboat with respect to the water is at an angle θ north of west. With respect to shore, Jack travels 250 m in 4.2 min, so his speed with respect to shore is

$$v_{RS} = \frac{250 \text{ m}}{4.2 \text{ min} \times 60 \text{ s/min}} = 0.992 \text{ m/s}$$

We can find the angle at which the rowboat should be headed by finding the tangent of the angle between $\vec{v}_{RW}$ and $\vec{v}_{RS}$:

$$\tan \theta = \frac{v_{WS}}{v_{RS}} = \frac{0.61 \text{ m/s}}{0.992 \text{ m/s}}$$

$$\theta = 32° \text{ N of W}$$

The speed at which Jack is able to row with respect to still water is the magnitude of $\vec{v}_{RW}$. Since $\vec{v}_{RS}$ and $\vec{v}_{WS}$ are *perpendicular*, the Pythagorean theorem yields

$$|\vec{v}_{RW}| = \sqrt{v_{WS}^2 + v_{RS}^2} = \sqrt{(0.61 \text{ m/s})^2 + (0.992 \text{ m/s})^2}$$

$$= 1.16 \text{ m/s}$$

Jack rows at a speed of 1.16 m/s with respect to the water.

⚠ **Discussion**   If $\vec{v}_{RS}$ and $\vec{v}_{WS}$ had not been perpendicular, we could not have used the Pythagorean

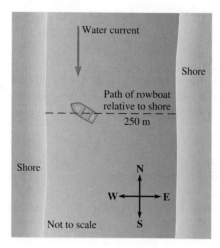

Figure 3.28

Rowing across a river.

theorem in this way. Rather, we would use the component method to add the two vectors.

If Jack had headed the rowboat directly west, the current would have carried him south, so he would have traveled in a direction south of west relative to shore. He has to compensate by heading upstream at just such an angle that his velocity relative to shore is directed west.

Practice Problem 3.11   Heading Straight Across

If Jack were to head straight across the river, in what direction with respect to shore would he travel? How long would it take him to cross? How far downstream would he be carried? Assume that he rows at the same speed with respect to the water as in Example 3.11.

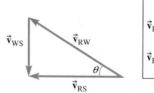

$\vec{v}_{WS}$ is velocity of water with respect to shore
$\vec{v}_{RS}$ is velocity of rowboat with respect to shore
$\vec{v}_{RW}$ is velocity of rowboat with respect to water

Figure 3.29

Graphical addition of the velocity vectors.

What does the path of the falling clam look like to the gull?

At the beginning of this chapter, we asked what the path followed by the falling clam looks like as seen by the gull flying through the air. With respect to a beachcomber on the ground and ignoring air resistance, the clam has a constant horizontal velocity component given to it by the gull and a changing vertical component of velocity due to gravity (Fig. 3.30a); the clam moves in a parabolic path. If the gull continues to fly at the same horizontal velocity after dropping the clam, it is directly overhead when the clam hits the rock because they both have the same constant horizontal component of velocity with respect to Earth.

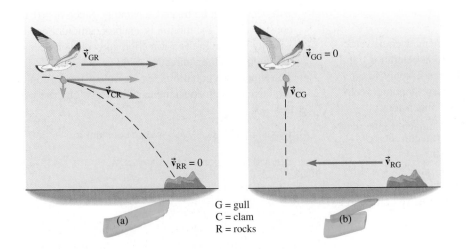

In its own reference frame—that is, using its own position as the origin of the coordinate axes—the gull sees the clam drop straight down toward the ground while rocks and other objects on the beach are moving horizontally (Fig. 3.30b). The bird sees a collision between the horizontally moving rocks and the vertically falling clam. At any instant, if the velocity of the clam with respect to the gull is $\vec{v}_{CG}$, the velocity of the gull with respect to the rocks is $\vec{v}_{GR}$, and the velocity of the clam with respect to the rocks is $\vec{v}_{CR}$, then $\vec{v}_{CG} + \vec{v}_{GR} = \vec{v}_{CR}$.

Master the Concepts

- Vectors are added graphically by drawing each vector so that its tail is placed at the tip of the previous vector. The sum is drawn as a vector arrow from the tail of the first vector to the tip of the last. Addition of vectors is commutative: $\vec{A} + \vec{B} = \vec{B} + \vec{A}$.

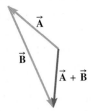

- Vectors are subtracted by adding the opposite of the second vector: $\vec{A} - \vec{B} = \vec{A} + (-\vec{B})$.
- Addition and subtraction of vectors algebraically using components is generally easier and more accurate than the graphical method. The graphical method is still a useful first step to get an approximate answer.
- To find the components of a vector, first draw a right triangle with the vector as the hypotenuse and the other two sides parallel to the x- and y-axes. Then use the trigonometric functions to find the magnitudes of the components. The correct algebraic sign must be determined for each component. The same triangle can be used to

find the magnitude and direction of a vector if its components are known.

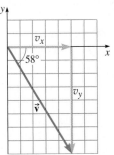

- To add vectors algebraically, add their components to find the components of the sum:

$$\vec{A} + \vec{B} = \vec{C} \text{ if and only if}$$
$$A_x + B_x = C_x \text{ and } A_y + B_y = C_y$$

- The x- and y-axes are chosen to make the problem easiest to solve. Any choice is valid as long as the two are perpendicular. If the direction of the acceleration is known, choose x- and y-axes so that the acceleration vector is parallel to one of the axes.
- Position, displacement, velocity, and acceleration are vector quantities with both magnitude and direction. They must be added and subtracted as vectors.

continued on next page

Master the Concepts continued

- The equations for position, displacement, average velocity, instantaneous velocity, average acceleration, and instantaneous acceleration in Chapter 2 apply to *each perpendicular component* of the corresponding vector quantities for motion in two or three dimensions.

- The instantaneous velocity vector is tangent to the path of motion.

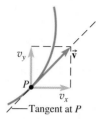

- The instantaneous acceleration vector does *not* have to be tangent to the path of motion, since velocities can change both in direction and in magnitude.

- For a projectile or any object moving with constant acceleration in the ± y-direction, the motion in the x- and y-directions can be treated separately. Since $a_x = 0$, v_x is constant. Thus, the motion is a superposition of constant velocity motion in the x-direction and constant acceleration motion in the y-direction.

- The kinematic equations for an object moving in two dimensions with constant acceleration along the y-axis are

x-axis: $a_x = 0$ y-axis: constant a_y

$\Delta v_x = 0$ (v_x is constant) $\Delta v_y = a_y \Delta t$ (3-19)

$\Delta x = v_x \Delta t$ $\Delta y = \frac{1}{2}(v_{fy} + v_{iy})\Delta t$ (3-20)

$\Delta y = v_{iy}\Delta t + \frac{1}{2}a_y(\Delta t)^2$ (3-21)

$v_{fy}^2 = v_{iy}^2 = 2a_y\Delta y$ (3-22)

- To relate the velocities of objects measured in different reference frames, use the vector equation

$$\vec{v}_{AC} = \vec{v}_{AB} + \vec{v}_{BC} \qquad (3\text{-}24)$$

where $\vec{v}_{AC}$ represents the velocity of A relative to C, and so forth.

Conceptual Questions

1. If two vectors have the same magnitude, are they necessarily equal? If not, why not? Can two vectors with different magnitudes ever be equal?

2. (a) Is it possible for the sum of two vectors to be smaller in magnitude than the magnitude of either vector? (b) Is it possible for the magnitude of the sum of two vectors to be larger than the sum of the magnitudes of the two vectors?

3. What is the distinction between a vector and a scalar quantity? Give two examples of each.

4. Is it possible for two identical projectiles with identical initial speeds, but with two different angles of elevation, to land in the same spot? Explain. Ignore air resistance and sketch the trajectories.

5. If the trajectory is parabolic in one reference frame, is it always, never, or sometimes parabolic in another reference frame that moves at constant velocity with respect to the first reference frame? If the trajectory can be other than parabolic, what else can it be?

6. You are standing on a balcony overlooking the beach. You throw a ball straight up into the air with speed v_i and throw an identical ball straight down with speed v_i. Ignoring air resistance, how do the speeds of the balls compare just before they hit the ground?

7. You throw a ball up with initial speed v_i and when it reaches its high point at height h, you throw another ball into the air with the same initial speed v_i. Will the two balls cross at half the height h, or more than half, or less than half? Explain.

8. If an object is traveling at a constant velocity, is it necessarily traveling in a straight line? Explain.

9. Can the average speed and the magnitude of the average velocity ever be equal? If so, under what circumstances?

10. Give an example of an object whose acceleration is (1) in the same direction as its velocity, (2) opposite its velocity, and (3) perpendicular to its velocity.

11. Name a situation where the speed of an object is constant while the velocity is not.

12. Tell whether or not each of the following objects has a constant velocity and explain your reasoning. (a) A car driving around a curve at constant speed on a flat road. (b) A car driving straight up a 6° incline at constant speed. (c) The Moon.

13. Explain how to add two displacement vectors of magnitudes $3L$ and $4L$ so that the vector sum has magnitude (a) L; (b) $7L$; (c) $5L$.

14. Compare the advantages and disadvantages of the two methods of vector addition (graphical and algebraic).

15. Can the x-component of a vector ever be greater than the magnitude of the vector? Explain.

16. Why is the muzzle of a rifle not aimed directly at the center of the target? Why is this more important at longer ranges?

17. Does the monkey, coconut, and hunter demonstration still work if the hunter is in a higher tree and the arrow is pointed *downward* at the monkey and coconut? Explain.

Multiple-Choice Questions

1. Vector $\vec{\mathbf{A}}$ in the drawing is equal to
   - (a) $\vec{\mathbf{C}} + \vec{\mathbf{D}}$
   - (b) $\vec{\mathbf{C}} + \vec{\mathbf{D}} + \vec{\mathbf{E}}$
   - (c) $\vec{\mathbf{C}} + \vec{\mathbf{F}}$
   - (d) $\vec{\mathbf{B}} + \vec{\mathbf{C}}$
   - (e) $\vec{\mathbf{B}} + \vec{\mathbf{F}}$

Multiple-Choice Questions 1 and 2

2. Which vector sum is *not* equal to zero?
   - (a) $\vec{\mathbf{C}} + \vec{\mathbf{D}} + \vec{\mathbf{E}}$
   - (b) $\vec{\mathbf{B}} + \vec{\mathbf{C}} + \vec{\mathbf{F}}$
   - (c) $\vec{\mathbf{D}} + \vec{\mathbf{F}}$
   - (d) $\vec{\mathbf{A}} + \vec{\mathbf{B}} + \vec{\mathbf{F}}$

3. A hunter spots a pheasant flying along horizontally. If he shoots the pheasant, the time interval between the bird being shot and the dead bird hitting the ground depends on
   - (a) the speed with which the bird was flying.
   - (b) the height of the bird above the ground.
   - (c) the speed of the bird and its height above the ground.

4. A runner moves along a circular track at a constant speed.
   - (a) Her acceleration is zero.
   - (b) Her velocity is constant.
   - (c) Both (a) and (b) are true.
   - (d) Both her acceleration and her velocity are changing.

5. A boy plans to cross a river in a rubber raft. The current flows from north to south at 1 m/s. In what direction should he head to get across the river to the east bank in the least amount of time if he is able to paddle the raft at 1.5 m/s in still water?
   - (a) directly to the east
   - (b) south of east
   - (c) north of east
   - (d) The three directions require the same time to cross the river.

6. A boy plans to paddle a rubber raft across a river to the east bank while the current flows downriver from north to south at 1 m/s. He is able to paddle the raft at 1.5 m/s in still water. In what direction should he head the raft to go straight east across the river to the opposite bank?
   - (a) directly to the east
   - (b) south of east
   - (c) north of east
   - (d) north
   - (e) south

7. A kicker kicks a football from the 5-yard line to the 45-yard line (both on the same half of the field). Ignoring air resistance, where along the trajectory is the speed of the football a minimum?
   - (a) at the 5-yard line, just after the football leaves the kicker's foot
   - (b) at the 45-yard line, just before the football hits the ground
   - (c) at the 15-yard line, while the ball is still going higher
   - (d) at the 35-yard line, while the ball is coming down
   - (e) at the 25-yard line, when the ball is at the top of its trajectory

8. Two balls, identical except for color, are projected horizontally from the roof of a tall building at the same instant. The initial speed of the red ball is twice the initial speed of the blue ball. Ignoring air resistance,
   - (a) the red ball reaches the ground first.
   - (b) the blue ball reaches the ground first.
   - (c) both balls land at the same instant with different speeds.
   - (d) both balls land at the same instant with the same speed.

9. A person stands on the roof garden of a tall building with one ball in each hand. If the red ball is thrown horizontally off the roof and the blue ball is simultaneously dropped over the edge, which statement is true?
   - (a) Both balls hit the ground at the same time, but the red ball has a higher speed just before it strikes the ground.
   - (b) The blue ball strikes the ground first, but with a lower speed than the red ball.
   - (c) The red ball strikes the ground first with a higher speed than the blue ball.
   - (d) Both balls hit the ground at the same time with the same speed.

10. A ball is thrown into the air and follows a parabolic trajectory. At the highest point in the trajectory,
    - (a) the velocity is zero, but the acceleration is not zero.
    - (b) both the velocity and the acceleration are zero.
    - (c) the acceleration is zero, but the velocity is not zero.
    - (d) neither the acceleration nor the velocity are zero.

11. A ball is thrown into the air and follows a parabolic trajectory. Point A is the highest point in the trajectory and point B is a point as the ball is falling back to the ground. Choose the correct relationship between the speeds and the magnitudes of the acceleration at the two points.
    - (a) $v_A > v_B$ and $a_A = a_B$
    - (b) $v_A < v_B$ and $a_A > a_B$
    - (c) $v_A = v_B$ and $a_A \neq a_B$
    - (d) $v_A < v_B$ and $a_A = a_B$

Questions 12–14. Two projectiles launched with the same initial speed but at different launch angles 30° and 60° land at the same spot (see Fig. 3.22). Ignore air resistance. Answer choices:
   - (a) projectile launched at 30°
   - (b) projectile launched at 60°
   - (c) They are equal.

12. Which has the larger horizontal velocity component v_x?

13. Which has a longer time of flight Δt (time interval between launch and hitting the ground)?

14. For which is the product $v_x \Delta t$ larger?

Problems

ⓒ   Combination conceptual/quantitative problem

♈   Biological or medical application

✦   Challenging problem

Blue #   Detailed solution in the Student Solutions Manual

① ②   Problems paired by concept

〰   Text website interactive or tutorial

3.1 Graphical Addition and Subtraction of Vectors

1. Displacement vector $\vec{A}$ is directed to the west and has magnitude 2.56 km. A second displacement vector is also directed to the west and has magnitude 7.44 km. (a) What are the magnitude and direction of $\vec{A} + \vec{B}$? (b) What are the magnitude and direction of $\vec{A} - \vec{B}$? (c) What are the magnitude and direction of $\vec{B} - \vec{A}$?

2. Vector $\vec{A}$ is directed along the positive x-axis and has magnitude 1.73 units. Vector $\vec{B}$ is directed along the negative x-axis and has magnitude 1.00 unit. (a) What are the magnitude and direction of $\vec{A} + \vec{B}$? (b) What are the magnitude and direction of $\vec{A} - \vec{B}$? (c) What are the magnitude and direction of $\vec{B} - \vec{A}$?

3. Two vectors have magnitudes 3.0 and 4.0. How are the directions of the two vectors related if (a) the sum has magnitude 7.0, or (b) if the sum has magnitude 5.0? (c) What relationship between the directions gives the smallest magnitude sum and what is this magnitude?

4. A runner is practicing on a circular track that is 300 m in circumference. From the point farthest to the west on the track, he starts off running due north and follows the track as it curves around toward the east. (a) If he runs halfway around the track and stops at the farthest eastern point of the track, what is the distance he traveled? (b) What is his displacement?

5. Two displacement vectors each have magnitude 20 km. One is directed 60° above the $+x$-axis; the other is directed 60° below the $+x$-axis. What is the vector sum of these two displacements? Use graph paper to find your answer.

6. Orville walks 320 m due east. He then continues walking along a straight line, but in a different direction, and stops 200 m northeast of his starting point. How far did he walk during the second portion of the trip and in what direction?

7. Vectors $\vec{A}$, $\vec{B}$, and $\vec{C}$ are shown in the figure. (a) Draw vectors $\vec{D}$ and $\vec{E}$, where $\vec{D} = \vec{A} + \vec{B}$ and $\vec{E} = \vec{A} + \vec{C}$. (b) Show that $\vec{A} + \vec{B} = \vec{B} + \vec{A}$ by graphical means.

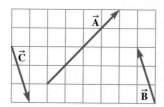

8. Two vectors, each of magnitude 4.0 cm, are directed at a small angle α below the horizontal as shown. (The grid is 1 cm on a side.) (a) Let $\vec{C} = \vec{A} + \vec{B}$. Sketch the direction of $\vec{C}$ and estimate its magnitude. (b) Let $\vec{D} = \vec{A} - \vec{B}$. Sketch the direction of $\vec{D}$ and estimate its magnitude. (〰 tutorial: vectors)

9. Michaela is planning a trip in Ireland from Killarney to Cork to visit Blarney Castle. (See Example 3.2.) She also wants to visit Mallow, which is located 39 km due east of Killarney and 22 km due north of Cork. Draw the displacement vectors for the trip when she travels from Killarney to Mallow to Cork. (a) What is the magnitude of her displacement once she reaches Cork? (b) How much additional distance does Michaela travel in going to Cork by way of Mallow instead of going directly from Killarney to Cork?

10. A scout troop is practicing its orienteering skills with map and compass. First they walk due east for 1.2 km. Next, they walk 45° west of north for 2.7 km. In what direction must they walk to go directly back to their starting point? How far will they have to walk? Use graph paper, ruler, and protractor to find a geometrical solution.

ⓒ11. Prove that the displacement for a trip is equal to the vector sum of the displacements for each leg of the trip. [*Hint:* Imagine a trip that consists of n segments. The trip starts at position $\vec{r}_1$, proceeds to $\vec{r}_2$, then to $\vec{r}_3, \ldots,$ then to $\vec{r}_{n-1}$, then finally to $\vec{r}_n$. Write an expression for each displacement as the difference of two position vectors and then add them.]

12. A sailboat sails from Marblehead Harbor directly east for 45 nautical miles, then 60° south of east for 20.0 nautical miles, returns to an easterly heading for 30.0 nautical miles, and sails 30° east of north for 10.0 nautical miles, then west for 62 nautical miles. At that time the boat becomes becalmed and the auxiliary engine fails to start. The crew decides to notify the Coast Guard of their position. Using graph paper, ruler, and protractor, sketch a graphical addition of the displacement vectors and estimate their position.

3.2 Vector Addition and Subtraction Using Components

13. A vector is 20.0 m long and makes an angle of 60.0° counterclockwise from the y-axis (on the side of the $-x$-axis). What are the x- and y-components of this vector?

14. Vector $\vec{A}$ has magnitude 4.0 units; vector $\vec{B}$ has magnitude 6.0 units. The angle between $\vec{A}$ and $\vec{B}$ is 60.0°. What is the magnitude of $\vec{A} + \vec{B}$?

15. Vector $\vec{A}$ is directed along the positive y-axis and has magnitude $\sqrt{3.0}$ units. Vector $\vec{B}$ is directed along the negative x-axis and has magnitude 1.0 unit. (a) What are

the magnitude and direction of $\vec{A} + \vec{B}$? (b) What are the magnitude and direction of $\vec{A} - \vec{B}$? (c) What are the *x*- and *y*-components of $\vec{B} - \vec{A}$?

16. Vector $\vec{a}$ has components $a_x = -3.0$ m/s^2 and $a_y = +4.0$ m/s^2. (a) What is the magnitude of $\vec{a}$? (b) What is the direction of $\vec{a}$? Give an angle with respect to one of the coordinate axes.

17. In Problem 8, let $\alpha = 10°$ and find the magnitude of vector $\vec{C}$ using the component method.

18. In Problem 8, let $\alpha = 10°$ and find the magnitude of vector $\vec{D}$ using the component method.

19. Find the *x*- and *y*-components of the four vectors shown in the drawing.

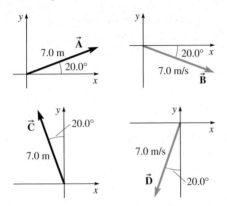

20. The velocity vector of a sprinting cheetah has *x*- and *y*-components $v_x = +16.4$ m/s and $v_y = -26.3$ m/s. (a) What is the magnitude of the velocity vector? (b) What angle does the velocity vector make with the +*x*- and −*y*-axes?

21. In each of these, the *x*- and *y*-components of a vector are given. Find the magnitude and direction of the vector. (a) $A_x = -5.0$ m/s, $A_y = +8.0$ m/s. (b) $B_x = +120$ m, $B_y = -60.0$ m. (c) $C_x = -13.7$ m/s, $C_y = -8.8$ m/s. (d) $D_x = 2.3$ m/s^2, $D_y = 6.5$ cm/s^2.

22. A vector $\vec{A}$ has a magnitude of 22.2 cm and makes an angle of 130.0° with the positive *x*-axis. What are the *x*- and *y*-components of this vector?

23. Vector $\vec{B}$ has magnitude 7.1 and direction 14° below the +*x*-axis. Vector $\vec{C}$ has *x*-component $C_x = -1.8$ and *y*-component $C_y = -6.7$. Compute (a) the *x*- and *y*-components of $\vec{B}$; (b) the magnitude and direction of $\vec{C}$; (c) the magnitude and direction of $\vec{C} + \vec{B}$; (d) the magnitude and direction of $\vec{C} - \vec{B}$; (e) the *x*- and *y*-components of $\vec{C} - \vec{B}$.

24. Margaret walks to the store using the following path: 0.500 miles west, 0.200 miles north, 0.300 miles east. What is her total displacement? That is, what is the length and direction of the vector that points from her house directly to the store? Use vector components to find the answer.

25. Jerry bicycles from his dorm to the local fitness center: 3.00 miles east and 2.00 miles north. Cindy's apartment is located 1.50 miles west of Jerry's dorm. If Cindy is able to meet Jerry at the fitness center by bicycling in a straight line, what is the length and direction she must travel?

26. Repeat Problem 10 using the component (algebraic) method.

27. Use the component method to obtain a more accurate description of the sailboat's location in Problem 12.

28. You will be hiking to a lake with some of your friends by following the trails indicated on a map at the trailhead. The map says that you will travel 1.6 mi directly north, then 2.2 mi in a direction 35° east of north, then finally 1.1 mi in a direction 15° north of east. At the end of this hike, how far will you be from where you started, and what direction will you be from your starting point?

3.3 Velocity

29. A runner times his speed around a circular track with a circumference of 0.478 mi. At the start he is running toward the east and the track starts bending toward the north. If he goes halfway around, he will be running toward the west. He finds that he has run a distance of 0.750 mi in 4.00 min. What is his (a) average speed and (b) average velocity in m/s?

30. A runner times his speed around a track with a circumference of 0.50 mi. He finds that he has run a distance of 1.00 mi in 4.0 min. What is his (a) average speed and (b) average velocity magnitude in m/s?

31. Peggy drives from Cornwall to Atkins Glen in 45 min. Cornwall is 73.6 km from Illium in a direction 25° west of south. Atkins Glen is 27.2 km from Illium in a direction 15° south of west. Using Illium as your origin, (a) draw the initial and final position vectors, (b) find the displacement during the trip, and (c) find Peggy's average velocity for the trip.

32. To get to a concert in time, a harpsichordist has to drive 122 mi in 2.00 h. (a) If he drove at an average speed of 55.0 mi/h in a due west direction for the first 1.20 h, what must be his average speed if he is heading 30.0° south of west for the remaining 48.0 min? (b) What is his average velocity for the entire trip?

33. A bicycle travels 3.2 km due east in 0.10 h, then 4.8 km at 15.0° east of north in 0.15 h, and finally another 3.2 km due east in 0.10 h to reach its destination. The time lost in turning is negligible. What is the average velocity for the entire trip?

34. A car travels east at 96 km/h for 1.0 h. It then travels 30.0° east of north at 128 km/h for 1.0 h. (a) What is the average speed for the trip? (b) What is the average velocity for the trip?

35. A speedboat heads west at 108 km/h for 20.0 min. It then travels at 60.0° south of west at 90.0 km/h for 10.0 min. (a) What is the average speed for the trip? (b) What is the average velocity for the trip?

36. See Problem 9. During Michaela's travel from Killarney to Cork via Mallow, her actual travel time in the car is

48 min. (a) What is her average speed in m/s? (b) What is the magnitude of her average velocity in m/s?

◆37. Geoffrey drives from his home town due east at 90.0 km/h for 80.0 min. After visiting a friend for 15.0 min, he drives in a direction 30.0° south of west at 76.0 km/h for 45.0 min to visit another friend. (a) How far is it to his home from the second town? (b) If it takes him 45.0 min to drive directly home, what is his average velocity on the third leg of the trip? (c) What is his average velocity during the first two legs of his trip? (d) What is his average velocity over the entire trip? (e) What is his average speed during the entire trip if he spent 55.0 min visiting the second friend?

3.4 Acceleration

38. A hawk is flying north at 2.0 m/s with respect to the ground; 10.0 s later, it is flying south at 5.0 m/s. What is its average acceleration during this time interval?

39. A skydiver is falling straight down at 55 m/s when he opens his parachute and slows to 8.3 m/s in 3.5 s. What is the average acceleration of the skydiver during those 3.5 s?

40. A car travels three quarters of the way around a circle of radius 20.0 m in a time of 3.0 s at a constant speed. The initial velocity is west and the final velocity is south. (a) Find its average velocity for this trip. (b) What is the car's average acceleration during these 3.0 s? (c) Explain how a car moving at constant speed has a nonzero average acceleration.

41. At $t = 0$, an automobile traveling north begins to make a turn. It follows one-quarter of the arc of a circle with a radius of 10.0 m until, at $t = 1.60$ s, it is traveling east. The car does not alter its speed during the turn. Find (a) the car's speed, (b) the change in its velocity during the turn, and (c) its average acceleration during the turn.

42. At the beginning of a 3.0-h plane trip, you are traveling due north at 192 km/h. At the end, you are traveling 240 km/h in the northwest direction (45° west of north). (a) Draw your initial and final velocity vectors. (b) Find the change in your velocity. (c) What is your average acceleration during the trip?

43. John drives 16 km directly west from Orion to Chester at a speed of 90 km/h, then directly south for 8.0 km to Seiling at a speed of 80 km/h, then finally 34 km southeast to Oakwood at a speed of 100 km/h. Assume he travels at constant velocity during each of the three segments. (a) What was the change in velocity during this trip? [*Hint:* Do not assume he starts from rest and stops at the end.] (b) What was the average acceleration during this trip?

44. A particle's constant acceleration is south at 2.50 m/s². At $t = 0$, its velocity is 40.0 m/s east. What is its velocity at $t = 8.00$ s?

45. A particle's constant acceleration is north at 100 m/s². At $t = 0$, its velocity vector is 60 m/s east. At what time will the magnitude of the velocity be 100 m/s?

3.5 Motion in a Plane with Constant Acceleration

46. A baseball is thrown horizontally from a height of 9.60 m above the ground with a speed of 30.0 m/s. Where is the ball after 1.40 s has elapsed?

47. A clump of soft clay is thrown horizontally from 8.50 m above the ground with a speed of 20.0 m/s. Where is the clay after 1.50 s? Assume it sticks in place when it hits the ground.

48. A tennis ball is thrown horizontally from an elevation of 14.0 m above the ground with a speed of 20.0 m/s. (a) Where is the ball after 1.60 s? (b) If the ball is still in the air, how long before it hits the ground and where will it be with respect to the starting point once it lands?

49. A ball is thrown from a point 1.0 m above the ground. The initial velocity is 19.6 m/s at an angle of 30.0° above the horizontal. (a) Find the maximum height of the ball above the ground. (b) Calculate the speed of the ball at the highest point in the trajectory.

50. An arrow is shot into the air at an angle of 60.0° above the horizontal with a speed of 20.0 m/s. (a) What are the x- and y-components of the velocity of the arrow 3.0 s after it leaves the bowstring? (b) What are the x- and y-components of the displacement of the arrow during the 3.0-s interval?

51. You are working as a consultant on a video game designing a bomb site for a World War I airplane. In this game, the plane you are flying is traveling horizontally at 40.0 m/s at an altitude of 125 m when it drops a bomb. (a) Determine how far horizontally from the target you should release the bomb. (b) What direction is the bomb moving just before it hits the target?

52. You have been employed by the local circus to plan their human cannonball performance. For this act, a spring-loaded cannon will shoot a human projectile, the Great Flyinski, across the big top to a net below. The net is located 5.0 m lower than the muzzle of the cannon from which the Great Flyinski is launched. The cannon will shoot the Great Flyinski at an angle of 35.0° above the horizontal and at a speed of 18.0 m/s. The ringmaster has asked that you decide how far from the cannon to place the net so that the Great Flyinski will land in the net and not be splattered on the floor, which would greatly disturb the audience. What do you tell the ringmaster? (🌐 interactive: projectile motion)

53. A cannonball is catapulted toward a castle. The cannonball's velocity when it leaves the catapult is 40 m/s at an angle of 37° with respect to the horizontal and the cannonball is 7.0 m above the ground at this time. (a) What is the maximum height above the ground reached by the cannonball? (b) Assuming the cannonball makes it over the castle walls and lands back down on the ground, at what horizontal distance from its release point will it land? (c) What are the x- and y-components of the cannonball's velocity just before it lands? The y-axis points up.

54. After being assaulted by flying cannonballs, the knights on the castle walls (12 m above the ground) respond by propelling flaming pitch balls at their assailants. One ball lands on the ground at a distance of 50 m from the castle walls. If it was launched at an angle of 53° above the horizontal, what was its initial speed?

55. From the edge of the rooftop of a building, a boy throws a stone at an angle 25.0° above the horizontal. The stone hits the ground 4.20 s later, 105 m away from the base of the building. (Ignore air resistance.) (a) For the stone's path through the air, sketch graphs of x, y, v_x, and v_y as functions of time. These need to be only *qualitatively* correct—you need not put numbers on the axes. (b) Find the initial velocity of the stone. (c) Find the initial height h from which the stone was thrown. (d) Find the maximum height H reached by the stone.

56. Two angles are complementary when their sum is 90.0°. Find the ranges for two projectiles launched with identical initial speeds of 36.2 m/s at angles of elevation above the horizontal that are complementary pairs. (a) For one trial, the angles of elevation are 36.0° and 54.0°. (b) For the second trial, the angles of elevation are 23.0° and 67.0°. (c) Finally, the angles of elevation are both set to 45.0°. (d) What do you notice about the range values for each complementary pair of angles? At which of these angles was the range greatest?

57. The range R of a projectile is defined as the magnitude of the horizontal displacement of the projectile *when it returns to its original altitude.* (In other words, the range is the distance between the launch point and the impact point on flat ground.) A projectile is launched at $t = 0$ with initial speed v_i at an angle θ above the horizontal. (a) Find the time t at which the projectile returns to its original altitude. (b) Show that the range is

$$R = \frac{v_i^2 \sin 2\theta}{g}$$

[*Hint:* Use the trigonometric identity $\sin 2\theta = 2 \sin \theta \cos \theta$.] (c) What value of θ gives the maximum range? What is this maximum range?

58. Use the expression in Problem 57 to find (a) the maximum range of a projectile with launch speed v_i and (b) the launch angle θ at which the maximum range occurs.

59. A projectile is launched at $t = 0$ with initial speed v_i at an angle θ above the horizontal. (a) What are v_x and v_y at the projectile's highest point? (b) Find the time t at which the projectile reaches its maximum height. (c) Show that the maximum height H of the projectile is

$$H = \frac{(v_i \sin \theta)^2}{2g}$$

60. A ballplayer standing at home plate hits a baseball that is caught by another player at the same height above the ground from which it was hit. The ball is hit with an initial velocity of 22.0 m/s at an angle of 60.0° above the horizontal. (tutorial: projectile) (a) How high will the ball rise? (b) How much time will elapse from the

time the ball leaves the bat until it reaches the fielder? (c) At what distance from home plate will the fielder be when he catches the ball?

61. You are planning a stunt to be used in an ice skating show. For this stunt a skater will skate down a frictionless ice ramp that is inclined at an angle of 15.0° above the horizontal. At the bottom of the ramp, there is a short horizontal section that ends in an abrupt drop off. The skater is supposed to start from rest somewhere on the ramp, then skate off the horizontal section and fly through the air a horizontal distance of 7.00 m while falling vertically for 3.00 m, before landing smoothly on the ice. How far up the ramp should the skater start this stunt?

62. A suspension bridge is 60.0 m above the level base of a gorge. A stone is thrown or dropped from the bridge. Ignore air resistance. At the location of the bridge g has been measured to be 9.83 m/s². (a) If you drop the stone, how long does it take for it to fall to the base of the gorge? (b) If you *throw* the stone straight down with a speed of 20.0 m/s, how long before it hits the ground? (c) If you throw the stone with a velocity of 20.0 m/s at 30.0° above the horizontal, how far from the point directly below the bridge will it hit the level ground?

63. A circus performer is shot out of a cannon and flies over a net that is placed horizontally 6.0 m from the cannon. When the cannon is aimed at an angle of 40° above the horizontal, the performer is moving in the horizontal direction and just barely clears the net as he passes over it. What is the muzzle speed of the cannon and how high is the net?

64. Show that for a projectile launched at an angle of 45° the maximum height of the projectile is one quarter of the range (the distance traveled on flat ground).

3.6 Velocity Is Relative; Reference Frames

65. Two cars are driving toward each other on a straight, flat Kansas road. The Jeep Wrangler is traveling at 82 km/h north and the Ford Taurus is traveling at 48 km/h south, both measured relative to the road. What is the velocity of the Jeep relative to an observer in the Ford?

66. Two cars are driving toward each other on a straight and level road in Alaska. The BMW is traveling at 100.0 km/h north and the VW is traveling at 42 km/h south, both velocities measured relative to the road. At a certain instant, the distance between the cars is 10.0 km. Approximately how long will it take from that instant for the two cars to meet? [*Hint:* Consider a reference frame in which one of the cars is at rest.]

67. A car is driving directly north on the freeway at a speed of 110 km/h and a truck is leaving the freeway driving 85 km/h in a direction that is 35° west of north. What is the velocity of the truck relative to the car?

68. A Nile cruise ship takes 20.8 h to go upstream from Luxor to Aswan, a distance of 208 km, and 19.2 h to

make the return trip downstream. Assuming the ship's speed relative to the water is the same in both cases, calculate the speed of the current in the Nile.

69. An airplane has a velocity relative to the ground of 210 m/s toward the east. The pilot measures his airspeed (the speed of the plane relative to the air) to be 160 m/s. What is the minimum wind velocity possible?

70. A small plane is flying directly west with an airspeed of 30.0 m/s. The plane flies into a region where the wind is blowing at 10.0 m/s at an angle of 30° to the south of west. (a) If the pilot does not change the heading of the plane, what will be the ground speed of the airplane? (b) What will be the new directional heading, relative to the ground, of the airplane? (🅦 tutorial: flight of crow)

71. A small plane is flying directly west with an airspeed of 30.0 m/s. The plane flies into a region where the wind is blowing at 10.0 m/s at an angle of 30° to the south of west. In that region, the pilot changes the directional heading to maintain her due west heading. (a) What is the change she makes in the directional heading to compensate for the wind? (b) After the heading change, what is the ground speed of the airplane?

72. A boat that can travel at 4.0 km/h in still water crosses a river with a current of 1.8 km/h. At what angle must the boat be pointed upstream to travel straight across the river? In other words, in what direction is the velocity of the boat relative to the water?

73. At an antique car rally, a Stanley Steamer automobile travels north at 40 km/h and a Pierce Arrow automobile travels east at 50 km/h. Relative to an observer riding in the Stanley Steamer, what are the *x*- and *y*-components of the velocity of the Pierce Arrow car? The *x*-axis is to the east and the *y*-axis is to the north.

74. Sheena can row a boat at 3.00 mi/h in still water. She needs to cross a river that is 1.20 mi wide with a current flowing at 1.60 mi/h. Not having her calculator ready, she guesses that to go straight across, she should head 60.0° upstream. (a) What is her speed with respect to the starting point on the bank? (b) How long does it take her to cross the river? (c) How far upstream or downstream from her starting point will she reach the opposite bank? (d) In order to go straight across, what angle upstream should she have headed?

75. A dolphin wants to swim directly back to its home bay, which is 0.80 km due west. It can swim at a speed of 4.00 m/s relative to the water, but a uniform water current flows with speed 2.83 m/s in the southeast direction. (a) What direction should the dolphin head? (b) How long does it take the dolphin to swim the 0.80-km distance home?

76. Demonstrate with a vector diagram that a displacement is the same when measured in two different reference frames that are at rest with respect to each other.

77. A boy is attempting to swim directly across a river; he is able to swim at a speed of 0.500 m/s relative to the water. The river is 25.0 m wide and the boy ends up at 50.0 m downstream from his starting point. (a) How fast is the current flowing in the river? (b) What is the speed of the boy relative to a friend standing on the riverbank?

78. An aircraft has to fly between two cities, one of which is 600.0 km north of the other. The pilot starts from the southern city and encounters a steady 100.0 km/h wind that blows from the northeast. The plane has a cruising speed of 300.0 km/h in still air. (a) In what direction (relative to east) must the pilot head her plane? (b) How long does the flight take?

Comprehensive Problems

79. Jason is practicing his tennis stroke by hitting balls against a wall. The ball leaves his racquet at a height of 60 cm above the ground at an angle of 80° with respect to the *vertical*. (a) The speed of the ball as it leaves the racquet is 20 m/s and it must travel a distance of 10 m before it reaches the wall. How far above the ground does the ball strike the wall? (b) Is the ball on its way up or down when it hits the wall?

80. Imagine a trip where you drive along an east-west highway at 80.0 km/h for 45.0 min and then you turn onto a highway that runs 38.0° north of east and travel at 60.0 km/h for 30.0 min. (a) What is your average velocity for the trip? (b) What is your average velocity on the return trip when you head the opposite way and drive 38.0° south of west at 60.0 km/h for the first 30.0 min and then west at 80.0 km/h for the last 45.0 min?

81. A jetliner flies east for 600.0 km, then turns 30.0° toward the south and flies another 300.0 km. (a) How far is the plane from its starting point? (b) In what direction could the jetliner have flown directly to the same destination (in a straight-line path)? (c) If the jetliner flew at a constant speed of 400.0 km/h, how long did the trip take? (d) Moving at the same speed, how long would the direct flight have taken?

82. An African swallow carrying a very small coconut is flying horizontally with a speed of 18 m/s. (a) If it drops the coconut from a height of 100 m above the Earth, how long will it take before the coconut strikes the ground? (b) At what horizontal distance from the release point will the coconut strike the ground?

83. A pilot starting from Athens, New York, wishes to fly to Sparta, New York, which is 320 km from Athens in the direction 20.0° N of E. The pilot heads directly for Sparta and flies at an airspeed of 160 km/h. After flying for 2.0 h, the pilot expects to be at Sparta, but instead he finds himself 20 km due west of Sparta. He has forgotten to correct for the wind. (a) What is the velocity of

the plane relative to the air? (b) Find the velocity (magnitude and direction) of the plane relative to the ground. (c) Find the wind speed and direction.

84. The citizens of Paris were terrified during World War I when they were suddenly bombarded with shells fired from a long-range gun known as Big Bertha. The barrel of the gun was 36.6 m long and it had a muzzle speed of 1.46 km/s. When the gun's angle of elevation was set to 55°, what would be the range? For the purposes of solving this problem, neglect air resistance. (The actual range at this elevation was 121 km; air resistance cannot be ignored for the high muzzle speed of the shells.)

85. You are serving as a consultant for the newest James Bond film. In one scene, Bond must fire a projectile from a cannon and hit the enemy headquarters located on the top of a cliff 75.0 m above and 350 m from the cannon. The cannon will shoot the projectile at an angle of 40.0° above the horizontal. The director wants to know what the speed of the projectile must be when it is fired from the cannon so that it will hit the enemy headquarters. What do you tell him? [*Hint:* Don't assume the projectile will hit the headquarters at the highest point of its flight.]

86. The pilot of a small plane finds that the airport where he intended to land is fogged in. He flies 55 mi west to another airport to find that conditions there are too icy for him to land. He flies 25 mi at 15° east of south and is finally able to land at the third airport. (a) How far and in what direction must he fly the next day to go directly to his original destination? (b) How many extra miles beyond his original flight plan has he flown?

87. A particle has a constant acceleration of 5.0 m/s² to the east. At time $t = 0$, it is 2.0 m east of the origin and its velocity is 20 m/s north. What are the components of its position vector at $t = 2.0$ s?

88. A baseball batter hits a long fly ball that rises to a height of 44 m. An outfielder on the opposing team can run at 7.6 m/s. What is the farthest the fielder can be from where the ball will land so that it is possible for him to catch the ball?

89. A locust jumps at an angle of 55.0° and lands 0.800 m from where it jumped. (a) What is the maximum height of the locust during its jump? Ignore air resistance. (b) If it jumps with the same initial speed at an angle of 45.0°, would the maximum height be larger or smaller? (c) What about the range? (d) Calculate the maximum height and range for this angle.

90. A helicopter is flying horizontally at 8.0 m/s and an altitude of 18 m when a package of emergency medical supplies is ejected horizontally backward with a speed of 12 m/s *relative to the helicopter*. Ignoring air resistance, what is the horizontal distance between the package and the helicopter when the package hits the ground?

91. An airplane is traveling from New York to Paris, a distance of 5.80×10^3 km. Ignore the curvature of the Earth. (a) If the cruising speed of the airplane is 350.0 km/h, how much time will it take for the airplane to make the round-trip on a calm day? (b) If a steady wind blows from New York to Paris at 60.0 km/h, how much time will the round-trip take? (c) How much time will it take if there is a crosswind of 60.0 km/h?

92. A gull is flying horizontally 8.00 m above the ground at 6.00 m/s. The bird is carrying a clam in its beak and plans to crack the clamshell by dropping it on some rocks below. Ignoring air resistance, (a) what is the horizontal distance to the rocks at the moment that the gull should let go of the clam? (b) With what speed relative to the rocks does the clam smash into the rocks? (c) With what speed relative to the gull does the clam smash into the rocks?

93. A beanbag is thrown horizontally from a dorm room window a height h above the ground. It hits the ground a horizontal distance h (the *same* distance h) from the dorm directly below the window from which it was thrown. Ignoring air resistance, find the direction of the beanbag's velocity just before impact.

94. In a plate glass factory, sheets of glass move along a conveyor belt at a speed of 15.0 cm/s. An automatic cutting tool descends at preset intervals to cut the glass to size. Since the assembly belt must keep moving at constant speed, the cutter is set to cut at an angle to compensate for the motion of the glass. If the glass is 72.0 cm wide and the cutter moves across the width at a speed of 24.0 cm/s, at what angle should the cutter be set?

95. A pilot wants to fly from Dallas to Oklahoma City, a distance of 330 km at an angle of 10.0° west of north. The pilot heads directly toward Oklahoma City with an air speed of 200 km/h. After flying for 1.0 h, the pilot finds that he is 15 km off course to the west of where he expected to be after one hour assuming there was no wind. (a) What is the velocity and direction of the wind? (b) In what direction should the pilot have headed his plane to fly directly to Oklahoma City without being blown off course?

96. A ball is thrown horizontally off the edge of a cliff with an initial speed of 20.0 m/s. (a) How long does it take for the ball to fall to the ground 20.0 m below? (b) How long would it take for the ball to reach the ground if it were dropped from rest off the cliff edge? (c) How long would it take the ball to fall to the ground if it were thrown at an initial velocity of 20.0 m/s but 18° below the horizontal?

97. A marble is rolled so that it is projected horizontally off the top landing of a staircase. The initial speed of the marble is 3.0 m/s. Each step is 0.18 m high and 0.30 m wide. Which step does the marble strike first?

98. A motor scooter rounds a curve on the highway at a constant speed of 20.0 m/s. The original direction of the

scooter was due east; after rounding the curve the scooter is heading 36° north of east. The radius of curvature of the road at the location of the curve is 150 m. What is the average acceleration of the scooter as it rounds the curve?

◆ 99. You want to make a plot of the trajectory of a projectile. That is, you want to make a plot of the height y of the projectile as a function of horizontal distance x. The projectile is launched from the origin with an initial speed v_i at an angle θ above the horizontal. Show that the equation of the trajectory followed by the projectile is

$$y = \left(\frac{v_{iy}}{v_{ix}}\right)x + \left(\frac{-g}{2v_{ix}^2}\right)x^2$$

◆ 100. A person climbs from a Paris metro station to the street level by walking up a stalled escalator in 94 s. It takes 66 s to ride the same distance when standing on the escalator when it is operating normally. How long would it take for him to climb from the station to the street by walking up the moving escalator?

Answers to Practice Problems

3.1 No; the checkbook balance may increase or decrease, but there is no spatial direction associated with it. When we say it "goes down," we do not mean that it moves in a direction toward the center of Earth! Rather, we really mean that it decreases. The balance is a scalar.

3.2 240 mi 20° W of S

3.3 $A_x = +16$ km; $A_y = -8.2$ km; $B_x = +17$ km; $B_y = 0$ km; $C_x = -11$ km; $C_y = +47$ km

3.4 $|\vec{v}_{av}|$ can never be greater than the average speed because the magnitude of the displacement cannot be greater than the distance traveled. $|\vec{v}_{av}|$ can be equal to the average speed if the magnitude of the displacement is equal to the distance traveled, which is true when the motion is along a straight line with no change in direction.

3.5 (a) 1.64 m/s directed 33° east of north; (b) 0.82 m/s² directed 33° east of north

3.6 2

3.7 $v_i^2/(4g)$

3.8 Ignoring air resistance, the two arrows have the same constant horizontal velocity component: $v_{2x} - v_{1x} = 0$ (choosing the x-axis horizontal and the y-axis up). Their vertical velocity components are different, but they *change at the same rate*, so $v_{2y} - v_{1y}$ stays constant. The difference in their velocities ($\vec{v}_2 - \vec{v}_1$) stays constant. This constant difference in their velocities makes the difference in their positions ($\vec{r}_2 - \vec{r}_1$) change with time

3.9 $v_{fx} = 500.0$ m/s; $v_{fy} = -19.8$ m/s; bullet enters the water at an angle of 2.27° below the horizontal

3.10 (a) 1.0 m/s; (b) 15 min

3.11 28° south of west; 3.6 min; 130 m

Answers to Checkpoints

3.2 $C_x = -5.5$ km and $C_y = -7.2$ km

3.4 Velocity is a vector quantity. The plane's speed does not change, but its velocity does. Therefore, $\Delta\vec{v} \neq 0$ and $\vec{a}_{av} = \Delta\vec{v}/\Delta t \neq 0$.

3.5 The horizontal velocity component does not change. The vertical component is zero at the highest point, so the velocity vector is directed horizontally. The acceleration is constant and directed vertically downward throughout the flight, including at the highest point.

3.6 (a) 19.5 m/s (b) 1.5 m/s

Force and Newton's Laws of Motion

A sailplane (or "glider") is a small, unpowered, high-performance aircraft. A sailplane must be initially towed a few thousand feet into the air by a small airplane, after which it relies on regions of upward-moving air such as thermals and ridge currents to ascend further. Suppose a small plane requires about 120 m of runway to take off by itself. When it is towing a sailplane, how much runway does it need? (See p. 120 for the answer.)

Concepts & Skills to Review

- addition of vectors (Sections 3.1 and 3.2)
- vector components (Section 3.2)
- acceleration (Sections 2.3 and 3.4)
- motion with constant acceleration (Sections 2.4 and 3.5)
- motion diagrams (Section 2.5)

4.1   FORCE

Just as human life would be dull without social interactions, the physical universe would be dull without physical interactions. Social interactions with friends and family change our behavior; physical interactions change the "behavior" (motion, temperature, etc.) of matter.

Force: a push or pull that one object exerts on another

An interaction between two objects can be described and measured in terms of two *forces*, one exerted on each of the two interacting objects. A **force** is a push or a pull. When you play soccer, your foot exerts a force on the ball while the two are in contact, thereby changing the speed and direction of the ball's motion. At the same time, the ball exerts a force on your foot, the effect of which you can feel. To understand the motion of an object, whether it be a soccer ball or the International Space Station, we need to analyze the forces acting on the object.

Long-Range Forces   Forces exerted on macroscopic objects—objects that are large enough for us to observe without instrumentation—can be either long-range forces or contact forces. **Long-range forces** do not require the two objects to be touching. These forces can exist even if the two objects are far apart and even if there are other objects between the two. For example, gravity is a long-range force. The gravitational force exerted on the Earth by the Sun keeps the Earth in orbit around the Sun, despite the great distance between them and despite other planets that occasionally come between them. The Earth also exerts a long-range gravitational force on objects on or near its surface. We call the magnitude of the gravitational force that a planet or moon exerts on a nearby object the object's **weight**.

The **weight** of an object near a planet or moon is the magnitude of the gravitational force exerted on it by that planet or moon.

PHYSICS AT HOME

Besides gravity, other long-range forces are electric or magnetic in nature. On a dry day, run a comb vigorously through your hair until you hear some crackling. Now hold the comb a few centimeters from small pieces of a torn paper napkin. Observe the long-range electrical interaction between the paper and the comb.

Now take a refrigerator magnet. Hold it near but not touching the refrigerator door. You can feel the effect of a long-range magnetic interaction.

Part 3 of this book treats electromagnetic forces in detail. Until then, you can safely assume that gravity is the only significant long-range interaction unless the statement of a problem indicates otherwise.

Contact forces exist only as long as the objects are touching one another.

Contact Forces   All forces exerted on macroscopic objects, other than long-range gravitational and electromagnetic forces, involve contact. **Contact forces** exist only as long as the objects are touching one another. Your foot has no noticeable effect on a soccer ball's motion until the two come into contact, and the force lasts only as long as they are in contact. Once the ball moves away from your foot, your foot has no further influence over the ball's motion.

The idea of contact is a useful simplification for macroscopic objects. What we call a single contact force is really the net effect of enormous numbers of electromagnetic forces between atoms on the surfaces of the two objects. On an atomic scale, the idea of "contact" breaks down. There is no way to define "contact" between two atoms—in other words, there is no unique distance between the atoms at which the forces they exert on one another suddenly become zero.

Measuring Forces

If the concept of force is to be useful in physics, there must be a way to measure forces. Consider a simple spring scale (Fig. 4.1). As the scale's pan is pulled down, a spring is stretched. The harder you pull, the more the spring stretches. As the spring stretches, an attached pointer moves. Then all we have to do to measure the applied force is to calibrate the scale so the amount of stretch measures the magnitude of the force. For many springs, the extension is approximately proportional to the force, which makes calibration easy.

In the United States, supermarket scales are generally calibrated to measure forces in pounds (lb). In the SI system, the unit of force is the **newton** (N). To convert pounds to newtons, use the approximate conversion factors

$$1 \text{ lb} = 4.448 \text{ N} \quad \text{or} \quad 1 \text{ N} = 0.2248 \text{ lb} \tag{4-1}$$

There are more sophisticated means for measuring forces than a supermarket scale. Even so, many operate on the same principle as the supermarket scale: a force is measured by the deformation—change of size or shape—it produces in some object.

Force Is a Vector Quantity

The magnitude of a force is *not* a complete description of the force. The *direction* of the force is equally important. The direction of the brief contact force exerted by a soccer player's foot on the ball can make the difference between scoring a goal or not (Fig. 4.2). Force is a vector quantity that must be added (or subtracted) using the same methods used for other vector quantities such as position, velocity, and acceleration.

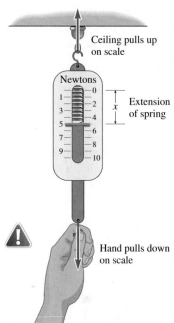

Figure 4.1 As the bottom of a spring scale is pulled downward, the spring stretches. We can measure the force by measuring the extension of the spring. For many springs, the extension is approximately proportional to the force, which makes calibration easy. Note that there is a pull on *both* ends of the scale. The ceiling pulls up on the scale and supports the scale from above.

Figure 4.2 A soccer player's foot exerts a force on the ball only when they are touching.

Example 4.1

Traction on a Foot

In a traction apparatus, three cords pull on the central pulley, each with magnitude 22.0 N, in the directions shown in Fig. 4.3. What is the sum of the forces exerted on the central pulley by the three cords? Give the magnitude and direction of the sum.

Strategy   First, we sketch the graphical addition of the three forces to get an estimate of the magnitude and direction of the sum. Then, to get an accurate answer, we resolve the three forces into their x- and y-components, sum the components, and then calculate the magnitude and direction of the sum.

Solution   Figure 4.4 shows the graphical addition of the three forces exerted on the central pulley by the cords. From this sketch, we can tell that the sum of the three forces is at a relatively small angle above the horizontal (roughly half of 45°) and has a magnitude a bit larger than 44 N.

To find an algebraic solution, we find the components along the x- and y-axes and add them (Fig. 4.5). The x-components of the forces are

$$F_{1x} = F_{2x} = (22.0 \text{ N}) \cos 45.0°$$

$$F_{3x} = (22.0 \text{ N}) \cos 30.0°$$

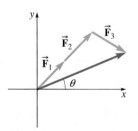
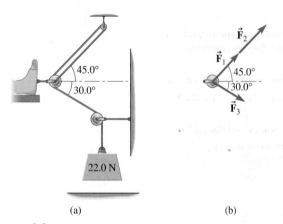

45.0°
30.0°

22.0 N

(a)

$\vec{F}_2$
$\vec{F}_1$
45.0°
30.0°
$\vec{F}_3$

(b)

Figure 4.3

(a) A foot in traction; (b) the three forces exerted on the central pulley by the cords.

The y-components of the forces are

$$F_{1y} = F_{2y} = (22.0 \text{ N}) \sin 45.0°$$

$$F_{3y} = (-22.0 \text{ N}) \sin 30.0°$$

The sum of the x-components is

$$F_x = F_{1x} + F_{2x} + F_{3x}$$
$$= 2 \times (22.0 \text{ N}) \cos 45.0° + (22.0 \text{ N}) \cos 30.0°$$
$$= 31.11 \text{ N} + 19.05 \text{ N} = 50.16 \text{ N}$$

We keep an extra decimal place for now to minimize round-off error. The sum of the y-components is

$$F_y = F_{1y} + F_{2y} + F_{3y}$$
$$= 2 \times (22.0 \text{ N}) \sin 45.0° + (-22.0 \text{ N}) \sin 30.0°$$
$$= 31.11 \text{ N} - 11.00 \text{ N} = 20.11 \text{ N}$$

The magnitude of the sum is (Fig. 4.6):

$$F = \sqrt{F_x^2 + F_y^2} = \sqrt{(50.16 \text{ N})^2 + (20.11 \text{ N})^2} = 54.0 \text{ N}$$

and the direction of the sum is

$$\theta = \tan^{-1} \frac{\text{opposite}}{\text{adjacent}} = \tan^{-1} \frac{20.11 \text{ N}}{50.16 \text{ N}} = 21.8°$$

The sum of the forces exerted on the pulley by the three cords is 54.0 N at an angle 21.8° above the $+x$-axis.

Discussion   To check the answer, look back at the graphical estimate. The magnitude of the sum (54 N) is somewhat larger than 44 N and the direction is at an angle very nearly half of 45° above the horizontal.

Practice Problem 4.1   Changing the Pulley Angles

The pulleys are moved, after which $\vec{F}_1$ and $\vec{F}_2$ are at an angle of 30.0° above the x-axis and $\vec{F}_3$ is 60.0° below the x-axis. (a) What is the sum of these three forces in component form? (b) What is the magnitude of the sum? (c) At what angle with the horizontal is the sum?

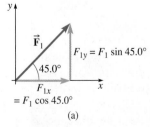

y
$\vec{F}_2$ $\vec{F}_3$
$\vec{F}_1$
θ
x

Figure 4.4

Graphical sum of the forces on the pulley due to the cords.

y
$\vec{F}_1$
$F_{1y} = F_1 \sin 45.0°$
45.0°
$F_{1x} = F_1 \cos 45.0°$
x
(a)

y
$F_{3x} = F_3 \cos 30.0°$
30.0° x
$F_{3y} = -F_3 \sin 30.0°$
$\vec{F}_3$
(b)

Figure 4.5

Finding the components of (a) $\vec{F}_1$ and (b) $\vec{F}_3$. For clarity, the vector arrows are drawn twice as long as they were in Fig. 4.4.

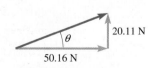

20.11 N
θ
50.16 N

Figure 4.6

Finding the sum from its components.

Net Force

When more than one force acts on an object, the subsequent motion of the object is determined by the *net force* acting on the object. The **net force** is the vector sum of all the forces acting on an object.

Definition of net force:

If $\vec{F}_1, \vec{F}_2, \ldots, \vec{F}_n$ are *all* the forces acting on an object, then the net force $\vec{F}_{net}$ acting on that object is the vector sum of those forces:

$$\vec{F}_{net} = \sum \vec{F} = \vec{F}_1 + \vec{F}_2 + \cdots + \vec{F}_n \qquad (4\text{-}2)$$

The symbol $\sum$ is a capital Greek letter sigma that stands for "sum."

Free-Body Diagrams

An essential tool used to find the net force acting on an object is a **free-body diagram** (FBD): a simplified sketch of a single object with force vectors drawn to represent *every* force *acting on that object*. (For example, the sum of three forces calculated in Example 4.1 is *not* the net force on the central pulley because the forces on the pulley due to the patient's leg and due to gravity are not included.) The net force must *not* include any forces that act on other objects. To draw an FBD:

- Draw the object in a simplified way—you don't have to be Michelangelo to solve physics problems! Almost any object can be represented as a box or a circle, or even a dot.
- Identify all the forces that are exerted on the object. Take care not to omit any forces that are exerted on the object. Consider that everything touching the object may exert one or more contact forces. Then identify long-range forces (for now, just gravity unless electric or magnetic forces are specified in the problem).
- Check your list of forces to make sure that each force is exerted *on* the object of interest *by* some other object. Make sure you have not included any forces that are exerted *on other objects*.
- Draw vector arrows representing all the forces acting on the object. We usually draw the vectors as arrows that start on the object and point away from it. Draw the arrows so they correctly illustrate the directions of the forces. If you have enough information to do so, draw the lengths of the arrows so they are proportional to the magnitudes of the forces.

Example 4.2

Net Force on an Airplane

The forces on an airplane in flight heading eastward are as follows: gravity = 16.0 kN (kilonewtons), downward; lift = 16.0 kN, upward; thrust = 1.8 kN, east; and drag = 0.8 kN, west. (Lift, thrust, and drag are three forces that the air exerts on the plane.) What is the net force on the plane?

Strategy All the forces acting on the plane are given in the statement of the problem. After drawing these forces in the FBD for the plane, we add the forces to find the net force. To resolve the force vectors into components, we choose x- and y-axes pointing east and north, respectively. All four forces are then lined up with the axes, so each will have only one nonzero component, with a sign that indicates the direction along that axis. For example, the drag force points in the $-x$-direction, so its x-component is negative and its y-component is zero.

continued on next page

Example 4.2 continued

Solution   Figure 4.7a is the FBD for the plane, using $\vec{L}, \vec{T}$, and $\vec{D}$ for the lift, thrust, and drag, respectively. $\vec{W}$ stands for the gravitational force on the plane; its magnitude is the plane's weight W. The sum of the x-components of the forces is

$$\sum F_x = L_x + T_x + W_x + D_x$$

$$= 0 + (1.8 \text{ kN}) + 0 + (-0.8 \text{ kN}) = 1.0 \text{ kN}$$

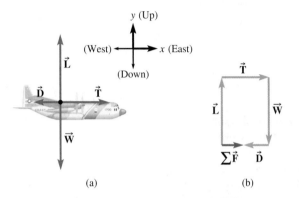

(a)

The sum of the y-components of the forces is

$$\sum F_y = L_y + T_y + W_y + D_y$$

$$= (16 \text{ kN}) + 0 + (-16 \text{ kN}) + 0 = 0$$

The net force is 1.0 kN east.

Discussion   A graphical check of the vector addition is a good idea. Figure 4.7b shows that the sum of the four forces is indeed in the +x-direction (east).

Practice Problem 4.2   New Forces on the Airplane

Find the net force on the airplane if the forces are gravity = 16.0 kN, downward; lift = 15.5 kN, upward; thrust = 1.2 kN, north; drag = 1.2 kN, south.

Figure 4.7

(a) FBD for the airplane. (b) Graphical addition of the four force vectors.

4.2   INERTIA AND EQUILIBRIUM: NEWTON'S FIRST LAW OF MOTION

In 1687, Isaac Newton (1643–1727) published one of the greatest scientific works of all time, his *Philosophiae Naturalis Principia Mathematica* (or *Principia* for short). The Latin title translates as *The Mathematical Principles of Natural Philosophy*. In the *Principia*, Newton stated three laws of motion that form the basis of classical physics.

To pre-Newtonian thinkers, it seemed that there must be two different sets of physical laws: one set to describe the motion of the heavenly bodies, thought to be perfect and enduring, and another to describe the motion of earthly bodies that always come to rest. Together with his law of universal gravitation, Newton's laws of motion showed for the first time that the motion of the heavenly bodies (the Sun, the planets, and their satellites) and the motion of earthly bodies can be understood using the same physical principles.

Newton's First Law of Motion

Newton's first law says that an object acted on by zero net force moves in a straight line with constant speed, or, if it is at rest, remains at rest. Using the concept of the velocity vector, which is a measure of both the speed *and the direction of motion* of an object, we can state the first law:

> **Newton's First Law of Motion**
>
> An object's velocity vector $\vec{v}$ remains constant if and only if the net force acting on the object is zero.

This concise statement of Newton's first law includes both the case of an object at rest (zero velocity) and a moving object (nonzero velocity). Certainly it makes sense that an object at rest remains at rest unless some force acts on it to make it start to move. On the

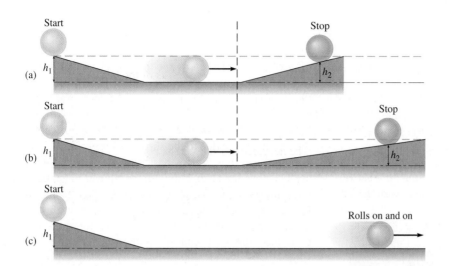

Start ... Stop

(a) h_1 ... h_2

Start ... Stop

(b) h_1 ... h_2

Start ... Rolls on and on

(c) h_1

Figure 4.8 (a) Galileo found that a ball rolled down an incline stops when it reaches *almost* the same height on the second incline. He decided that it would reach the *same* height if resistive forces could be eliminated. (b) As the second incline is made less and less steep, the ball rolls farther and farther before stopping. (c) If the second incline is horizontal, and there are no resistive forces, the ball would never stop.

other hand, it may not be obvious that an object can continue to move with constant speed in a straight line without forces acting to keep it moving. In our experience, most moving objects come to rest because of forces that oppose motion, like friction and air resistance. A hockey puck can slide the entire length of a rink with very little change in speed or direction because the ice is slippery (frictional forces are small). If we could remove *all* the resistive forces, including friction and air resistance, the puck would slide without changing its speed or direction at all.

No force is required to keep an object in motion if there are no forces opposing its motion. When a hockey player strikes the puck with his stick, the brief contact force exerted on the puck by the stick changes the puck's velocity, but once the puck loses contact with the stick, it slides along the ice even though the stick no longer exerts a force on it.

Inertia Newton's first law is also called the **law of inertia**. In physics, **inertia** means resistance to *changes* in velocity. It does *not* mean resistance to the continuation of motion (or the tendency to come to rest). Newton based the law of inertia on the ideas of some of his predecessors, including Galileo Galilei (1564–1642) and René Descartes (1596–1650). In a series of clever experiments in which he rolled a ball up inclines of different angles, Galileo postulated that, if he could eliminate all resistive forces, a ball rolling on a horizontal surface would never stop (Fig. 4.8). Galileo made a brilliant conceptual leap from the real world with friction to an imagined, ideal world, free of friction. The law of inertia contradicted the view of the Greek philosopher Aristotle (384–322 B.C.E.). Almost 2000 years before Galileo, Aristotle had formulated his view that the natural state of an object is to be at rest; and, for an object to remain in motion, a force would have to act on it continuously. Galileo conjectured that, in the absence of friction and other resistive forces, no continued force is needed to keep an object moving.

However, Galileo thought that the sustained motion of an object would be in a great circle around the Earth. Shortly after Galileo's death, Descartes argued that the motion of an object free of any forces should be along a straight line rather than a circle. Newton acknowledged his debt to Galileo, Descartes, and others when he wrote: "If I have seen farther, it is because I was standing on the shoulders of giants."

Conceptual Example 4.3

Snow Shoveling

The task of shoveling newly fallen snow from the driveway can be thought of as a struggle against the inertia of the snow. Without the application of a net force, the snow remains at rest on the ground. However, there is an important way that the inertia of the snow makes it *easier* to shovel. Explain.

continued on next page

Conceptual Example 4.3 continued

Strategy   Think about the physical motions used when shoveling snow. (If you live where there is no snow, think about shoveling gravel from a wheelbarrow to line a garden path.) In order for the shoveling to be facilitated by the snow's inertia, there must be a time when the snow is moving on its own, without the shovel pushing it.

Solution and Discussion   Imagine scooping up a shovelful of snow and swinging the shovel forward toward the side

of the driveway. The snow and the shovel are both in motion. Then suddenly the forward motion of the shovel stops, but the snow continues to move forward because of its inertia; it slides forward off the shovel, to be pulled down to the ground by gravity. The snow does not stop moving forward when the forward force due to the shovel is removed.

This procedure works best with fairly dry snow. Wet sticky snow tends to cling to the shovel. The frictional force on the snow due to the shovel keeps it from moving forward and makes the job far more difficult. In this case, it might help to give the shovel a thin coating of cooking oil to reduce the frictional force the shovel exerts on the snow.

Conceptual Practice Problem 4.3   **Inertia on the Subway**

Negar, a college student, stands on a subway car, holding on to an overhead strap. As the train starts to pull out of the station, she feels thrust toward the rear of the car; as the train comes to a stop at the next station, she feels thrust forward. Explain the role played by inertia in this situation.

PHYSICS AT HOME

For an easy demonstration of inertia, place a quarter on top of an index card, or a credit card, balanced on top of a drinking glass (Fig. 4.9a). With your thumb and forefinger, flick the card so it flies out horizontally from under the quarter. What happens to the quarter? The horizontal force on the coin due to friction is small. With a negligibly small horizontal force, the coin tends to remain motionless while the card slides out from under it (Fig. 4.9b). Once the card is gone, gravity pulls the coin down into the glass (Fig. 4.9c).

Equilibrium

An object in translational equilibrium has a net force of zero acting on it.

When the net force acting on an object is zero, the object is said to be in **translational equilibrium**. *Equilibrium* conveys the idea that the forces are in balance; there is as much force upward as there is downward, as much to the right as to the left, and so forth. Any object moving with a constant velocity, whether at rest or moving in a

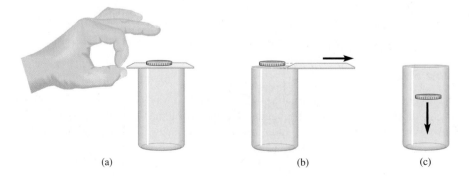

Figure 4.9 A demonstration of inertia.

(a)   (b)   (c)

straight line at constant speed, is in translational equilibrium. A vector can only have zero magnitude if all of its components are zero, so

For an object in equilibrium,

$$\sum F_x = 0 \quad \text{and} \quad \sum F_y = 0 \quad (\text{and} \quad \sum F_z = 0) \qquad (4\text{-}3)$$

In an equilibrium problem, choose x- and y-axes so the fewest number of force vectors have both x- and y-components. It is always good practice to make a conscious *choice* of axes and then to draw them in the FBDs and any other sketches that you make in solving the problem.

Example 4.4

Sliding a Chest

In order to slide a chest that weighs 750 N across the floor at constant velocity, you must push it horizontally with a force of 450 N (Fig. 4.10). Find the contact force that the floor exerts on the chest.

Strategy The chest moves with constant velocity, so it is in equilibrium. The net force acting on it is zero. We will identify all the forces acting on the chest, draw an FBD, do a graphical addition of the forces, choose x- and y-axes, resolve the forces into their x- and y-components, and then set $\Sigma F_x = 0$ and $\Sigma F_y = 0$.

Solution There are three forces acting on the chest. The gravitational force $\vec{\mathbf{W}}$ has magnitude 750 N and is directed downward. Your push $\vec{\mathbf{F}}$ has magnitude 450 N and its direction is horizontal. The contact force due to the floor $\vec{\mathbf{C}}$ has unknown magnitude and direction. However, remembering that the chest is in equilibrium, upward and downward force components must balance, as must the horizontal force components. Therefore, $\vec{\mathbf{C}}$ must be roughly in the direction shown in the FBD (Fig. 4.11a), as is confirmed by adding the three forces graphically (Fig. 4.11b). The sum is zero because the tip of the last vector ends up at the tail of the first one.

Figure 4.10

Sliding a chest across the floor.

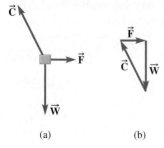

Figure 4.11

(a) An FBD for the chest; (b) graphical addition of the three forces showing that the sum is zero.

(a)   (b)

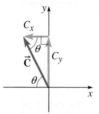

Figure 4.12

Finding the magnitude and direction of the contact force.

Choosing the x-axis to the right and the y-axis up means that two of the three force vectors, $\vec{\mathbf{W}}$ and $\vec{\mathbf{F}}$, have one component that is zero:

$$W_x = 0 \quad \text{and} \quad W_y = -750 \text{ N}$$
$$F_x = 450 \text{ N} \quad \text{and} \quad F_y = 0$$

Now we set the x- and y-components of the net force each equal to zero because the chest is in equilibrium.

$$\sum F_x = W_x + F_x + C_x = 0 + 450 \text{ N} + C_x = 0$$
$$\sum F_y = W_y + F_y + C_y = -750 \text{ N} + 0 + C_y = 0$$

These equations tell us the components of $\vec{\mathbf{C}}$: $C_x = -450$ N and $C_y = +750$ N. Then the magnitude of the contact force is (Fig. 4.12)

$$C = \sqrt{C_x^2 + C_y^2} = \sqrt{(-450 \text{ N})^2 + (750 \text{ N})^2} = 870 \text{ N}$$

$$\theta = \tan^{-1} \frac{\text{opposite}}{\text{adjacent}} = \tan^{-1} \frac{750 \text{ N}}{450 \text{ N}} = 59°$$

The contact force due to the floor is 870 N, directed 59° above the leftward horizontal ($-x$-axis).

Discussion The x- and y-components of the contact force and its magnitude and direction are all reasonable based on the graphical addition, so we can be confident that we did not make an error such as a sign error with one of the components.

Practice Problem 4.4   The Chest at Rest

Suppose the same chest is at rest. You push it horizontally with a force of 110 N but it does not budge. What is the contact force on the chest due to the floor during the time you are pushing?

Using Newton's first law, we can understand how a spring scale can be used to measure weight (the magnitude of the gravitational force exerted on an object). If a melon remains at rest in the pan of the scale, the net force on the melon must be zero. There are only two forces acting on the melon: gravity pulls down and the scale pulls up. Then these two forces must be equal in magnitude and opposite in direction. The scale measures the magnitude of the force it exerts on the melon, which is equal to the weight of the melon.

4.3   NET FORCE, MASS, AND ACCELERATION: NEWTON'S SECOND LAW OF MOTION

When a *nonzero* net force acts on an object, the object's velocity changes. Newton's second law says that the *rate of change of the object's velocity*—that is, the object's acceleration—is proportional to the net force acting on it and inversely proportional to its mass:

Newton's Second Law

$$\vec{\mathbf{a}} = \frac{1}{m}\sum\vec{\mathbf{F}} \quad \text{or} \quad \sum\vec{\mathbf{F}} = m\vec{\mathbf{a}} \tag{4-4}$$

If the net force is zero, then the acceleration is zero, in accordance with Newton's first law. If the net force is not zero, then the acceleration has the same direction as the net force. When the net force is constant, the acceleration is also constant. In component form, Newton's second law is

$$\sum F_x = ma_x \quad \text{and} \quad \sum F_y = ma_y \tag{4-5}$$

If all the forces acting on an object are known, then Eq. (4-4) can be used to calculate its acceleration. Alternatively, sometimes we know the object's acceleration but we have incomplete information about the forces acting on it; then Eq. (4-4) provides information about the unknown forces.

SI Unit of Force

The SI unit of force, the newton, is *defined* so that a net force of 1 N gives a 1-kg mass an acceleration of 1 m/s^2:

$$1\,\text{N} = 1\,\text{kg·m/s}^2 \tag{4-6}$$

Defining the unit of force in this way makes it possible to write Eqs. (4-4) and (4-5) without needing a constant of proportionality to convert between the force unit and kg·m/s^2.

What Is Mass?

The acceleration of an object is proportional to the net force on it and is in the same direction (Fig. 4.13). A larger net force causes a more rapid change in the velocity vector. Newton's second law also says that the acceleration is inversely proportional to the object's mass. The same net force acting on two different objects causes a smaller acceleration on the object with greater mass (Fig. 4.14). Mass is a measure of an object's inertia—the amount of resistance to *changes in velocity*. Newton's second law serves as our *definition* of mass.

In everyday language mass and weight are sometimes used as synonyms, but in physics, mass and weight are different physical properties. The mass of an object is a measure of its inertia, while weight is the magnitude of the gravitational force acting on it. Imagine taking a shuffleboard puck to the Moon. Since the Moon's surface gravity is

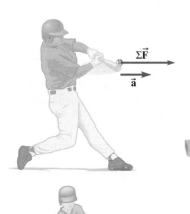

$\sum\vec{\mathbf{F}}$

$\vec{\mathbf{a}}$

$\sum\vec{\mathbf{F}}$

$\vec{\mathbf{a}}$

Figure 4.13 The acceleration of a baseball is proportional to the net force acting on it.

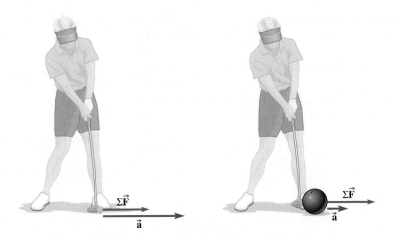

Figure 4.14 The same net force acting on two different objects produces accelerations in inverse proportion to the masses.

weaker than the Earth's, the puck's weight would be smaller on the Moon, but the puck's *mass* would be the same as on Earth. Ignoring the effects of friction, an astronaut playing shuffleboard on the Moon would have to exert the same horizontal force on the puck as on Earth to give it the same acceleration (Fig. 4.15).

4.4 INTERACTION PAIRS: NEWTON'S THIRD LAW OF MOTION

Forces always exist in pairs. Every force is part of an interaction between two objects and each of the interacting objects exerts a force on the other. We call the two forces an **interaction pair**; each force is the **interaction partner** of the other. When you push open a door, the door pushes you. When two cars collide, each exerts a force on the other. Note that interaction partners *act on different objects*—the two objects that are interacting.

Earth		Moon	
(a)	(b)	(c)	(d)

Figure 4.15 An astronaut playing shuffleboard (a) on Earth and (c) on the Moon. FBDs for a puck of mass m being given the same acceleration $\vec{a}$ on a *frictionless* court on (b) Earth and (d) on the Moon. The contact force on the puck due to the *pushing stick* ($\vec{F}_{stick}$) must be the same since the mass of the puck is the same: $\Sigma\vec{F} = \vec{F}_{stick} = m\vec{a}$.

Newton's third law of motion says that interaction partners always have the *same magnitude* and are in *opposite directions*.

Newton's Third Law of Motion

In an interaction between two objects, each object exerts a force on the other. These two forces are equal in magnitude and opposite in direction.

Conceptual Example 4.5

An Orbiting Satellite

Earth exerts a gravitational force on an orbiting communications satellite. What is the interaction partner of this force?

Strategy   The question concerns a gravitational interaction between two objects: Earth and the satellite. In this interaction, each object exerts a gravitational force on the other.

Solution   The interaction partner is the gravitational force exerted on the Earth by the satellite.

Discussion   Does the satellite really exert a force on the Earth with the same magnitude as the force Earth exerts on the satellite? If so, why does the satellite orbit Earth rather than Earth orbiting the satellite? Newton's third law says that the interaction partners are equal in magnitude, but does not say that these two forces have equal *effects*. The effect of a net force on an object's motion depends on the object's mass. These two forces of equal magnitude have

vastly different effects due to the great discrepancy between the masses of the Earth and the satellite.

On the other hand, if a massive planet orbits a star in a relatively small orbit, the gravitational force that the planet exerts on the star can make the star wobble enough to be observed. The wobble enables astronomers to discover planets orbiting stars other than the Sun. The planets do not reflect enough light toward Earth to be seen, but their presence can be inferred from the effect they have on the star's motion.

Conceptual Practice Problem 4.5   Interaction Partner of a Surface Contact Force

In Example 4.4, the contact force exerted on the chest by the floor was 870 N, directed 59° above the leftward horizontal (−*x*-axis). Describe the interaction partner of this force—in other words, what object exerts it on what other object? What are the magnitude and direction of the interaction partner?

  Do not assume that Newton's third law is involved *every* time two forces *happen to be equal and opposite*—*it ain't necessarily so!* You will encounter many situations in which two equal and opposite forces act *on a single object*. These forces cannot be *interaction partners* because they act on the same object. Interaction partners act on *different objects*, one on each of the two objects that are interacting.

✓ CHECKPOINT 4.4

In the photo, two children are pulling on a toy. If they are exerting equal and opposite forces on the toy, are these two forces interaction partners?

The forces exerted by these two children on a toy cannot be interaction partners because they act on the same object (the toy). The interaction of the force exerted by a child on the toy is the force that the *toy* exerts *on that child.*

PHYSICS AT HOME

The next time you go swimming, notice that you use Newton's third law to get the water to push you forward. When you push down and backward on the water with your arms and legs, the water pushes up and forward on you. The various swimming strokes are devised so that you exert as large a force as possible backward on the water during the power part of the stroke, and then as small a force as possible forward on the water during the return part of the stroke.

Internal and External Forces

When we say that a baseball has interactions with the Earth (gravity), with a baseball bat, and with the air, we are treating the baseball as a single entity. But the ball really consists of an enormous number of protons, neutrons, and electrons, all interacting with each other. The protons and neutrons interact with each other to form atomic nuclei; the nuclei interact with electrons to form atoms; interactions between atoms form molecules; and the molecules interact to form the structure of the thing we call a baseball. It would be difficult to have to deal with all of these interactions to predict the motion of a baseball.

Defining a System   Let us call the set of particles comprising the baseball a **system**. Once we have defined a system, we can classify all the interactions that affect the system as either **internal** or **external** to the system. For an internal interaction, *both* interacting objects are part of the system. When we add up all the forces acting on the system to find the net force, every internal interaction contributes two forces—an interaction pair—that always add to zero. For an external interaction, *only one of the two interaction partners is exerted on the system.* The other partner is exerted on an object outside the system and does not contribute to the net force on the system. Therefore, to find the net force on the system, we can ignore all the internal forces and just add the external forces. The insight that internal forces always add to zero is particularly powerful because the choice of what constitutes a system is completely arbitrary. We can choose *any* set of objects and define it to be a system. In one problem, it may be convenient to think of the baseball as a system; in another, we may choose a system consisting of both the baseball and the bat. The second choice might be useful if we do not have detailed information about the interaction between the bat and the ball.

4.5 GRAVITATIONAL FORCES

Newton's Law of Universal Gravitation

Now we turn our attention to learning about some forces in more detail, beginning with gravity. According to **Newton's law of universal gravitation,** any two objects exert gravitational forces on each other that are proportional to the masses (m_1 and m_2) of the two objects and inversely proportional to the square of the distance (r) between their centers. Strictly speaking, the law of gravitation as presented here only applies to point particles and symmetrical spheres. (The *point particle* is a common model in physics used when the size of an object is negligibly small and the internal structure is irrelevant.) Nevertheless, the law of gravitation is *approximately* true for any two objects if the distance between their centers is large compared with their sizes.

In mathematical language, the magnitude of the gravitational force is written:

$$F = \frac{Gm_1m_2}{r^2} \qquad (4\text{-}7)$$

where the constant of proportionality ($G = 6.674 \times 10^{-11}$ N·m^2/kg^2) is called the **universal gravitational constant.** Equation (4-7) is only part of the law of universal gravitation because it gives only the magnitudes of the gravitational forces that each object exerts on the other. The directions are equally important: each object is pulled toward the other's center (Fig. 4.16). In other words, gravity is an attractive force. The forces on the two objects are equal in magnitude and the directions are opposite, as they must be since they form an interaction pair.

Gravitational forces exerted *by* ordinary objects on each other are so small as to be negligible in most cases (see Practice Problem 4.6). Gravitational forces exerted by Earth, on the other hand, are much larger due to Earth's large mass.

Figure 4.16 Gravity is always an attractive force. The force that each body exerts on the other is equal in magnitude, even though the masses may be very different. The force exerted *on the Moon by the Earth* is of the same magnitude as the force exerted on the Earth by the Moon. The directions are opposite.

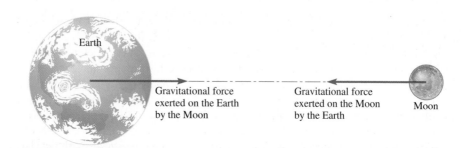

Earth

Gravitational force exerted on the Earth by the Moon

Gravitational force exerted on the Moon by the Earth

Moon

Example 4.6

Weight at High Altitude

When you are in a commercial airliner cruising at an altitude of 6.4 km, by what percentage has your weight (as well as the weight of the airplane) changed compared with your weight on the ground?

Strategy Your weight is the magnitude of Earth's gravitational force exerted on you. Newton's law of universal gravitation gives the magnitude of the gravitational force at a distance r from the center of the Earth. For your weight on the ground W_1, we can use the mean radius of the Earth R_E as the distance between the Earth's center and you: $r_1 = R_E = 6.37 \times 10^6$ m (Fig. 4.17). At an altitude of $h = 6.4 \times 10^3$ m above the surface, your weight is W_2 and your distance from Earth's center is $r_2 = R_E + h$. Your mass m, the mass of the Earth M_E (= 5.97×10^{24} kg), and G are the same in the two cases, so it is efficient to write a ratio of the weights and let those factors cancel out.

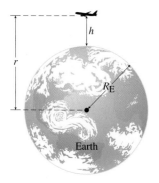

R_E

Earth

Figure 4.17

The gravitational force depends on the distance r to the *center* of the Earth.

Solution The ratio of your weight in the airplane to your weight on the ground is

$$\frac{W_2}{W_1} = \frac{\dfrac{GM_E m}{r_2^2}}{\dfrac{GM_E m}{r_1^2}} = \frac{r_1^2}{r_2^2} = \frac{R_E^2}{(R_E + h)^2}$$

$$= \left(\frac{6.37 \times 10^6 \text{ m}}{6.37 \times 10^6 \text{ m} + 6.4 \times 10^3 \text{ m}}\right)^2 = 0.998$$

Since $0.998 = 1 - 0.002$ and $0.002 = 0.2/100$, your weight decreases by 0.2%.

Discussion Although 6400 m may seem like a significant altitude to us, it's a small fraction of the Earth's radius (0.10%), so the weight change is a small percentage. When judging whether a quantity is small or large, always ask: "Small (or large) compared to what?"

Practice Problem 4.6   A Creative Defense

After an automobile collision, one driver claims that the gravitational force between the two cars caused the collision. Estimate the magnitude of the gravitational force exerted by one car on another when they are driving side-by-side in parallel lanes and comment on the driver's claim.

Gravitational Field Strength

For an object near Earth's surface, the distance between the object and the Earth's center is very nearly equal to the Earth's mean radius, $R_E = 6.37 \times 10^6$ m. The mass of the Earth is $M_E = 5.97 \times 10^{24}$ kg, so the weight of an object of mass m near Earth's surface is

$$W = \frac{GM_E m}{R_E^2} = m\left(\frac{GM_E}{R_E^2}\right) \tag{4-8}$$

Notice that for objects near Earth's surface, the constants in the parentheses are always the same and the weight of the object is proportional to its mass. Rather than recalculate that combination of constants over and over, we call the combination the **gravitational field strength** g near Earth's surface:

$$g = \frac{GM_E}{R_E^2} = \frac{6.674 \times 10^{-11}\ \text{N·m}^2\text{·kg}^{-2} \times (5.97 \times 10^{24}\ \text{kg})}{(6.37 \times 10^6\ \text{m})^2} \approx 9.8\ \text{N/kg} \quad (4\text{-}9)$$

The units *newtons per kilogram* reinforce the conclusion that weight is proportional to mass: g tells us how many newtons of gravitational force are exerted on an object for every kilogram of the object's mass. The weight of a 1.0-kg object near Earth's surface is 9.8 N (2.2 lb). Using g, the weight of an object of mass m near Earth's surface is usually written

Relationship between mass and weight:

$$W = mg \qquad\qquad (4\text{-}10)$$

Variations in Earth's Gravitational Field   The Earth is not a perfect sphere; it is slightly flattened at the poles. Since the distance from the surface to the center of the Earth is smaller there, the field strength at sea level is greatest at the poles (9.832 N/kg) and smallest at the equator (9.814 N/kg). Altitude also matters; as you climb above sea level, your distance from Earth's center increases and the field strength decreases. Tiny local variations in the field strength are also caused by geologic formations. On top of dense bedrock, g is a little greater than above less dense rock. Geologists and geophysicists measure these variations to study Earth's structure and also to locate deposits of various minerals, water, and oil. The device they use, a *gravimeter*, is essentially a mass hanging on a spring. As the gravimeter is carried from place to place, the extension of the spring increases where g is larger and decreases where g is smaller. The mass hanging from the spring does not change, but its weight does ($W = mg$).

Furthermore, due to Earth's rotation, the *effective* value of g that we measure in a coordinate system attached to Earth's surface is slightly less than the true value of the field strength. This effect is greatest at the equator, where the effective value of g is 9.784 N/kg, about 0.3% smaller than the true value of g. The effect gradually decreases with latitude to zero at the poles. We learn more about this effect in Chapter 5.

The most important thing to remember from this discussion is that, unlike G, g is *not* a universal constant. The value of g is a function of position. Near Earth's surface, the variations are small, so we can adopt an average value $g = 9.80$ N/kg as a default.

Gravitational Field and Free-Fall Acceleration

An object in free fall is assumed to have only one force acting on it: gravity. Other forces, such as air resistance, must be negligibly small for this approximation to be valid. We can write the gravitational force on the object as $\vec{\mathbf{W}} = m\vec{\mathbf{g}}$, where the gravitational field vector $\vec{\mathbf{g}}$ has magnitude g and is directed downward (in the direction of the gravitational force). From Newton's second law,

$$\vec{\mathbf{F}}_{net} = m\vec{\mathbf{g}} = m\vec{\mathbf{a}}$$

Dividing by the mass yields

$$\vec{\mathbf{a}} = \vec{\mathbf{g}} \qquad\qquad (4\text{-}11)$$

Therefore, the acceleration of an object in free fall is $\vec{\mathbf{g}}$, regardless of the object's mass. Since 1 N = 1 kg·m/s², 9.80 N/kg = 9.80 m/s²—the magnitude of the free-fall acceleration near Earth's surface has average value 9.80 m/s².

More massive objects have the same free-fall acceleration as less massive objects. True, a more massive object is harder to accelerate: the acceleration of an object subjected to a given force is inversely proportional to its mass. However, the stronger gravitational force on a more massive object compensates for its greater inertia, giving it the same free-fall acceleration as a less massive object.

Gravitational Field Strength on Other Planets

Equation (4-10) can be used to find the weight of an object at or above the surface of *any* planet or moon, but the value of *g* will be different due to the different mass *M* of the planet or moon and the different distance *r* from the planet's center:

$$g = \frac{GM}{r^2} \tag{4-12}$$

✓ CHECKPOINT 4.5

If you climb Mt. McKinley, what happens to the weight of your gear? What happens to its mass?

Example 4.7

"Weighing" Figs in Kilograms

In most countries other than the United States, produce is sold in mass units (grams or kilograms) rather than in force units (pounds or newtons). The scale still measures a force, but the scale is calibrated to show the mass of the produce instead of its weight. What is the weight of 350 g of fresh figs, in newtons and in pounds?

Strategy   Weight is mass times the gravitational field strength. We will assume $g = 9.80$ N/kg. The weight in newtons can be converted to pounds using the conversion factor 1 N = 0.2248 lb.

Solution   The weight of the figs in newtons is

$$W = mg = 0.35 \text{ kg} \times 9.80 \text{ N/kg} = 3.43 \text{ N}$$

Converting to pounds,

$$W = 3.43 \text{ N} \times 0.2248 \text{ lb/N} = 0.771 \text{ lb}$$

The figs weigh 3.4 N or 0.77 lb.

Discussion   This is the weight of the figs at a location where *g* has its average value of 9.80 N/kg. The figs would weigh a little more in the northern city of St. Petersburg, Russia, and a little less in Quito, Ecuador, which is near the equator.

Practice Problem 4.7   Figs on the Moon

What would those figs weigh on the surface of the Moon, where $g = 1.62$ N/kg?

CONNECTION:

In Example 4.4, we resolved the contact force on a sliding chest into components perpendicular to and parallel to the contact surface. It is often convenient to think of these components as two separate but related contact forces: the *normal force* and the *frictional force*.

Normal force: a contact force between two solid objects that is perpendicular to the contact surfaces. Each object pushes the other one away.

4.6   CONTACT FORCES

We have already solved some problems involving forces exerted between two solid objects in contact. Now we look at contact forces in more detail.

Normal Force

A contact force perpendicular to the contact surface that prevents two objects from passing through one another is called the **normal force**. (In geometry, the word *normal* means *perpendicular*.) Consider a book resting on a horizontal table surface. The normal force due to the table must have just the right magnitude to keep the book from falling through the table. If no other vertical forces act, the normal force on the book is equal in magnitude to the book's weight because the book is in equilibrium (Fig. 4.18a).

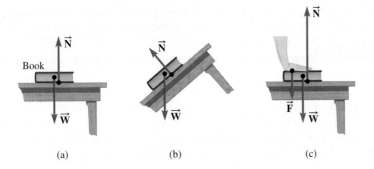

(a)   (b)   (c)

Figure 4.18 (a) The normal force is equal in magnitude to the weight of the book; the two forces sum to zero. (b) On an incline, the normal force is smaller than the weight of the book and is not vertical. (c) If you push down on the book ($\vec{F}$), the normal force on the book due to the table is larger than the book's weight.

According to Newton's third law, there is also a normal force exerted on the table by the book; this normal force acts downward and is of equal magnitude. In *everyday* language, we might say that the table "feels the book's weight." That is not an accurate statement in the language of physics. The table cannot "feel" the gravitational force on the book; the table can only feel forces exerted *on the table*. What the table does "feel" is the normal force—a *contact* force—exerted on the table by the book.

If the table's surface is horizontal, the normal force on the book will be vertical and equal in magnitude to the book's weight. If the surface of the table is *not* horizontal, the normal force is not vertical and is not equal in magnitude to the weight of the book. Remember that the normal force is *perpendicular to the contact surface* (Fig. 4.18b). Even on a horizontal surface, if there are other vertical forces acting on the book, then the normal force is *not* equal in magnitude to the book's weight (Fig. 4.18c). Never *assume* anything about the magnitude of the normal force. In general, we can figure out what the magnitude of the normal force must be in various situations if we have enough information about other forces.

What Causes Normal Forces? How does the table "know" how hard to push on the book? First imagine putting the book on a bathroom scale instead of the table. A spring inside the scale provides the upward force. The spring "knows" how hard to push because, as it is compressed, the force it exerts increases. When the book reaches equilibrium, the spring is exerting just the right amount of force, so there is no tendency to compress it further. The spring is compressed until it pushes up with a force equal to the book's weight. If the spring were stiffer, it would exert the same upward force but with less compression.

The forces that bind atoms together in a rigid solid, like the table, act like extremely stiff springs that can provide large forces with little compression—so little that it's usually not noticed. The book makes a tiny indentation in the surface of the table (Fig. 4.19); a heavier book would make a slightly larger indentation. If the book were to be placed on a soft foam surface, the indentation would be much more noticeable.

✓ CHECKPOINT 4.6

Your laptop is resting on the surface of your desk, which stands on four legs on the floor. Identify the normal forces acting on the desk and give their directions.

Friction

A contact force *parallel* to the contact surface is called **friction**. We distinguish two types: **static friction** and **kinetic (or sliding) friction**. When the two objects are slipping or sliding across one another, as when a loose shingle slides down a roof, the friction is kinetic. When no slipping or sliding occurs, such as between the tires of a car parked on a hill and the road surface, the friction is called static. Static friction acts to prevent objects from *starting* to slide; kinetic friction acts to try to make sliding objects

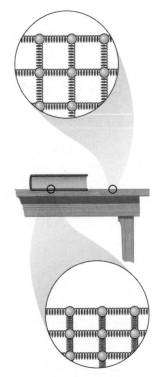

Figure 4.19 The book compresses the "atomic springs" in the table until they push up on the book to hold it up. The slight decrease in the distance between atoms is greatly exaggerated here.

stop sliding. Note that two objects in contact with one another that move with the same velocity exert *static* frictional forces on one another, because there is no *relative* motion between the two. For example, if a conveyor belt carries an air freight package up an incline and the package is not sliding, the two move with the same velocity and the friction is *static*.

Static Friction   Frictional forces are complicated on the microscopic level and are an active field of current research. Despite the complexities, we can make some approximate statements about the frictional forces between dry, solid surfaces. In a simplified model, the maximum magnitude of the force of static friction $f_{s,max}$ that can occur in a particular situation is proportional to the magnitude of the normal force N acting between the two surfaces.

$$f_{s,max} \propto N$$

If you want better traction between the tires of a rear-wheel-drive car and the road, it helps to put something heavy in the trunk to increase the normal force between the tires and the road.

The constant of proportionality is called the **coefficient of static friction** (symbol μ_s):

Maximum force of static friction:
$$f_{s,max} = \mu_s N \qquad (4\text{-}13)$$

Since $f_{s,max}$ and N are both magnitudes of forces, μ_s is a dimensionless number. Its value depends on the condition and nature of the surfaces. Equation (4-13) provides only an *upper limit* on the force of static friction in a particular situation. The actual force of friction in a given situation is not necessarily the maximum possible. It tells us only that, if sliding does not occur, the magnitude of the static frictional force is less than or equal to this upper limit:

$$f_s \leq \mu_s N \qquad (4\text{-}14)$$

Kinetic (Sliding) Friction   For sliding or kinetic friction, the force of friction is only weakly dependent on the speed and is roughly proportional to the normal force. In the simplified model we will use, the force of kinetic friction is assumed to be proportional to the normal force and independent of speed:

Force of kinetic (sliding) friction:
$$f_k = \mu_k N \qquad (4\text{-}15)$$

where f_k is the magnitude of the force of kinetic friction and μ_k is called the **coefficient of kinetic friction.** The coefficient of static friction is always larger than the coefficient of kinetic friction for an object on a given surface. On a horizontal surface, a larger force is required to start the object moving than is required to keep it moving at a constant velocity.

Direction of Frictional Forces   Equations (4-13) through (4-15) relate only the *magnitudes* of the frictional and normal forces on an object. Remember that the frictional force is perpendicular to the normal force between the same two surfaces. Friction is always parallel to the contact surface, but there are many directions parallel to a given contact surface. Here are some rules of thumb for determining the direction of a frictional force.

- The static frictional force acts in whatever direction necessary to prevent the objects from beginning to slide or slip.

- Kinetic friction acts in a direction that tends to make the sliding stop. If a book slides to the left along a table, the table exerts a kinetic frictional force on the book to the right, in the direction opposite to the motion of the book.
- From Newton's third law, frictional forces come in interaction pairs. If the table exerts a frictional force on the sliding book to the right, the book exerts a frictional force on the table to the *left* with the same magnitude.

Example 4.8

Coefficient of Kinetic Friction for the Sliding Chest

Example 4.4 involved sliding a 750-N chest to the right at constant velocity by pushing it with a horizontal force of 450 N. We found that the contact force on the chest due to the floor had components $C_x = -450$ N and $C_y = +750$ N, where the x-axis points to the right and the y-axis points up (see Fig. 4.20). What is the coefficient of kinetic friction for the chest-floor surface?

Strategy To find the coefficient of friction, we need to know what the normal and frictional forces are. They are the

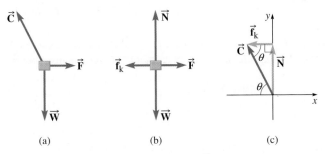

Figure 4.20 (a) FBD for the chest. $\vec{C}$ is the contact force due to the floor. (b) FBD in which the contact force is replaced by two perpendicular forces, the normal force $\vec{N}$ and the kinetic frictional force $\vec{f}_k$. (c) Resolving $\vec{C}$ into normal and frictional components.

components of the contact force that are perpendicular and parallel to the contact surface. Since the surface is horizontal (in the x-direction), the x-component of the contact force is friction and the y-component is the normal force.

Solution The magnitude of the force due to sliding friction is $f_k = |C_x| = 450$ N. The magnitude of the normal force is $N = |C_y| = 750$ N. Now we can calculate the coefficient of kinetic friction from $f_k = \mu_k N$:

$$\mu_k = \frac{f_k}{N} = \frac{450 \text{ N}}{750 \text{ N}} = 0.60$$

Discussion If we had written $f_k = C_x = -450$ N, we would have ended up with a negative coefficient of friction. The coefficient of friction is a relationship between the *magnitudes* of two forces, so it cannot be negative.

Practice Problem 4.8   Chest at Rest

Suppose the same chest is at rest. You push to the right with a force of 110 N but the chest does not budge. What are the normal and frictional forces on the chest due to the floor while you are pushing? Explain why you do not need to know the coefficient of static friction to answer this question.

Conceptual Example 4.9

Horse and Sleigh

A horse pulls a sleigh to the right at constant velocity on level ground. The horse exerts a horizontal force $\vec{F}_{sh}$ on the sleigh. (The subscripts indicate the force on the sleigh due to the horse.) (a) Draw three FBDs, one for the horse, one for the sleigh, and one for the system horse + sleigh. (b) To make the sleigh increase its velocity, there must be a nonzero net force to the right acting on the sleigh. Suppose the horse pulls harder (F_{sh} increases in magnitude).

According to Newton's third law, the sleigh always pulls back on the horse with a force of *the same* magnitude as the force with which the horse pulls the sleigh. Does this mean that no matter how hard it pulls, the horse can never make the net force on the sleigh nonzero? Explain. (c) Identify the interaction partner of each force acting on the sleigh.

continued on next page

Conceptual Example 4.9 continued

$\vec{v}$ (constant)

s = sleigh
g = ground
h = horse
E = Earth

$\vec{N}_{sg}$

$\vec{f}_{sg}$ $\vec{F}_{sh}$

$\vec{F}_{sE}$

Figure 4.21

FBD for the sleigh.

in magnitude and opposite in direction. From the FBDs, $\vec{f}_{hg} = -\vec{F}_{hs}$ and $\vec{f}_{sg} = -\vec{F}_{sh}$. Because $\vec{F}_{hs}$ and $\vec{F}_{sh}$ are interaction partners, they are equal and opposite. Therefore, $\vec{f}_{hg}$ and $\vec{f}_{sg}$ are equal and opposite. The system is in equilibrium.

(b) The FBD for the sleigh (see Fig. 4.21) shows that if the horse pulls the sleigh with a force greater in magnitude than the force of friction on the sleigh ($F_{sh} > f_{sg}$), then the net force on the sleigh is nonzero and to the right. From Fig. 4.22, we need $f_{hg} > F_{hs}$ to have a nonzero net force to the right on the horse. So the frictional force on the horse would have to increase to enable it to pull the sleigh with a greater force. Then in Fig. 4.23, the two frictional forces are no longer equal in magnitude. The forward frictional force on the horse is greater than the backward frictional force on the sleigh, so the net force on the system horse + sleigh is to the right.

Strategy (a) In each FBD, we include only the *external* forces acting on that system. All three systems move with constant velocity, so the net force on each is zero. (b) Looking at the FBD for the sleigh, we can determine the conditions under which the net force on the sleigh can be nonzero. (c) For a force exerted on the sleigh by X, its interaction partner must be the force exerted on X by the sleigh.

Solution and Discussion (a) If we think of the normal and frictional forces as separate forces, then there are four forces acting on the sleigh: the force exerted by the horse $\vec{F}_{sh}$, the gravitational force due to Earth $\vec{F}_{sE}$, the normal force on the sleigh due to the ground $\vec{N}_{sg}$, and kinetic (sliding) friction due to the ground $\vec{f}_{sg}$. Figure 4.21 shows the FBD for the sleigh. The net force is zero, so its horizontal and vertical components must each be zero: $\vec{F}_{sh} + \vec{f}_{sg} = 0$ and $\vec{N}_{sg} + \vec{F}_{sE} = 0$.

Similarly, four forces are acting on the horse: the force exerted by the sleigh $\vec{F}_{hs}$, the gravitational force $\vec{F}_{hE}$, the normal force due to the ground $\vec{N}_{hg}$, and friction due to the ground $\vec{f}_{hg}$. Newton's third law says that $\vec{F}_{hs} = -\vec{F}_{sh}$; the sleigh pulls back on the horse with a force equal in magnitude to the forward pull of the horse on the sleigh. Therefore, $\vec{F}_{hs}$ is to the left and has the same magnitude as $\vec{F}_{sh}$. The horse is in equilibrium, so $\vec{F}_{hs} + \vec{f}_{hg} = 0$ and $\vec{N}_{hg} + \vec{F}_{hE} = 0$. The first of these equations means that the frictional force has to be to the *right*. How does the horse get friction to push it *forward*? By pushing *backward* on the ground with its feet. We all do the same thing when taking a step; by pushing backward on the ground, we get the ground to push forward on us. This is *static* friction because the horse's hoof is not sliding along the ground. If there were no friction (imagine the ground to be icy), the hoof might slide backward. Static friction acts to prevent sliding, so the frictional force on the hoof is forward. Figure 4.22 shows the FBD for the horse.

Of the eight forces acting either on the horse or on the sleigh, two are internal forces for the horse + sleigh system: $\vec{F}_{sh}$ and $\vec{F}_{hs}$. They add to zero since they are interaction partners, so we can omit them from the FBD for the system (Fig. 4.23). The two frictional forces on the system horse + sleigh are *not* interaction partners, but they are equal

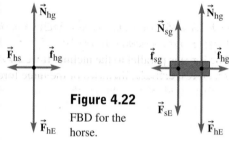

$\vec{N}_{hg}$

$\vec{F}_{hs}$ $\vec{f}_{hg}$

$\vec{F}_{hE}$

Figure 4.22

FBD for the horse.

$\vec{N}_{sg}$

$\vec{f}_{sg}$ $\vec{f}_{hg}$

$\vec{F}_{sE}$

$\vec{N}_{hg}$

$\vec{f}_{hg}$

$\vec{F}_{hE}$

Figure 4.23

FBD for the system, horse and sleigh. The internal forces $\vec{F}_{sh}$ and $\vec{F}_{hs}$ are omitted—they form an interaction pair, so they add to zero.

(c)

Force Exerted on Sleigh	Interaction Partner
Force on the sleigh due to the horse $\vec{F}_{sh}$	Force on the horse due to the sleigh $\vec{F}_{hs}$
Gravitational force on the sleigh due to Earth $\vec{F}_{sE}$	Gravitational force on Earth due to the sleigh $\vec{F}_{Es}$
Normal force on the sleigh due to the ground $\vec{N}_{sg}$	Normal force on the ground due to the sleigh $\vec{N}_{gs}$
Friction on the sleigh due to the ground $\vec{f}_{sg}$	Friction on the ground due to the sleigh $\vec{f}_{gs}$

Practice Problem 4.9   Passing a Truck

A car is moving north and speeding up to pass a truck on a level road. The combined contact force exerted *on the road by all four tires* has vertical component 11.0 kN downward and horizontal component 3.3 kN southward. The drag force exerted on the car by the air is 1.2 kN southward. (a) Draw the FBD for the car. (b) What is the weight of the car? (c) What is the net force acting on the car?

Microscopic Origin of Friction   What looks like the smooth surface of a solid to the unaided eye is generally quite rough on a microscopic scale (Fig. 4.24). Friction is caused by atomic or molecular bonds between the "high points" on the surfaces of the two objects. These bonds are formed by microscopic electromagnetic forces that hold the atoms or molecules together. If the two objects are pushed together harder, the surfaces deform a little more, enabling more "high points" to bond. That is why the force of kinetic friction and the maximum force of static friction are proportional to the normal force. A bit of lubricant drastically decreases the frictional forces, because the two surfaces can float past one another without many of the "high points" coming into contact.

In static friction, when these molecular bonds are stretched, they pull back harder. The bonds have to be broken before sliding can begin. Once sliding begins, molecular bonds are continually made and broken as "high points" come together in a hit-or-miss fashion. These bonds are generally not as strong as those formed in the absence of sliding, which is why $\mu_s > \mu_k$.

For dry, solid surfaces, the amount of friction depends on how smooth the surfaces are and how many contaminants are present on the surface. Does polishing two steel surfaces decrease the frictional forces when they slide across each other? Not necessarily. In an extreme case, if the surfaces are extremely smooth and all surface contaminants are removed, the steel surfaces form a "cold weld"—essentially, they become one piece of steel. The atoms bond as strongly with their new neighbors as they do with the old.

Figure 4.24 Friction is caused by bonds between atoms that form between the "high points" of the two surfaces that come into contact.

Equilibrium on an Inclined Plane

Suppose we wish to pull a large box up a *frictionless* incline to a loading dock platform. Figure 4.25 shows the three forces acting on the box. $\vec{\mathbf{F}}_a$ represents the applied force with which we pull. The force is parallel to the incline. If we choose the x- and y-axes to be horizontal and vertical, respectively, then two of the three forces have both x- and y-components. On the other hand, if we choose the x-axis parallel to the incline and the y-axis perpendicular to it, then only one of the three forces has both x- and y-components (the gravitational force).

With axes chosen, the weight of the box is then resolved into two perpendicular components (Fig. 4.26a). To find the x- and y-components of the gravitational force $\vec{\mathbf{W}}$, we must determine the angle that $\vec{\mathbf{W}}$ makes with one of the axes. The right triangle of Fig. 4.26b shows that $\alpha + \phi = 90°$, since the interior angles of a triangle add up to $180°$. The x- and y-axes are perpendicular, so $\alpha + \beta = 90°$. Therefore, $\beta = \phi$.

The y-component of $\vec{\mathbf{W}}$ is perpendicular to the surface of the incline. From Fig. 4.26a, the side parallel to the y-axis is adjacent to angle β, so

$$\cos \beta = \frac{\text{adjacent}}{\text{hypotenuse}} = \frac{|W_y|}{|W|}$$

Since W_y is negative and $W = mg$,

$$W_y = -mg \cos \beta = -mg \cos \phi$$

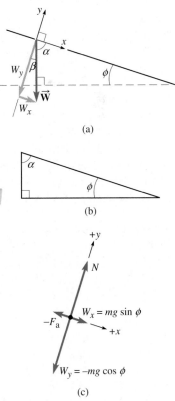

Figure 4.26 (a) Resolving the weight into components parallel to and perpendicular to the incline. (b) A right triangle shows that $\alpha + \phi = 90°$. (c) FBD for the box on the incline.

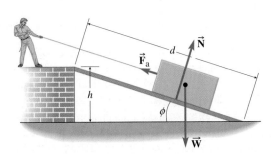

Figure 4.25 A box of mass m pulled up an incline.

The x-component of the weight tends to make the box slide down the incline (in the positive x-direction). Using the same triangle,

$$W_x = +mg \sin \phi$$

When the box is pulled with a force equal in magnitude to W_x up the incline (in the negative x-direction), it will slide up with constant velocity. The component of the box's weight perpendicular to the incline is supported by the normal force $\vec{N}$ that pushes the box away from the incline. Figure 4.26c is an FBD in which the gravitational force is separated into its x- and y-components.

If the box is in equilibrium, whether at rest or moving along the incline at constant velocity, the force components along each axis sum to zero:

$$\sum F_x = (-F_a) + mg \sin \phi = 0$$

and

$$\sum F_y = N + (-mg \cos \phi) = 0$$

On an incline, the normal force is *not* equal in magnitude to the weight and it does not point straight up. If the applied force has magnitude $mg \sin \phi$, we can pull the box up the incline at constant velocity. If friction acts on the box, we must pull with a force greater than $mg \sin \phi$ to slide the box up the incline at constant velocity.

Example 4.10

Pushing a Safe up an Incline

A new safe is being delivered to the Corner Book Store. It is to be placed in the wall at a height of 1.5 m above the floor. The delivery people have a portable ramp, which they plan to use to help them push the safe up and into position. The mass of the safe is 510 kg, the coefficient of static friction along the incline is $\mu_s = 0.42$, and the coefficient of kinetic friction along the incline is $\mu_k = 0.33$. The ramp forms an angle $\theta = 15°$ above the horizontal. (a) How hard do the movers have to push to start the safe moving up the incline? Assume that they push in a direction parallel to the incline. (b) To slide the safe up at a constant speed, with what magnitude force must the movers push?

Strategy   (a) When the safe *starts* to move, its velocity is changing, so the safe is *not* in equilibrium. Nevertheless, to find the minimum applied force to start the safe moving, we can find the *maximum* applied force for which the safe *remains at rest*—an equilibrium situation. (b) The safe is in equilibrium

as it slides with a constant velocity. Both parts of the problem can be solved by drawing the FBD, choosing axes, and setting the x- and y-components of the net force equal to zero.

Solution   First we draw a diagram to show forces acting (Fig. 4.27). Before resolving the forces into components, we must choose x- and y-axes. To use the coefficient of friction, we have to resolve the contact force on the safe due to the incline into components *parallel and perpendicular to the incline*—friction and the normal force, respectively—rather than into horizontal and vertical components. Therefore, we choose x- and y-axes parallel and perpendicular to the incline so friction is along the x-axis and the normal force is along the y-axis.

The gravitational force $\vec{W}$ can be resolved into its components: $W_x = -mg \sin \theta$ and $W_y = -mg \cos \theta$ (Fig. 4.28a).

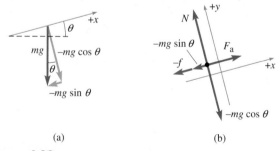

(a)   (b)

Figure 4.28

(a) Resolving the weight into x- and y-components, and (b) an FBD in which the weight is replaced with its x- and y-components.

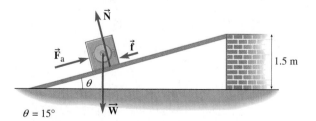

$\theta = 15°$

Figure 4.27

Forces acting on the safe as it is moved up the incline.

continued on next page

Example 4.10 continued

Now we draw the FBD with $\vec{W}$ replaced by its components (Fig. 4.28b).

(a) Suppose that the safe is initially at rest. As the movers start to push, F_a gets larger and the force of static friction gets larger to "try" to keep the safe from sliding. Eventually, at some value of F_a, static friction reaches its maximum possible value $\mu_s N$. If the movers continue to push harder, increasing F_a further, the force of static friction cannot increase past its maximum value $\mu_s N$, so the safe starts to slide. The direction of the frictional force is along the incline and downward since friction is "trying" to keep the safe from sliding *up* the incline.

The normal force is *not* equal in magnitude to the weight of the safe. To find the normal force, sum the y-components of the forces:

$$\sum F_y = N + (-mg \cos \theta) = 0$$

Then $N = mg \cos \theta$. The normal force is *less than the weight* since $\cos \theta < 1$.

When the movers push with the largest force for which the safe does *not* slide,

$$\sum F_x = F_{ax} + f_x + W_x = 0$$

The applied force is in the $+x$-direction, so $F_{ax} = +F_a$. The frictional force has its maximum magnitude and is in the $-x$-direction, so $f_x = -f_{s,max} = -\mu_s N = -\mu_s mg \cos \theta$. From the FBD, $W_x = -mg \sin \theta$. Then,

$$\sum F_x = F_a - \mu_s mg \cos \theta - mg \sin \theta = 0$$

Solving for F_a,

$$F_a = mg (\mu_s \cos \theta + \sin \theta)$$
$$= 510 \text{ kg} \times 9.80 \text{ m/s}^2 \times (0.42 \times \cos 15° + \sin 15°)$$
$$= 3300 \text{ N}$$

An applied force that *exceeds* 3300 N starts the box moving up the incline.

(b) Once the safe is sliding, the movers need only push hard enough to make the net force on the safe equal to zero if they want the safe to slide at constant velocity. We are now dealing with sliding friction, so the frictional force is now $f_x = -\mu_k N = -\mu_k mg \cos \theta$.

$$\sum F_x = F_{ax} + f_x + W_x$$
$$= F_a - \mu_k mg \cos \theta - mg \sin \theta$$
$$= 0$$
$$F_a = mg (\mu_k \cos \theta + \sin \theta)$$
$$= 510 \text{ kg} \times 9.80 \text{ m/s}^2 \times (0.33 \times \cos 15° + \sin 15°)$$
$$= 2900 \text{ N}$$

The movers push with a force $\vec{F}_a$ of magnitude 2900 N directed up the incline.

Discussion In (b), the expression $F_a = mg (\mu_k \cos \theta + \sin \theta)$ shows that the applied force up the incline has to balance the sum of two forces down the incline: the frictional force ($\mu_k mg \cos \theta$) and the component of the gravitational force down the incline ($mg \sin \theta$). This balance of forces is shown graphically in the FBD (Fig. 4.28b).

Practice Problem 4.10   Smoothing the Infield Dirt

During the seventh-inning stretch of a baseball game, groundskeepers drag mats across the infield dirt to smooth it. A groundskeeper is pulling a mat at a constant velocity by applying a force of 120 N at an angle of 22° above the horizontal. The coefficient of kinetic friction between the mat and the ground is 0.60. Find (a) the magnitude of the frictional force between the dirt and the mat and (b) the weight of the mat.

PHYSICS AT HOME

To estimate the coefficient of static friction between a penny and the cover of your physics book, place the penny on the book and slowly lift the cover. Note the angle of the cover when the penny starts to slide. Explain how you can use this angle to find the coefficient of static friction. Can you devise an experiment to find the coefficient of kinetic friction?

4.7   TENSION

Consider a heavy chandelier hanging by a chain from the ceiling (Fig. 4.29a). The chandelier is in equilibrium, so the upward force on it due to the chain is equal in magnitude to the chandelier's weight. With what force does the chain pull downward on the ceiling? The ceiling has to pull up with a force equal to the total weight of the chain and the chandelier. The interaction partner of this force—the force the chain exerts on the ceiling—is

Figure 4.29 (a) The chain pulls up on the chandelier and pulls down on the ceiling. (b) The chain is under tension. Each link is pulled in opposite directions by its neighbors.

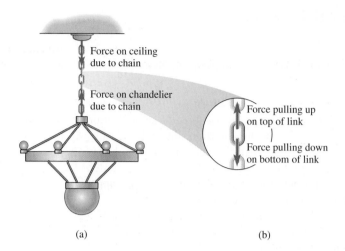

Force on ceiling due to chain

Force on chandelier due to chain

Force pulling up on top of link

Force pulling down on bottom of link

(a)   (b)

equal in magnitude and opposite in direction. Therefore, if the weight of the chain is negligibly small compared with the weight of the chandelier, then the chain exerts forces of equal magnitude at its two ends. The forces at the ends would *not* be equal, however, if you grabbed the chain in the middle and pulled it up or down or if we could not neglect the weight of the chain. We can generalize this observation:

> An *ideal* cord (or rope, string, tendon, cable, or chain) pulls in the direction of the cord with forces of equal magnitude on the objects attached to its ends as long as no external force is exerted on it anywhere between the ends. An ideal cord has zero mass and zero weight.

A single link of the chain (Fig. 4.29b) is pulled at both ends by the neighboring links. The magnitude of these forces is called the **tension** in the chain. Similarly, a little segment of a cord is pulled at both its ends by the tension in the neighboring pieces of the cord. If the segment is in equilibrium, then the net force acting on it is zero. As long as there are no other forces exerted on the segment, the forces exerted by its neighbors must be equal in magnitude and opposite in direction. Therefore, the tension has the same value everywhere and is equal to the force that the cord exerts on the objects attached to its ends.

Example 4.11

Archery Practice

Figure 4.30 shows the bowstring of a bow and arrow just before it is released. The archer is pulling back on the midpoint of the bowstring with a horizontal force of 162 N. What is the tension in the bowstring?

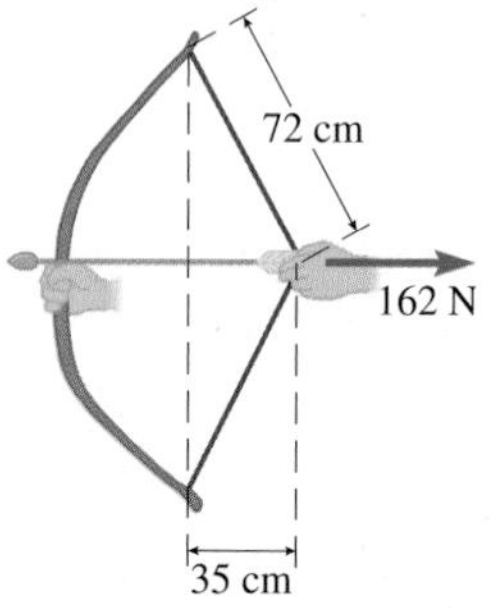

72 cm

162 N

35 cm

Figure 4.30

The force applied to the bowstring by an archer.

Strategy Consider a small segment of the bowstring that touches the archer's finger. That piece of the string is in equilibrium, so the net force acting on it is zero. We draw the FBD, choose coordinate axes, and apply the equilibrium condition: $\Sigma F_x = 0$ and $\Sigma F_y = 0$. We know the force exerted on the

segment of string by the archer's fingers. That segment is also pulled on each end by the tension in the string. Can we assume the tension in the string is the same everywhere? The weight of the string is small compared with the other forces acting on it. The archer pulls sideways on the bowstring, exerting little or no *tangential* force, so we can assume the tension is the same everywhere.

Solution Figure 4.31a is an FBD for the segment of bowstring being considered. The forces are labeled with their magnitudes: F_a for the force applied by the archer's finger and T for each of the tension forces. Figure 4.31b shows these three forces adding to zero. From this sketch, we expect the tension T to be roughly the same as F_a. We choose the x-axis to the right and the y-axis upward. To find the

continued on next page

Example 4.11 continued

components of the forces due to tension in the string, we draw a triangle (Fig. 4.31c). From the measurements given, we can find the angle θ.

$$\sin \theta = \frac{\text{opposite}}{\text{hypotenuse}} = \frac{35 \text{ cm}}{72 \text{ cm}} = 0.486$$

$$\theta = \sin^{-1} 0.486 = 29.1°$$

The x-component of the tension force exerted on the upper end of the segment is

$$T_x = -T \sin \theta$$

The x-component of the force exerted on the lower end of the string is the same. Therefore,

$$\Sigma F_x = -2T \sin \theta + F_a = 0$$

Solving for T,

$$T = \frac{F_a}{2 \sin \theta} = \frac{162 \text{ N}}{2 \times 0.486} = 170 \text{ N}$$

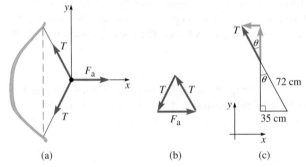

(a)   (b)   (c)

Figure 4.31

(a) FBD for a point on the bowstring with the magnitudes of the forces labeled. (b) Graphical addition of the three forces showing that the sum is zero. (c) The angle θ is used to find the x- and y-components of the forces exerted at each end of the bowstring.

Discussion The tension is only slightly larger than F_a, a reasonable result given the picture of graphical vector addition in Fig. 4.31b.

In this problem, only the x-components of the forces had to be used. The y-components must also add to zero. At the upper end of the string, the y-component of the force exerted by the bow is $+T \cos \theta$, while at the lower end it is $-T \cos \theta$. Therefore, $\Sigma F_y = 0$.

The expression $T = F_a/(2 \sin \theta)$ can be evaluated for limiting values of θ to make sure that the expression is correct. As θ approaches 90°, the tension approaches

$$\frac{F_a}{2 \sin 90°} = \frac{1}{2} F_a$$

That is correct because the archer would be pulling to the right with a force F_a, while each side of the bowstring would pull to the left with a force of magnitude T. For equilibrium, $F_a = 2T$ or $T = \frac{1}{2} F_a$.

As θ gets smaller, $\sin \theta$ decreases and the tension increases (for a fixed value of F_a). That agrees with our intuition. The larger the tension, the smaller the angle the string needs to make in order to supply the necessary horizontal force.

Practice Problem 4.11 Tightrope Practice

Jorge decides to rig up a tightrope in the backyard so his children can develop a good sense of balance (Fig. 4.32). For safety reasons, he positions a horizontal cable only 0.60 m above the ground. If the 6.00-m-long cable sags by 0.12 m from its taut horizontal position when Denisha (weight 250 N) is standing on the middle of it, what is the tension in the cable? Ignore the weight of the cable.

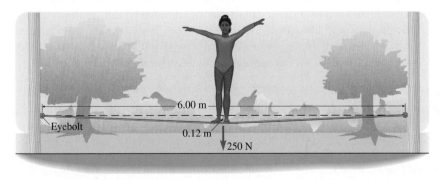

Figure 4.32

Tightrope for balancing practice.

Application: Tensile Forces in the Body Tensile forces are central in the study of animal motion, or biomechanics. Muscles are usually connected by tendons, one at each end of the muscle, to two different bones, which in turn are linked at a joint (Fig. 4.33). Usually one of the bones is more easily moved than the other. When the muscle contracts, the tension in the tendons increases, pulling on both of the bones.

Figure 4.33 A muscle contracts, increasing the tension in the attached tendons. The tendons exert forces on two different bones.

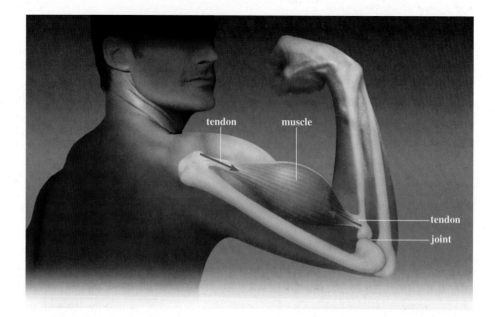

PHYSICS AT HOME

Sit with your arm bent at the elbow with a heavy object on the palm of your hand. You can feel the contraction of the biceps muscle. With your other hand, feel the tendon that connects the biceps muscle to your forearm.

Now place your hand palm down on the desktop and push down. Now it is the triceps muscle that contracts, pulling up on the bone on the other side of the elbow joint. Muscles and tendons cannot push; they can only pull. The biceps muscle cannot push the forearm downward, but the triceps muscle can pull on the other side of the joint. In both cases, the arm acts as a lever.

Figure 4.34 Using a pulley to lift an object by pulling *downward* on a rope with force $\vec{F}$.

Application: Ideal Pulleys   A pulley can change the direction of the force exerted by a cord under tension. To lift something heavy, it is easier to stand on the ground and pull *down* on the rope than to get above the weight on a platform and pull up on the rope (Fig. 4.34).

An *ideal* pulley has no mass and no friction. An ideal pulley exerts no forces on the cord that are *tangent* to the cord—it is not pulling in either direction along the cord. As a result, the tension of an ideal cord that runs through an ideal pulley is the same on both sides of the pulley. An ideal pulley changes the direction of the force exerted by a cord without changing its magnitude. As long as a real pulley has a small mass and negligible amount of friction, we can approximate it as an ideal pulley.

Example 4.12

A Two-Pulley System

A 1804-N engine is hauled upward at constant speed (Fig. 4.35). What are the tensions in the three ropes labeled A, B, and C? Assume the ropes and the pulleys labeled L and R are ideal.

Strategy   The engine and pulley L move up at constant speed, so the net force on each of them is zero. Pulley R is at rest, so the net force on it is also zero. We can draw the FBD for any or all of these objects and then apply the equilibrium condition. If the pulleys are ideal, the tension in the rope is the same on both sides of the pulley. Therefore, rope C—which is attached to the ceiling, passes around both pulleys, and is pulled downward at the other end—has the same tension

continued on next page

Example 4.12 continued

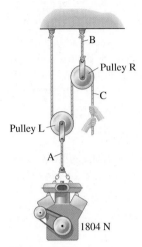

Figure 4.35

A system of pulleys used to raise a heavy weight.

throughout. Call the tensions in the three ropes T_A, T_B, and T_C. To analyze the forces exerted on a pulley, we define our system so the part of the rope wrapped around the pulley is considered part of the pulley. Then there are two cords pulling on the pulley, each with the same tension.

Solution   There are two forces acting on the engine: the gravitational force (1804 N, downward) and the upward pull of rope A. These must be equal and opposite (Fig. 4.36a), since the net force is zero. Therefore $T_A = 1804$ N.

The FBD for pulley L (Fig. 4.36b) shows rope A pulling down with a force of magnitude T_A and rope C pulling upward on *each side*. The rope has the same tension throughout, so all forces labeled T_C in Fig. 4.36b,c have the same magnitude. For the net force to equal zero,

$$2T_C = T_A$$
$$T_C = \tfrac{1}{2}T_A = 902.0 \text{ N}$$

Figure 4.36c is the FBD for pulley R. Rope B pulls upward on it with a force of magnitude T_B. On *each side* of the pulley, rope C pulls downward. For the net force to equal zero,

$$T_B = 2T_C = 1804 \text{ N}$$

Discussion   The engine is raised by pulling *down* on a rope—the pulleys change the direction of the applied force needed to lift the engine. In this case they also change the *magnitude* of the required force. They do that by making the rope pull up on the engine twice, so the person pulling the rope only needs to exert a force equal to half the engine's weight.

Practice Problem 4.12   System of Ropes, Pulleys, and Engine

Consider the entire collection of ropes, pulleys, and the engine to be a single system. Draw the FBD for this system and show that the net force on the system is zero. [*Hint:* Remember that only forces exerted by objects *external* to the system are included in the FBD.]

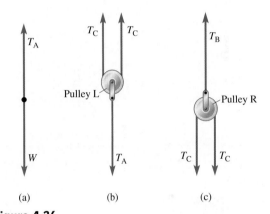

Figure 4.36

(a) FBD for the engine. (b) FBD for pulley L and (c) FBD for pulley R.

4.8   APPLYING NEWTON'S SECOND LAW

We can now apply Newton's second law to a great variety of situations involving the forces we have encountered so far—gravity, contact forces, and tension. The following steps are helpful in most problems that involve Newton's second law.

Problem-Solving Strategy for Newton's Second Law

- Decide what object will have Newton's second law applied to it.
- Identify all the *external* forces acting on that object.
- Draw an FBD to show all the forces acting on the object.
- Choose a coordinate system. If the direction of the net force is known, choose axes so that the net force (and the acceleration) are along one of the axes.
- Find the net force by adding the forces as vectors.
- Use Newton's second law to relate the net force to the acceleration.
- Relate the acceleration to the change in the velocity vector during a time interval of interest.

Example 4.13

The Broken Suitcase

The wheels fall off Beatrice's suitcase, so she ties a rope to it and drags it along the floor of the airport terminal (Fig. 4.37). The rope makes a 40.0° angle with the horizontal. The suitcase has a mass of 36.0 kg and Beatrice pulls on the rope with a force of 65.0 N. (a) What is the magnitude of the normal force acting on the suitcase due to the floor? (b) If the coefficient of kinetic friction between the suitcase and the marble floor is $\mu_k = 0.13$, find the frictional force acting on the suitcase. (c) What is the acceleration of the suitcase while Beatrice pulls with a 65.0 N force at 40.0°? (d) Starting from rest, for how long a time must she pull with this force until the suitcase reaches a comfortable walking speed of 0.5 m/s?

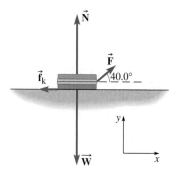

Figure 4.37

Beatrice dragging her suitcase.

Strategy Since the suitcase is dragged horizontally along the floor, the vertical component of its velocity is always zero. The vertical acceleration component of the suitcase is zero because the vertical velocity component does not change. (If it did have a vertical acceleration component, the suitcase would begin to move either down through the floor or up into the air.) If we choose the $+y$-axis up and the $+x$-axis to be horizontal, then $a_y = 0$. We resolve the forces acting on the suitcase into their components, draw a free-body diagram for the suitcase, and apply Newton's second law.

Solution (a) Figure 4.38 shows the forces acting on the suitcase, where $\vec{F}$ is the force exerted by Beatrice. All the other forces are either parallel or perpendicular to the floor, so only $\vec{F}$ needs to be resolved into x- and y-components.

$$F_x = F \cos 40.0° = 65.0 \text{ N} \times 0.766 = 49.8 \text{ N}$$

$$F_y = F \sin 40.0° = 65.0 \text{ N} \times 0.643 = 41.8 \text{ N}$$

Figure 4.39 is an FBD in which $\vec{F}$ is replaced by its components. The vertical force components add to zero since $a_y = 0$.

$$\sum F_y = ma_y = 0$$

$$N + F \sin 40.0° - W = 0$$

We can solve this equation for the magnitude of the normal force. The magnitude of the gravitational force is $W = mg$, so

$$N = mg - F \sin 40.0°$$

$$= (36.0 \text{ kg} \times 9.80 \text{ N/kg}) - (65.0 \text{ N} \times \sin 40.0°)$$

$$= 352.8 \text{ N} - 41.8 \text{ N} = 311 \text{ N}$$

(b) The magnitude of the kinetic frictional force is

$$f_k = \mu_k N = 0.13 \times 311 \text{ N} = 40.43 \text{ N}$$

Rounding to two significant figures, the frictional force is 40 N in the $-x$-direction (opposite the motion of the suitcase).

(c) The y-component of the acceleration is zero. To find the x-component, we apply Newton's second law to the x-components of the forces acting on the suitcase:

$$\sum F_x = +F \cos 40.0° + (-f_k)$$

$$= 49.79 \text{ N} - 40.43 \text{ N} = 9.36 \text{ N}$$

$$a_x = \frac{\sum F_x}{m} = \frac{9.36 \text{ N}}{36.0 \text{ kg}} = 0.260 \text{ m/s}^2$$

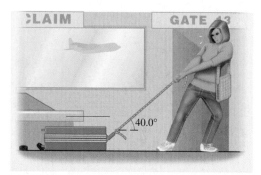

Figure 4.38

Forces acting on a suitcase dragged along the floor. The lengths of the vector arrows are not to scale.

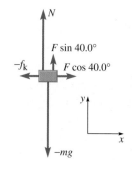

Figure 4.39

FBD for the suitcase, with the forces represented by their x- and y-components.

continued on next page

Example 4.13 continued

Here we have replaced newtons per kilogram with the equivalent meters per second squared, the usual way to write the SI units of acceleration. The acceleration is 0.3 m/s^2 in the +x-direction.

(d) With constant a_x,

$$\Delta v_x = a_x \Delta t$$

The suitcase starts from rest so $v_{ix} = 0$ and $\Delta v_x = v_{fx} - v_{ix} = v_{fx}$. Then,

$$\Delta t = \frac{v_{fx}}{a_x} = \frac{0.5 \text{ m/s}}{0.260 \text{ m/s}^2} = 2 \text{ s}$$

Discussion What Beatrice probably wants to do is to drag the suitcase along at constant velocity. To do that, she must first accelerate the suitcase from rest. Once the suitcase is moving at the desired velocity, she pulls a little less hard, so the net force is zero and the suitcase slides at constant speed. She would do so without thinking much about it, of course!

Practice Problem 4.13 The Continuing Story . . .

(a) How hard does Beatrice pull at a 40.0° angle while the suitcase slides along the floor at constant velocity? [*Hint:* Do *not* assume that the normal force is the same as in the previous discussion.] (b) The suitcase is moving at 0.50 m/s. Beatrice changes the force to 42 N at 40.0°. How long does it take the suitcase to come to rest?

Sometimes two or more objects are constrained to have the same acceleration by the way they are connected. In Example 4.14, we look at a train engine pulling five freight cars. The couplings maintain a fixed distance between the cars, so at any instant the cars move with the same velocity; if they didn't, the distance between them would change. The velocities don't have to be constant, they just have to change in exactly the same way, which implies that the accelerations must also be the same at any instant.

Example 4.14

Coupling Force on First and Last Freight Cars

A train engine pulls out of a station along a straight horizontal track with five identical freight cars behind it, each of which weighs 90.0 kN. The train reaches a speed of 15.0 m/s within 5.00 min of starting out. Assuming the engine pulls with a constant force during this interval, with what magnitude of force does the coupling between cars pull forward on the first and last of the freight cars? Ignore air resistance and friction on the freight cars.

Strategy A sketch of the situation is shown in Fig. 4.40. To find the force exerted by the first coupling, we consider all five cars to be one system so we do not have to worry about the force exerted on the first car by the second car. The only *external* forces on the group of five cars are the normal force,

gravity, and the pull of the first coupling. To find the force exerted by the fifth coupling, we consider car five by itself to be a system. In each case, once we identify a system, we draw a free-body diagram, choose a coordinate system, and then apply Newton's second law.

As discussed previously, the engine and the cars must all have the same acceleration at any instant. We expect the acceleration to be *constant* because the engine pulls with a constant force. We can calculate the acceleration of the train from the initial and final velocities and the elapsed time.

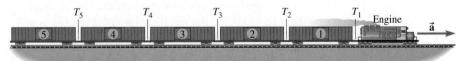

Figure 4.40
An engine pulling five identical freight cars. The entire train has a constant acceleration $\vec{a}$ to the right.

continued on next page

Example 4.14 continued

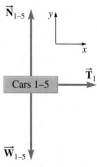

Figure 4.41

FBD for the system consisting of cars 1–5 (but not the engine).

Solution   For the tension T_1 in the first coupling, we consider the five cars as *one system* of mass M. Figure 4.41 shows the FBD in which cars 1 to 5 are treated as a single object. We choose the x-axis in the direction of motion of the train and the y-axis up. Since the train moves along the x-axis, the acceleration vector is along the x-axis. Therefore, $a_y = 0$. Using the y-component of Newton's second law, the vertical forces add to zero:

$$\sum F_y = Ma_y = N_{1-5} - W_{1-5} = 0$$

The only external horizontal force is the force $\vec{T}_1$ due to the tension in the first coupling. This force is constant according to the problem statement, so we know that the acceleration a_x is constant:

$$\sum F_x = T_1 = Ma_x$$

The mass of the system M is five times the mass of one car m. We are given the *weight* of one car ($W = 90.0$ kN $= 9.00 \times 10^4$ N). From the relation between mass and weight, $W = mg$, the mass of one car is $m = W/g$ and the mass of five cars is $M = 5W/g$.

The constant acceleration of the train is

$$a_x = \frac{\Delta v_x}{\Delta t} = \frac{v_{fx} - v_{ix}}{t_f - t_i} = \frac{15.0 \text{ m/s} - 0}{300 \text{ s} - 0} = 0.0500 \text{ m/s}^2$$

Therefore,

$$T_1 = Ma_x = \frac{5W}{g} \times \frac{\Delta v_x}{\Delta t} = \frac{5 \times 9.00 \times 10^4 \text{ N}}{9.80 \text{ m/s}^2} \times \frac{15.0 \text{ m/s}}{300 \text{ s}}$$

$$= 2.30 \text{ kN}$$

Now consider the last freight car (car 5). If we ignore friction and air resistance, the only external forces acting are the force $\vec{T}_5$ due to the tension in the fifth coupling, the normal force $\vec{N}_5$, and the gravitational force $\vec{W}_5$; the FBD is shown in Fig. 4.42. Since $\vec{N}_5 + \vec{W}_5 = 0$, the net force is equal to $\vec{T}_5$. From Newton's second law,

$$\sum F_x = T_5 = ma_x = \frac{W}{g} a_x$$

$$T_5 = \frac{W}{g} \times \frac{\Delta v_x}{\Delta t} = \frac{9.00 \times 10^4 \text{ N}}{9.80 \text{ m/s}^2} \times \frac{15.0 \text{ m/s}}{300 \text{ s}} = 459 \text{ N}$$

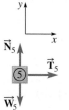

Figure 4.42

FBD for car 5. (Vector lengths are not to the same scale as those in Fig. 4.41.)

Discussion   We considered two systems (cars 1 to 5 and car 5) that have the same acceleration and different masses. As expected, the net force is proportional to the mass: the net force on five cars is five times the net force on one car.

The solution to this problem is much simpler when Newton's second law is applied to a system comprised of all five cars, rather than to each car individually. Although the problem can be solved by looking at individual cars, to find the tension in the first coupling you would have to draw five FBDs (one for each car) and apply Newton's second law five times. That's because each car, except the fifth, is acted on by the unequal tensions in the couplings on either side. You'd have to first find the tension in the fifth coupling, then the fourth, then the third, and so on.

Practice Problem 4.14   Coupling Force Between First and Second Freight Cars

With what force does the coupling between the first and second cars pull forward on the second car? [*Hint:* Try two methods. One of them is to draw the FBD for the first car and apply Newton's *third* law as well as the second.]

Example 4.15 deals with two objects connected by an ideal cord. Although it may have a nonzero acceleration, the net force on an *ideal* cord is still zero because it has *zero mass*: if $m = 0$, then $\sum \vec{F} = m\vec{a} = 0$. As a result, the tension is the same at the two ends as long as no external force acts on the cord between the ends (Fig. 4.43a). An ideal cord that passes over an ideal pulley has the same tension at its ends. The pulley exerts an external force on part of the cord, but this force is everywhere *perpendicular to the cord*. As Fig. 4.43b shows, an external force that has no component tangent to the cord does not affect the tension in the cord.

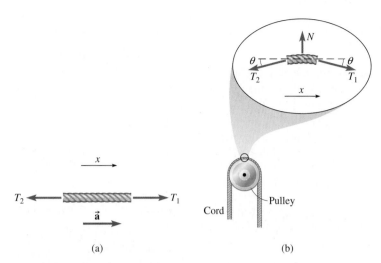

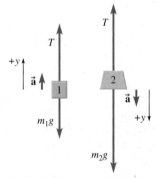

(a)

(b)

Figure 4.43 (a) FBD for an ideal cord with acceleration $\vec{a}$. Applying Newton's second law along the x-axis: $\Sigma F_x = T_1 - T_2 = ma_x$. The ideal cord has mass $m = 0$, so $T_1 = T_2$: the tensions at the ends are equal. (b) An ideal cord passing around an ideal pulley and the FBD for a short segment of the cord at the top of the pulley. Choosing the x-axis to be horizontal, the normal force has no x-component. Applying Newton's second law along the x-axis: $\Sigma F_x = T_1 \cos \theta - T_2 \cos \theta = ma_x$. With $m = 0$, $T_1 = T_2$. The same reasoning can be applied to any segment of cord in contact with the pulley to show that the tensions are the same on either side of the pulley.

Example 4.15

Two Blocks Hanging on a Pulley

In Fig. 4.44, two blocks are connected by an ideal cord that does not stretch; the cord passes over an ideal pulley. If the masses are $m_1 = 26.0$ kg and $m_2 = 42.0$ kg, what are the accelerations of each block and the tension in the cord?

Strategy Since m_2 is greater than m_1, the downward force of gravity is stronger on the right side than on the left. We expect block 2's acceleration to be downward and block 1's to be upward.

The cord does not stretch, so blocks 1 and 2 move at the same speed at any instant (in opposite directions). Therefore, the accelerations of the two blocks are equal in magnitude and opposite in direction. If the accelerations had different magnitudes, then soon the two blocks would be moving with different speeds. That could only happen if the cord either stretches or contracts.

The tension in the cord must be the same everywhere along the cord since the masses of the cord and pulley are negligible and the pulley turns without friction.

We treat each block as a separate system, draw FBDs for each, and then apply Newton's second law to each. It is convenient to choose the positive y-direction differently for the two blocks since we know their accelerations are in opposite directions. For each block, we choose the $+y$-axis in the direction of the acceleration of that block: upward for block m_1

and downward for m_2. Doing so means that a_y has the same magnitude *and sign* (both positive) for the two blocks.

Solution Figure 4.45 shows FBDs for the two blocks. Two forces act on each: gravity and the pull of the cord. The

Figure 4.44

Two hanging blocks connected on either side of a frictionless pulley by a massless, flexible cord that does not stretch.

Figure 4.45

FBDs for the hanging blocks. We draw the acceleration vector *next to* each FBD as a guide—the net force has to be in the direction of the acceleration. However, the acceleration vector is not *part of* the FBD (it is not a force to be added to the others).

continued on next page

Example 4.15 continued

acceleration vectors are drawn *next to* the FBDs. Thus, we know the direction of the net force: it is always the same as the direction of the acceleration. Then we know that the tension must be greater than $m_1 g$ to give block 1 an upward acceleration and less than $m_2 g$ to give block 2 a downward acceleration. The +y-axes are drawn for each block to be in the direction of the acceleration.

From the FBD of block 1, the pull of the cord is in the +y-direction and the gravitational force is in the −y-direction. Then Newton's second law for block 1 is

$$\sum F_{1y} = T - m_1 g = m_1 a_{1y}$$

For block 2, the pull of the cord is in the −y-direction and the gravitational force is in the +y-direction. Newton's second law for block 2 is

$$\sum F_{2y} = m_2 g - T = m_2 a_{2y}$$

The tension T in the cord is the same in the two equations. Also a_{1y} and a_{2y} are identical, so we write them simply as a_y. We then have a system of two equations with two unknowns. We can add the equations to obtain

$$m_2 g - m_1 g = m_2 a_y + m_1 a_y$$

Solving for a_y, we find

$$a_y = \frac{(m_2 - m_1)g}{m_2 + m_1}$$

Substituting numerical values,

$$a_y = \frac{(42.0 \text{ kg} - 26.0 \text{ kg}) \times 9.80 \text{ N/kg}}{42.0 \text{ kg} + 26.0 \text{ kg}}$$

$$= 2.31 \text{ m/s}^2$$

since

$$1 \frac{\text{N}}{\text{kg}} = 1 \frac{\text{kg} \cdot \text{m/s}^2}{\text{kg}} = 1 \text{ m/s}^2$$

The blocks have the same magnitude acceleration. For block 1 the acceleration points upward and for block 2 it points downward.

To find T we can substitute the expression for a_y into either of the two original equations. Using the first equation,

$$T - m_1 g = m_1 \frac{(m_2 - m_1)g}{m_2 + m_1}$$

Solving for T yields

$$T = \frac{2m_1 m_2}{m_1 + m_2} g$$

Substituting,

$$T = \frac{2 \times 26.0 \text{ kg} \times 42.0 \text{ kg}}{68.0 \text{ kg}} \times 9.80 \text{ N/kg} = 315 \text{ N}$$

Discussion   A few quick checks:

- a_y is positive, which means that the accelerations are in the directions we expect.
- The tension (315 N) is between $m_1 g$ (255 N) and $m_2 g$ (412 N), as it must be for the accelerations to be in opposite directions.
- The units and dimensions are correct for all equations.
- We can check algebraic expressions in special cases for which we have some intuition. For example, if the masses had been *equal*, we expect the blocks to hang in equilibrium (either at rest or moving at constant velocity) due to the equal pull of gravity on the two blocks. Substituting $m_1 = m_2$ into the expressions for a_y and T gives $a_y = 0$ and $T = m_1 g = m_2 g$, which is just what we expect.

⚠️ Note that we did *not* find out which way the blocks move. We found the directions of their *accelerations.* If the blocks start out at rest, then the block of mass m_2 moves downward and the block of mass m_1 moves upward. However, if initially m_2 is moving up and m_1 down, they continue to move in those directions, slowing down since their accelerations are opposite to their velocities. Eventually they come to rest and then reverse directions.

Practice Problem 4.15   Another Check

Using the numerical values of the tension and the acceleration calculated in Example 4.15, verify Newton's second law directly for each of the two blocks.

Examples 4.16, 4.17, and 4.18 illustrate how different concepts and problem-solving techniques from Chapters 2–4 can be brought together to find the solution to a physics problem.

Example 4.16

Hauling a Crate up to a Third-Floor Window

A student is moving into a dorm room on the third floor and he decides to use a block and tackle arrangement (Fig. 4.46) to move a crate of mass 91 kg from the ground up to his window. If the breaking strength of the available rope is 550 N, what is the minimum time required to haul the crate to the level of the window, 30.0 m above the ground, without breaking the rope?

Strategy The tension in the rope is T and is the same at both ends or anywhere along the rope, assuming the rope and pulleys are ideal. Two pieces of rope support the lower pulley, each pulling upward with a force of magnitude T. The gravitational force acts downward. We draw an FBD for the system consisting of the crate and the lower pulley and set the tension equal to the breaking force of the rope to find the maximum possible acceleration of the crate. Then we use the maximum acceleration to find the minimum time to move the required distance to the third-floor window. We choose the y-axis to be upward. Known: $m = 91$ kg; $\Delta y = 30.0$ m; $T_{max} = 550$ N; $v_{iy} = 0$. To find: Δt, the time to raise the crate 30.0 m with the maximum tension in the cable.

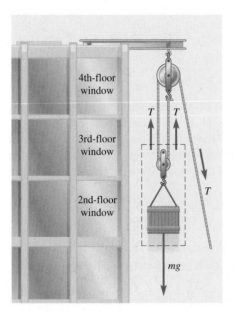

Figure 4.46
Block and tackle setup.

Solution From the FBD (Fig. 4.47), if the forces acting up are greater than the force acting down, the net force is upward and the crate's acceleration is upward. In terms of components, with the +y-direction chosen to be upward,

$$\sum F_y = T + T - mg = ma_y$$

Solving for the acceleration,

$$a_y = \frac{T + T - mg}{m}$$

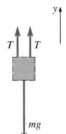

Figure 4.47
FBD for the crate and lower pulley. (This system is outlined by dashed lines in Fig. 4.46.)

Setting $T = 550$ N, the maximum possible value before the cable breaks, and substituting the other known values:

$$a_y = \frac{550 \text{ N} + 550 \text{ N} - 91 \text{ kg} \times 9.80 \text{ m/s}^2}{91 \text{ kg}} = 2.288 \text{ m/s}^2$$

The time to move the crate up a distance Δy starting from rest can be found from

$$\Delta y = v_{iy} \Delta t + \tfrac{1}{2}a_y(\Delta t)^2 \tag{3-21}$$

Setting $v_{iy} = 0$ and solving for Δt, we find

$$\Delta t = \pm\sqrt{\frac{2\,\Delta y}{a_y}}$$

Our equation applies only for $\Delta t \geq 0$ (the crate reaches the window *after* it leaves the ground). Taking the positive root and substituting numerical values,

$$\Delta t = \sqrt{\frac{2 \times 30.0 \text{ m}}{2.288 \text{ m/s}^2}} = 5.1 \text{ s}$$

This is the minimum possible to haul the crate up without breaking the rope.

Discussion In reality, the student is not likely to achieve this *minimum possible* time. To do so would mean pulling the rope at an unrealistic speed. At the end of the 5.1-s interval, $v_{fy} = 2.288$ m/s$^2 \times 5.1$ s = 12 m/s! More likely, the student would hoist the crate at a roughly constant velocity (except at the beginning, to get it moving, and at the end, to let it come to rest). For motion with a constant velocity, the tension in the rope would be equal to half the weight of the crate (450 N).

Practice Problem 4.16   Hauling the Crate with a Single Pulley

If only a single pulley, attached to the beam above the fourth floor, were available and if the student had a few friends to help him pull on the cable, could they haul the crate up to the third-floor window using the same rope? If so, what is the minimum time required to do so?

Example 4.17

Towing a Glider

What length runway does the plane need?

A small plane of mass 760 kg requires 120 m of runway to take off by itself. (120 m is the horizontal displacement of the plane just before it lifts off the runway, not the entire length of the runway.) As a simplified model, ignore friction and drag forces and assume the plane's engine exerts a constant forward force on the plane. (a) When the plane is towing a 330-kg glider, how much runway does it need? (b) If the final speed of the plane just before it lifts off the runway is 28 m/s, what is the tension in the tow cable while the plane and glider are moving along the runway?

Strategy   We draw FBDs for the two cases: plane alone, then plane + glider. The motion in both cases is horizontal (along the runway), because we are told the displacement *before it lifts off the runway.* Until the plane begins to lift off the runway, its vertical acceleration component is zero. We need not be concerned with the vertical forces (gravity, the normal force, and lift—the upward force on the plane's wings due to the air) since they cancel one another to produce zero vertical acceleration. We use Newton's second law to compare the accelerations in the two cases and then use the accelerations to compare the displacements.

Solution   (a) When the plane takes off by itself, four forces act on it (see Fig. 4.48). Three are vertical and the third—the thrust due to the engine—is horizontal. Choosing the x-axis to be horizontal, Newton's second law says

$$\sum F_{1x} = F = m_1 a_{1x}$$

where F is the thrust, m_1 is the plane's mass, and a_{1x} is its horizontal acceleration component.

When the glider is towed, we can consider the plane, glider, and cable to be a single system (see Fig. 4.49). There

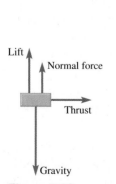

Figure 4.48

FBD for the plane.

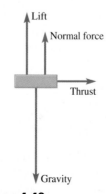

Figure 4.49

FBD for the system plane + glider.

is still only one horizontal external force and it is the same thrust as before. The tension in the cable is an *internal* force. Therefore,

$$\sum F_{2x} = F = (m_1 + m_2)a_x$$

where $m_1 + m_2$ is the total mass of the system (plane mass m_1 plus glider mass m_2) and a_x is the horizontal acceleration component of plane and glider. We ignore the mass of the cable.

The problem statement gives neither the thrust nor either of the accelerations. We can continue by setting the thrusts equal and finding the ratio of the accelerations:

$$m_1 a_{1x} = (m_1 + m_2)a_x \Rightarrow \frac{a_x}{a_{1x}} = \frac{m_1}{m_1 + m_2}$$

The magnitude of the acceleration is inversely proportional to the mass of the system for the same net force.

How is the acceleration related to the runway distance? The plane must get to the same final speed in order to lift off the runway. From our two basic constant acceleration equations

$$\Delta v_x = v_{fx} - v_{ix} = a_x \Delta t \qquad (2\text{-}9)$$
$$\Delta x = \tfrac{1}{2}(v_{fx} + v_{ix})\,\Delta t \qquad (2\text{-}11)$$

we can substitute $v_{ix} = 0$ and eliminate Δt to find

$$\Delta x = \tfrac{1}{2}(v_{fx} + 0)\left(\frac{v_{fx}}{a_x}\right) = \frac{v_{fx}^2}{2a_x}$$

In both cases, the displacement is inversely proportional to the acceleration and the acceleration is inversely proportional to the mass of the system. Therefore, the displacement is *directly* proportional to the mass. Letting $\Delta x_1 = 120$ m be the displacement of the plane without the glider, we can set up a proportion:

$$\frac{\Delta x}{\Delta x_1} = \frac{a_{1x}}{a_x} = \frac{m_1 + m_2}{m_1} = \frac{1090 \text{ kg}}{760 \text{ kg}} = 1.434$$

$$\Delta x = 1.434 \times 120 \text{ m} = 172.08 \text{ m} \rightarrow 170 \text{ m}$$

(b) The final speed given enables us to find the acceleration:

$$\Delta x = \frac{v_{fx}^2}{2a_x} \quad \text{or} \quad a_x = \frac{v_{fx}^2}{2\Delta x}$$

With $v_{fx} = 28$ m/s, $v_{ix} = 0$, and $\Delta x = 172.08$ m,

$$a_x = \frac{(28 \text{ m/s})^2}{2 \times 172.08 \text{ m}} = 2.278 \text{ m/s}^2$$

The tension in the cable is the only horizontal force acting on the glider. Therefore,

$$\sum F_x = T = m_2 a_x = 330 \text{ kg} \times 2.278 \text{ m/s}^2 = 751.7 \text{ N} \rightarrow 750 \text{ N}$$

continued on next page

Example 4.17 continued

Discussion   This solution is based on a simplified model, so we can only regard the answers as approximate. Nevertheless, it illustrates Newton's second law. The same net force produces an acceleration inversely proportional to the mass of the object upon which it acts. Here we have the same net force acting on two different objects: first the plane alone, then the plane and glider together.

Alternatively, we can look at forces acting only on the plane. When towing the glider, the cable pulls backward on the plane. The net force *on the plane* is smaller, so its acceleration is smaller. The smaller acceleration means that it takes more time to reach takeoff speed and travels a longer distance before lifting off the runway.

Practice Problem 4.17   Engine Thrust

What is the thrust provided by the airplane's engines in Example 4.17?

Example 4.18

A Pulley, an Incline, and Two Blocks

A block of mass $m_1 = 2.60$ kg rests on an incline that is angled at $30.0°$ above the horizontal (Fig. 4.50). An ideal cord is connected from block 1 over an ideal, frictionless pulley to another block of mass $m_2 = 2.20$ kg that is hanging 2.00 m above the ground. The coefficient of kinetic friction between the incline and block 1 is 0.180. The blocks are initially at rest. (a) How long does it take for block 2 to reach the ground? (b) Sketch a motion diagram for block 2 with a time interval of 0.5 s.

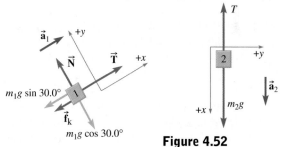

Figure 4.51
FBD for block 1.

Figure 4.52
FBD for block 2 with the downward direction chosen as +x.

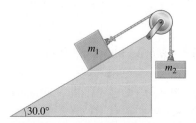

Figure 4.50
Block on an incline connected to a hanging block by a cord passing over a pulley.

Strategy   The problem says that the blocks start from rest and that block 2 hits the floor, so block 2's acceleration is downward and block 1's is up the incline. For block 1, we choose axes parallel and perpendicular to the incline so that its acceleration has only one nonzero component. The magnitudes of the accelerations of the two blocks are equal since they are connected by an ideal cord that does not stretch. Since the cord and pulley are ideal, the tension is the same at the two ends.

Solution   (a) We start by drawing separate FBDs for each block (Figs. 4.51 and 4.52). Since block 1 slides up the incline, the frictional force $\vec{f}_k$ acts down the incline to oppose the sliding. The gravitational force on block 1 is resolved into two components, one along the incline and one perpendicular to the incline.

Using the FBDs, we write Newton's second law in component form for each block. Block 1 has no acceleration component perpendicular to the incline. It does not sink into the incline or rise above it; it can only slide along the incline. Thus, the net force on block 1 in the direction perpendicular to the incline—the direction we have chosen as the y-axis for block 1—is zero.

$$\sum F_y = N - m_1 g \cos \theta = 0$$

or

$$N = m_1 g \cos \theta$$

Here $\theta = 30.0°$. Along the incline, in the x-direction for block 1, the acceleration is nonzero:

$$\sum F_x = T - m_1 g \sin \theta - f_k = m_1 a_x$$

The kinetic frictional force is related to the normal force:

$$f_k = \mu_k N = \mu_k m_1 g \cos \theta$$

By substitution,

$$T - m_1 g \sin \theta - \mu_k m_1 g \cos \theta = m_1 a_x \qquad (1)$$

continued on next page

Example 4.18 continued

For block 2, we choose an x-axis pointing downward. Doing so simplifies the solution, since then the two blocks have the same a_x. Applying Newton's second law,

$$\sum F_x = m_2 g - T = m_2 a_x \tag{2}$$

The tension in the cord T and the x-component of acceleration a_x are both unknown in Eqs. (1) and (2). We solve for T in Eq. (2) and substitute into Eq. (1):

$$T = m_2 g - m_2 a_x = m_2(g - a_x)$$

$$m_2(g - a_x) - m_1 g \sin\theta - \mu_k m_1 g \cos\theta = m_1 a_x$$

Rearranging and solving for a_x yields

$$a_x = \frac{m_2 - m_1(\sin\theta + \mu_k \cos\theta)}{m_1 + m_2} g \tag{3}$$

Substituting the known and given values,

$$a_x = \frac{2.20 \text{ kg} - 2.60 \text{ kg} \times (0.50 + 0.180 \times 0.866)}{2.60 \text{ kg} + 2.20 \text{ kg}} \times 9.80 \text{ m/s}^2$$

$$= 1.01 \text{ m/s}^2$$

Block 2 has a distance of 2.00 m to travel starting from rest with a constant downward acceleration of 1.01 m/s². From Eq. (2-12) with $v_{ix} = 0$,

$$\Delta x = \tfrac{1}{2} a_x (\Delta t)^2$$

The time to travel that distance is

$$\Delta t = \sqrt{\frac{2\Delta x}{a_x}} = \sqrt{\frac{2 \times 2.00 \text{ m}}{1.01 \text{ m/s}^2}} = 2.0 \text{ s}$$

(b) Figure 4.53 shows the motion diagram for block 2. Choosing $x_i = 0$ and $t_i = 0$, the position as a function of time is $x = \tfrac{1}{2} a_x t^2$.

Discussion   One advantage to solving for a_x algebraically in Eq. (3) before substituting numerical values is that

dimensional analysis can easily be used to check for errors. In Eq. (3), the quantity in parentheses is dimensionless—the values of trigonometric functions are pure numbers as are coefficients of friction. Therefore, the numerator is the sum of two quantities with dimensions of force, the denominator is the sum of two masses, and force divided by mass gives an acceleration.

What if the problem did not tell us the directions of the blocks' accelerations? We could figure it out by comparing the force with which gravity pulls down on block 2 ($m_2 g$) with the component of the gravitational force pulling block 1 down the incline ($m_1 g \sin\theta$). Whichever is greater "wins the tug-of-war," assuming that static friction doesn't prevent the blocks from starting to slide. Once we know the direction of block 1's acceleration, we can determine the direction of the kinetic frictional force. If block 1 is not initially at rest, the kinetic frictional force opposes the direction of sliding, even though that may be opposite to the direction of the acceleration.

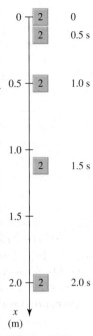

Figure 4.53
Motion diagram for block 2.

t (s)	x (m)
0	0
0.5	0.125
1.0	0.50
1.5	1.125
2.0	2.0

Practice Problem 4.18   More Fun with a Pulley and an Incline

Suppose that $m_1 = 3.8$ kg and $m_2 = 1.2$ kg and the coefficient of kinetic friction is 0.18. The blocks are released from rest and block 1 starts to slide. (a) Does block 1 slide up or down the incline? (b) In which direction does the kinetic frictional force act? (c) Find the acceleration of block 1.

✓ CHECKPOINT 4.8

Is it ever useful to choose the x- and y-axes so the x-axis is not horizontal? If yes, give an example.

4.9   REFERENCE FRAMES

Imagine a train moving at constant velocity with respect to the ground (Fig. 4.54). Suppose Tim does some experiments using the train's reference frame for his measurements. Greg does similar experiments using the reference frame of the ground. Tim and Greg disagree about the numerical value of an object's velocity, but since their velocity

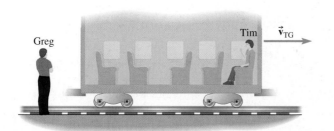

Figure 4.54 Greg's frame of reference is that of the ground; Tim's is that of the train, which moves at constant velocity $\vec{v}_{TG}$ with respect to the ground.

measurements *differ by a constant*, they will always agree about *changes* in velocity and about accelerations. Both observers can use Newton's second law to relate the net force to the acceleration. The basic laws of physics, such as Newton's laws of motion, work equally well in any two reference frames if they move with a constant relative velocity.

Newton's First Law Defines an Inertial Reference Frame   You might wonder why we need Newton's first law—isn't it just a special case of the second law when $\sum \vec{F} = 0$? No, the first law *defines* what kind of reference frame we can use when applying the second law. For the second law to be valid, we must use an *inertial reference frame*—a reference frame in which the law of inertia holds—to observe the motion of objects. The law of inertia is a *postulate* of classical mechanics—an assumption that is used as a starting point. It is not something we can prove experimentally.

Is a reference frame attached to Earth's surface truly inertial? No, but it is close enough in many circumstances. When analyzing the motion of a soccer ball, the fact that Earth rotates about its axis does not have much effect. But if we want to analyze the motion of a meteor falling from a great distance toward Earth, Earth's rotation must be considered. We will take a closer look at the effect of Earth's rotation in Chapter 5.

4.10 APPARENT WEIGHT

Imagine being in an elevator when the cable snaps. Assume that some safety mechanism brings you to rest after you have been in free fall for a while. While you are in free fall, you *seem* to be "weightless," but your weight has not changed; the Earth still pulls downward with the same gravitational force. In free fall, gravity gives the elevator and everything in it a downward acceleration equal to $\vec{g}$. If you jump up from the elevator floor, you seem to "float" up to the ceiling of the elevator. Your *weight* hasn't changed, but your *apparent* weight is zero while you are in free fall.

Similarly, astronauts in a space station in orbit around the Earth are in free fall (their acceleration is equal to the local value of $\vec{g}$). Earth exerts a gravitational force on them so they are not weightless; their *apparent* weight is zero.

Imagine an object that appears to be resting on a bathroom scale. The scale measures the object's *apparent* weight W', which is equal to the true weight only if the object and the scale have zero acceleration. Newton's second law requires that

$$\sum \vec{F} = \vec{N} + m\vec{g} = m\vec{a}$$

where $\vec{N}$ is the normal force of the scale pushing up. The apparent weight W' is the reading of the scale—that is, the magnitude of $\vec{N}$:

$$W' = |\vec{N}| = N$$

In Fig. 4.55a, the acceleration of the elevator is upward. The normal force must be larger than the weight for the net force to be upward (Fig. 4.55b). Writing the forces in component form where the +y-direction is upward

$$\sum F_y = N - mg = ma_y$$

or

$$N = mg + ma_y$$

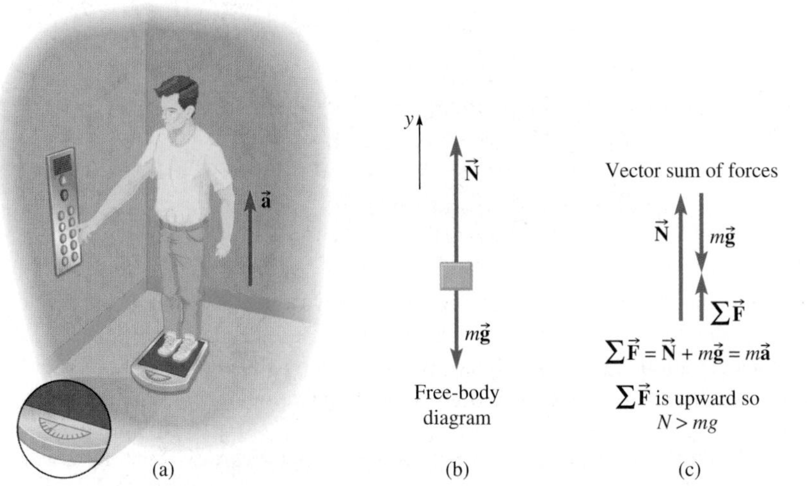

Figure 4.55 (a) Apparent weight in an elevator with acceleration upward. (b) FBD for the passenger. (c) The normal force must be greater than the weight to have an upward net force.

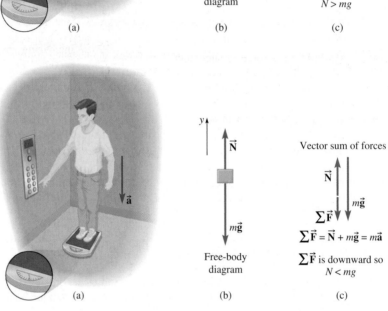

Free-body diagram

Vector sum of forces

$$\sum \vec{F} = \vec{N} + m\vec{g} = m\vec{a}$$

$\sum \vec{F}$ is upward so
$N > mg$

(a)   (b)   (c)

Figure 4.56 (a) Apparent weight in an elevator with acceleration downward. (b) FBD for the passenger. (c) The normal force must be less than the weight to have a downward net force.

Free-body diagram

Vector sum of forces

$$\sum \vec{F} = \vec{N} + m\vec{g} = m\vec{a}$$

$\sum \vec{F}$ is downward so
$N < mg$

(a)   (b)   (c)

Therefore,

$$W' = N = m(g + a_y) \tag{4-16}$$

Since the elevator's acceleration is upward, $a_y > 0$; the apparent weight is greater than the true weight (Fig. 4.55c).

In Fig. 4.56a, the acceleration is downward. Then the net force must also point downward. The normal force is still upward, but it must be smaller than the weight in order to produce a downward net force (Fig. 4.56b). It is still true that $W' = m(g + a_y)$, but now the acceleration is downward ($a_y < 0$). The apparent weight is less than the true weight (Fig. 4.56c). If the elevator is in free fall, then $a_y = -g$ and the apparent weight of the unfortunate passenger is zero.

Example 4.19

Apparent Weight in an Elevator

A passenger weighing 598 N rides in an elevator. What is the apparent weight of the passenger in each of the following situations? In each case, the magnitude of the elevator's acceleration is 0.500 m/s². (a) The passenger is on the first floor and has pushed the button for the fifteenth floor; the elevator is beginning to move upward. (b) The elevator is slowing down as it nears the fifteenth floor.

continued on next page

Example 4.19 continued

Strategy In each case, we sketch the FBD for the passenger. The apparent weight is equal to the magnitude of the normal force exerted by the floor on the passenger. The only other force acting is gravity. Newton's second law lets us find the normal force from the weight and the acceleration. Known: $W = 598$ N; magnitude of the acceleration is $a = 0.500$ m/s^2. To find: W'.

Solution (a) Let the +y-axis be upward. When the elevator starts up from the first floor it has acceleration in the upward direction as its speed increases. Since the elevator's acceleration is upward, $a_y > 0$ (as in Fig. 4.55). We expect the apparent weight $W' = N$ to be greater than the true weight—the floor must push up with a force greater than W to cause an upward acceleration. Figure 4.57 is the FBD. Newton's second law says

$$\sum F_y = N - W = ma_y$$

Since $W = mg$, we can substitute $m = W/g$.

$$W' = N = W + ma_y = W + \frac{W}{g}a_y = W\left(1 + \frac{a_y}{g}\right)$$

$$= 598 \text{ N} \times \left(1 + \frac{0.500 \text{ m/s}^2}{9.80 \text{ m/s}^2}\right) = 629 \text{ N}$$

Figure 4.57
FBD for the passenger in an elevator with upward acceleration.

(b) When the elevator approaches the fifteenth floor, it is slowing down while still moving upward; its acceleration is downward ($a_y < 0$) as in Fig. 4.56. The apparent weight is less than the true weight. Figure 4.58 is the FBD. Again, $\sum F_y = N - W = ma_y$, but this time $a_y = -0.500$ m/s^2.

$$N = W\left(1 + \frac{a_y}{g}\right)$$

$$= 598 \text{ N} \times \left(1 + \frac{-0.500 \text{ m/s}^2}{9.80 \text{ m/s}^2}\right) = 567 \text{ N}$$

Discussion The apparent weight is greater when the direction of the elevator's acceleration is upward. That can happen in two cases: either the elevator is moving up with increasing speed, or it is moving down with decreasing speed.

Practice Problem 4.19 Elevator Descending

What is the apparent weight of a passenger of mass 42.0 kg traveling in an elevator in each of the following situations? In each case, the magnitude of the elevator's acceleration is 0.460 m/s^2. (a) The passenger is on the fifteenth floor and has pushed the button for the first floor; the elevator is beginning to move downward. (b) The elevator is slowing down as it nears the first floor.

Figure 4.58
FBD for the passenger in an elevator with downward acceleration.

PHYSICS AT HOME

Take a bathroom scale to an elevator. Stand on the scale inside the elevator and push a button for a higher floor. When the elevator's acceleration is upward, you can feel the increase in your apparent weight and can see the increase by the reading on the scale. When the elevator slows down to stop, the elevator's acceleration is downward and your apparent weight is less than your true weight.

What is happening in your body while the elevator accelerates? The inertia principle means that your blood and internal organs cannot have the same acceleration as the elevator until the correct net force acts on them. Blood tends to collect in the lower extremities during acceleration upward and in the upper body during acceleration downward until the forces exerted on the blood by the body readjust to give the blood the same acceleration as the elevator. Likewise, the internal organs shift position within the body cavity, resulting in a funny feeling in the gut as the elevator starts and stops. To avoid this problem, high-speed express elevators in skyscrapers keep the acceleration relatively small, but maintain that acceleration long enough to reach high speeds. That way, the elevator can travel quickly to the upper floors without making the passengers feel too uncomfortable.

✓ CHECKPOINT 4.10

You are standing on a bathroom scale in an elevator that is moving downward. Nearing your stop, the elevator's speed is decreasing. Is the scale reading greater or less than your weight?

4.11   AIR RESISTANCE

So far we have ignored the effect of air resistance on falling objects and projectiles. A skydiver relies on a parachute to provide a large force of air resistance (also called **drag**). Even with the parachute closed, drag is not negligible when the skydiver is falling rapidly. The drag force is similar to friction between two solid surfaces in that the direction of the force *opposes the motion* of the object through the air. However, in contrast to the force of friction, the magnitude of the drag force is strongly dependent on the speed of the object. In many cases, air drag is proportional to the square of the speed. Drag also depends on the size and shape of the object.

Since the drag force increases as the speed increases, a falling object approaches an equilibrium situation in which the drag force is equal in magnitude to the weight but opposite in direction. The velocity at which this equilibrium occurs is called the object's *terminal velocity*. (See text website for a more detailed treatment of drag.)

PHYSICS AT HOME

Drop a basket-style paper coffee filter (or a cupcake paper) and a penny simultaneously from as close to the ceiling as you can safely do so. Air resistance on the penny is negligible unless it is dropped from a very high balcony. At the other extreme, the effect of air resistance on the coffee filter is very noticeable; it reaches its terminal speed almost immediately. Stack several (two to four) coffee filters together and drop them simultaneously with a single coffee filter. Why is the terminal speed higher for the stack? Crumple a coffee filter into a ball and drop it simultaneously with the penny. Air resistance on the coffee filter is now reduced, but still noticeable.

4.12   FUNDAMENTAL FORCES

One of the main goals of physics has been to understand the immense variety of forces in the universe in terms of the fewest number of fundamental laws. Physics has made great progress in this quest for *unification*; today all forces are understood in terms of just four fundamental interactions (Fig. 4.59). At the high temperatures present in the early universe, two of these interactions—the electromagnetic and weak forces—are now understood as the effects of a single electroweak interaction. The ultimate goal is to describe all forces in terms of a single interaction.

Gravity   You may be surprised to learn that gravity is by far the *weakest* of the fundamental forces. Any two objects exert gravitational forces on one another, but the force is tiny unless at least one of the masses is large. We tend to notice the relatively large gravitational forces exerted by planets and stars, but not the feeble gravitational

Figure 4.59 All forces result from just four fundamental forces: gravity, electromagnetism, and the weak and strong forces.

forces exerted by smaller objects, such as the gravitational force this book exerts on your body.

Gravity has an unlimited range. The force gets weaker as the distance between two objects increases, but it never drops exactly to zero, no matter how far apart the objects get.

Newton's law of gravity is an early example of unification. Before Newton, people did not understand that the same kind of force that makes an apple fall from a tree also keeps the planets in their orbits around the Sun. A single law—Newton's law of universal gravitation—describes both.

Electromagnetism   The electromagnetic force is unlimited in range, like gravity. It acts on particles with electric charge. The electric and magnetic forces were unified into a single theoretical framework in the nineteenth century. We study electromagnetic forces in detail in Part 3 of this book.

Electromagnetism is the fundamental interaction that binds electrons to nuclei to form atoms and binds atoms together in molecules and solids. It is responsible for the properties of solids, liquids, and gases and forms the basis of the sciences of chemistry and biology. It is the fundamental interaction behind all macroscopic contact forces such as the frictional and normal forces between surfaces and forces exerted by springs, muscles, and the wind.

The electromagnetic force is *much* stronger than gravity. For example, the electrical repulsion of two electrons at rest is about 10^{43} times as strong as the gravitational attraction between them. Macroscopic objects have a nearly perfect balance of positive and negative electric charge, resulting in a nearly perfect balance of attractive and repulsive electromagnetic forces between the objects. Therefore, despite the fundamental strength of the electromagnetic forces, the net electromagnetic force between two macroscopic objects is often negligibly small except when atoms on the two surfaces come very close to each other—what we think of as *in contact*. On a microscopic level, there is no fundamental difference between contact forces and other electromagnetic forces.

The Strong Force   The strong force holds protons and neutrons together in the atomic nucleus. The same force binds quarks (a family of elementary particles) in combinations so they can form protons and neutrons and many more exotic subatomic particles. The strong force is the strongest of the four fundamental forces—hence its name—but its range is short: its effect is negligible at distances much larger than the size of an atomic nucleus (about 10^{-15} m).

The Weak Force   The range of the weak force is even shorter than that of the strong force (about 10^{-17} m). It is manifest in many radioactive decay processes.

Master the Concepts

- A *force* is a push or a pull. Gravity and electromagnetic forces have unlimited range. All other forces exerted on macroscopic objects involve contact. Force is a vector quantity.
- The SI unit of force is the newton: $1 \text{ N} = 1 \text{ kg·m/s}^2$.
- The *net force* on a system is the vector sum of all the forces acting on it:

$$\vec{\mathbf{F}}_{\text{net}} = \sum \vec{\mathbf{F}} = \vec{\mathbf{F}}_1 + \vec{\mathbf{F}}_2 + \cdots + \vec{\mathbf{F}}_n \qquad (4\text{-}2)$$

Since all the internal forces form interaction pairs, we need only sum the external forces.

- *Newton's first law of motion*: If zero net force acts on an object, then the object's velocity does not change. Velocity is a vector whose magnitude is the speed at which the object moves and whose direction is the direction of motion.
- *Newton's second law of motion* relates the net force acting on an object to the object's acceleration and its mass:

$$\vec{\mathbf{a}} = \frac{\sum \vec{\mathbf{F}}}{m} \quad \text{or} \quad \sum \vec{\mathbf{F}} = m\vec{\mathbf{a}} \qquad (4\text{-}4)$$

The acceleration is always in the same direction as the net force. Many problems involving Newton's second law—whether equilibrium or nonequilibrium—can be solved by treating the *x*- and *y*-components of the forces and the acceleration separately:

$$\sum F_x = ma_x \quad \text{and} \quad \sum F_y = ma_y \qquad (4\text{-}5)$$

- *Newton's third law of motion*: In an interaction between two objects, each object exerts a force on the other. These two forces are equal in magnitude and opposite in direction.
- A *free-body diagram* (FBD) includes vector arrows representing every force acting on the chosen object

due to some other object, but no forces acting on other objects.

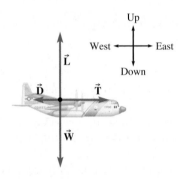

- The magnitude of the *gravitational force* between two objects is

$$F = \frac{Gm_1 m_2}{r^2} \qquad (4\text{-}7)$$

where r is the distance between their centers. Each object is pulled toward the other's center.

- The *weight* of an object is the magnitude of the gravitational force acting on it. An object's weight is proportional to its mass: $W = mg$ [Eq. (4-10)], where g is the gravitational field strength. Near Earth's surface, $g \approx 9.80$ N/kg.
- The *normal force* is a contact force perpendicular to the contact surfaces that pushes each object away from the other.

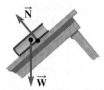

- *Friction* is a contact force parallel to the contact surfaces. In a simplified model, the kinetic frictional force and the maximum static frictional force are proportional

continued on next page

to the normal force acting between the same contact surfaces.

$$f_s \leq \mu_s N \qquad (4\text{-}14)$$

$$f_k = \mu_k N \qquad (4\text{-}15)$$

The static frictional force acts in the direction that tends to keep the surfaces from beginning to slide. The direction of the kinetic frictional force is in the direction that would tend to make the sliding stop.

- An ideal cord pulls in the direction of the cord with forces of equal magnitude on the objects attached to its ends as long as no external force tangent to the cord is exerted on it anywhere between the ends. The tension of an ideal cord that runs through an ideal pulley is the same on both sides of the pulley.

- An object that is accelerating has an apparent weight that differs from its true weight. The apparent weight is equal to the normal force exerted by a supporting surface with the same acceleration. A helpful trick is to think of the apparent weight as the reading of a bathroom scale that supports the object.

- The drag force exerted on an object moving through air opposes the motion of the object but, unlike kinetic friction, is strongly dependent on the object's speed. When an object falls at its terminal velocity, the drag force is equal and opposite to the gravitational force, so the acceleration is zero.

- At the fundamental level, there are four interactions: gravity, the strong and weak interactions, and the electromagnetic interaction. Contact forces are large-scale manifestations of many microscopic electromagnetic interactions.

Conceptual Questions

1. Explain the need for automobile seat belts in terms of Newton's first law.

2. An American visitor to Finland is surprised to see heavy metal frames outside of all the apartment buildings. On Saturday morning the purpose of the frames becomes evident when several apartment dwellers appear, carrying rugs and carpet beaters to each frame. What role does the principle of inertia play in the rug beating process? Do you see a similarity to the role the principle of inertia plays when you throw a baseball?

3. You are lying on the beach after a dip in the ocean where the waves were buffeting you around. Is it true that there are now no forces acting on you? Explain.

4. A dog goes swimming at the beach and then shakes himself all over to get dry. What principle of physics aids in the drying process? Explain.

5. In an attempt to tighten the loosened steel head of a hammer, a carpenter holds the hammer vertically, raises it up, and then brings it down rapidly, hitting the bottom end of the wood handle on a two-by-four board. Explain how this tightens the head back onto the handle.

6. When a car begins to move forward, what force makes it do so? Remember that it has to be an *external* force; the internal forces all add to zero. How does the engine facilitate the propelling force?

7. Two cars are headed toward each other in opposite directions along a narrow country road. The cars collide head-on, crumpling up the hoods of both. Describe what happens to the car bodies in terms of the principle of inertia. Does the rear end of the car stop at the same time as the front end?

8. Can a body in free fall be in equilibrium? Explain.

9. (a) What assumptions do you make when you call the reading of a bathroom scale your "weight"? What does the scale really tell you? (b) Under what circumstances might the reading of the scale *not* be equal to your weight?

10. A freight train consists of an engine and several identical cars on level ground. Determine whether each of these statements is correct or incorrect and explain why. (a) If the train is moving at constant speed, the engine must be pulling with a force greater than the train's weight. (b) If the train is moving at constant speed, the engine's pull on the first car must exceed that car's backward pull on the engine. (c) If the train is coasting, its inertia makes it slow down and eventually stop.

11. (a) Does a man weigh more at the North Pole or at the equator? (b) Does he weigh more at the top of Mt. Everest or at the base of the mountain?

12. What is the acceleration of an object thrown straight up into the air at the highest point of its motion? Does the answer depend on whether air resistance is negligible or not? Explain.

13. If a wagon starts at rest and pulls back on you with a force equal to the force you pull on it, as required by Newton's third law, how is it possible for you to make the wagon start to move? Explain.

14. You are standing on a bathroom scale in an elevator. In which of these situations must the scale read the same as when the elevator is at rest? Explain. (a) Moving up at constant speed. (b) Moving up with increasing speed. (c) In free fall (after the elevator cable has snapped).

15. A heavy ball hangs from a string attached to a sturdy wooden frame. A second string is attached to a hook on the bottom of the lead ball. You pull slowly and steadily on the lower string. Which string do you think will break first? Explain.

16. An SUV collides with a Mini Cooper convertible. Is the force exerted on the Mini by the SUV greater than, equal to, or less than the force exerted on the SUV by the Mini? Explain.

17. You are standing on one end of a light wooden raft that has floated 3 m away from the pier. If the raft is 6 m long by 2.5 m wide and you are standing on the raft end nearest to the pier, can you propel the raft back toward the pier where a friend is standing with a pole and hook trying to reach you? You have no oars. Make suggestions of what to do without getting yourself wet.

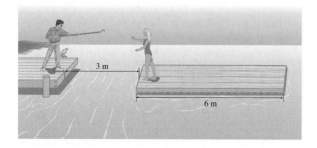

18. What does it mean when we refer to a cord as an "ideal cord" and a pulley as an "ideal pulley"?

19. If a feather and a lead brick are dropped simultaneously from the top of a ladder, the lead brick hits the ground first. What would happen if the experiment is repeated on the surface of the Moon?

20. A baseball is tossed straight up. Taking into consideration the force of air resistance, is the magnitude of the baseball's acceleration zero, less than g, equal to g, or greater than g on the way up? At the top of the flight? On the way down? Explain. [*Hint:* The force of air resistance is directed opposite to the velocity. Assume in this case that its magnitude is less than the weight.]

21. Why might an elevator cable break during acceleration when lifting a lighter load than it normally supports at rest or at constant velocity?

22. If air resistance is ignored, what force(s) act on an object in free fall?

23. The net force acting on an object is constant. Under what circumstances does the object move along a straight line? Under what circumstances does the object move along a curved path?

24. Pulleys and inclined planes are examples of *simple machines.* Explain what these machines do in Examples 4.10, 4.12, and 4.16 to make a task easier to perform.

25. For a problem about a crate sliding along an inclined plane, is it possible to choose the x-axis so that it is parallel to the incline?

26. A bird sits on a stretched clothesline, causing it to sag slightly. Is the tension in the line greatest where the bird sits, greater at either end of the line where it is attached to poles, or the same everywhere along the line? Treat the line as an ideal cord with negligible weight.

27. You decide to test your physics knowledge while going over a waterfall in a barrel. You take a baseball into the barrel with you and as you are falling vertically downward, you let go of the ball. What do you expect to see for the motion of the ball relative to the barrel? Will the ball fall faster than you and move toward the bottom of the barrel? Will it move slower than you and approach the top of the barrel, or will it hover apparently motionless within the falling barrel? Explain. [*Warning:* Do not try this.]

Multiple-Choice Questions

1. Interaction partners
   (a) are equal in magnitude and opposite in direction and act on the same object.
   (b) are equal in magnitude and opposite in direction and act on different objects.
   (c) appear in an FBD for a given object.
   (d) always involve gravitational force as one partner.
   (e) act in the same direction on the same object.

2. Within a given system, the internal forces
   (a) are always balanced by the external forces.
   (b) all add to zero.
   (c) are determined only by subtracting the external forces from the net force on the system.
   (d) determine the motion of the system.
   (e) can never add to zero.

3. A friction force is
   (a) a contact force that acts parallel to the contact surfaces.
   (b) a contact force that acts perpendicular to the contact surfaces.
   (c) a scalar quantity since it can act in any direction along a surface.
   (d) always proportional to the weight of an object.
   (e) always equal to the normal force between the objects.

4. When a force is called a "normal" force, it is
   (a) the usual force expected given the arrangement of a system.
   (b) a force that is perpendicular to the surface of the Earth at any given location.
   (c) a force that is always vertical.
   (d) a contact force perpendicular to the contact surfaces between two solid objects.
   (e) the net force acting on a system.

5. Your car won't start, so you are pushing it. You apply a horizontal force of 300 N to the car, but it doesn't budge. What force is the interaction partner of the 300 N force you exert?
   (a) the frictional force exerted on the car by the road
   (b) the force exerted on you by the car
   (c) the frictional force exerted on you by the road
   (d) the normal force on you by the road
   (e) the normal force on the car by the road

6. Which of these is *not* a long-range force?
   (a) the force that makes raindrops fall to the ground
   (b) the force that makes a compass point north
   (c) the force that a person exerts on a chair while sitting
   (d) the force that keeps the Moon in its orbital path around the Earth

7. When an object is in translational equilibrium, which of these statements is *not* true?
   (a) The vector sum of the forces acting on the object is zero.
   (b) The object must be stationary.
   (c) The object has a constant velocity.
   (d) The speed of the object is constant.

8. To make an object start moving on a surface with friction requires
   (a) less force than to keep it moving on the surface.
   (b) the same force as to keep it moving on the surface.
   (c) more force than to keep it moving on the surface.
   (d) a force equal to the weight of the object.

9. A thin string that can support a weight of 35.0 N, but breaks under any larger weight, is attached to the ceiling of an elevator. How large a mass can be attached to the string if the initial acceleration as the elevator starts to ascend is 3.20 m/s^2?
   (a) 3.57 kg
   (b) 2.69 kg
   (c) 4.26 kg
   (d) 2.96 kg
   (e) 5.30 kg

10. A woman stands on a bathroom scale in an elevator that is not moving. The scale reads 500 N. The elevator then moves downward at a constant velocity of 4.5 m/s. What does the scale read while the elevator descends with constant velocity?
    (a) 100 N
    (b) 250 N
    (c) 450 N
    (d) 500 N
    (e) 750 N

11. A 70.0-kg man stands on a bathroom scale in an elevator. What does the scale read if the elevator is slowing down at a rate of 3.00 m/s^2 while descending?
    (a) 70 kg   (b) 476 N   (c) 686 N
    (d) 700 N   (e) 896 N

12. A space probe leaves the solar system to explore interstellar space. Once it is far from any stars, when must it fire its rocket engines?
    (a) All the time, in order to keep moving.
    (b) Only when it wants to speed up.
    (c) When it wants to speed up or slow down.
    (d) Only when it wants to turn.
    (e) When it wants to speed up, slow down, or turn.

13. A small plane climbs with a constant velocity of 250 m/s at an angle of 28° with respect to the horizontal. Which statement is true concerning the magnitude of the net force on the plane?
    (a) It is equal to zero.
    (b) It is equal to the weight of the plane.
    (c) It is equal to the magnitude of the force of air resistance.
    (d) It is less than the weight of the plane but greater than zero.
    (e) It is equal to the component of the weight of the plane in the direction of motion.

14. Two blocks are connected by a light string passing over a pulley (see the figure and [W] tutorial: pulley). The block with mass m_1 slides on the frictionless horizontal surface, while the block with mass m_2 hangs vertically. ($m_1 > m_2$.) The tension in the string is
    (a) zero.
    (b) less than $m_2 g$.
    (c) equal to $m_2 g$.
    (d) greater than $m_2 g$, but less than $m_1 g$.
    (e) equal to $m_1 g$.
    (f) greater than $m_1 g$.

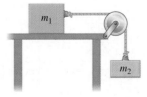

Problems

  ⊙ Combination conceptual/quantitative problem
  ⚕ Biological or medical application
  ✦ Challenging problem
Blue #   Detailed solution in the Student Solutions Manual
 ① ② Problems paired by concept
  ⓦ Text website interactive or tutorial

4.1 Force

1. A person is standing on a bathroom scale. Which of the following is *not* a force exerted *on the scale*: a contact force due to the floor, a contact force due to the person's feet, the weight of the person, the weight of the scale?

2. A sack of flour has a weight of 19.8 N. What is its weight in pounds?

3. An astronaut weighs 175 lb. What is his weight in newtons?

4. Does the concept of a contact force apply to both a macroscopic scale and an atomic scale? Explain.

5. A force of 20 N is directed at an angle of 60° above the *x*-axis. A second force of 20 N is directed at an angle of 60° below the *x*-axis. What is the vector sum of these two forces?

6. Juan is helping his mother rearrange the living room furniture. Juan pushes on the armchair with a force of 30 N directed at an angle of 15° above a horizontal line while his mother pushes with a force of 40 N directed at an angle of 20° below the same horizontal. What is the vector sum of these two forces?

7. In the drawing, what is the vector sum of forces $\vec{A} + \vec{B} + \vec{C}$ if each grid square is 2 N on a side?

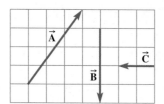

8. In the drawing, what is the vector sum of forces $\vec{D} + \vec{E} + \vec{F}$ if each grid square is 2 N on a side?

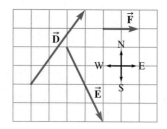

9. Two of Robin Hood's men are pulling a sledge loaded with some gold along a path that runs due north to their hideout. One man pulls his rope with a force of 62 N at an angle of 12° east of north and the other pulls with the same force at an angle of 12° west of north. Assume the ropes are parallel to the ground. What is the sum of these two forces on the sledge?

10. A barge is hauled along a straight-line section of canal by two horses harnessed to tow ropes and walking along the tow paths on either side of the canal. Each horse pulls with a force of 560 N at an angle of 15° with the centerline of the canal. Find the sum of the two forces exerted by the horses on the barge.

11. On her way to visit Grandmother, Red Riding Hood sat down to rest and placed her 1.2-kg basket of goodies beside her. A wolf came along, spotted the basket, and began to pull on the handle with a force of 6.4 N at an angle of 25° with respect to vertical. Red was not going to let go easily, so she pulled on the handle with a force of 12 N. If the net force on the basket is straight up, at what angle was Red Riding Hood pulling?

12. A parked automobile slips out of gear, rolls unattended down a slight incline, and then along a level road until it hits a stone wall. Draw an FBD to show the forces acting on the car while it is in contact with the wall.

13. Two objects, *A* and *B*, are acted on by the forces shown in the FBDs. Is the magnitude of the net force acting on object *B* greater than, less than, or equal to the magnitude of the net force acting on object *A*? Make a scale drawing on graph paper and explain the result.

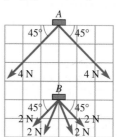

14. Find the magnitude and direction of the net force on the object in each of the FBDs for this problem.

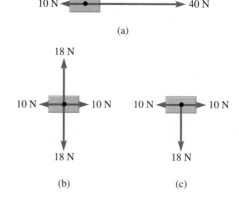

15. A truck driving on a level highway is acted on by the following forces: a downward gravitational force of 52 kN (kilonewtons); an upward contact force due to the road of 52 kN; another contact force due to the road of 7 kN, directed east; and a drag force due to air resistance of 5 kN, directed west. What is the net force acting on the truck?

4.2 Inertia and Equilibrium: Newton's First Law of Motion; 4.3 Net Force, Mass, and Acceleration: Newton's Second Law of Motion

16. A sailboat, tied to a mooring with a line, weighs 820 N. The mooring line pulls horizontally toward the west on the sailboat with a force of 110 N. The sails are stowed away and the wind blows from the west. The boat is moored on a still lake—no water currents push on it. Draw an FBD for the sailboat and indicate the magnitude of each force.

17. A hummingbird is hovering motionless beside a flower. The blur of its wings shows that they are rapidly beating up and down. If the air pushes upward on the bird with a force of 0.30 N, what is the weight of the hummingbird?

18. You are pulling a suitcase through the airport at a constant speed. The handle of the suitcase makes an angle of 60° with respect to the horizontal direction. If you pull with a force of 5.0 N parallel to the handle, what is the contact force due to the floor acting on the suitcase?

19. A model sailboat is slowly sailing west across a pond at 0.33 m/s. A gust of wind blowing at 28° south of west gives the sailboat a constant acceleration of magnitude 0.30 m/s^2 during a time interval of 2.0 s. (a) If the net force on the sailboat during the 2.0-s interval has magnitude 0.375 N, what is the sailboat's mass? (b) What is the new velocity of the boat after the 2.0-s gust of wind?

20. A man is lazily floating on an air mattress in a swimming pool. If the weight of the man and air mattress together is 806 N, what is the upward force of the water acting on the mattress?

21. A bag of potatoes with weight 39.2 N is suspended from a string that exerts a force of 46.8 N. If the bag's acceleration is upward at 1.90 m/s^2, what is the mass of the potatoes?

22. A 2010-kg elevator moves with an upward acceleration of 1.50 m/s^2. What is the force exerted by the cable on the elevator?

23. While an elevator of mass 2530 kg moves upward, the force exerted by the cable is 33.6 kN. (a) What is the acceleration of the elevator? (b) If at some point in the motion the velocity of the elevator is 1.20 m/s upward, what is the elevator's velocity 4.00 s later?

24. The vertical component of the acceleration of a sailplane is zero when the air pushes up against its wings with a force of 3.0 kN. (a) Assuming that the only forces on the sailplane are that due to gravity and that due to the air pushing against its wings, what is the gravitational force on the Earth due to the sailplane? (b) If the wing stalls and the upward force decreases to 2.0 kN, what is the acceleration of the sailplane?

25. A man lifts a 2.0-kg stone vertically with his hand at a constant upward velocity of 1.5 m/s. What is the magnitude of the total force of the man's hand on the stone?

26. A man lifts a 2.0-kg stone vertically with his hand at a constant upward *acceleration* of 1.5 m/s^2. What is the magnitude of the total force of the man's hand on the stone?

27. What is the acceleration of an automobile of mass 1.40×10^3 kg when it is subjected to a forward force of 3.36×10^3 N?

28. A large wooden crate is pushed along a smooth, frictionless surface by a force of 100 N. The acceleration of the crate is measured to be 2.5 m/s^2. What is the mass of the crate?

29. The forces on a small airplane (mass 1160 kg) in horizontal flight heading eastward are as follows: gravity = 16.000 kN downward, lift = 16.000 kN upward, thrust = 1.800 kN eastward, and drag = 1.400 kN westward. At $t = 0$, the plane's speed is 60.0 m/s. If the forces remain constant, how far does the plane travel in the next 60.0 s?

30. While an elevator of mass 832 kg moves downward, the tension in the supporting cable is a constant 7730 N. Between $t = 0$ and $t = 4.00$ s, the elevator's displacement is 5.00 m downward. What is the elevator's speed at $t = 4.00$ s?

4.4 Interaction Pairs: Newton's Third Law of Motion

31. A hanging potted plant is suspended by a cord from a hook in the ceiling. Draw an FBD for each of these: (a) the system consisting of plant, soil, and pot; (b) the cord; (c) the hook; (d) the system consisting of plant, soil, pot, cord, and hook. Label each force arrow using subscripts (for example, $\vec{F}_{ch}$ would represent the force exerted on the cord by the hook).

32. A bike is hanging from a hook in a garage. Consider the following forces: (a) the force of the Earth pulling down on the bike, (b) the force of the bike pulling up on the Earth, (c) the force of the hook pulling up on the bike, and (d) the force of the hook pulling down on the ceiling. Which two forces are equal and opposite because of Newton's third law? Which two forces are equal and opposite because of Newton's first law?

33. A woman who weighs 600 N sits on a chair with her feet on the floor and her arms resting on the chair's armrests. The chair weighs 100 N. Each armrest exerts an upward force of 25 N on her arms. The seat of the chair exerts an upward force of 500 N. (a) What force does the floor exert on her feet? (b) What force does the floor exert on the chair? (c) Consider the woman and the chair to be a single system. Draw an FBD for this system that includes all of the *external* forces acting on it.

34. A fisherman is holding a fishing rod with a large fish suspended from the line of the rod. Identify the forces acting on the rod and their interaction partners.

35. A fish is suspended by a line from a fishing rod. Choose two forces acting on the fish and describe the interaction partner of each.

Problems 34 and 35

36. A skydiver, who weighs 650 N, is falling at a constant speed with his parachute open. Consider the apparatus that connects the parachute to the skydiver to be part of the parachute. The parachute pulls upward with a force of 620 N. (a) What is the force of the air resistance acting on the skydiver? (b) Identify the forces and the interaction partners of each force exerted on the skydiver. (c) Identify the forces and interaction partners of each force exerted on the parachute.

37. Margie, who weighs 543 N, is standing on a bathroom scale that weighs 45 N. (a) With what force does the scale push up on Margie? (b) What is the interaction partner of that force? (c) With what force does the Earth push up on the scale? (d) Identify the interaction partner of that force.

38. Refer to Problem 36. Consider the skydiver and parachute to be a single system. What are the external forces acting on this system?

4.5 Gravitational Forces

39. (a) Calculate your weight in newtons. (b) What is the weight in newtons of 250 g of cheese? (c) Name a common object whose weight is about 1 N.

40. A young South African girl has a mass of 40.0 kg. (a) What is her weight in newtons? (b) If she came to the United States, what would her weight be in pounds as measured on an American scale? Assume $g = 9.80$ N/kg in both locations.

41. A man weighs 0.80 kN on Earth. What is his mass in kilograms?

42. An astronaut stands at a position on the Moon such that Earth is directly over head and releases a Moon rock that was in her hand. (a) Which way will it fall? (b) What is the gravitational force exerted by the Moon on a 1.0-kg rock resting on the Moon's surface? (c) What is the gravitational force exerted by the Earth on the same 1.0-kg rock resting on the surface of the Moon? (d) What is the net gravitational force on the rock?

43. Alex is on stage playing his bass guitar. Estimate the magnitude of the *gravitational* attraction between Alex and Pat, a fan who is standing 8 m from Alex. Alex has a mass of 55 kg and Pat has a mass of 40 kg.

44. The Space Shuttle carries a satellite in its cargo bay and places it into orbit around the Earth. Find the ratio of the Earth's gravitational force on the satellite when it is on a launch pad at the Kennedy Space Center to the gravitational force exerted when the satellite is orbiting 6.00×10^3 km above the launch pad.

45. How far above the surface of the Earth does an object have to be in order for it to have the same weight as it would have on the surface of the Moon? (Ignore any effects from the Earth's gravity for the object on the Moon's surface or from the Moon's gravity for the object above the Earth.)

46. Find and compare the weight of a 65-kg man on Earth with the weight of the same man on (a) Mars, where $g = 3.7$ N/kg; (b) Venus, where $g = 8.9$ N/kg; and (c) Earth's Moon, where $g = 1.6$ N/kg.

47. Find the altitudes above the Earth's surface where Earth's gravitational field strength would be (a) two thirds and (b) one third of its value at the surface. [*Hint:* First find the radius for each situation; then recall that the altitude is the distance from the *surface* to a point above the surface. Use proportional reasoning.]

48. During a balloon ascension, wearing an oxygen mask, you measure the weight of a calibrated 5.00-kg mass and find that the value of the gravitational field strength at your location is 9.792 N/kg. How high above sea level, where the gravitational field strength was measured to be 9.803 N/kg, are you located?

49. At what altitude above the Earth's surface would your weight be half of what it is at the Earth's surface?

50. (a) What is the magnitude of the gravitational force that the Earth exerts on the Moon? (b) What is the magnitude of the gravitational force that the Moon exerts on the Earth? See the inside front and back covers for necessary information.

51. What is the approximate magnitude of the gravitational force between the Earth and the Voyager spacecraft when they are separated by 15 billion km? Each spacecraft has a mass of approximately 825 kg during the mission, although the mass at launch was 2100 kg because of expendable Titan-Centaur rockets.

✦52. In free fall, we assume the acceleration to be constant. Not only is air resistance ignored, but the gravitational field strength is assumed to be constant. From what height can an object fall to the Earth's surface such that the gravitational field strength changes less than 1.000% during the fall?

4.6 Contact Forces

53. A book rests on the surface of the table. Consider the following four forces that arise in this situation: (a) the force of the Earth pulling on the book, (b) the force of the table pushing on the book, (c) the force of the book pushing on the table, and (d) the force of the book pulling on the Earth. The book is not moving. Which pair of forces must be equal in magnitude and opposite in direction even though they are *not* an interaction pair?

54. A crate full of artichokes rests on a ramp that is inclined 10.0° above the horizontal. Give the direction of the normal force and the friction force acting on the crate in each of these situations. (a) The crate is at rest. (b) The crate is being pushed and is sliding up the ramp. (c) The crate is being pushed and is sliding down the ramp.

55. Mechanical advantage is the ratio of the force required without the use of a simple machine to that needed when using the simple machine. Compare the force to lift an object with that needed to slide the same object up a frictionless incline and show that the mechanical advantage of the inclined plane is the length of the incline divided by the height of the incline (d/h in Fig. 4.25).

56. An 80.0-N crate of apples sits at rest on a ramp that runs from the ground to the bed of a truck. The ramp is inclined at 20.0° to the ground. (a) What is the normal force exerted on the crate by the ramp? (b) The interaction partner of this normal force has what magnitude and direction? It is exerted *by* what object *on* what object? Is it a contact or a long-range force? (c) What is the static frictional force exerted on the crate by the ramp? (d) What is the minimum possible value of the coefficient of static friction? (e) The normal and frictional forces are perpendicular components of the contact force exerted on the crate by the ramp. Find the magnitude and direction of the contact force.

57. An 85-kg skier is sliding down a ski slope at a constant velocity. The slope makes an angle of 11° above the horizontal direction. (a) Ignoring any air resistance, what is the force of kinetic friction acting on the skier? (b) What is the coefficient of kinetic friction between the skis and the snow?

Problems 58–60. A crate of potatoes of mass 18.0 kg is on a ramp with angle of incline 30° to the horizontal. The coefficients of friction are $\mu_s = 0.75$ and $\mu_k = 0.40$. Find the frictional force (magnitude and direction) on the crate if

58. the crate is at rest.

59. the crate is sliding down the ramp.

60. the crate is sliding *up* the ramp.

61. You grab a book and give it a quick push across the top of a horizontal table. After a short push, the book slides across the table, and because of friction, comes to a stop. (a) Draw an FBD of the book while you are pushing it. (b) Draw an FBD of the book after you have stopped pushing it, while it is sliding across the table. (c) Draw an FBD of the book after it has stopped sliding. (d) In which of the preceding cases is the net force on the book not equal to zero? (e) If the book has a mass of 0.50 kg and the coefficient of friction between the book and the table is 0.40, what is the net force acting on the book in part (b)? (f) If there were no friction between the table and the book, what would the free-body diagram for part (b) look like? Would the book slow down in this case? Why or why not?

62. (a) In Example 4.10, if the movers stop pushing on the safe, can static friction hold the safe in place without having it slide back down? (b) If not, what minimum force needs to be applied to hold the safe in place?

✦63. A 3.0-kg block is at rest on a horizontal floor. If you push horizontally on the 3.0-kg block with a force of 12.0 N, it just starts to move. (a) What is the coefficient of static friction? (b) A 7.0-kg block is stacked on top of the 3.0-kg block. What is the magnitude F of the force, acting horizontally on the 3.0-kg block as before, that is required to make the two blocks start to move?

64. A horse is trotting along pulling a sleigh through the snow. To move the sleigh, of mass m, straight ahead at a constant speed, the horse must pull with a force of magnitude T. (a) What is the net force acting on the sleigh? (b) What is the coefficient of kinetic friction between the sleigh and the snow?

✦65. Before hanging new William Morris wallpaper in her bedroom, Brenda sanded the walls lightly to smooth out some irregularities on the surface. The sanding block weighs 2.0 N and Brenda pushes on it with a force of 3.0 N at an angle of 30.0° with respect to the vertical, and angled toward the wall. Draw an FBD for the sanding block as it moves straight up the wall at a constant speed. What is the coefficient of kinetic friction between the wall and the block?

66. Four separate blocks are placed side by side in a left-to-right row on a table. A horizontal force, acting toward the right, is applied to the block on the far left end of the row. Draw FBDs for (a) the second block on the left and for (b) the system of four blocks.

✦67. A box sits on a horizontal wooden ramp. The coefficient of static friction between the box and the ramp is 0.30. You grab one end of the ramp and lift it up, keeping the other end of the ramp on the ground. What is the angle between the ramp and the horizontal direction when the box begins to slide down the ramp? (🎧 tutorial: crate on ramp)

✦68. In a playground, two slides have different angles of incline θ_1 and θ_2 ($\theta_2 > \theta_1$). A child slides down the first at constant speed; on the second, his acceleration down the slide is a. Assume the coefficient of kinetic friction is the same for both slides. (a) Find a in terms of θ_1, θ_2, and g. (b) Find the numerical value of a for $\theta_1 = 45°$ and $\theta_2 = 61°$.

4.7 Tension

69. A sailboat is tied to a mooring with a horizontal line. The wind is from the southwest. Draw an FBD and identify all the forces acting on the sailboat.

70. A towline is attached between a car and a glider. As the car speeds due east along the runway, the towline exerts a horizontal force of 850 N on the glider. What is the magnitude and direction of the force exerted by the glider on the towline?

71. In Example 4.14, find the tension in the coupling between cars 2 and 3. (🎧 tutorial: towing a train)

72. A 200.0-N sign is suspended from a horizontal strut of negligible weight. The force exerted on the strut by the wall is horizontal. Draw an FBD to show the forces acting on the strut. Find the tension T in the diagonal cable supporting the strut.

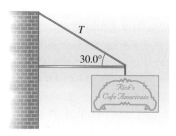

73. Two boxes with different masses are tied together on a frictionless ramp surface. What is the tension in each of the cords?

74. A pulley is attached to the ceiling. Spring scale A is attached to the wall and a rope runs horizontally from it and over the pulley. The same rope is then attached to spring scale B. On the other side of scale B hangs a 120-N weight. What are the readings of the two scales A and B? The weights of the scales are negligible.

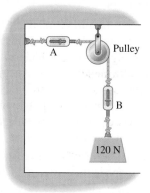

75. Spring scale A is attached to the floor and a rope runs vertically upward, loops over the pulley, and runs down on the other side to a 120-N weight. Scale B is attached to the ceiling and the pulley is hung below it. What are the readings of the two spring scales, A and B? Neglect the weights of the pulley and scales.

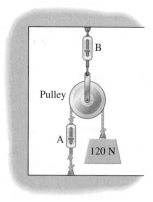

76. Two springs are connected in series so that spring scale A hangs from a hook on the ceiling and a second spring scale, B, hangs from the hook at the bottom of scale A.

Apples weighing 120 N hang from the hook at the bottom of scale B. What are the readings on the upper scale A and the lower scale B? Ignore the weights of the scales.

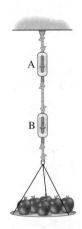

77. A pulley is hung from the ceiling by a rope. A block of mass M is suspended by another rope that passes over the pulley and is attached to the wall. The rope fastened to the wall makes a right angle with the wall. Ignore the masses of the rope and the pulley. Find (a) the tension in the rope from which the pulley hangs and (b) the angle θ that the rope makes with the ceiling.

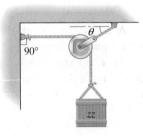

78. A 2.0-kg ball tied to a string fixed to the ceiling is pulled to one side by a force $\vec{\mathbf{F}}$. Just before the ball is released and allowed to swing back and forth, (a) how large is the force $\vec{\mathbf{F}}$ that is holding the ball in position and (b) what is the tension in the string?

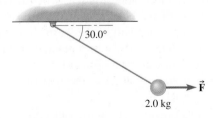

79. A 45-N lithograph is supported by two wires. One wire makes a 25° angle with the vertical and the other makes a 15° angle with the vertical. Find

the tension in each wire. (tutorial: hanging picture)

◆80. A crow perches on a clothesline midway between two poles. Each end of the rope makes an angle of θ below the horizontal where it connects to the pole. If the weight of the crow is W, what is the tension in the rope? Ignore the weight of the rope.

◆81. The drawing shows an elastic cord attached to two back teeth and stretched across a front tooth. The purpose of this arrangement is to apply a force $\vec{\mathbf{F}}$ to the front tooth. (The figure has been simplified by running the cord straight from the front tooth to the back teeth.) If the tension in the cord is 1.2 N, what are the magnitude and direction of the force $\vec{\mathbf{F}}$ applied to the front tooth?

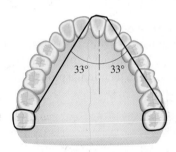

◆82. A cord, with a spring balance to measure forces attached midway along, is hanging from a hook attached to the ceiling. A mass of 10 kg is hanging from the lower end of the cord. The spring balance indicates a reading of 98 N for the force. Then two people hold the opposite ends of the same cord and pull against each other horizontally until the balance in the middle again reads 98 N. With what force must each person pull to attain this result?

◆83. Two blocks, masses m_1 and m_2, are connected by a massless cord. If the two blocks are pulled with a constant tension on a frictionless surface by applying a force of magnitude T_2 to a second cord connected to m_2, what is the ratio of the tensions in the two cords T_1/T_2 in terms of the masses?

4.8 Applying Newton's Second Law

84. A 6.0-kg block, starting from rest, slides down a frictionless incline of length 2.0 m. When it arrives at the bottom of the incline, its speed is v_f. At what distance from the top of the incline is the speed of the block $0.50\ v_f$?

85. The coefficient of static friction between a block and a horizontal floor is 0.40, while the coefficient of kinetic friction is 0.15. The mass of the block is 5.0 kg. A horizontal force is applied to the block and slowly increased. (a) What is the value of the applied horizontal force at the instant that the block starts to slide? (b) What is the net force on the block after it starts to slide?

86. A 2.0-kg toy locomotive is pulling a 1.0-kg caboose. The frictional force of the track on the caboose is 0.50 N backward along the track. If the train's acceleration forward is $3.0\ \text{m/s}^2$, what is the magnitude of the force exerted by the locomotive on the caboose?

87. A block of mass $m_1 = 3.0$ kg rests on a frictionless horizontal surface. A second block of mass $m_2 = 2.0$ kg hangs from an ideal cord of negligible mass that runs over an ideal pulley and then is connected to the first block. The blocks are released from rest. (a) Find the acceleration of the two blocks after they are released. (b) What is the velocity of the first block 1.2 s after the release of the blocks, assuming the first block does not run out of room on the table and the second block does not land on the floor? (c) How far has block 1 moved during the 1.2-s interval? (d) What is the displacement of the blocks from their initial positions 0.40 s after they are released?

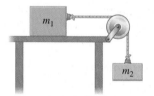

Problems 87 and 153

88. An engine pulls a train of 20 freight cars, each having a mass of 5.0×10^4 kg with a constant force. The cars move from rest to a speed of 4.0 m/s in 20.0 s on a straight track. Ignoring friction, what is the force with which the 10th car pulls the 11th one (at the middle of the train)? (🅦 tutorial: school bus)

89. In Fig. 4.44, two blocks are connected by a lightweight, flexible cord that passes over a frictionless pulley. (a) If $m_1 = 3.0$ kg and $m_2 = 5.0$ kg, what are the accelerations of each block? (b) What is the tension in the cord?

90. A rope is attached from a truck to a 1400-kg car. The rope will break if the tension is greater than 2500 N. Ignoring friction, what is the maximum possible acceleration of the truck if the rope does not break? Should

the driver of the truck be concerned that the rope might break?

91. Two blocks are connected by a lightweight, flexible cord that passes over a frictionless pulley. If $m_1 = 3.6$ kg and $m_2 = 9.2$ kg, and block 2 is initially at rest 140 cm above the floor, how long does it take block 2 to reach the floor?

92. A 10.0-kg watermelon and a 7.00-kg pumpkin are attached to each other via a cord that wraps over a pulley, as shown. Friction is negligible everywhere in this system. (🅦 tutorial: pulley) (a) Find the accelerations of the pumpkin and the watermelon. Specify magnitude and direction. (b) If the system is released from rest, how far along the incline will the pumpkin travel in 0.30 s? (c) What is the speed of the watermelon after 0.20 s?

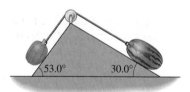

93. In the physics laboratory, a glider is released from rest on a frictionless air track inclined at an angle. If the glider has gained a speed of 25.0 cm/s in traveling 50.0 cm from the starting point, what was the angle of inclination of the track? Draw a graph of $v_x(t)$ when the positive x-axis points down the track.

◆94. A 10.0-kg block is released from rest on a frictionless track inclined at an angle of 55°. (a) What is the net force on the block after it is released? (b) What is the acceleration of the block? (c) If the block is released from rest, how long will it take for the block to attain a speed of 10.0 m/s? (d) Draw a motion diagram for the block. (e) Draw a graph of $v_x(t)$ for values of velocity between 0 and 10 m/s. Let the positive x-axis point down the track.

◆95. A box full of books rests on a wooden floor. The normal force the floor exerts on the box is 250 N. (a) You push horizontally on the box with a force of 120 N, but it refuses to budge. What can you say about the coefficient of static friction between the box and the floor?

(b) If you must push horizontally on the box with a force of at least 150 N to start it sliding, what is the coefficient of static friction? (c) Once the box is sliding, you only have to push with a force of 120 N to keep it sliding. What is the coefficient of kinetic friction?

◆ 96. A helicopter is lifting two crates simultaneously. One crate with a mass of 200 kg is attached to the helicopter by a cable. The second crate with a mass of 100 kg is hanging below the first crate and attached to the first crate by a cable. As the helicopter accelerates upward at a rate of 1.0 m/s^2, what is the tension in each of the two cables?

4.10 Apparent Weight

97. Oliver has a mass of 76.2 kg. He is riding in an elevator that has a downward acceleration of 1.37 m/s^2. With what magnitude force does the elevator floor push upward on Oliver?

98. While on an elevator, Jaden's apparent weight is 550 N. When he is on the ground, the scale reading is 600 N. What is Jaden's acceleration?

99. When on the ground, Ian's weight is measured to be 640 N. When Ian is on an elevator, his apparent weight is 700 N. What is the net force on the system (Ian and the elevator) if their combined mass is 1050 kg?

100. Refer to Example 4.19. What is the apparent weight of the same passenger (weighing 598 N) in the following situations? In each case, the magnitude of the elevator's acceleration is 0.50 m/s^2. (a) After having stopped at the 15th floor, the passenger pushes the 8th floor button; the elevator is beginning to move downward. (b) The elevator is moving downward and is slowing down as it nears the 8th floor.

101. You are standing on a bathroom scale inside an elevator. Your weight is 140 lb, but the reading of the scale is 120 lb. (a) What is the magnitude and direction of the acceleration of the elevator? (b) Can you tell whether the elevator is speeding up or slowing down?

102. Yolanda, whose mass is 64.2 kg, is riding in an elevator that has an upward acceleration of 2.13 m/s^2. What force does she exert on the floor of the elevator?

103. Felipe is going for a physical before joining the swim team. He is concerned about his weight, so he carries his scale into the elevator to check his weight while heading to the doctor's office on the 21st floor of the building. If his scale reads 750 N while the elevator has an upward acceleration of 2.0 m/s^2, what does the nurse measure his weight to be?

◆ 104. Luke stands on a scale in an elevator that has a constant acceleration upward. The scale reads 0.960 kN. When Luke picks up a box of mass 20.0 kg, the scale reads 1.200 kN. (The acceleration remains the same.) (a) Find the acceleration of the elevator. (b) Find Luke's weight.

4.12 Fundamental Forces

105. Which of the fundamental forces has the shortest range, yet is responsible for producing the sunlight that reaches Earth?

106. Which of the fundamental forces governs the motion of planets in the solar system? Is this the strongest or the weakest of the fundamental forces? Explain.

107. Which of the following forces have an unlimited range: strong force, contact force, electromagnetic force, gravitational force?

108. Which of the following forces bind electrons to nuclei to form atoms: strong force, contact force, electromagnetic force, gravitational force?

109. Which of the fundamental forces binds quarks together to form protons, neutrons, and many exotic subatomic particles?

Comprehensive Problems

110. A car is driving on a straight, level road at constant speed. Draw an FBD for the car, showing the significant forces that act upon it.

111. A skier with a mass of 63 kg starts from rest and skis down an icy (frictionless) slope that has a length of 50 m at an angle of 32° with respect to the horizontal. At the bottom of the slope, the path levels out and becomes horizontal, the snow becomes less icy, and the skier begins to slow down, coming to rest in a distance of 140 m along the horizontal path. (a) What is the speed of the skier at the bottom of the slope? (b) What is the coefficient of kinetic friction between the skier and the horizontal surface?

112. You want to push a 65-kg box up a 25° ramp. The coefficient of kinetic friction between the ramp and the box is 0.30. With what magnitude force parallel to the ramp should you push on the box so that it moves up the ramp at a constant speed?

113. An airplane is cruising along in a horizontal level flight at a constant velocity, heading due west. (a) If the weight of the plane is 2.6×10^4 N, what is the net force on the plane? (b) With what force does the air push upward on the plane?

▼ 114. A young boy with a broken leg is undergoing traction. (a) Find the magnitude of the total force of the traction apparatus applied to the leg, assuming the weight of the leg is 22 N and the weight hanging from the traction apparatus is also 22 N. (b) What is the horizontal component of the traction force acting on the leg? (c) What is the magnitude of the force exerted on the femur by the lower leg?

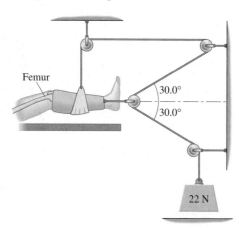

◉ 115. When you hold up a 100-N weight in your hand, with your forearm horizontal and your palm up, the force exerted by your biceps is much larger than 100 N— perhaps as much as 1000 N. How can that be? What other forces are acting on your arm? Draw an FBD for the forearm, showing all of the forces. Assume that all the forces exerted on the forearm are purely vertical— either up or down.

116. In the sport of curling, popular in Canada and Ireland, a player slides a 20.0-kg granite stone down a 38-m-long ice rink. Draw FBDs for the stone (a) while it sits at rest on the ice; (b) while it slides down the rink; (c) during a head-on collision with an opponent's stone that was at rest on the ice.

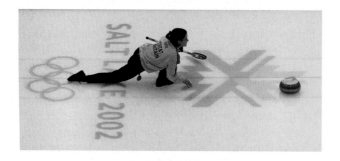

117. A truck is towing a 1000-kg car at a constant speed up a hill that makes an angle of $\alpha = 5.0°$ with respect to the horizontal. A rope is attached from the truck to the car at an angle of $\beta = 10.0°$ with respect to horizontal. Ignore any friction in this problem. (a) Draw an FBD showing all the forces on the car. Indicate the angle

that each force makes with either the vertical or horizontal direction. (b) What is the tension in the rope?

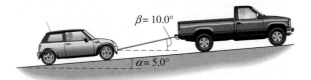

◉ 118. The readings of the two spring scales shown in the drawing are the same. (a) Explain why they are the same. [*Hint:* Draw free-body diagrams.] (b) What is the reading?

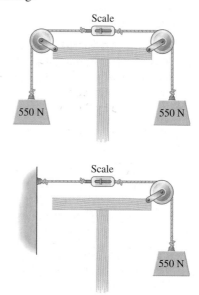

119. The tallest spot on Earth is Mt. Everest, which is 8850 m above sea level. If the radius of the Earth to sea level is 6370 km, how much does the gravitational field strength change between the sea level value at that location (9.826 N/kg) and the top of Mt. Everest?

120. By what percentage does the weight of an object change when it is moved from the equator at sea level, where the effective value of g is 9.784 N/kg, to the North Pole where $g = 9.832$ N/kg?

121. Two canal workers pull a barge along the narrow waterway at a constant speed. One worker pulls with a force of 105 N at an angle of 28° with respect to the forward motion of the barge and the other worker, on the opposite tow path, pulls at an angle of 38° relative to the barge motion. Both ropes are parallel to the ground. (a) With what magnitude force should the second worker pull to make the sum of the two forces be in the forward direction? (b) What is the magnitude of the force on the barge from the two tow ropes?

122. A large wrecking ball of mass m is resting against a wall. It hangs from the end of a cable that is attached at its upper end to a crane that is just touching the wall.

The cable makes an angle of θ with the wall. Ignoring friction between the ball and the wall, find the tension in the cable.

123. The figure shows the quadriceps and the patellar tendons attached to the patella (the kneecap). If the tension T in each tendon is 1.30 kN, what is (a) the magnitude and (b) the direction of the contact force $\vec{\mathbf{F}}$ exerted on the patella by the femur?

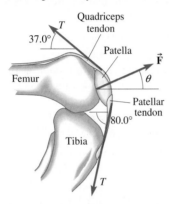

124. The coefficient of static friction between a block and a horizontal floor is 0.35, while the coefficient of kinetic friction is 0.22. The mass of the block is 4.6 kg and it is initially at rest. (a) What is the minimum horizontal applied force required to make the block start to slide? (b) Once the block is sliding, if you keep pushing on it with the same minimum starting force as in part (a), does the block move with constant velocity or does it accelerate? (c) If it moves with constant velocity, what is its velocity? If it accelerates, what is its acceleration?

125. Two blocks lie side by side on a frictionless table. The block on the left is of mass m; the one on the right is of mass $2m$. The block on the right is pushed to the left with a force of magnitude F, pushing the other block in turn. What force does the block on the left exert on the block to its right?

126. A locomotive pulls a train of 10 identical cars, on a track that runs east-west, with a force of 2.0×10^6 N directed east. What is the force with which the *last* car to the west pulls on the rest of the train?

127. The coefficient of static friction between a brick and a wooden board is 0.40 and the coefficient of kinetic friction between the brick and board is 0.30. You place the brick on the board and slowly lift one end of the board off the ground until the brick starts to slide down the board. (a) What angle does the board make with the ground when the brick starts to slide? (b) What is the acceleration of the brick as it slides down the board?

128. A woman of mass 51 kg is standing in an elevator. (a) If the elevator floor pushes up on her feet with a force of 408 N, what is the acceleration of the elevator? (b) If the elevator is moving at 1.5 m/s as it passes the fourth floor on its way down, what is its speed 4.0 s later?

129. In Fig. 4.15 an astronaut is playing shuffleboard on Earth. The puck has a mass of 2.0 kg. Between the board and puck the coefficient of static friction is 0.35 and of kinetic friction is 0.25. (a) If she pushes the puck with a force of 5.0 N in the forward direction, does the puck move? (b) As she is pushing, she trips and the force in the forward direction suddenly becomes 7.5 N. Does the puck move? (c) If so, what is the acceleration of the puck along the board if she maintains contact between puck and stick as she regains her footing while pushing steadily with a force of 6.0 N on the puck? (d) She carries her game to the Moon and again pushes a moving puck with a force of 6.0 N forward. Is the acceleration of the puck during contact more, the same, or less than on Earth? Explain. (tutorial: rough table)

130. You want to hang a 15-N picture as in part (a) using some very fine twine that will break with more than 12 N of tension. Can you do this? What if you have it as illustrated in part (b) of the figure?

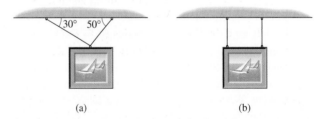

(a)   (b)

131. A roller coaster is towed up an incline at a steady speed of 0.50 m/s by a chain parallel to the surface of the incline. The slope is 3.0%, which means that the elevation increases by 3.0 m for every 100.0 m of horizontal distance. The mass of the roller coaster is 400.0 kg. Ignoring friction, what is the magnitude of the force exerted on the roller coaster by the chain?

132. A 320-kg satellite is in orbit around the Earth 16 000 km above the Earth's surface. (a) What is the weight of the satellite when in orbit? (b) What was its weight when it was on the Earth's surface, before being launched? (c) While it orbits the Earth, what force does the satellite exert on the Earth?

133. The mass of the Moon is 0.0123 times that of the Earth. A spaceship is traveling along a line connecting the centers of the Earth and the Moon. At what distance from the Earth does the spaceship find the gravitational pull of the Earth equal in magnitude to that of the Moon? Express your answer as a percentage of the distance between the centers of the two bodies.

134. A model rocket is fired vertically from rest. It has a net acceleration of 17.5 m/s^2. After 1.5 s, its fuel is exhausted and its only acceleration is that due to gravity. (a) Ignoring air resistance, how high does the rocket travel? (b) How long after liftoff does the rocket return to the ground?

135. The model rocket in Problem 134 has a mass of 87 g and you may assume the mass of the fuel is much less than 87 g. (a) What was the net force on the rocket during the first 1.5 s after liftoff? (b) What force was exerted on the rocket by the burning fuel? (c) What was the net force on the rocket after its fuel was spent? (d) The rocket's vertical velocity was zero instantaneously when it was at the top of its trajectory. What were the net force and acceleration on the rocket at this instant?

136. A toy freight train consists of an engine and three identical cars. The train is moving to the right at constant speed along a straight, level track. Three spring scales are used to connect the cars as follows: spring scale A is located between the engine and the first car; scale B is between the first and second cars; scale C is between the second and third cars. (a) If air resistance and friction are negligible, what are the relative readings on the three spring scales A, B, and C? (b) Repeat part (a), taking air resistance and friction into consideration this time. [Hint: Draw an FBD for the car in the middle.] (c) If air resistance and friction together cause a force of magnitude 5.5 N on each car, directed toward the left, find the readings of scales A, B, and C.

137. Four *identical* spring scales, A, B, C, and D are used to hang a 220.0-N sack of potatoes. (a) Assume the scales have negligible weights and all four scales show the same reading. What is the reading of each scale? (b) Suppose that each scale has a weight of 5.0 N. If scales B and D show the same reading, what is the reading of each scale?

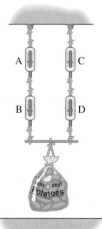

138. A computer weighing 87 N rests on the horizontal surface of your desk. The coefficient of friction between the computer and the desk is 0.60. (a) Draw an FBD for the computer. (b) What is the magnitude of the frictional force acting on the computer? (c) How hard would you have to push on it to get it to start to slide across the desk?

139. A refrigerator magnet weighing 0.14 N is used to hold up a photograph weighing 0.030 N. The magnet attracts the refrigerator door with a magnetic force of 2.10 N. (a) Identify the interactions between the magnet and other objects. (b) Draw an FBD for the magnet, showing all the forces that act on it. (c) Which of these forces are long-range and which are contact forces? (d) Find the magnitudes of all the forces acting on the magnet.

140. A 50.0-kg crate is suspended between the floor and the ceiling using two spring scales, one attached to the ceiling and one to the floor. If the lower scale reads 120 N, what is the reading of the upper scale? Ignore the weight of the scales.

141. Spring scale A is attached to the ceiling. A 10.0-kg mass is suspended from the scale. A second spring scale, B, is hanging from a hook at the bottom of the 10.0-kg mass and a 4.0-kg mass hangs from the second spring scale. (a) What are the readings of the two scales if the masses of the scales are negligible? (b) What are the readings if each scale has a mass of 1.0 kg?

142. A crate of oranges weighing 180 N rests on a flatbed truck 2.0 m from the back of the truck. The coefficients of friction between the crate and the bed are $\mu_s = 0.30$ and $\mu_k = 0.20$. The truck drives on a straight, level highway at a constant 8.0 m/s. (a) What is the force of friction acting on the crate? (b) If the truck speeds up with an acceleration of 1.0 m/s^2, what is the force of the friction on the crate? (c) What is the maximum acceleration the truck can have without the crate starting to slide?

143. A crate of books is to be put on a truck by rolling it up an incline of angle θ using a dolly. The total mass of the crate and the dolly is m. Assume that rolling the dolly up the incline is the same as sliding it up a frictionless surface. (a) What is the magnitude of the *horizontal* force that must be applied just to hold the crate in place on the incline? (b) What horizontal force must be applied to roll the crate up at constant speed? (c) In order to start the dolly moving, it must be accelerated from rest. What horizontal force must be applied to give the crate an acceleration up the incline of magnitude a? (tutorial: cart on ramp)

144. A toy cart of mass m_1 moves on frictionless wheels as it is pulled by a string under tension T. A block of mass m_2 rests on top of the cart. The coefficient of static friction between the cart and the block is μ. Find the maximum tension T that will not cause the block to slide on the

cart if the cart rolls on (a) a horizontal surface; (b) up a ramp of angle θ above the horizontal. In both cases, the string is parallel to the surface on which the cart rolls.

145. A helicopter of mass M is lowering a truck of mass m onto the deck of a ship. (a) At first, the helicopter and the truck move downward together (the length of the cable doesn't change). If their downward speed is decreasing at a rate of $0.10g$, what is the tension in the cable? (b) As the truck gets close to the deck, the helicopter stops moving downward. While it hovers, it lets out the cable so that the truck is still moving downward. If the truck's downward speed is decreasing at a rate of $0.10g$, while the helicopter is at rest, what is the tension in the cable?

146. The coefficient of static friction between block A and a horizontal floor is 0.45 and the coefficient of static friction between block B and the floor is 0.30. The mass of each block is 2.0 kg and they are connected together by a cord. (a) If a horizontal force $\vec{F}$ pulling on block B is slowly increased, in a direction parallel to the connecting cord, until it is barely enough to make the two blocks start moving, what is the magnitude of $\vec{F}$ at the instant that they start to slide? (b) What is the tension in the cord connecting blocks A and B at that same instant?

147. Tamar wants to cut down a large, dead poplar tree with her chain saw, but she does not want it to fall onto the nearby gazebo. Yoojin comes to help with a long rope. Yoojin, a physicist, suggests they tie the rope taut from the poplar to the oak tree and then pull *sideways* on the rope as shown in the figure. If the rope is 40.0 m long and Yoojin pulls sideways at the midpoint of the rope with a force of 360.0 N, causing a 2.00-m sideways displacement of the rope at its midpoint, what force will the rope exert on the poplar tree? Compare this with pulling the rope directly away from the poplar with a force of 360.0 N and explain why the values are different. [*Hint:* Until the poplar is cut through enough to start falling, the rope is in equilibrium.]

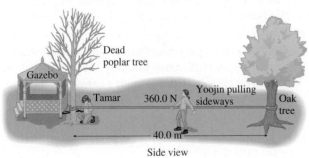
Side view

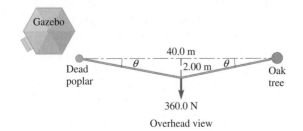
Overhead view

148. A student's head is bent over her physics book. The head weighs 50.0 N and is supported by the muscle force $\vec{F}_m$ exerted by the neck extensor muscles and by the contact force $\vec{F}_c$ exerted at the atlantooccipital joint. Given that the magnitude of $\vec{F}_m$ is 60.0 N and is directed 35° below the horizontal, find (a) the magnitude and (b) the direction of $\vec{F}_c$.

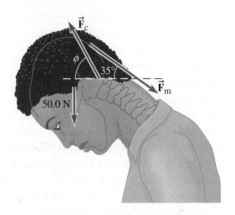

149. (a) If a spacecraft moves in a straight line between the Earth and the Sun, at what point would the force of gravity on the spacecraft due to the Sun be as large as that due to the Earth? (b) If the spacecraft is close to, but not at, this equilibrium point, does the net force on the spacecraft tend to push it toward or away from the equilibrium point? [*Hint:* Imagine the spacecraft a small distance d closer to the Earth and find out which gravitational force is stronger.]

150. While trying to decide where to hang a framed picture, you press it against the wall to keep it from falling. The picture weighs 5.0 N and you press against the frame with a force of 6.0 N at an angle of 40° from the vertical. (a) What is the direction of the normal force exerted on the picture by your hand? (b) What is the direction of the normal force exerted on the picture by the wall? (c) What is the coefficient of static friction between the wall and the picture? The frictional force exerted on the picture by the wall can have two possible directions. Explain why.

151. In a movie, a stuntman places himself on the front of a truck as the truck accelerates. The coefficient of friction between the stuntman and the truck is 0.65. The stuntman is not standing on anything but can "stick" to

the front of the truck as long as the truck continues to accelerate. What minimum forward acceleration will keep the stuntman on the front of the truck?

+152. An airplane of mass 2800 kg has just lifted off the runway. It is gaining altitude at a constant 2.3 m/s while the horizontal component of its velocity is increasing at a rate of 0.86 m/s². Assume $g = 9.81$ m/s². (a) Find the direction of the force exerted on the airplane by the air. (b) Find the horizontal and vertical components of the plane's acceleration if the force due to the air has the same magnitude but has a direction 2.0° closer to the vertical than its direction in part (a).

+153. In the figure with Problem 87, the block of mass m_1 slides to the right with coefficient of kinetic friction μ_k on a horizontal surface. The block is connected to a hanging block of mass m_2 by a light cord that passes over a light, frictionless pulley. (a) Find the acceleration of each of the blocks and the tension in the cord. (b) Check your answers in the special cases $m_1 \ll m_2$, $m_1 \gg m_2$, and $m_1 = m_2$. (c) For what value of m_2 (if any) do the two blocks slide at constant velocity? What is the tension in the cord in that case?

Answers to Practice Problems

4.1 (a) $F_x = 49.1$ N, $F_y = 2.9$ N; (b) $F = 49.2$ N; (c) 3.4° above the horizontal

4.2 0.5 kN downward

4.3 In the first case, the principle of inertia says that Negar tends to stay at rest with respect to the ground as the subway car begins to move forward, until forces acting on her (exerted by the strap and the floor) make her move forward. In the second case, Negar keeps moving forward with respect to the ground with constant speed as the subway car slows down, until forces acting on her make her slow down as well.

4.4 760 N, 81.7° above the –x-axis or 8.3° to the left of the +y-axis

4.5 The contact force exerted on the floor by the chest; 870 N, 59° below the rightward horizontal (+x-axis)

4.6 For $m_1 = m_2 = 1000$ kg and $r = 4$ m, $F \approx 4$ μN, which is about the same magnitude as the weight of a mosquito. The claim that this tiny force caused the collision is ridiculous.

4.7 0.57 N or 0.13 lb

4.8 The chest is in equilibrium, so the net force on it is zero. Setting the net force equal to zero separately for the

horizontal and vertical components gives the answer: the normal force is 750 N, up, and the frictional force is 110 N, to the left. The quantity $\mu_s N$ is the *maximum* possible magnitude of the force of static friction for a surface. In this problem, the frictional force does not necessarily have the maximum possible magnitude.

4.9 (a)

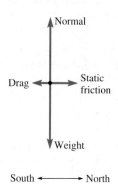

(b) Weight of the car = 11.0 kN; (c) 2.1 kN northward

4.10 (a) 110 N; (b) 230 N

4.11 3100 N

4.12

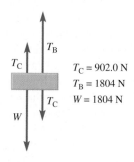

$T_C = 902.0$ N
$T_B = 1804$ N
$W = 1804$ N

4.13 (a) 54 N; (b) 1.8 s

4.14 1.84 kN

4.15 Block 1: $\Sigma F_{1y} = T - m_1 g = 315$ N $- 255$ N $= 60$ N; $m_1 a_{1y} = 60$ N. Block 2: $\Sigma F_{2y} = m_2 g - T = 412$ N $- 315$ N $= 97$ N; $m_2 a_{2y} = 97$ N.

4.16 Impossible to pull the crate up with a single pulley. The entire weight of the crate would be supported by a single strand of cable and that weight exceeds the breaking strength of the cable.

4.17 2500 N

4.18 (a) down the incline; (b) up the incline; (c) 0.2 m/s² down the incline

4.19 (a) 392 N; (b) 431 N

Answers to Checkpoints

4.4 The two forces exerted by the two children on a toy cannot be interaction partners because they act on the *same* object (the toy), not on two different objects. Interaction partners act on different objects, one on each of the two objects that are interacting. The interaction partner of the force exerted by one child on the toy is the force that the toy exerts on that child.

4.5 The weight of the gear decreases as the value of *g* decreases. The mass of the gear does not change.

4.6 One upward normal force on each leg due to the floor and one downward normal force on the desktop due to the laptop.

4.8 Yes. For motion along an incline, it simplifies the problem to choose one axis parallel to the incline and the other perpendicular to the incline.

4.10 Your velocity is downward and decreasing in magnitude, so your acceleration is upward. Then the upward normal force exerted on you by the scale must be greater than your weight. The scale reading is greater than your weight.

Circular Motion

German athlete Susanne Keil throws the hammer during the German Athletics championships. Keil qualified for the 2004 Olympics in Athens with a 67.77-m throw.

In the track and field event called the *hammer throw*, the "hammer" is actually a metal ball (mass 4.00 kg for women or 7.26 kg for men) attached by a cable to a grip. The athlete whirls the hammer several times around while not leaving a circle of radius 2.1 m and then releases it. The winner is the athlete whose hammer lands the greatest distance away. How large a force does an athlete have to exert on the grip to whirl the massive hammer around in a circle? What kind of path does the hammer follow once it is released? (See pp. 155–156 for the answer.)

- gravitational forces (Section 4.5)
- Newton's second law: force and acceleration (Sections 4.3 and 4.8)
- velocity and acceleration (Sections 2.2 and 2.3)
- apparent weight (Section 4.10)
- normal and frictional forces (Section 4.6)

5.1   DESCRIPTION OF UNIFORM CIRCULAR MOTION

Ask someone to name the most important machine ever invented by humans and you are likely to get the *wheel* as a response. Rotating objects are so important to modern— and even not-so-modern—technology that we barely notice them. Examples include wheels on cars, bicycles, trains, and lawnmowers; propellers on airplanes and helicopters; CDs and DVDs; computer hard drives; the gears and hands of an analog clock; amusement park rides and centrifuges—the list is endless.

Rotation of a Rigid Body   To describe circular motion, we could use the familiar definitions of displacement, velocity, and acceleration. But much of the circular motion around us occurs in the rotation of a rigid object. A **rigid body** is one for which the distance between any two points of the body remains the same when the body is translated or rotated. When such an object rotates, every point on the object moves in a circular path. The radius of the path for any point is the distance between that point and the axis of rotation. When a compact disk spins inside a CD player, different points on the CD have different velocities and accelerations. The velocity and acceleration of a given point keep changing direction as the CD spins. It would be clumsy to describe the rotation of the CD by talking about the motion of arbitrary points on it. However, some quantities are the *same* for every point on the CD. It is much simpler, for instance, to say "the CD spins at 210 rpm" instead of saying "a point 6.0 cm from the rotation axis of the CD is moving at 1.3 m/s."

In a **rigid body,** the distance between any two points is constant.

The abbreviation rpm means *revolutions per minute.*

Angular Displacement and Angular Velocity   To simplify the description of circular motion, we concentrate on *angles* instead of distances. If a CD spins through $\frac{1}{4}$ of a turn, every point moves through the same angle (90°), but points at different radii move different linear distances. On the CD shown in Fig. 5.1, point 1 near the axis of rotation moves through a smaller distance than point 4 on the circumference. For this reason we define a set of variables that are analogous to displacement, velocity, and acceleration, but use angular measure instead of linear distance. Instead of displacement, we speak of **angular displacement** $\Delta\theta$, the angle through which the CD turns. A point on the CD moves along the circumference of a circle. As the point moves from the angular position θ_i to the angular position θ_f, a radial line drawn between the center of the circle and that point sweeps out an angle $\Delta\theta = \theta_f - \theta_i$, which is the angular displacement of the CD during that time interval (Fig. 5.2).

Definition of angular displacement:

$$\Delta\theta = \theta_f - \theta_i \tag{5-1}$$

The sign of the angular displacement indicates the sense of the rotation. The usual convention is that a positive angular displacement represents counterclockwise rotation and a negative angular displacement represents clockwise rotation. Counterclockwise and clockwise are only well defined for a particular viewing direction; counterclockwise rotation viewed from above is clockwise when viewed from below.

CONNECTION:

Equations (5-1) through (5-3) have a familiar form because ω is the *rate of change* of θ, just as velocity is the rate of change of position.

+ Counterclockwise
− Clockwise

Remember that the notation $\lim_{\Delta t \to 0}$ indicates that $\Delta \theta$ is the angular displacement during a *very short* time interval Δt (short enough that the ratio $\Delta \theta / \Delta t$ doesn't change significantly if we make the time interval even shorter).

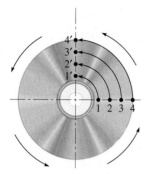

Figure 5.1 A CD rotates through $\frac{1}{4}$ turn; points 1, 2, 3, and 4 travel through the same angle but different distances to reach their new positions, marked 1′, 2′, 3′, and 4′, respectively.

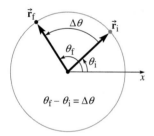

Figure 5.2 Angular positions such as θ_i and θ_f are measured counterclockwise from a reference axis (usually the x-axis).

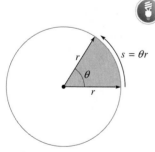

Figure 5.3 Definition of the radian: angle θ in radians is the arc length s divided by the radius r. The angle shown is 1 rad $\approx 57.3°$.

The **average angular velocity** ω_{av} is the average rate of change of the angular displacement.

Definition of average angular velocity:

$$\omega_{av} = \frac{\Delta \theta}{\Delta t} \tag{5-2}$$

If we let the time interval Δt become shorter and shorter, we are averaging over smaller and smaller time intervals. In the limit $\Delta t \to 0$, ω_{av} becomes the **instantaneous angular velocity** ω.

Definition of instantaneous angular velocity:

$$\omega = \lim_{\Delta t \to 0} \frac{\Delta \theta}{\Delta t} \tag{5-3}$$

The angular velocity also indicates—through its algebraic sign—in what direction the CD is spinning. Since angular displacements can be measured in degrees or radians, angular velocities have units such as degrees/second, radians/second, degrees/day, and the like.

Radian Measure   You may be most familiar with measuring angles in degrees, but in many situations the most convenient measure is the **radian**. One such situation is when we relate the angular displacement or angular velocity of a rotating object with the distance traveled by, or the speed of, some point on the object.

In Fig. 5.3, an angle θ between two radii of a circle define an arc of length s. We say that θ is the angle *subtended* by the arc. The arc length is proportional to both the radius of the circle and to the angle subtended. The angle θ in radians is *defined* as

$$\theta \text{ (in radians)} = \frac{s}{r} \tag{5-4}$$

where r is the radius of the circle. Since an angle in radians is defined by the ratio of two lengths, it is dimensionless (a pure number). We use the term radians, abbreviated "rad," to keep track of the angular measure used. Since "rad" is not a physical unit like meters or kilograms, it does not have to balance in Eq. (5-4). For the same reason, we can drop "rad" whenever there is no chance of being misunderstood. We can write $\omega = 23 \text{ s}^{-1}$ as long as context makes it clear that we mean 23 radians per second.

In equations that relate linear variables to angular variables [such as Eq. (5-4)], think of r as the number of meters of arc length per radian of angle subtended. In other words, think of r as having units of meters per radian. Doing so, the radians cancel out in these equations. For example, if $\theta = 2.0 \text{ rad}$ and $r = 1.2 \text{ m}$, then the arc length is

$$s = \theta r = 2.0 \text{ rad} \times 1.2 \frac{\text{m}}{\text{rad}} = 2.4 \text{ m}$$

Since the arc length for an angle of 360° is the circumference of the circle, the radian measure of an angle of 360° is

$$\theta = \frac{s}{r} = \frac{2\pi r}{r} = 2\pi \text{ rad}$$

Therefore, the conversion factor between degrees and radians is

$$360° = 2\pi \text{ rad} \tag{5-5}$$

Example 5.1

Angular Speed of Earth

Earth is rotating about its axis. What is its angular speed in rad/s? (The question asks for angular *speed*, so we do not have to worry about the direction of rotation.)

Strategy The Earth's angular velocity is constant, or nearly so. Therefore, we can calculate the average angular velocity for any convenient time interval and, in turn, the Earth's instantaneous angular speed $|\omega|$.

Solution It takes the Earth 1 day to complete one rotation, during which the angular displacement is 2π rad. More formally, during a time interval $\Delta t = 1$ day, the angular displacement of the Earth is $\Delta\theta = 2\pi$ rad. So the angular speed of the Earth is 2π rad/day, and then convert days to seconds.

$$1 \text{ day} = 24 \text{ h} = 24 \text{ h} \times 3600 \text{ s/h} = 86\,400 \text{ s}$$

$$|\omega| = \frac{2\pi \text{ rad}}{86\,400 \text{ s}} = 7.3 \times 10^{-5} \text{ rad/s}$$

Discussion Notice that this problem is analogous to a problem in linear motion such as: "A car travels in a straight line at constant speed. In 3 h, it has traveled 192 mi. What is its velocity in m/s?" Just about everything in circular motion and rotation has this kind of analog—which means we can draw heavily on what we have already learned.

Earth actually completes one rotation in 23.9345 h (see inside back cover) rather than in 24 h due to Earth's motion around the Sun. This distinction would be important only if we needed a more precise value of $|\omega|$ (more than two significant figures).

Practice Problem 5.1   Angular Speed of Venus

Venus completes one rotation about its axis every 5816 h. What is the angular speed of the rotation of Venus in rad/s?

Relation Between Linear and Angular Speed

For a point moving in a circular path of radius r, the linear distance traveled along the circular path during an angular displacement of $\Delta\theta$ (in radians) is the arc length s where

$$s = r|\Delta\theta| = r|\theta_f - \theta_i| \quad \text{(angles in radians)} \tag{5-6}$$

The point in question could be a point particle moving in a circular path, or it could be any point on a rotating rigid object. Since Eq. (5-6) comes directly from the definition of the radian, any equation derived from Eq. (5-6) is valid only when the angles are measured in radians.

What is the linear speed at which the point moves? The average linear speed is the distance traveled divided by the time interval,

$$v_{av} = \frac{s}{\Delta t} = \frac{r|\Delta\theta|}{\Delta t} \quad (\Delta\theta \text{ in radians})$$

We recognize $\Delta\theta/\Delta t$ as the average angular velocity ω_{av}. If we take the limit as Δt approaches zero, both average quantities (v_{av} and ω_{av}) become instantaneous quantities. Therefore, the relationship between linear speed and angular speed is

$$v = r|\omega| \quad (\omega \text{ in radians per unit time}) \tag{5-7}$$

Equation (5-7) relates only the *magnitudes* of the linear and angular speeds. The direction of the velocity vector $\vec{v}$ is tangent to the circular path. For a rotating object, points farther from the axis move at higher linear speeds; they have a circle of bigger radius to travel and, therefore, cover more distance in the same time interval. For example, a person standing at the equator has a much higher linear speed due to Earth's rotation than does a person standing at the Arctic Circle (see Fig. 5.4).

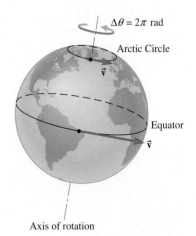

Figure 5.4 A person standing at the Equator is moving much faster than another person standing at the Arctic Circle, but their *angular* speeds are the same.

⚠ In uniform circular motion, speed is constant but velocity is *not* constant because the *direction* of the velocity is changing.

Period and Frequency

When the speed of a point moving in a circle is constant, its motion is called **uniform circular motion**. Even though the speed of the point is constant, the velocity is not: the direction of the velocity is changing. This distinction is important when we find the acceleration of an object in uniform circular motion (Section 5.2). The time for the point to travel completely around the circle is called the **period** of the motion, T. The **frequency** of the motion, which is the number of revolutions per unit time, is defined as

$$f = \frac{1}{T} \tag{5-8}$$

since

$$\frac{\text{revolutions}}{\text{second}} = \frac{1}{\text{second/revolution}}$$

✓ CHECKPOINT 5.1

If it takes $\frac{1}{7200}$ of a second for a computer hard drive to spin around once, what is its frequency?

The speed is the total distance traveled divided by the time taken,

$$v = \frac{2\pi r}{T} = 2\pi r f$$

Then, for uniform circular motion

$$|\omega| = \frac{v}{r} = 2\pi f \quad (\omega \text{ in radians per unit time}) \tag{5-9}$$

SI unit of frequency: 1 Hz = 1 rev/s

where, in SI units, angular velocity ω is measured in rad/s and frequency f is measured in hertz (Hz). The hertz is a derived unit equal to 1 rev/s. The dimensions of Eq. (5-9) are correct since both revolutions and radians are pure numbers. The physical dimensions on both sides are a number per second (s^{-1}).

Example 5.2

Speed in a Centrifuge

A centrifuge is spinning at 5400 rpm. (a) Find the period (in s) and frequency (in Hz) of the motion. (b) If the radius of the centrifuge is 14 cm, how fast (in m/s) is an object at the outer edge moving?

Strategy Remember that rpm means *revolutions per minute*. 5400 rpm *is* the frequency, but in a unit other than Hz. After a unit conversion, the other quantities can be found using the relations already discussed.

Solution (a) First convert rpm to Hz:

$$f = 5400 \, \frac{\text{rev}}{\text{min}} \times \frac{1 \text{ min}}{60 \text{ s}} = 90 \text{ rev/s}$$

so the frequency is $f = 90$ Hz $= 90 \, s^{-1}$. The period is

$$T = 1/f = 0.011 \text{ s}$$

(b) To find the linear speed, we first find the angular speed in rad/s:

$$|\omega| = 90 \, \frac{\text{rev}}{\text{s}} \times 2\pi \, \frac{\text{rad}}{\text{rev}} = 180\pi \text{ rad/s}$$

continued on next page

Example 5.2 continued

So $|\omega| = 2\pi f = 180\,\pi$ rad/s. The linear speed is

$$v = |\omega|r = 180\pi\,\text{s}^{-1} \times 0.14\text{ m} = 79\text{ m/s}$$

Discussion Notice that much of this problem was done with unit conversions. Instead of memorizing a formula such as $|\omega| = 2\pi f$, an understanding of where the formula came from (in this case, that 2π radians correspond to one revolution) is more useful and less prone to error.

Practice Problem 5.2   Clothing in the Drier

An automatic clothing drier spins at 51.6 rpm. If the radius of the drier drum is 30.5 cm, how fast is the outer edge of the drum moving?

Rolling Without Slipping: Rotation and Translation Combined

When an object is rolling, it is both rotating and translating. The wheel rotates about an axle, but the axle is not at rest; it moves forward or backward. What is the relationship between the angular speed of the wheel and the linear speed of the axle? You might guess that $v = |\omega|r$ is the answer. You would be right, as long as the object rolls without slipping or skidding.

There is no fixed relationship between the linear and angular speeds of a wheel if it is allowed to skid or slip. When an impatient driver guns the engine the instant a traffic light turns green, the automobile wheels are likely to slip. The rubber sliding against the road surface makes the squealing sound and leaves tracks on the road. The driver could actually make the acceleration of the car greater by giving the engine *less* gas. When the wheels are skidding or slipping, *kinetic* friction propels the car forward instead of the potentially larger force of *static* friction.

For a wheel that rolls *without* slipping, as the wheel turns through one complete rotation, the axle moves a distance equal to the circumference of the wheel (Fig. 5.5). Think of a paint roller leaving a line of paint as it rolls along a wall. After one complete

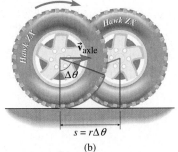

13.0 m/s

$2\pi r$

65.0 cm

Tire position after one revolution

(a)

$s = r\Delta\theta$

(b)

Figure 5.5 (a) As a wheel of radius r that rolls without slipping turns through one complete revolution, the distance its axle moves is equal to the circumference of the wheel ($2\pi r$). (b) As a wheel rolls without slipping through an angle $\Delta\theta$, the distance the axle moves is equal to the arc length s.

rotation, the same point on the roller wheel is touching the wall as was initially touching it. The length of the line of paint is $2\pi r$. The elapsed time is T, so the axle's speed is

$$v_{\text{axle}} = \frac{2\pi r}{T}$$

while the angular speed of the roller is

$$|\omega| = \frac{2\pi}{T}$$

Thus,

$$v_{\text{axle}} = |\omega| r \quad (\omega \text{ in radians per unit time}) \tag{5-10}$$

Example 5.3

Angular Speed of a Rolling Wheel

Kevin is riding his motorcycle at a speed of 13.0 m/s. If the diameter of the rear tire is 65.0 cm, what is the angular speed of the rear wheel? Assume that it rolls without slipping.

Strategy   The given diameter of the tire enables us to find the circumference and, thus, the distance traveled in one revolution of the wheel. From the speed of the motorcycle we can find how many revolutions the tire must make per second.

Solution   During one revolution of the wheel, the motorcycle travels a distance equal to the tire's circumference $2\pi r$ (Fig. 5.5). Then the time to make one revolution is T and the speed v is

$$v = \frac{\text{distance}}{\text{time}} = \frac{2\pi r}{T}$$

Therefore, $T = 2\pi r/v$. For each revolution there is an angular displacement of $\Delta\theta = 2\pi$ radians, so

$$|\omega| = \frac{|\Delta\theta|}{\Delta t} = \frac{2\pi}{T}$$

Substituting $T = 2\pi r/v$ and remembering that the radius is half the diameter,

$$|\omega| = \frac{2\pi}{2\pi r/v} = \frac{v}{r} = \frac{13.0 \text{ m/s}}{(0.650 \text{ m})/2} = 40.0 \frac{\text{rad}}{\text{s}}$$

Discussion   Check: Time for one revolution is

$$\frac{2\pi \text{ rad}}{40.0 \text{ rad/s}} = 0.157 \text{ s.}$$

Time to travel a distance $2\pi r = 2.04$ m is

$$\frac{2.04 \text{ m}}{13.0 \text{ m/s}} = 0.157 \text{ s.}$$

Looks good.

You could have obtained this answer immediately by looking back through the text for the equation $|\omega| = v/r$ and plugging in numbers, but the solution here shows that you can re-create that equation. Here, and in many cases, there is no need to memorize a formula if you understand the concepts behind the formula. You are then less apt to make a mistake by forgetting a factor or constant in the equation, or by using an inappropriate formula. For another example, if an object moves along a straight line at a constant velocity, you know instantly that the displacement is the velocity times the time interval—not because you have memorized an equation ($\Delta\vec{r} = \vec{v}\,\Delta t$), but because you understand the concepts of displacement and velocity. This is the sort of internalization of scientific thinking that you will develop with more and more practice in problem solving.

Practice Problem 5.3   Rolling Drum

A cylindrical steel drum is tipped over and rolled along the floor of a warehouse. If the drum has a radius of 0.40 m and makes one complete turn every 8.0 s, how long does it take to roll the drum 36 m?

5.2   RADIAL ACCELERATION

For a particle undergoing uniform circular motion, the *magnitude* of the velocity vector is constant, but its direction is continuously changing. At any instant of time, the direction of the instantaneous velocity is tangent to the path, as discussed in Section 3.2. Since the *direction* of the velocity continually changes, the particle has a nonzero acceleration.

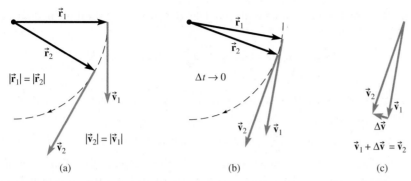

Figure 5.6 Uniform circular motion at constant speed. (a) The velocity vector is always tangent to the circular path and perpendicular to the radius at that point. (b) As the time interval between two velocity measurements decreases, the angle between the velocity vectors decreases. (c) The change in velocity ($\Delta \vec{\mathbf{v}}$) is found by placing the tails of the two velocity vectors together. Then $\Delta \vec{\mathbf{v}}$ is drawn from the tip of the initial velocity ($\vec{\mathbf{v}}_1$) to the tip of the final velocity ($\vec{\mathbf{v}}_2$) so that $\vec{\mathbf{v}}_1 + \Delta \vec{\mathbf{v}} = \vec{\mathbf{v}}_2$.

In Fig. 5.6a, two velocity vectors of equal magnitude are drawn tangent to a circular path of radius r, representing the velocity at two different times of an object moving around a circular path with constant speed. At any instant, the velocity vector is perpendicular to a radius drawn from the center of the circle to the position of the object. As the time between velocity measurements approaches zero, the radii become closer together (Fig. 5.6b). To find the acceleration, $\vec{\mathbf{a}} = \lim\limits_{\Delta t \to 0} \dfrac{\Delta \vec{\mathbf{v}}}{\Delta t}$, we must first find the change in velocity $\Delta \vec{\mathbf{v}}$ for a very short time interval. Figure 5.6c shows that as the time interval Δt approaches zero, the angle between the two velocities also approaches zero and $\Delta \vec{\mathbf{v}}$ becomes perpendicular to the velocity.

Since $\Delta \vec{\mathbf{v}}$ is perpendicular to the velocity, it is directed along a radius of the circle. Inspection of Figs. 5.6b and 5.6c shows that $\Delta \vec{\mathbf{v}}$ is radially *inward* (toward the center of the circle). Since the acceleration $\vec{\mathbf{a}}$ has the same direction as $\Delta \vec{\mathbf{v}}$ (in the limit $\Delta t \to 0$), the acceleration is also directed radially inward (Fig. 5.7)—that is, along a radius of the circular path toward the center of the circle. The acceleration of an object undergoing *uniform* circular motion is often called the **radial acceleration $\vec{\mathbf{a}}_r$**. The word *radial* here just reminds us of the direction of the acceleration. (A synonym for radial acceleration is *centripetal acceleration*. *Centripetal* means "toward the center.")

CONNECTION:

Radial acceleration is not a new kind of acceleration. The acceleration vector for an object moving in uniform circular motion is directed radially inward toward the center of the circle.

In uniform circular motion, the direction of the acceleration is radially inward (toward the center of the circular path).

√ **CHECKPOINT 5.2**

Does a radial acceleration mean that the speed of the object is changing?

Magnitude of the Radial Acceleration

To find the magnitude of the radial acceleration for uniform circular motion, we must find the change in velocity $\Delta \vec{\mathbf{v}}$ for a time interval Δt in the limit $\Delta t \to 0$. The velocity keeps the same magnitude but changes direction at a steady rate, equal to the angular velocity ω. In a time interval Δt, the velocity $\vec{\mathbf{v}}$ rotates through an angle equal to the angular displacement $\Delta \theta = \omega \, \Delta t$. During this time interval, the velocity vector sweeps out an arc of a circle of "radius" v (Fig. 5.8). In the limit $\Delta t \to 0$, the

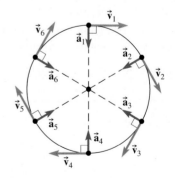

Figure 5.7 In uniform circular motion, the acceleration is always directed toward the center of the circle, perpendicular to the velocity (see [interactive] interactive: circular motion).

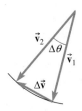

Figure 5.8 The velocity vector sweeps out an arc of a circle whose "length" is nearly equal to that of the chord $\Delta \vec{v}$.

magnitude of $\Delta \vec{v}$ becomes equal to the arc length, since a very short arc approaches a straight line. Then

$$|\Delta \vec{v}| = \text{arc length} = \text{radius of circle} \times \text{angle subtended}$$
$$= v\,|\Delta \theta| = v\,|\omega|\,\Delta t$$

Acceleration is the rate of change of velocity, so the magnitude of the radial acceleration is

$$a_r = |\vec{a}| = \frac{|\Delta \vec{v}|}{\Delta t} = v|\omega| \quad (\omega \text{ in radians per unit time}) \tag{5-11}$$

where absolute value symbols are used with the vector quantities to indicate their magnitudes. Velocity and angular velocity are not independent; $v = |\omega|r$. It is usually most convenient to write the magnitude of the radial acceleration in terms of one or the other of these two quantities. So we write the radial acceleration in two other equivalent ways using $v = |\omega|r$:

$$a_r = \frac{v^2}{r} \quad \text{or} \quad a_r = \omega^2 r \quad (\omega \text{ in radians per unit time}) \tag{5-12}$$

Note that Eqs. (5-11) and (5-12) assume that ω is expressed in *radians* per unit time (normally rad/s, but rad/min or rad/h would be correct).

Example 5.4

A Spinning CD

If a CD spins at 210 rpm, what is the radial acceleration of a point on the outer rim of the CD? The CD is 12 cm in diameter.

Strategy   From the number of revolutions per minute, we can find the frequency and the angular velocity. The angular velocity and the radius of the CD enable us to calculate the radial acceleration.

Solution   We convert 210 rpm into a frequency in revolutions per second (Hz).

$$f = 210\,\frac{\text{rev}}{\text{min}} \times \frac{1}{60}\,\frac{\text{min}}{\text{s}} = 3.5\,\frac{\text{rev}}{\text{s}} = 3.5\,\text{Hz}$$

For each revolution, the CD rotates through an angle of 2π radians. The angular velocity is

$$|\omega| = 2\pi f = 2\pi\,\frac{\text{radians}}{\text{rev}} \times 3.5\,\frac{\text{rev}}{\text{s}} = 7.0\pi\,\text{rad/s}$$

Then using Eq. (5-12), the radial acceleration is

$$a_r = \omega^2 r = (7.0\pi\,\text{rad/s})^2 \times 0.060\,\text{m} = 29\,\text{m/s}^2$$

Discussion   When finding the radial acceleration, use whichever form of Eq. (5-12) is more convenient. For rotating objects such as the spinning CD, it's usually easiest to think in terms of the angular velocity. For an object moving around a circle, such as a satellite in orbit whose speed is known, it might be easier to use v^2/r. Since the two equations are equivalent, either can be used in any situation.

Practice Problem 5.4   **Radial Acceleration of a Point on an Old Record**

What is the radial acceleration of a point 25.4 cm from the center of a record that is rotating at 78 rpm on a turntable?

Applying Newton's Second Law to Uniform Circular Motion

Now that we know the magnitude and direction of the acceleration of any object in uniform circular motion, we can use Newton's second law to relate the net force acting on the object to the speed and radius of its motion. The net force is found in the usual way: each of the individual forces acting on the object is identified and then the forces are added as vectors. Every force acting must be exerted *by some other object*. Resist the temptation to add in a new, separate force just because something moves in a circle. For an object to move in a circle at constant speed, real, physical forces such as gravity, tension, normal forces, and friction must act on it; these forces combine to produce a net force that has the correct magnitude and is always perpendicular to the velocity of the object.

Problem-Solving Strategy for an Object in Uniform Circular Motion

1. Begin as for any Newton's second law problem: identify all the forces acting on the object and draw an FBD.
2. Choose perpendicular axes at the point of interest so that one is radial and the other is tangent to the circular path.
3. Find the radial component of each force.
4. Apply Newton's second law as follows:

$$\sum F_r = ma_r$$

where $\sum F_r$ is the radial component of the net force and the radial component of the acceleration is

$$a_r = \frac{v^2}{r} = \omega^2 r$$

(For uniform circular motion, neither the net force nor the acceleration has a tangential component.)

Example 5.5

The Hammer Throw

 What force does the athlete exert on the grip? What path does the hammer follow after release?

An athlete whirls a 4.00-kg hammer six or seven times around and then releases it. Although the purpose of whirling it around several times is to increase the hammer's speed, assume that *just before* the hammer is released, it moves at constant speed along a circular arc of radius 1.7 m. At the instant she releases the hammer, it is 1.0 m above the ground and its velocity is directed 40° above the horizontal. The hammer lands a horizontal distance of 74.0 m away. What force does the athlete apply to the grip just before she releases it? Ignore air resistance.

Strategy After release, the only force acting on the hammer is gravity. The hammer moves in a parabolic trajectory like any other projectile. By analyzing the projectile motion of the hammer, we can find the speed of the hammer just

after its release. Just *before* release, the forces acting on the hammer are the tension in the cable and gravity. We can relate the net force on the hammer to its radial acceleration, calculated from the speed and radius of its path. The problem becomes two subproblems, one dealing with circular motion and the other with projectile motion. The final velocity for the circular motion is the initial velocity for the projectile motion.

Solution During its projectile motion, the initial velocity has magnitude v_i (to be determined) and direction $\theta = 40°$ above the horizontal. Choosing the +y-axis pointing up, the displacement of the hammer (in component form) is $\Delta x = 74.0$ m and $\Delta y = -1.0$ m (Fig. 5.9), the acceleration of the hammer is $a_x = 0$ and $a_y = -g$, and the initial velocity is $v_{ix} = v_i \cos \theta$ and $v_{iy} = v_i \sin \theta$. Then, from Eqs. (4-8) and (4-9),

$$\Delta x = (v_i \cos \theta) \Delta t \quad \text{and} \quad \Delta y = (v_i \sin \theta) \Delta t - \tfrac{1}{2} g(\Delta t)^2$$

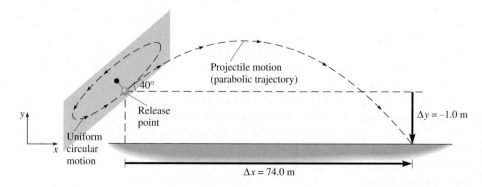

Figure 5.9

Path of the hammer from just before its release until it hits the ground. (Distances are *not* to scale.)

continued on next page

Example 5.5 continued

Solving the left equation for Δt and substituting into the right equation gives

$$\Delta y = v_i \sin \theta \, \frac{\Delta x}{v_i \cos \theta} - \frac{1}{2} g \left(\frac{\Delta x}{v_i \cos \theta} \right)^2$$

After a bit of algebra, we can solve for v_i. First we multiply through by $2v_i^2 \cos^2 \theta$:

$$2v_i^2 \cos^2 \theta \, \Delta y = 2v_i^2 \cos^2 \theta \, \frac{\Delta x \sin \theta}{\cos \theta}$$

$$- \frac{2 \, v_i^2 \cos^2 \theta}{2} g \left(\frac{\Delta x}{v_i \cos \theta} \right)^2$$

Subtracting the first term on the right side from both sides and factoring out v_i^2,

$$v_i^2 (2 \, \Delta y \cos^2 \theta - 2 \, \Delta x \cos \theta \sin \theta) = -g(\Delta x)^2$$

Now we solve for v_i:

$$v_i = \sqrt{\frac{g(\Delta x)^2}{2\Delta x \cos\theta \, \sin\theta - 2\Delta y \cos^2 \theta}}$$

$$= \sqrt{\frac{9.80 \text{ m/s}^2 \times (74.0 \text{ m})^2}{2(74.0 \text{ m}) \cos 40° \sin 40° - 2(-1.0 \text{ m}) \cos^2 40°}}$$

$$= 26.9 \text{ m/s}$$

The net force on the hammer can be found from Newton's second law. The two forces acting on the hammer are due to the tension in the cable and to gravity (Fig. 5.10). We ignore the gravitational force, assuming that the hammer's weight is small compared with the tension in the cable. Then the tension in the cable is the only significant force acting on the hammer. Assuming uniform circular motion, the cable pulls radially inward and causes a radial

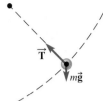

Figure 5.10

FBD for the hammer just before its release. (Not to scale.)

acceleration of magnitude v^2/r. Newton's second law in the radial direction is

$$\sum F_r = T = ma_r = \frac{mv^2}{r}$$

Substituting numerical values,

$$T = \frac{4.00 \text{ kg} \times (26.9 \text{ m/s})^2}{1.7 \text{ m}} = 1700 \text{ N}$$

The tension is much larger than the weight of the hammer (≈ 40 N), so the assumption that we could ignore the weight is justified. The athlete must apply a force of magnitude 1700 N—almost 400 lb—to the grip.

Discussion   This example demonstrates the cumulative nature of physics concepts. The basic concepts keep reappearing, to be used over and over and to be extended for use in new contexts. Part of the problem involves new concepts (radial acceleration); the rest of the problem involves old material (Newton's second law, projectile motion, and tension in a cord).

Practice Problem 5.5   Rotating Carousel

A horse located 8.0 m from the central axis of a rotating carousel moves at a speed of 6.0 m/s. The horse is at a fixed height (it does not move up and down). What is the net force acting on a child seated on this horse? The child's weight is 130 N.

Example 5.6

Conical Pendulum

Suppose you whirl a stone in a horizontal circle at a slow speed so that the weight of the stone is *not* negligible compared with the tension in the cord. Then the cord cannot be horizontal—the tension must have a vertical component to cancel the weight and leave a horizontal net force (Fig. 5.11). If the cord has length L, the stone has mass m, and the cord makes an angle ϕ with the vertical direction, what is the constant angular speed of the stone?

Strategy   The net force must point toward the center of the circle, since the stone is in uniform circular motion.

With the stone in the position depicted in Fig. 5.11a, the direction of the net force is along the $+x$-axis. This time the tension in the cord does not pull toward the center, but the *net* force does.

Solution   Start by drawing an FBD (Fig. 5.11b). Now apply Newton's second law in component form. The acceleration has components $a_x = \omega^2 r$ and $a_y = 0$. For the x-components,

$$\sum F_x = T \sin \phi = ma_x = m\omega^2 r$$

continued on next page

Example 5.6 continued

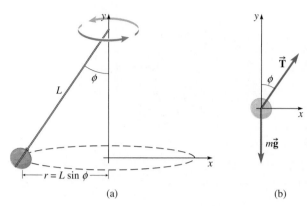

Figure 5.11

(a) A stone is whirled in a horizontal circle of radius $r = L \sin \phi$.
(b) An FBD for the stone.

Since the problem does not specify r, we must express r in terms of L and ϕ. In Fig. 5.11a, the radius forms a right triangle with the cord and the y-axis. Then

$$r = L \sin \phi$$

and

$$\sum F_x = T \sin \phi = m\omega^2 L \sin \phi$$

Therefore, $T = m\omega^2 L$. For the y-components,

$$\sum F_y = T \cos \phi - mg = ma_y = 0 \quad \Rightarrow \quad T \cos \phi = mg$$

Now we eliminate the tension:

$$(m\omega^2 L) \cos \phi = mg$$

Solving for $|\omega|$,

$$|\omega| = \sqrt{\frac{g}{L \cos \phi}}$$

Discussion We should check the dimensions of the final expression. Since $\cos \phi$ is dimensionless,

$$\sqrt{\frac{[L/T^2]}{[L]}} = \frac{1}{[T]}$$

which is correct for ω (SI unit rad/s).

Another check is to ask how ω and ϕ are related for a given length cord. As ϕ increases toward 90°, the cord gets closer to horizontal and the radius increases. In our expression, as ϕ increases, $\cos \phi$ decreases and, therefore, ω increases, in accordance with experience: the stone would have to be whirled faster and faster to make the cord more nearly horizontal.

Conceptual Practice Problem 5.6   Conical Pendulum on the Moon

Examine the result of Example 5.6 to see how ω depends on g, all other things being equal. Where the gravitational field is weaker, do you have to whirl the stone faster or more slowly to keep the cord at the same angle ϕ? Is that in accord with your intuition?

5.3   UNBANKED AND BANKED CURVES

Unbanked Curves   When you drive an automobile in a circular path along an unbanked roadway, friction acting on the tires due to the pavement acts to keep the automobile moving in a curved path. This frictional force acts *sideways*, toward the center of the car's circular path (Fig. 5.12). The frictional force might also have a tangential component; for example, if the car is braking, a component of the frictional force makes the car slow down by acting backward (opposite to the car's velocity). For now we assume that the car's speed is constant and that the forward or backward component of the frictional force is negligibly small.

As long as the tires roll without slipping, there is no relative motion between the bottom of the tires and the road, so it is the force of *static* friction that acts (see Section 4.6). If the car is in a skid, then it is the smaller force of kinetic friction that acts as the bottom portion of the tire slides along the pavement. As the speed of the car increases, or for slippery surfaces with low coefficients of friction, the static frictional force may not be enough to hold the car in its curved path.

Banked Curves   To help prevent cars from going into a skid or losing control, the roadway is often banked (tilted at a slight angle) around curves so that the outer portion of the road—the part farthest from the center of curvature—is higher than the inner portion. Banking changes the angle and magnitude of the normal force, $\vec{N}$, so that it has a horizontal component N_x directed toward the center of curvature (in the

Application of radial acceleration and contact forces: banked roadways

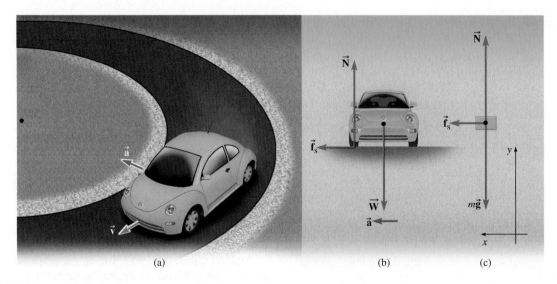

Figure 5.12 (a) A car negotiating a curve at constant speed on an *unbanked* roadway. The car's acceleration is toward the center of the circular path. (b) A head-on view of the same car. The center of the circular path is to the left as viewed here. The force vectors $\vec{\mathbf{N}}$ and $\vec{\mathbf{f}}_s$ are shown acting on one tire, but they represent the *total* normal and frictional forces acting on all four tires. (c) FBD for the car.

radial direction—see Fig. 5.13). Then we need no longer rely solely on friction to keep the car moving in a circular path as it negotiates the curve; this component of the normal force acts to help the car remain on the curved path. Figure 5.13 shows a banked road with the normal force, the gravitational force, and, in parts (b) and (c), the radial component of the normal force N_x. We choose the axes so that the x-axis is in the direction of the acceleration, which is to the left; the axes are *not* parallel and perpendicular to the incline.

Figure 5.13 (a) Head-on view of a car negotiating a curve at constant speed on a *banked* roadway. The car's acceleration is toward the center of the circular path (to the left as viewed here). $\vec{\mathbf{N}}$ represents the *total* normal force acting on all four tires. The car moves at just the right speed so that the frictional force is zero. (b) Resolving the normal force into x- and y-components. (c) FBD for the car with the normal force represented by its components.

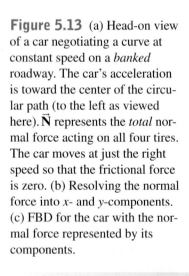

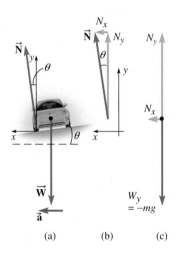

Example 5.7

A Possible Skid: Unbanked and Banked Curves

A car is going around an unbanked curve at the recommended speed of 11 m/s (see Fig. 5.12). (a) If the radius of curvature of the path is 25 m and the coefficient of static friction between the rubber and the road is $\mu_s = 0.70$, does the car skid as it goes around the curve? (b) What happens if the driver ignores the highway speed limit sign and travels at 18 m/s? (c) What speed is safe for traveling around the curve if the road surface is wet from a recent rainstorm and the

continued on next page

Example 5.7 continued

coefficient of static friction between the wet road and the rubber tires is $\mu_s = 0.50$? (d) For a car to safely negotiate the curve in icy conditions at a speed of 13 m/s, what banking angle would be required (see Fig. 5.13)?

Strategy The force of static friction is the only horizontal force acting on the car when the curve is not banked. The maximum force of static friction, which depends on road conditions, determines the maximum possible radial acceleration of the car. Therefore, we can compare the radial acceleration necessary to go around the curve at the specified speeds with the maximum possible radial acceleration determined by the coefficient of static friction. For part (d), in icy conditions we cannot rely much on friction, but the normal force has a horizontal component when the road is banked.

Solution (a) We find the radial acceleration required for a speed of 11 m/s:

$$a_r = \frac{v^2}{r} = \frac{(11 \text{ m/s})^2}{25 \text{ m}} = 4.8 \text{ m/s}^2$$

In order to have that acceleration, the component of the net force acting toward the center of curvature must be

$$\sum F_r = ma_r = m \frac{v^2}{r}$$

The only force with a horizontal component is the static frictional force acting on the tires due to the road (see the FBD in Fig. 5.12c). Therefore,

$$\sum F_r = f_s = m \frac{v^2}{r}$$

We must check to make sure that the maximum frictional force is not exceeded:

$$f_s \leq \mu_s N$$

Since $N = mg$, the car can go around the curve without skidding as long as

$$\cancel{m} \frac{v^2}{r} \leq \mu_s \cancel{m} g$$

Thus, the radial acceleration cannot exceed $\mu_s g$. That limits the car to speeds satisfying

$$v \leq \sqrt{\mu_s g r}$$

Substituting numerical values,

$$v \leq \sqrt{0.70 \times 9.80 \text{ m/s}^2 \times 25 \text{ m}} = 13 \text{ m/s}$$

Since 11 m/s is less than the maximum safe speed of 13 m/s, the car safely negotiates the curve.

(b) At 18 m/s, the car moves at a speed higher than the maximum safe speed of 13 m/s. The frictional force cannot supply the radial acceleration needed for the car to go around the curve—the car goes into a skid.

(c) In part (a), we found that the car is limited to speeds satisfying

$$v \leq \sqrt{\mu_s g r}$$

With $\mu_s = 0.50$, the maximum safe speed is

$$v_{\text{max}} = \sqrt{\mu_s g r} = \sqrt{0.50 \times 9.80 \text{ m/s}^2 \times 25 \text{ m}} = 11 \text{ m/s}$$

which is the same maximum speed recommended by the road sign. The highway engineer knew what she was doing when she had the sign placed along the road.

(d) Finally, we find the banking angle that would enable cars to travel around the curve at 13 m/s in icy conditions. Assuming that friction is negligible, the horizontal component of the normal force is the only horizontal force. With the x-axis pointing toward the center of curvature and the y-axis vertical (Fig. 5.13),

$$\sum F_x = N \sin \theta = mv^2/r \qquad (1)$$

and

$$\sum F_y = N \cos \theta - mg = 0 \qquad (2)$$

Dividing Eq. (1) by Eq. (2) gives

$$\frac{N \sin \theta}{N \cos \theta} = \tan \theta = \frac{mv^2/r}{mg} = \frac{v^2}{rg}$$

$$\theta = \tan^{-1} \frac{v^2}{rg} = \tan^{-1} \frac{(13 \text{ m/s})^2}{25 \text{ m} \times 9.80 \text{ m/s}^2} = 35° \qquad (3)$$

Discussion Notice that the mass of the car does not appear in Eq. (3); the same banking angle holds for a scooter, motorcycle, car, or tractor-trailer. Notice also that the banking angle depends on the square of the speed. Automobile racetracks and bicycle racetracks have highly banked road surfaces at hairpin curves to minimize skidding of the high-speed vehicles. However, a banking angle of 35° is far greater than those used in practice along public roadways. Careful drivers would not try to drive around this curve in icy conditions at 13 m/s. What do you think might happen in icy conditions to a car that is traveling *very slowly* along a road banked at such a steep angle?

Highway curves are banked at slight angles to help drivers who are driving at reasonable speeds for the road conditions. They are not banked to save speed demons from their folly.

Practice Problem 5.7 A Bobsled Race

A bobsled races down an icy hill and then comes on a horizontal curve, located 60.0 m from the bottom of the hill. The sled is traveling at 22.4 m/s (50 mph) as it approaches the curve that has a radius of curvature of 50.0 m. The curve is banked at an angle of 45° and the frictional force on the sled runners is negligible. Does the sled make it safely around the curve?

Application of radial acceleration: banking angle of an airplane

Figure 5.14 The lift force $\vec{L}$ is perpendicular to the wings of the plane. To turn, the pilot tilts the wings so a component of the lift force is directed toward the center of the circular path of the plane.

Application of radial acceleration: circular orbits

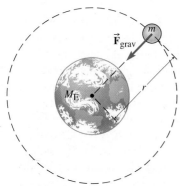

Figure 5.15 Satellite in orbit around Earth.

If there is *no* friction between the road and the tires, then there is only one speed at which it is safe to drive around a given curve. *With* friction, there is a *range* of safe speeds. The static frictional force can have any magnitude from 0 to $\mu_s N$ and it can be directed either up or down the bank of the road.

When an airplane pilot makes a turn in the air, the pilot makes use of a banking angle. The airplane itself is tilted as if it were traveling over an inclined surface. Because of the shape of the wings, an aerodynamic force called *lift* acts upward when the plane is in level flight. To go around a turn, the wings are tilted; the lift force stays perpendicular to the wings and, therefore, now has a horizontal component (Fig. 5.14), just as the normal force has a horizontal component for a car on a banked curve. This component supplies the necessary radial acceleration, while the vertical component of the lift holds the plane up. Therefore,

$$L_x = ma_r = \frac{mv^2}{r} \quad \text{and} \quad L_y = mg$$

where the *x*-axis is horizontal and the *y*-axis is vertical. The lift force is different in its physical origin from the normal force, but its components split up the same way, so a plane in a turn banks its wings at the same angle that a road would be banked for the same speed and radius of curvature. Of course, planes usually move much faster than cars and use large radii of curvature when they turn.

✓ CHECKPOINT 5.3

A plane can't make a turn without tilting its wings. Why can a car turn on a flat road?

5.4   CIRCULAR ORBITS OF SATELLITES AND PLANETS

A satellite can orbit Earth in a circular path because of the long-range gravitational force on the satellite due to the Earth. The magnitude of the gravitational force on the satellite is

$$F = \frac{Gm_1 m_2}{r^2} \tag{2-6}$$

where the universal gravitational constant is $G = 6.67 \times 10^{-11}$ N·m²/kg². We can use Newton's second law to find the speed of a satellite in circular orbit at constant speed. Let m be the mass of the satellite and M_E be the mass of the Earth. The direction of the gravitational force on the satellite is always toward the center of the Earth, which is the center of the orbit (Fig. 5.15). Since gravity is the only force acting on the satellite,

$$\sum F_r = G\frac{mM_E}{r^2}$$

where r is the distance from the *center* of the Earth to the satellite. Then, from Newton's second law,

$$\sum F_r = ma_r = \frac{mv^2}{r}$$

Setting these equal,

$$G\frac{mM_E}{r^2} = \frac{mv^2}{r}$$

Solving for the speed yields

$$v = \sqrt{\frac{GM_E}{r}} \tag{5-13}$$

Notice that the mass of the satellite does not appear in the equation for speed; it has been algebraically canceled. The greater inertia of a more massive satellite is overcome by a proportionally greater gravitational force acting on it. Thus, the speed of a satellite in a circular orbit does not depend on the mass of the satellite. Equation (5-13) also shows that satellites in lower orbits (smaller radii) have greater speeds.

We have been discussing satellites orbiting Earth, but the same principles apply to the circular orbits of satellites around other planets and to the orbits of the planets around the Sun. For planetary orbits, the mass of the Sun would appear in Eq. (5-13) instead of the Earth's mass, because the *Sun's* gravitational pull keeps the planets in their orbits. The planetary orbits are actually ellipses (Fig. 5.16) instead of circles, although for most of the planets in the solar system the ellipses are nearly circular. Mercury is the exception; its orbit is markedly different from a circle.

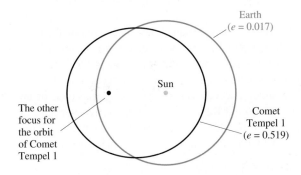

Figure 5.16 The shapes of two elliptical orbits around the Sun. (The *sizes* of the orbits are not to scale.) An ellipse looks like an elongated circle. The degree of elongation is measured by a quantity called the eccentricity *e*. A circle is a special case of an ellipse with *e* = 0. Most of the planetary orbits are nearly circular, with the exception of Mercury. The sum of the distances from any point on an ellipse to each of two fixed points (called the *foci*) is constant. The Sun is at one focus of each orbit. Since Earth's orbit is nearly circular, the second focus is very near the Sun.

Example 5.8

Speed of a Satellite

The Hubble Space Telescope is in a circular orbit 613 km above Earth's surface. The average radius of the Earth is 6.37×10^3 km and the mass of Earth is 5.97×10^{24} kg. What is the speed of the telescope in its orbit?

Strategy We first need to find the orbital radius of the telescope. It is not 613 km; that is the distance from the *surface* of Earth to the telescope. We must add the radius of the Earth to 613 km to find the orbital radius, which is measured from the center of the Earth to the telescope. Then we use Newton's second law, along with what we know about radial acceleration.

Solution The radius of the telescope's orbit is

$$r = 6.13 \times 10^2 \text{ km} + 6.37 \times 10^3 \text{ km} = (0.613 + 6.37) \times 10^3 \text{ km}$$

$$= 6.98 \times 10^3 \text{ km}$$

The net force on the telescope is equal to the gravitational force, given by Newton's law of gravity. Newton's second law relates the net force to the acceleration. Both are directed radially inward.

$$\sum F_r = \frac{GmM_E}{r^2} = \frac{mv^2}{r}$$

where *m* is the mass of the telescope. Solving for the speed, we find

$$v = \sqrt{\frac{GM_E}{r}}$$

$$v = \sqrt{\frac{6.67 \times 10^{-11} \text{ N} \cdot \text{m}^2/\text{kg}^2 \times 5.97 \times 10^{24} \text{ kg}}{6.98 \times 10^6 \text{ m}}}$$

$$v = 7550 \text{ m/s} = 27\,200 \text{ km/h}$$

Discussion *Any* satellite orbiting Earth at an altitude of 613 km has this same speed, regardless of its mass.

Practice Problem 5.8 Speed of Earth in Its Orbit

What is the speed of Earth in its approximately circular orbit about the Sun? The average Earth–Sun distance is 1.50×10^{11} m and the mass of the Sun is 1.987×10^{30} kg. Once you find the speed, use it along with the distance traveled by the Earth during one revolution about the Sun to calculate the time in seconds for one orbit.

Kepler's Laws of Planetary Motion

At the beginning of the seventeenth century, Johannes Kepler (1571–1630) proposed three laws to describe the motion of the planets. These laws predated Newton's laws of motion and his law of gravity. They offered a far simpler description of planetary motion than anything that had been proposed previously. We turn history on its head and look at one of Kepler's laws as a consequence of Newton's laws. The fact that Newton could derive Kepler's laws from his own work on gravity was seen as a confirmation of Newtonian mechanics.

Kepler's laws of planetary motion are

- The planets travel in elliptical orbits (Fig. 5.16) with the Sun at one focus of the ellipse.
- A line drawn from a planet to the Sun sweeps out equal areas in equal time intervals.
- The square of the orbital period is proportional to the cube of the average distance from the planet to the Sun.

Kepler's first law can be derived from the inverse square law of gravitational attraction. The derivation is a bit complicated, but for any two objects that have such an attraction, the orbit of one about the other is an ellipse, with the stationary object located at one focus. (Planetary orbits are also affected by gravitational interactions with other planets; Kepler's laws ignore these small effects.) The circle is a special case of an ellipse where the two foci coincide. We discuss Kepler's second law in Chapter 8.

We can derive Kepler's third law from Newton's law of universal gravitation for the special case of a circular orbit. The gravitational force gives rise to the radial acceleration:

Application of radial acceleration: Kepler's third law for a circular orbit

$$\sum F_r = \frac{GmM_{Sun}}{r^2} = \frac{mv^2}{r}$$

Solving for v yields

$$v = \sqrt{\frac{GM_{Sun}}{r}}$$

The distance traveled during one revolution is the circumference of the circle, which is equal to $2\pi r$. The speed is the distance traveled during one orbit divided by the period:

$$v = \sqrt{\frac{GM_{Sun}}{r}} = \frac{2\pi r}{T}$$

Now we solve for T:

$$T = 2\pi \sqrt{\frac{r^3}{GM_{Sun}}}$$

Squaring both sides yields

$$T^2 = \frac{4\pi^2}{GM_{Sun}} r^3 = \text{constant} \times r^3 \tag{5-14}$$

Equation (5-14) is Kepler's third law: the square of the period of a planet is directly proportional to the cube of the average orbital radius.

Application of radial acceleration: geostationary orbits

Although Kepler's laws were derived for the motion of planets, they apply to satellites orbiting the Earth as well. Many satellites, such as those used for communications, are placed in a *geostationary* (or *geosynchronous*) orbit—a circular orbit in Earth's equatorial plane whose period is equal to Earth's rotational period (Fig. 5.17). A satellite in geostationary orbit remains directly above a particular point on the equator; to observers on the ground, it seems to hover above that point without moving. Due to their fixed positions with respect to Earth's surface, geostationary satellites are used as relay stations for communication signals. In Example 5.9, we find the speed of a geostationary satellite.

√ **CHECKPOINT 5.4**

Do all geostationary satellites, no matter their masses, have to be the same height above Earth? Explain.

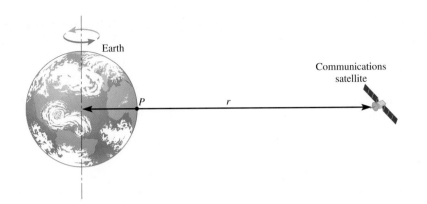

Figure 5.17 Geostationary satellite orbiting the Earth. The satellite has the same angular velocity as Earth, so it is always directly above point *P*.

Example 5.9

Geostationary Satellite

A 300.0-kg communications satellite is placed in a geostationary orbit 35,800 km above a relay station located in Kenya. What is the speed of the satellite in orbit?

Strategy The period of the satellite is 1 d or approximately 24 h. To find the speed of the satellite in orbit we use Newton's law of gravity and his second law of motion along with what we know about radial acceleration.

Solution Let *m* be the mass of the satellite and let M_E be the mass of the Earth. Gravity is the only force acting on the satellite in its orbit. From Newton's law of universal gravitation, Newton's second law, and the expression for radial acceleration,

$$\sum F_r = \frac{GmM_E}{r^2} = \frac{mv^2}{r}$$

Solving for the speed yields

$$v = \sqrt{\frac{GM_E}{r}}$$

We must add the mean radius of the Earth, $R_E = 6.37 \times 10^6$ m, to the height of the satellite above the Earth's surface to find the orbital radius.

$$r = h + R_E = 3.58 \times 10^7 \text{ m} + 0.637 \times 10^7 \text{ m}$$

$$= 4.217 \times 10^7 \text{ m}$$

Substituting numerical values into the speed equation,

$$v = \sqrt{\frac{6.67 \times 10^{-11} \text{ N·m}^2/\text{kg}^2 \times 5.97 \times 10^{24} \text{ kg}}{4.217 \times 10^7 \text{ m}}}$$

$$= \sqrt{9.443 \times 10^6 \text{ m}^2/\text{s}^2}$$

$$v = 3.07 \times 10^3 \text{ m/s}$$

Discussion This result, an orbital speed of 3.07 km/s and a distance above Earth's surface of 35,800 km, applies to *all*

geostationary satellites. The mass of the satellite does not matter; it cancels out of the equations for orbital radius and for speed.

If we were actually putting a satellite into orbit, we would use a more accurate value for the period. We should use a time of 23 h and 56 min, which is the length of a *sidereal day*—the time for Earth to complete one rotation about its axis relative to the fixed stars. The solar day, 24 h, is the period of time between the daily appearances of the Sun at its highest point in the sky. The fact that Earth moves around the Sun is what causes the difference between these two ways of measuring the length of a day. The error introduced by using the longer time is negligible in this problem.

We can use Kepler's third law to check the result. Examples 5.8 and 5.9 both concern circular orbits around the Earth. Is the square of the period proportional to the cube of the orbital radius? From Example 5.8, $r_1 = 6.98 \times 10^3$ km and

$$T_1 = \frac{2\pi r_1}{v} = \frac{2\pi \times 6.98 \times 10^3 \text{ km}}{7.55 \text{ km/s}} = 5810 \text{ s}$$

From the present example, $r_2 = 4.22 \times 10^7$ m and

$$T_2 = 24 \text{ h} \times \frac{3600 \text{ s}}{1 \text{ h}} = 86\,400 \text{ s}$$

The ratio of the squares of the periods is

$$\left(\frac{T_2}{T_1}\right)^2 = \left(\frac{86\,400 \text{ s}}{5810 \text{ s}}\right)^2 = 221$$

The ratio of the cubes of the radii is

$$\left(\frac{r_2}{r_1}\right)^3 = \left(\frac{4.22 \times 10^7 \text{ m}}{6.98 \times 10^6 \text{ m}}\right)^3 = 221$$

Practice Problem 5.9 Orbital Radius of Venus

The period of the orbit of Venus around the Sun is 0.615 Earth years. Using this information, find the radius of its orbit in terms of *R*, the radius of Earth's orbit around the Sun.

Orbiting Satellites

A satellite revolves about Earth with an orbital radius of r_1 and speed v_1. If an identical satellite were set into circular orbit with the same speed about a planet of mass three times that of Earth, what would its orbital radius be?

Strategy   We can apply Newton's law of universal gravitation and set up a ratio to solve for the new orbital radius.

Solution   From Newton's second law, the magnitude of the gravitational force on the satellite is equal to the satellite's mass times the magnitude of its radial acceleration:

$$\frac{Gm M_E}{r_1^2} = \frac{m v_1^2}{r_1}$$

where M_E and m are the masses of Earth and of the satellite, respectively. Solving for r_1 yields

$$r_1 = \frac{GM_E}{v_1^2}$$

Now we apply Newton's second law to the orbit of the second satellite about the planet of mass $3M_E$:

$$\frac{Gm \times 3M_E}{r_2^2} = \frac{m v_1^2}{r_2}$$

$$r_2 = \frac{G \times 3M_E}{v_1^2}$$

The ratio of r_2 to r_1 is

$$\frac{r_2}{r_1} = \frac{G \times 3M_E/v_1^2}{GM_E/v_1^2} = 3$$

Thus, $r_2 = 3r_1$.

Discussion   Notice that we did not rush to substitute numerical values for the constants G and M_E into the equations. We took the ratio r_2/r_1 so that these constants cancel.

Practice Problem 5.10   Period of Lunar Lander

A lunar lander is orbiting about the Moon. If the radius of its orbit is $\frac{1}{3}$ the radius of Earth, what is the period of its orbit?

5.5   NONUNIFORM CIRCULAR MOTION

So far we have focused on *uniform* circular motion. Now we can extend the discussion to nonuniform circular motion, where the angular velocity changes with time.

Figure 5.18a shows the velocity vectors $\vec{v}_1$ and $\vec{v}_2$ at two different times for an object moving in a circle with changing speed. In this case, the speed is increasing ($v_2 > v_1$). In Fig. 5.18b, we subtract $\vec{v}_1$ from $\vec{v}_2$ to find the change in velocity. In the limit $\Delta t \to 0$, $\Delta\vec{v}$ does *not* become perpendicular to the velocity, as it did for uniform circular motion. Thus, the direction of the acceleration is *not* radial if the speed is changing. However, we can resolve the acceleration into tangential and radial components

Figure 5.18 Motion along a circular path with a changing speed: (a) the magnitude of velocity $\vec{v}_2$ is greater than the magnitude of velocity $\vec{v}_1$, (b) the direction of $\Delta\vec{v}$ is not radial when the speed is changing, and (c) components of $\vec{a}$ can be taken along a tangent to the curved path (a_t) and along a radius (a_r).

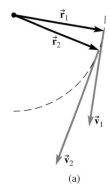 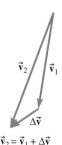 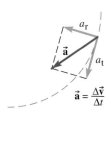

(a)   (b)   (c)

(Fig. 5.18c). The radial component a_r changes the *direction* of the velocity, and the tangential component a_t changes the *magnitude* of the velocity. Since these are perpendicular components of the acceleration, the magnitude of the acceleration is

$$a = \sqrt{a_r^2 + a_t^2}$$

Using the same method as in Section 5.2 to find the radial acceleration, but working here with only the radial *component* of the acceleration, we find that

$$a_r = \frac{v^2}{r} = \omega^2 r \quad (\omega \text{ in radians per unit time}) \tag{5-12}$$

For circular motion, *whether uniform or nonuniform*, the radial component of the acceleration is given by Eq. (5-12). However, in *uniform* circular motion the radial component of the acceleration a_r is constant in magnitude, but for nonuniform circular motion a_r changes as the speed changes.

Also still true for nonuniform circular motion is the relationship between speed and angular speed:

$$v = r\,|\omega| \tag{5-7}$$

Many problems involving nonuniform circular motion are solved in the same way as for uniform circular motion. We find the *radial component of the net force* and then apply Newton's second law along the radial direction:

$$\sum F_r = ma_r$$

Problem-Solving Strategy for an Object in Nonuniform Circular Motion

1. Begin as for any Newton's second law problem: Identify all the forces acting on the object and draw an FBD.
2. Choose perpendicular axes at the point of interest so that one axis is radial and the other is tangent to the circular path.
3. Find the radial component of each force.
4. Apply Newton's second law along the radial direction:

$$\sum F_r = ma_r$$

where

$$a_r = \frac{v^2}{r} = \omega^2 r$$

5. If necessary, apply Newton's second law to the tangential force components:

$$\sum F_t = ma_t$$

The tangential acceleration component a_t determines how the speed of the object changes.

✓ CHECKPOINT 5.5

For an object in circular motion, what is it about the radial acceleration that distinguishes between uniform and nonuniform circular motion?

Example 5.11

Vertical Loop-the-Loop

A roller coaster includes a vertical circular loop of radius 20.0 m (Fig. 5.19a). What is the minimum speed at which the car must move at the top of the loop so that it doesn't lose contact with the track?

Strategy A roller coaster car moving around a vertical loop is in nonuniform circular motion; its speed decreases on the way up and increases on the way back down. Nevertheless, it is moving in a circle and has a radial acceleration component as given in Eq. (5-12) as long as it moves in a circle. The only forces acting on the car are gravity and the normal force of the track pushing the car. Even if frictional or drag forces are present, at the top of the loop they act in the tangential direction and, thus, do not contribute to the radial component of the net force. At the top of the loop, the track exerts a normal force on the car as long as the car

moves with a speed great enough to stay on the track. If the car moves too slowly, it loses contact with the track and the normal force is then zero.

Solution The normal force exerted by the track on the car at the top pushes the car *away* from the track (downward); the normal force cannot pull up on the car. Then, at the top of the loop, the gravitational force and the normal force both point straight down toward the center of the loop. Figure 5.19b is an FBD for the car. From Newton's second law,

$$\sum F_r = N + mg = ma_r = \frac{mv_{top}^2}{r}$$

or

$$N = \frac{mv_{top}^2}{r} - mg$$

continued on next page

Figure 5.19 (a) A roller coaster car on a vertical circular loop. At the bottom of the loop, the car's acceleration $\vec{a}_{bottom}$ points upward toward the center of the circle. At the top of the loop, the car's acceleration $\vec{a}_{top}$ points downward. The magnitude of $\vec{a}_{top}$ is smaller than that of $\vec{a}_{bottom}$ because the speed is smaller at the top than at the bottom. (b) FBD for the car at the top of the loop. The track is above the car, so the normal force on the car due to the track is *downward*. (c) FBD for the car at the bottom of the loop.

Example 5.11 continued

where v_{top} stands for the speed at the top. In this expression, N stands for the magnitude of the normal force. Since $N \geq 0$,

$$m \left(\frac{v_{top}^2}{r} - g \right) \geq 0$$

or

$$v_{top} \geq \sqrt{gr}$$

Imagine sending a roller coaster car around the loop many times with a slightly smaller speed at the top each time. As v_{top} approaches $\sqrt{gr}$, the normal force at the top gets smaller and smaller. When $v_{top} = \sqrt{gr}$, the normal force just becomes zero at the top of the loop. Any slower and the car loses contact with the track *before* getting to the highest point and would fall off the track unless prevented from falling by a backup safety mechanism. Therefore, the minimum speed at the top is

$$v_{top} = \sqrt{gr} = \sqrt{9.80 \text{ m/s}^2 \times 20.0 \text{ m}} = 14.0 \text{ m/s}$$

Discussion   If the car is going faster than 14 m/s at the top, its radial acceleration is larger. The track pushing on the car provides the additional net force component that results in a larger radial acceleration. The minimum speed occurs when gravity alone provides the radial acceleration at the top of the loop. In other words, $a_r = g$ at the top of the loop for minimum speed.

Practice Problem 5.11   Normal Force at the Bottom of the Track

If the speed of the roller coaster at the *bottom* of the loop is 25 m/s, what is the normal force exerted on the car by the track in terms of the car's weight mg? (See Fig. 5.19c.)

PHYSICS AT HOME

Go outside on a warm day and fill a bucket with water. Swing the bucket around in a vertical circle over your head. What, if anything, keeps the water in the bucket when the bucket is upside down over your head? Why doesn't the water spill out? Do any upward forces act on the water at that point? [*Hint:* The FBD for the water when it is directly overhead is similar to the FBD for a roller coaster car at the top of a loop.]

Conceptual Example 5.12

Acceleration of a Pendulum Bob

A pendulum is released from rest at point A (Fig. 5.20). Sketch qualitatively an FBD and the acceleration vector for the pendulum bob at points B and C.

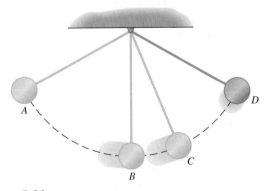

Figure 5.20
A pendulum swings to the right, starting from rest at point A.

Strategy   Two forces appear on each FBD: gravity and the force due to the cord. The gravitational force is the same at both points (magnitude mg, direction down), but the force due to the cord varies in magnitude and in direction. Its direction is always along the cord. The net force on the bob is the sum of these two forces and its direction is the same as the direction of the acceleration. We can use what we know about the acceleration to guide us in drawing the forces.

The pendulum bob moves along the arc of a circle, but not at constant speed. At any point, the radial component of the acceleration is $a_r = v^2/r$. Unless $v = 0$, the radial acceleration component is nonzero. As the pendulum bob swings toward the bottom (from A to B), its speed is increasing; as it rises on the other side, its speed is decreasing. When the speed is increasing, the tangential component of the acceleration a_t is in the same direction as the velocity. From B to D, the speed is decreasing and a_t is in the direction *opposite* to

continued on next page

Conceptual Example 5.12 continued

the velocity. At point B, the speed is neither increasing nor decreasing and $a_t = 0$.

Solution and Discussion   At point B, the tangential acceleration is zero, so the acceleration points in the radial direction: straight up (Fig. 5.21). The tension in the cord pulls straight up and gravity pulls down, so the tension must be larger than the weight of the bob to give an upward net force.

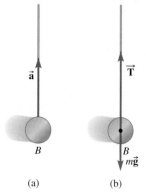

Figure 5.21
(a) Acceleration of the bob at point B. (b) FBD for the bob at B.

The acceleration at point C has both tangential and radial components. The tangential acceleration is opposite to the velocity because the bob is slowing down. Figure 5.22 shows the tangential and radial acceleration components added to form the acceleration vector $\vec{a}$ and the FBD for the bob.

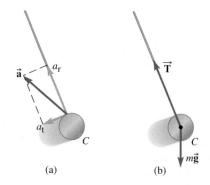

Figure 5.22
(a) At point C, the bob has both tangential and radial acceleration components. (b) FBD for the bob at C.

When the two forces are added, they give a net force in the same direction as the acceleration vector.

Conceptual Practice Problem 5.12   Analysis of the Bob at Point D

Sketch the FBD and the acceleration vector for the pendulum bob at point D, the highest point in its swing to the right.

5.6   TANGENTIAL AND ANGULAR ACCELERATION

An object in nonuniform circular motion has a changing speed and a changing angular velocity. To describe how the angular velocity changes, we define an angular acceleration. If the angular velocity is ω_1 at time t_1 and is ω_2 at time t_2, the change in angular velocity is

$$\Delta\omega = \omega_2 - \omega_1$$

The time interval during which the angular velocity changes is $\Delta t = t_2 - t_1$. The average rate at which the angular velocity changes is called the **average angular acceleration**, α_{av}.

$$\alpha_{av} = \frac{\omega_2 - \omega_1}{t_2 - t_1} = \frac{\Delta\omega}{\Delta t} \tag{5-15}$$

As we let the time interval become shorter and shorter, α_{av} approaches the **instantaneous angular acceleration**, α.

$$\alpha = \lim_{\Delta t \to 0} \frac{\Delta\omega}{\Delta t} \tag{5-16}$$

If ω is in units of rad/s, α is in units of rad/s^2.

The angular acceleration is closely related to the tangential component of the acceleration. The tangential component of velocity is

$$v_t = r\,|\omega| \tag{5-7}$$

Equation (5-7) gives us a way to relate tangential acceleration to the angular acceleration. The tangential acceleration is the rate of change of the tangential velocity, so

$$a_t = \frac{\Delta v_t}{\Delta t} = r\left|\frac{\Delta\omega}{\Delta t}\right| \quad \text{(in the limit } \Delta t \to 0)$$

Therefore,

$$a_t = r\,|\alpha| \tag{5-17}$$

Table 5.1	Relationships Between θ, ω, and α for Constant Angular Acceleration		

Constant Acceleration Along x-Axis		**Constant Angular Acceleration**	
$\Delta v_x = v_{fx} - v_{ix} = a_x \, \Delta t$	(2-9)	$\Delta \omega = \omega_f - \omega_i = \alpha \, \Delta t$	(5-18)
$\Delta x = \frac{1}{2}(v_{fx} + v_{ix}) \, \Delta t$	(2-11)	$\Delta \theta = \frac{1}{2}(\omega_f + \omega_i) \, \Delta t$	(5-19)
$\Delta x = v_{ix} \, \Delta t + \frac{1}{2} a_x (\Delta t)^2$	(2-12)	$\Delta \theta = \omega_i \, \Delta t + \frac{1}{2} \alpha (\Delta t)^2$	(5-20)
$v_{fx}^2 - v_{ix}^2 = 2 a_x \, \Delta x$	(2-13)	$\omega_f^2 - \omega_i^2 = 2 \alpha \, \Delta \theta$	(5-21)

> **CONNECTION:**
>
> Because α is the rate of change of ω, and ω is the rate of change of θ, the equations for constant α have the same form as those for constant a_x.

Constant Angular Acceleration

The mathematical relationships between θ, ω, and α are the same as the mathematical relationships between x, v_x, and a_x that we developed in Chapter 2. Each quantity is the instantaneous rate of change of the preceding quantity. For example, a_x is the rate of change of v_x and α is the rate of change of ω. Because the mathematical relationships are the same, we can draw upon the skills and equations we developed to solve problems with constant acceleration a_x. All we have to do is take the equations for constant acceleration and replace x with θ, v_x with ω, and a_x with α (see Table 5.1).

Equation (5-18) is the definition of average angular acceleration, with α_{av} replaced by α since the angular acceleration is constant. Constant α means that ω changes linearly with time; therefore, the average angular velocity is halfway between the initial and final angular velocities for any time interval $\omega_{av} = \frac{1}{2}(\omega_i + \omega_f)$. Using this form for ω_{av} along with the definition of ω_{av} ($\omega_{av} = \Delta\theta/\Delta t$) yields Eq. (5-19). Equations (5-20) and (5-21) can be derived from the preceding two relations in a manner analogous to the derivations of Eqs. (2-12) and (2-13) in Section 2.4.

✓ CHECKPOINT 5.6

A centrifuge is "spinning up" with a constant angular acceleration. Can the radial acceleration of a sample in the centrifuge be constant? Explain.

Example 5.13

A Rotating Potter's Wheel

A potter's wheel rotates from rest to 210 rpm in a time of 0.75 s. (a) What is the angular acceleration of the wheel during this time, assuming constant angular acceleration? (b) How many revolutions does the wheel make during this time interval? (c) Find the tangential and radial components of the acceleration of a point 12 cm from the rotation axis when the wheel is spinning at 180 rpm.

Strategy We know the initial and final frequencies, so we can find the initial and final angular velocities. We also know the time it takes for the wheel to get to the final angular velocity. That is all we need to find the average angular

acceleration that, for constant angular acceleration, is equal to the instantaneous angular acceleration. To find the number of revolutions, we can find the angular displacement $\Delta\theta$ in radians and then divide by 2π rad/rev. We can find the angular velocity at $t = 0.75$ s and use it to find the radial acceleration component. The tangential acceleration is calculated from α.

continued on next page

Example 5.13 continued

Solution (a) Initially the wheel is at rest, so the initial angular velocity is zero.

$$\omega_i = 0 \text{ rad/s}$$

Converting 210 rpm to rad/s gives the final angular velocity:

$$\omega_f = 210 \frac{\text{rev}}{\text{min}} \times \frac{1}{60} \frac{\text{min}}{\text{s}} \times 2\pi \frac{\text{rad}}{\text{rev}} = 7.0\pi \text{ rad/s}$$

The angular acceleration is the rate of change of the angular velocity. Since α is constant, we can calculate it by finding the *average* angular acceleration for the time interval:

$$\alpha = \frac{\omega_f - \omega_i}{t_f - t_i} = \frac{7.0\pi \text{ rad/s} - 0}{0.75 \text{ s} - 0} = \frac{7.0\pi \text{ rad/s}}{0.75 \text{ s}} = 29 \text{ rad/s}^2$$

(b) The angular displacement is

$$\Delta\theta = \tfrac{1}{2}(\omega_f + \omega_i)\,\Delta t = \tfrac{1}{2}(7.0\pi \text{ rad/s} + 0)(0.75 \text{ s}) = 8.25 \text{ rad}$$

Since 2π rad = one revolution, the number of revolutions is

$$\frac{8.25 \text{ rad}}{2\pi \text{ rad/rev}} = 1.3 \text{ rev}$$

(c) At 180 rpm, the angular velocity is

$$\omega = 180 \frac{\text{rev}}{\text{min}} \times \frac{1}{60} \frac{\text{min}}{\text{s}} \times 2\pi \frac{\text{rad}}{\text{rev}} = 6.0\pi \text{ rad/s}$$

The radial acceleration component is

$$a_r = \omega^2 r = (6.0\pi \text{ rad/s})^2 \times 0.12 \text{ m} = 43 \text{ m/s}^2$$

and the tangential acceleration component is

$$a_t = \alpha r = 29 \text{ rad/s}^2 \times 0.12 \text{ m} = 3.5 \text{ m/s}^2$$

Discussion A quick check involves another of the equations for constant acceleration:

$$\omega_f^2 - \omega_i^2 = 2\alpha\Delta\theta$$

Since $\omega_i = 0$, we can check

$$\omega_f = \sqrt{2\alpha\Delta\theta}$$

From the answers to (a) and (b),

$$\sqrt{2\alpha\Delta\theta} = \sqrt{2 \times 29 \text{ rad/s}^2 \times 8.25 \text{ rad}} = 22 \text{ rad/s}$$

The original value for ω_f in rad/s was 7.0π rad/s. Since $\pi \approx 22/7$, the check is successful.

Practice Problem 5.13   The London Eye

The London Eye, a Ferris wheel on the banks of the Thames, has radius 67.5 m. At its cruising angular speed, it takes 30.0 min to make one complete revolution. Suppose that it takes 20.0 s to bring the wheel from rest to its cruising speed and that the angular acceleration is constant during startup. (a) What is the angular acceleration during startup? (b) What is the angular displacement of the wheel during startup?

The London Eye

5.7   APPARENT WEIGHT AND ARTIFICIAL GRAVITY

Application of apparent weight and circular motion: apparent weightlessness of orbiting astronauts

You are no doubt familiar with pictures of astronauts "floating" while in orbit around the Earth. It seems as if the astronauts are weightless. To be truly weightless, the force of gravity acting on the astronauts due to Earth would have to be zero, or at least close to zero. Is it? We can calculate the weight of an astronaut in orbit. The orbital altitude for the space shuttle is typically about 600 km above the Earth. Then the orbital radius is 600 km + 6400 km = 7000 km. Comparing the astronaut's weight in orbit to his or her weight on Earth's surface,

$$\frac{W_{\text{orbit}}}{W_{\text{surface}}} = \frac{\dfrac{GM_E m}{(R_E + h)^2}}{\dfrac{GM_E m}{R_E^2}} = \frac{R_E^2}{(R_E + h)^2} = \frac{(6400 \text{ km})^2}{(7000 \text{ km})^2} = 0.84$$

The weight in orbit is 0.84 times the weight on the surface. The astronaut weighs less but certainly isn't *weightless*! Then why does the astronaut *seem* to be weightless?

Recall Section 4.10 on the apparent weightlessness of someone unfortunate enough to be in an elevator when the cable snaps. In that situation, the elevator and the passenger both have the same acceleration ($\vec{\mathbf{a}} = \vec{\mathbf{g}}$). Similarly, the astronaut has the same acceleration as the space shuttle, which is equal to the *local* gravitational field $\vec{\mathbf{g}}$. Apparent weightlessness occurs when $\vec{\mathbf{a}} = \vec{\mathbf{g}}$, where $\vec{\mathbf{g}}$ is the *local* gravitational field.

Application: Artificial Gravity In order for astronauts to spend long periods of time living in a space station without the deleterious effects of apparent weightlessness, *artificial gravity* would have to be created on the station. Many science fiction novels and movies feature ring-shaped space stations that rotate in order to create artificial gravity for the occupants. In a rotating space station, the acceleration of an astronaut is inward (toward the rotation axis), but the apparent gravitational field is outward. Therefore, the ceiling of rooms on the station are closest to the rotation axis and the floor is farthest away (Fig. 5.23).

The centrifuge is a device that creates artificial gravity on a smaller scale. Centrifuges are common not only in scientific and medical laboratories but also in everyday life. The first successful centrifuge was used to separate cream from milk in the 1880s. Water drips out of sopping wet clothes due to the pull of gravity when the clothes are hung on a clothesline, but the water is removed much faster by the artificial gravity created in the spin cycle of a washing machine.

The human body can be adversely affected not only by too little artificial gravity, but also by too much. Stunt pilots have to be careful about the accelerations to which they subject their bodies. An acceleration of about $3g$ can cause temporary blindness due to an inadequate supply of oxygen to the retina; the heart has difficulty pumping blood up to the head due to the blood's increased apparent weight. Larger accelerations can cause unconsciousness. Pressurized flight suits enable pilots to sustain accelerations up to about $5g$.

Figure 5.23 A rotating space station from the movie *2001: A Space Odyssey*. Note jogger in the upper half running on the floor.

Example 5.14

Stunt Pilot

Dave wants to practice vertical circles for a flying show exhibition. (a) What must the minimum radius of the circle be to ensure that his acceleration at the bottom does not exceed $3.0g$? The speed of the plane is 78 m/s at the bottom of the circle. (b) What is Dave's apparent weight at the bottom of the circular path? Express your answer in terms of his true weight.

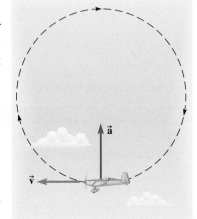

Figure 5.24

Velocity and acceleration vectors for the plane at the bottom of the circle.

Strategy For the *minimum* radius, we use the maximum possible radial acceleration since $a_r = v^2/r$. For the maximum radial acceleration, the *tangential* acceleration must be zero (Fig. 5.24)—the magnitude of the acceleration is $a = \sqrt{a_r^2 + a_t^2}$. Therefore, the radial acceleration component has magnitude $3.0g$ at the bottom. To find Dave's apparent weight, we do not need to use the numerical value of the radius found in part (a); we already know that his acceleration is upward and has magnitude $3.0g$.

Solution (a) The magnitude of the radial acceleration is

$$a_r = v^2/r$$

Solving for the radius,

$$r = \frac{v^2}{a_r} = \frac{v^2}{3.0g}$$

$$= \frac{(78 \text{ m/s})^2}{3.0 \times 9.8 \text{ m/s}^2} = 210 \text{ m}$$

continued on next page

Example 5.14 continued

(b) Dave's apparent weight is the magnitude of the normal force of the plane pushing up on him. Let the y-axis point upward. The normal force is up and the gravitational force is down (Fig. 5.25). Then

$$\sum F_y = N - mg = ma_y$$

where $a_y = +3.0g$. Therefore,

$$W' = N = m(g + a_y) = 4.0mg$$

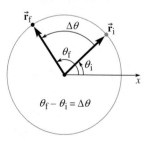

Figure 5.25 FBD for Dave.

His apparent weight is 4.0 times his true weight.

Discussion   It might have been tempting to jump to the conclusion that an acceleration of $3.0g$ means that his apparent weight is $3.0mg$. But is his apparent weight zero when his acceleration is zero? No.

Practice Problem 5.14   Astronaut's Apparent Weight

What is the apparent weight of a 730-N astronaut when her spaceship has an acceleration of magnitude $2.0g$ in the following two situations: (a) just above the surface of Earth, acceleration straight up; (b) far from any stars or planets?

Application of Apparent Weight to Objects at Rest with Respect to Earth's Surface   Due to Earth's rotation, the *effective* value of g measured in a coordinate system attached to Earth's surface is slightly less than the true value of the gravitational field strength (see Section 4.5). The net force of an object placed on a scale is *not* zero because the object has a radial acceleration $a_r = \omega^2 r$ directed toward Earth's axis of rotation (Fig. 5.26). This relatively small effect is greatest where r is greatest—at the equator, where the effective value of g is about 0.3% smaller than the true value of g.

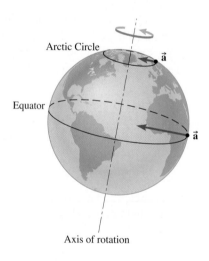

Figure 5.26 An object at rest with respect to Earth's surface has a radial acceleration due to Earth's rotation. The angular frequency ω is the same everywhere, so the radial acceleration $a_r = \omega^2 r$ is proportional to the distance from the axis of rotation.

Master the Concepts

- The angular displacement $\Delta\theta$ is the angle through which an object has turned. Positive and negative angular displacements indicate rotation in different directions. Conventionally, positive represents counterclockwise motion.

- Average angular velocity:

$$\omega_{av} = \frac{\theta_2 - \theta_1}{t_2 - t_1} = \frac{\Delta\theta}{\Delta t} \tag{5-2}$$

- Average angular acceleration:

$$\alpha_{av} = \frac{\omega_2 - \omega_1}{t_2 - t_1} = \frac{\Delta\omega}{\Delta t} \tag{5-15}$$

continued on next page

Master the Concepts continued

- The instantaneous angular velocity and acceleration are the limits of the average quantities as $\Delta t \to 0$.
- A useful measure of angle is the radian:

$$2\pi \text{ radians} = 360°$$

Using radian measure for θ, the arc length s of a circle of radius r subtended by an angle θ is

$$s = \theta r \quad (\theta \text{ in radian measure}) \qquad (5\text{-}4)$$

- Using radian measure for ω, the speed of an object in circular motion (including a point on a rotating object) is

$$v = r|\omega| \quad (\omega \text{ in radians per unit time}) \qquad (5\text{-}7)$$

- Using radian measure for α, the tangential acceleration component is related to the angular acceleration by

$$a_t = r|\alpha| \quad (\alpha \text{ in radians per time}^2) \qquad (5\text{-}17)$$

- An object moving in a circle has a radial acceleration component given by

$$a_r = \frac{v^2}{r} = \omega^2 r \quad (\omega \text{ in radians per unit time}) \qquad (5\text{-}12)$$

- The tangential and radial acceleration components are two perpendicular components of the acceleration vector. The radial acceleration component changes the direction of the velocity and the tangential acceleration component changes the speed.

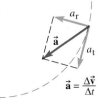

- Uniform circular motion means that v and ω are constant. In uniform circular motion, the time to complete one revolution is constant and is called the period T. The frequency f is the number of revolutions completed per second.

$$f = 1/T \qquad (5\text{-}8)$$

$$|\omega| = v/r = 2\pi f \qquad (5\text{-}9)$$

where the SI unit of angular velocity is rad/s and that of frequency is rev/s = Hz.

- A rolling object is both rotating and translating. If the object rolls without skidding or slipping, then

$$v_{\text{axle}} = r|\omega| \qquad (5\text{-}10)$$

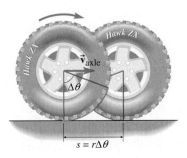

- Kepler's third law says that the square of the period of a planetary orbit is proportional to the cube of the orbital radius:

$$T^2 = \text{constant} \times r^3 \qquad (5\text{-}14)$$

- For constant angular acceleration, we can use equations analogous to those we developed for constant acceleration a_x:

$$\Delta \omega = \omega_f - \omega_i = \alpha \, \Delta t \qquad (5\text{-}18)$$

$$\Delta \theta = \tfrac{1}{2}(\omega_f + \omega_i) \, \Delta t \qquad (5\text{-}19)$$

$$\Delta \theta = \omega_i \, \Delta t + \tfrac{1}{2}\alpha (\Delta t)^2 \qquad (5\text{-}20)$$

$$\omega_f^2 - \omega_i^2 = 2\alpha \, \Delta \theta \qquad (5\text{-}21)$$

Conceptual Questions

1. Is depressing the "accelerator" (gas pedal) of a car the only way that the driver can make the car accelerate (in the physics sense of the word)? If not, what else can the driver do to give the car an acceleration?

2. Two children ride on a merry-go-round. One is 2 m from the axis of rotation and the other is 4 m from it. Which child has the larger (a) linear speed, (b) acceleration, (c) angular speed, and (d) angular displacement?

3. Explain why the orbital radius and the speed of a satellite in circular orbit are not independent.

4. In uniform circular motion, is the velocity constant? Is the acceleration constant? Explain.

5. In uniform circular motion, the net force is perpendicular to the velocity and changes the direction of the velocity but not the speed. If a projectile is launched horizontally, the net force (ignoring air resistance) is perpendicular to the initial velocity, and yet the projectile gains speed as it falls. What is the difference between the two situations?

6. The speed of a satellite in circular orbit around a planet does not depend on the mass of the satellite. Does it depend on the mass of the planet? Explain.

7. A flywheel (a massive disk) rotates with constant angular acceleration. For a point on the rim of the flywheel, is the tangential acceleration component constant? Is the radial acceleration component constant?

8. Explain why the force of gravity due to the Earth does not pull the Moon in closer and closer on an inward spiral until it hits Earth's surface.

9. When a roller coaster takes a sharp turn to the right, it feels as if you are pushed toward the left. Does a force push you to the left? If so, what is it? If not, why does there *seem* to be such a force?

10. Is there anywhere on Earth where a bathroom scale reads your true weight? If so, where? Where does your apparent weight due to Earth's rotation differ most from your true weight?

11. A physics teacher draws a cut-away view of a car rounding a banked curve as a rectangle atop a right triangle. A student draws a coordinate system based on the drawing. Is there another choice of axes that would make the problem easier to solve?

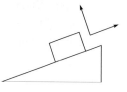

12. A bridal party is at a rehearsal dinner. The best man challenges the bride-groom to pick up an olive using only a brandy snif-ter. How does the groom accomplish this task?

Brandy snifter

Olive

Multiple-Choice Questions

1. A spider sits on a turntable that is rotating at a constant 33 rpm. The acceleration $\vec{a}$ of the spider is
   (a) greater the closer the spider is to the central axis.
   (b) greater the farther the spider is from the central axis.
   (c) nonzero and independent of the location of the spider on the turntable.
   (d) zero.

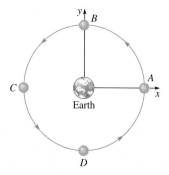

Multiple-Choice Questions 2–5 and Problem 36

Questions 2–5: A satellite in orbit travels around the Earth in uniform circular motion. In the figure, the satellite moves counterclockwise (*ABCDA*). Answer choices:

(a) $+x$ (b) $+y$ (c) $-x$ (d) $-y$
(e) 45° above $+x$ (toward $+y$)
(f) 45° below $+x$ (toward $-y$)
(g) 45° above $-x$ (toward $+y$)
(h) 45° below $-x$ (toward $-y$)

2. What is the direction of the satellite's instantaneous velocity at point *D*?

3. What is the direction of the satellite's average velocity for one quarter of an orbit, starting at *C* and ending at *D*?

4. What is the direction of the satellite's average acceleration for one half of an orbit, starting at *C* and ending at *A*?

5. What is the direction of the satellite's instantaneous acceleration at point *C*?

6. Two satellites are in orbit around Mars with the same orbital radius. Satellite 2 has twice the mass of satellite 1. The radial acceleration of satellite 2 has
   (a) twice the magnitude of the radial acceleration of satellite 1.
   (b) the same magnitude as the radial acceleration of satellite 1.
   (c) half the magnitude of the radial acceleration of satellite 1.
   (d) four times the magnitude of the radial acceleration of satellite 1.

Questions 7–8: A boy swings in a tire swing. Answer choices:
   (a) At the highest point of the motion
   (b) At the lowest point of the motion
   (c) At a point neither highest nor lowest
   (d) It is constant.

7. When is the tension in the rope the greatest?

8. When is the tangential acceleration the greatest?

Questions 9–10 concern these three statements:
   (1) Its acceleration is constant.
   (2) Its radial acceleration component is constant in magnitude.
   (3) Its tangential acceleration component is constant in magnitude.

9. An object is in uniform circular motion. Identify the correct statement(s).
   (a) 1 only   (b) 2 only   (c) 3 only
   (d) 1, 2, and 3   (e) 2 and 3   (f) 1 and 2
   (g) 1 and 3   (h) None of them

10. An object is in nonuniform circular motion with constant angular acceleration. Identify the correct statement(s). (Use the same answer choices as Question 9.)

11. An astronaut is out in space far from any large bodies. He uses his jets to start spinning, then releases a baseball he has been holding in his hand. Ignoring the gravitational force between the astronaut and the baseball, how would you describe the path of the baseball after it leaves the astronaut's hand?
   (a) It continues to circle the astronaut in a circle with the same radius it had before leaving the astronaut's hand.
   (b) It moves off in a straight line.
   (c) It moves off in an ever-widening arc.

12. An object moving in a circle at a constant speed has an acceleration that is
    (a) in the direction of motion
    (b) toward the center of the circle
    (c) away from the center of the circle
    (d) zero

Problems

◉   Combination conceptual/quantitative problem
♼   Biological or medical application
✦   Challenging problem
Blue #   Detailed solution in the Student Solutions Manual
① ②   Problems paired by concept
🌀   Text website interactive or tutorial

5.1 Description of Uniform Circular Motion

1. A carnival swing is fixed on the end of an 8.0-m-long beam. If the swing and beam sweep through an angle of 120°, what is the distance through which the riders move?

2. A soccer ball of diameter 31 cm rolls without slipping at a linear speed of 2.8 m/s. Through how many revolutions has the soccer ball turned as it moves a linear distance of 18 m?

3. Find the average angular speed of the second hand of a clock.

4. Convert these to radian measure: (a) 30.0°, (b) 135°, (c) $\frac{1}{4}$ revolution, (d) 33.3 revolutions.

5. A bicycle is moving at 9.0 m/s. What is the angular speed of its tires if their radius is 35 cm? (🌀 tutorial: car tire)

6. An elevator cable winds on a drum of radius 90.0 cm that is connected to a motor. (a) If the elevator is moving down at 0.50 m/s, what is the angular speed of the drum? (b) If the elevator moves down 6.0 m, how many revolutions has the drum made?

7. Grace is playing with her dolls and decides to give them a ride on a merry-go-round. She places one of them on an old record player turntable and sets the angular speed at 33.3 rpm. (a) What is their angular speed in rad/s? (b) If the doll is 13 cm from the center of the spinning turntable platform, how fast (in m/s) is the doll moving?

8. A wheel is rotating at a rate of 2.0 revolutions every 3.0 s. Through what angle, in radians, does the wheel rotate in 1.0 s?

✦ 9. In the construction of railroads, it is important that curves be gentle, so as not to damage passengers or freight. Curvature is not measured by the radius of curvature, but in the following way. First a 100.0-ft-long chord is measured. Then the curvature is reported as the angle subtended by two radii at the endpoints of the

chord. (The angle is measured by determining the angle between two tangents 100 ft apart; since each tangent is perpendicular to a radius, the angles are the same.) In modern railroad construction, track curvature is kept below 1.5°. What is the radius of curvature of a "1.5° curve"? [*Hint:* Since the angle is small, the length of the chord is approximately equal to the arc length along the curve.]

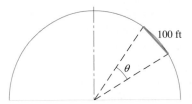

5.2 Radial Acceleration

10. Verify that all three expressions for radial acceleration ($v\omega$, v^2/r, and $\omega^2 r$) have the correct dimensions for an acceleration.

♼ 11. An apparatus is designed to study insects at an acceleration of magnitude 980 m/s² (= 100*g*). The apparatus consists of a 2.0-m rod with insect containers at either end. The rod rotates about an axis perpendicular to the rod and at its center. (a) How fast does an insect move when it experiences a radial acceleration of 980 m/s²? (b) What is the angular speed of the insect? (🌀 tutorial: centrifuge)

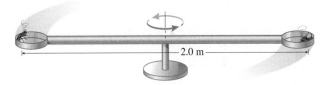

12. The rotor is an amusement park ride where people stand against the inside of a cylinder. Once the cylinder is spinning fast enough, the floor drops out. (a) What force keeps the people from falling out the bottom of the cylinder? (b) If the coefficient of friction is 0.40 and the cylinder has a radius of 2.5 m, what is the minimum angular speed of the cylinder so that the people don't fall out? (Normally the operator runs it considerably faster as a safety measure.)

13. Objects that are at rest relative to Earth's surface are in circular motion due to Earth's rotation. What is the radial acceleration of an African baobab tree located at the equator?

✦14. Earth's orbit around the Sun is nearly circular. The period is 1 yr = 365.25 d. (a) In an elapsed time of 1 d, what is Earth's angular displacement? (b) What is the change in Earth's velocity, $\Delta\vec{v}$? (c) What is Earth's average acceleration during 1 d? (d) Compare your answer for (c) to the magnitude of Earth's instantaneous radial acceleration. Explain.

15. A 0.700-kg ball is on the end of a rope that is 1.30 m in length. The ball and rope are attached to a pole and the entire apparatus, including the pole, rotates about the pole's symmetry axis. The rope makes an angle of 70.0° with respect to the vertical. What is the tangential speed of the ball?

70.0°

Axis of rotation

16. A child's toy has a 0.100-kg ball attached to two strings, A and B. The strings are also attached to a stick and the ball swings around the stick along a circular path in a horizontal plane. Both strings are 15.0 cm long and make an angle of 30.0° with respect to the horizontal. (a) Draw an FBD for the ball showing the tension forces and the gravitational force. (b) Find the magnitude of the tension in each string when the ball's angular speed is 6.00π rad/s.

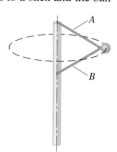

A

B

✦17. A child swings a rock of mass m in a horizontal circle using a rope of length L. The rock moves at constant speed v. (a) Ignoring gravity, find the tension in the rope. (b) Now include gravity (the weight of the rock is no longer negligible, although the weight of the rope still is negligible). What is the tension in the rope? Express the tension in terms of m, g, v, L, and the angle θ that the rope makes with the horizontal. (tutorial: skip rope)

✦18. A *conical pendulum* consists of a bob (mass m) attached to a string (length L) swinging in a horizontal circle (Fig. 5.11). As the string moves, it sweeps out the area of a cone. The angle that the string makes with the vertical is ϕ. (a) What is the tension in the string? (b) What is the period of the pendulum?

5.3 Unbanked and Banked Curves

19. A curve in a stretch of highway has radius R. The road is unbanked. The coefficient of static friction between the tires and road is μ_s. (a) What is the fastest speed that a car can safely travel around the curve? (b) Explain what happens when a car enters the curve at a speed greater than the maximum safe speed. Illustrate with an FBD. (interactive: banked curve)

20. A highway curve has a radius of 825 m. At what angle should the road be banked so that a car traveling at 26.8 m/s (60 mph) has no tendency to skid sideways on the road? [*Hint:* No tendency to skid means the frictional force is zero.]

21. A curve in a highway has radius of curvature 120 m and is banked at 3.0°. On a day when the road is icy, what is the safest speed to go around the curve?

22. A roller coaster car of mass 320 kg (including passengers) travels around a horizontal curve of radius 35 m. Its speed is 16 m/s. What is the magnitude and direction of the total force exerted on the car by the track?

23. A velodrome is built for use in the Olympics. The radius of curvature of the surface is 20.0 m. At what angle should the surface be banked for cyclists moving at 18 m/s? (Choose an angle so that no frictional force is needed to keep the cyclists in their circular path. Large banking angles *are* used in velodromes.)

24. A car drives around a curve with radius 410 m at a speed of 32 m/s. The road is not banked. The mass of the car is 1400 kg. (a) What is the frictional force on the car? (b) Does the frictional force necessarily have magnitude $\mu_s N$? Explain.

✦25. A car drives around a curve with radius 410 m at a speed of 32 m/s. The road is banked at 5.0°. The mass of the car is 1400 kg. (a) What is the frictional force on the car? (b) At what speed could you drive around this curve so that the force of friction is zero?

✦26. A curve in a stretch of highway has radius R. The road is banked at angle θ to the horizontal. The coefficient of static friction between the tires and road is μ_s. What is the fastest speed that a car can travel through the curve?

27. An airplane is flying at constant speed v in a horizontal circle of radius r. The lift force on the wings due to the air is perpendicular to the wings. At what angle to the vertical must the wings be banked to fly in this circle? (tutorial: plane in turn)

✦28. A road with a radius of 75.0 m is banked so that a car can navigate the curve at a speed of 15.0 m/s without any friction. When a car is going 20.0 m/s on this curve, what minimum coefficient of static friction is needed if the car is to navigate the curve without slipping?

5.4 Circular Orbits of Satellites and Planets

29. What is the average linear speed of the Earth about the Sun?

30. The orbital speed of Earth about the Sun is 3.0×10^4 m/s and its distance from the Sun is 1.5×10^{11} m. The mass of Earth is approximately 6.0×10^{24} kg and that of the Sun is 2.0×10^{30} kg. What is the magnitude of the force exerted by the Sun on Earth? [*Hint:* Two different methods are possible. Try both.]

31. Two satellites are in circular orbits around Jupiter. One, with orbital radius r, makes one revolution every 16 h. The other satellite has orbital radius $4.0r$. How long does the second satellite take to make one revolution around Jupiter?

32. The Hubble Space Telescope orbits Earth 613 km above Earth's surface. What is the period of the telescope's orbit?

33. Io, one of Jupiter's satellites, has an orbital period of 1.77 d. Europa, another of Jupiter's satellites, has an orbital period of about 3.54 d. Both moons have nearly circular orbits. Use Kepler's third law to find the distance of each satellite from Jupiter's center. Jupiter's mass is 1.9×10^{27} kg.

34. A spy satellite is in circular orbit around Earth. It makes one revolution in 6.00 h. (a) How high above Earth's surface is the satellite? (b) What is the satellite's acceleration?

35. Mars has a mass of about 6.42×10^{23} kg. The length of a day on Mars is 24 h and 37 min, a little longer than the length of a day on Earth. Your task is to put a satellite into a circular orbit around Mars so that it stays above one spot on the surface, orbiting Mars once each Mars day. At what distance from the center of the planet should you place the satellite?

✦36. A satellite travels around Earth in uniform circular motion at an altitude of 35 800 km above Earth's surface. The satellite is in geosynchronous orbit (that is, the time for it to complete one orbit is exactly 1 d). In the figure with Multiple-Choice Questions 2–5, the satellite moves counterclockwise (*ABCDA*). State directions in terms of the *x*- and *y*-axes. (a) What is the satellite's instantaneous velocity at point *C*? (b) What is the satellite's average velocity for one quarter of an orbit, starting at *A* and ending at *B*? (c) What is the satellite's average acceleration for one quarter of an orbit, starting at *A* and ending at *B*? (d) What is the satellite's instantaneous acceleration at point *D*?

✦37. A spacecraft is in orbit around Jupiter. The radius of the orbit is 3.0 times the radius of Jupiter (which is $R_J = 71\,500$ km). The gravitational field at the surface of Jupiter is 23 N/kg. What is the period of the spacecraft's orbit? [*Hint:* You don't need to look up any more data about Jupiter to solve the problem.]

5.5 Nonuniform Circular Motion

38. A roller coaster has a vertical loop with radius 29.5 m. With what minimum speed should the roller coaster car be moving at the top of the loop so that the passengers do not lose contact with the seats?

39. A pendulum is 0.80 m long and the bob has a mass of 1.0 kg. At the bottom of its swing, the bob's speed is 1.6 m/s. (a) What is the tension in the string at the bottom of the swing? (b) Explain why the tension is greater than the weight of the bob.

40. A 35.0-kg child swings on a rope with a length of 6.50 m that is hanging from a tree. At the bottom of the swing, the child is moving at a speed of 4.20 m/s. What is the tension in the rope?

41. A car approaches the top of a hill that is shaped like a vertical circle with a radius of 55.0 m. What is the fastest speed that the car can go over the hill without losing contact with the ground?

5.6 Tangential and Angular Acceleration

42. A child pushes a merry-go-round from rest to a final angular speed of 0.50 rev/s with constant angular acceleration. In doing so, the child pushes the merry-go-round 2.0 revolutions. What is the angular acceleration of the merry-go-round?

43. A cyclist starts from rest and pedals so that the wheels make 8.0 revolutions in the first 5.0 s. What is the angular acceleration of the wheels (assumed constant)?

44. During normal operation, a computer's hard disk spins at 7200 rpm. If it takes the hard disk 4.0 s to reach this angular velocity starting from rest, what is the average angular acceleration of the hard disk in rad/s²?

45. Derive Eq. 5-20 from Eqs. 5-18 and 5-19. [*Hint:* See the derivation of Eq. (2-12) in Section 2.4.]

46. Derive Eq. 5-21 from Eqs. 5-18 and 5-19.

✦47. A pendulum is 0.800 m long and the bob has a mass of 1.00 kg. When the string makes an angle of $\theta = 15.0°$ with the vertical, the bob is moving at 1.40 m/s. Find the tangential and radial acceleration components and the tension in the string. [*Hint:* Draw an FBD for the bob. Choose the *x*-axis to be tangential to the motion of the bob and the *y*-axis to be radial. Apply Newton's second law.]

Problems 47 and 48

48. Find the tangential acceleration of a freely swinging pendulum when it makes an angle θ with the vertical.

49. A turntable reaches an angular speed of 33.3 rpm in 2.0 s, starting from rest. (a) Assuming the angular acceleration is constant, what is its magnitude? (b) How many revolutions does the turntable make during this time interval?

50. A wheel's angular acceleration is constant. Initially its angular velocity is zero. During the first 1.0-s time interval, it rotates through an angle of 90.0°. (a) Through what angle does it rotate during the next 1.0-s time interval? (b) Through what angle during the third 1.0-s time interval?

51. A car that is initially at rest moves along a circular path with a constant tangential acceleration component of 2.00 m/s². The circular path has a radius of 50.0 m. The initial position of the car is at the far west location on the circle and the initial velocity is to the north. (a) After the car has traveled $\frac{1}{4}$ of the circumference, what is the speed of the car? (b) At this point, what is the radial acceleration component of the car? (c) At this same point, what is the total acceleration of the car?

52. A disk rotates with constant angular acceleration. The initial angular speed of the disk is 2π rad/s. After the disk rotates through 10π radians, the angular speed is 7π rad/s. (a) What is the magnitude of the angular acceleration? (b) How much time did it take for the disk to rotate through 10π radians? (c) What is the tangential acceleration of a point located at a distance of 5.0 cm from the center of the disk?

53. In a Beams ultracentrifuge, the rotor is suspended magnetically in a vacuum. Since there is no mechanical connection to the rotor, the only friction is the air resistance due to the few air molecules in the vacuum. If the rotor is spinning with an angular speed of 5.0×10^5 rad/s and the driving force is turned off, its spinning slows down at an angular rate of 0.40 rad/s². (a) How long does the rotor spin before coming to rest? (b) During this time, through how many revolutions does the rotor spin?

54. The rotor of the Beams ultracentrifuge (see Problem 53) is 20.0 cm long. For a point at the end of the rotor, find the (a) initial speed, (b) tangential acceleration component, and (c) maximum radial acceleration component.

5.7 Apparent Weight and Artificial Gravity

55. If a washing machine's drum has a radius of 25 cm and spins at 4.0 rev/s, what is the strength of the artificial gravity to which the clothes are subjected? Express your answer as a multiple of g.

56. A space station is shaped like a ring and rotates to simulate gravity. If the radius of the space station is 120 m, at what frequency must it rotate so that it simulates Earth's gravity? [*Hint:* The apparent weight of the astronauts must be the same as their weight on Earth.] (tutorial: space station)

57. A biologist is studying growth in space. He wants to simulate Earth's gravitational field, so he positions the plants on a rotating platform in the spaceship. The distance of each plant from the central axis of rotation is $r = 0.20$ m. What angular speed is required?

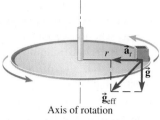

58. A biologist is studying plant growth and wants to simulate a gravitational field twice as strong as Earth's. She places the plants on a horizontal rotating table in her laboratory on Earth at a distance of 12.5 cm from the axis of rotation. What angular speed will give the plants an effective gravitational field $\vec{g}_{eff}$, whose magnitude is $2.0g$? [*Hint:* Remember to account for Earth's gravitational field as well as the artificial gravity when finding the apparent weight.]

59. Objects that are at rest relative to the Earth's surface are in circular motion due to Earth's rotation. (a) What is the radial acceleration of an object at the equator? (b) Is the object's apparent weight greater or less than its weight? Explain. (c) By what percentage does the apparent weight differ from the weight at the equator? (d) Is there any place on Earth where a bathroom scale reading is equal to your true weight? Explain.

60. A person of mass M stands on a bathroom scale inside a Ferris wheel compartment. The Ferris wheel has radius R and angular velocity ω. What is the apparent weight of the person (a) at the top and (b) at the bottom?

61. A person rides a Ferris wheel that turns with constant angular velocity. Her weight is 520.0 N. At the top of the ride her apparent weight is 1.5 N different from her true weight. (a) Is her apparent weight at the top 521.5 N or 518.5 N? Why? (b) What is her apparent weight at the bottom of the ride? (c) If the angular speed of the Ferris wheel is 0.025 rad/s, what is its radius?

62. Objects that are at rest relative to Earth's surface are in circular motion due to Earth's rotation. What is the radial acceleration of a painting hanging in the Prado Museum in Madrid, Spain, at a latitude of 40.2° North? (Note that the object's radial acceleration is not directed toward the center of the Earth.)

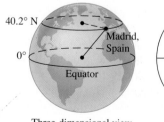

Three-dimensional view

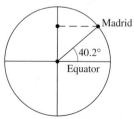

Rotation axis
Cross-sectional view

63. A rotating flywheel slows down at a constant rate due to friction in its bearings. After 1 min, its angular velocity has diminished to 0.80 of its initial value ω. At the end of the third minute, what is the angular velocity in terms of the initial value?

Comprehensive Problems

64. The Earth rotates on its own axis once per day (24.0 h). What is the tangential speed of the summit of Mt. Kilimanjaro (elevation 5895 m above sea level), which is located approximately on the equator, due to the rotation of the Earth? The equatorial radius of Earth is 6378 km.

65. A trimmer for cutting weeds and grass near trees and borders has a nylon cord of 0.23-m length that whirls about an axle at 660 rad/s. What is the linear speed of the tip of the nylon cord?

66. A high-speed dental drill is rotating at 3.14×10^4 rad/s. Through how many degrees does the drill rotate in 1.00 s?

67. A jogger runs counterclockwise around a path of radius 90.0 m at constant speed. He makes 1.00 revolution in 188.4 s. At $t = 0$, he is heading due east. (a) What is the jogger's instantaneous velocity at $t = 376.8$ s? (b) What is his instantaneous velocity at $t = 94.2$ s?

68. Two gears A and B are in contact. The radius of gear A is twice that of gear B. (a) When A's angular velocity is 6.00 Hz counterclockwise, what is B's angular velocity? (b) If A's radius to the tip of the teeth is 10.0 cm, what is the linear speed of a point on the tip of a gear tooth? What is the linear speed of a point on the tip of B's gear tooth?

Problems 68 and 69

69. If gear A in Problem 68 has an initial frequency of 0.955 Hz and an angular acceleration of 3.0 rad/s^2, how many rotations does each gear go through in 2.0 s?

70. The time to sunset can be estimated by holding out your arm, holding your fingers horizontally in front of your eyes, and counting the number of fingers that fit between the horizon and the setting Sun. (a) What is the angular speed, in radians per second, of the Sun's apparent circular motion around the Earth? (b) Estimate the angle subtended by one finger held at arm's length. (c) How long in minutes does it take the Sun to "move" through this same angle?

71. In the professional videotape recording system known as quadriplex, four tape heads are mounted on the circumference of a drum of radius 2.5 cm that spins at 1500 rad/s. (a) At what speed are the tape heads moving? (b) Why are moving tape heads used instead of stationary ones, as in audiotape recorders? [*Hint:* How fast would the tape have to move if the heads were stationary?]

72. The Milky Way galaxy rotates about its center with a period of about 200 million yr. The Sun is 2×10^{20} m from the center of the galaxy. How fast is the Sun moving with respect to the center of the galaxy?

73. A small body of mass 0.50 kg is attached by a 0.50-m-long cord to a pin set into the surface of a frictionless table top. The body moves in a circle on the horizontal surface with a speed of 2.0π m/s. (a) What is the magnitude of the radial acceleration of the body? (b) What is the tension in the cord?

74. Two blocks, one with mass $m_1 = 0.050$ kg and one with mass $m_2 = 0.030$ kg, are connected to one another by a string. The inner block is connected to a central pole by another string as shown in the figure with $r_1 = 0.40$ m and $r_2 = 0.75$ m. When the blocks are spun around on a horizontal frictionless surface at an angular speed of 1.5 rev/s, what is the tension in each of the two strings?

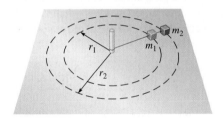

75. What's the fastest way to make a U-turn at constant speed? Suppose that you need to make a 180° turn on a circular path. The minimum radius (due to the car's steering system) is 5.0 m, while the maximum (due to the width of the road) is 20.0 m. Your acceleration must never exceed 3.0 m/s^2 or else you will skid. Should you use the smallest possible radius, so the distance is small, or the largest, so you can go faster without skidding, or something in between? What is the minimum possible time for this U-turn?

76. The Milky Way galaxy rotates about its center with a period of about 200 million yr. The Sun is 2×10^{20} m from the center of the galaxy. (a) What is the Sun's radial acceleration? (b) What is the net gravitational force on the Sun due to the other stars in the Milky Way?

77. Bacteria swim using a corkscrew-like helical flagellum that rotates. For a bacterium with a flagellum that has a pitch of 1.0 μm that rotates at 110 rev/s, how fast could it swim if there were no "slippage" in the medium in which it is swimming? The pitch of a helix is the distance between "threads."

78. You place a penny on a turntable at a distance of 10.0 cm from the center. The coefficient of static friction between the penny and the turntable is 0.350. The turntable's

angular acceleration is 2.00 rad/s². How long after you turn on the turntable will the penny begin to slide off of the turntable?

79. A coin is placed on a turntable that is rotating at 33.3 rpm. If the coefficient of static friction between the coin and the turntable is 0.1, how far from the center of the turntable can the coin be placed without having it slip off?

80. Grace, playing with her dolls, pretends the turntable of an old phonograph is a merry-go-round. The dolls are 12.7 cm from the central axis. She changes the setting from 33.3 rpm to 45.0 rpm. (a) For this new setting, what is the linear speed of a point on the turntable at the location of the dolls? (b) If the coefficient of static friction between the dolls and the turntable is 0.13, do the dolls stay on the turntable?

81. Your car's wheels are 65 cm in diameter and the wheels are spinning at an angular velocity of 101 rad/s. How fast is your car moving in kilometers per hour (assume no slippage)?

82. In an amusement park rocket ride, cars are suspended from 4.25-m cables attached to rotating arms at a distance of 6.00 m from the axis of rotation. The cables swing out at an angle of 45.0° when the ride is operating. What is the angular speed of rotation?

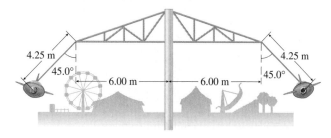

83. Centrifuges are commonly used in biological laboratories for the isolation and maintenance of cell preparations. For cell separation, the centrifugation conditions are typically 1.0×10^3 rpm using an 8.0-cm-radius rotor. (a) What is the radial acceleration of material in the centrifuge under these conditions? Express your answer as a multiple of g. (b) At 1.0×10^3 rpm (and with a 8.0-cm rotor), what is the net force on a red blood cell whose mass is 9.0×10^{-14} kg? (c) What is the net force on a virus particle of mass 5.0×10^{-21} kg under the same conditions? (d) To pellet out virus particles and even to separate large molecules such as proteins, super-high-speed centrifuges called ultracentrifuges are used in which the rotor spins in a vacuum to reduce heating due to friction. What is the radial acceleration inside an ultracentrifuge at 75 000 rpm with an 8.0-cm rotor? Express your answer as a multiple of g.

84. You take a homemade "accelerometer" to an amusement park. This accelerometer consists of a metal nut attached to a string and connected to a protractor, as

shown in the figure. While riding a roller coaster that is moving at a uniform speed around a circular path, you hold up the accelerometer and notice that the string is making an angle of 55° with respect to the vertical with the nut pointing away from the center of the circle, as shown. (a) What is the radial acceleration of the roller coaster? (b) What is your radial acceleration expressed as a multiple of g? (c) If the roller coaster track is turning in a radius of 80.0 m, how fast are you moving?

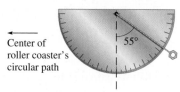

Center of roller coaster's circular path

55°

85. Massimo, a machinist, is cutting threads for a bolt on a lathe. He wants the bolt to have 18 threads per inch. If the cutting tool moves parallel to the axis of the would-be bolt at a linear velocity of 0.080 in./s, what must the rotational speed of the lathe chuck be to ensure the correct thread density? [*Hint:* One thread is formed for each complete revolution of the chuck.]

86. In Chapter 19 we will see that a charged particle can undergo uniform circular motion when acted on by a magnetic force and no other forces. (a) For that to be true, what must the angle between the magnetic force and the particle's velocity? (b) The magnitude of the magnetic force on a charged particle is proportional to the particle's speed, $F = kv$. Show that two identical charged particles moving in circles at different speeds in the same magnetic field must have the same period. (c) Show that the radius of the particle's circular path is proportional to the speed.

87. Find the orbital radius of a geosynchronous satellite. Do not assume the speed found in Example 5.9. Start by writing an equation that relates the period, radius, and speed of the orbiting satellite. Then apply Newton's second law to the satellite. You will have two equations with two unknowns (the speed and radius). Eliminate the speed algebraically and solve for the radius.

Answers to Practice Problems

5.1 3.001×10^{-7} rad/s

5.2 1.65 m/s

5.3 1.9 min

5.4 17 m/s²

5.5 60 N toward the center of the circular path

5.6 More slowly

5.7 No

5.8 29.7 km/s; 3.17×10^7 s

5.9 0.723R

5.10 2.44 h

5.11 4.2mg

5.12 Acceleration is purely tangential:

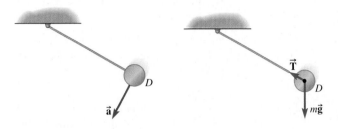

5.13 (a) 1.75×10^{-4} rad/s²; (b) 0.0349 rad (2.00°)

5.14 (a) 2200 N; (b) 1500 N

Answers to Checkpoints

5.1 7200 Hz

5.2 No, for uniform circular motion the *direction* of the velocity vector is continuously changing but the magnitude of the velocity (the speed) is unchanged.

5.3 The car has friction between the road and the tires to exert a horizontal force that causes the radial acceleration.

5.4 To be geosynchronous the satellites must have an orbital period of 1 d. The only quantities that affect the period are the mass of Earth and the radial distance from Earth's center. These quantities are the same for all satellites no matter the mass.

5.5 For nonuniform circular motion, the direction and the magnitude of the velocity are both changing. There are tangential and radial components to the acceleration. The magnitude of the radial component changes as the speed changes. For uniform circular motion, the magnitude of the velocity is constant but the direction changes. The radial acceleration is constant in magnitude (and the tangential acceleration is zero).

5.6 The radial acceleration cannot be constant because the radius r is constant but the angular velocity ω is changing $a_r = \omega^2 r$.

Review & Synthesis: Chapters 1–5

Review Exercises

1. From your knowledge of Newton's second law and dimensional analysis, find the units (in SI base units) of the spring constant k in the equation $F = kx$, where F is a force and x is a distance.

2. Harrison traveled 2.00 km west, then 5.00 km in a direction 53.0° south of west, then 1.00 km in a direction 60.0° north of west. (a) In what direction, and for how far, should Harrison travel to return to his starting point? (b) If Harrison returns directly to his starting point with a speed of 5.00 m/s, how long will the return trip take?

3. (a) How many center-stripe road reflectors, separated by 17.6 yd, are required along a 2.20-mile section of curving mountain roadway? (b) Solve the same problem for a road length of 3.54 km with the markers placed every 16.0 m. Would you prefer to be the highway engineer in a country with a metric system or U.S. customary units?

4. A baby was spitting up after nursing and the pediatrician prescribed Zantac syrup to reduce the baby's stomach acid. The prescription called for 0.75 mL to be taken twice a day for a month. The pharmacist printed a label for the bottle of syrup that said "3/4 tsp. twice a day." By what factor was the baby overmedicated before the error was discovered at the baby's next office visit two weeks later? [*Hint:* 1 tsp = 4.9 mL.]

5. Mike swims 50.0 m with a speed of 1.84 m/s, then turns around and swims 34.0 m in the opposite direction with a speed of 1.62 m/s. (a) What is his average speed? (b) What is his average velocity?

6. You are watching a television show about Navy pilots. The narrator says that when a Navy jet takes off, it accelerates because the engines are at full throttle and because there is a catapult that propels the jet forward. You begin to wonder how much force is supplied by the catapult. You look on the Web and find that the flight deck of an aircraft carrier is about 90 m long, that an F-14 has a mass of 33 000 kg, that each of the two engines supplies 27 000 lb of force, and that the takeoff speed of such a plane is about 160 mi/h. Estimate the average force on the jet due to the catapult.

7. On April 15, 1999, a Korean cargo plane crashed due to a confusion over units. The plane was to fly from Shanghai, China, to Seoul, Korea. After take-off the plane climbed to 900 m. Then the first officer was instructed by the Shanghai tower to climb to 1500 m and maintain that altitude. The captain, after reaching 1450 m, twice asked the first officer at what altitude they should fly. He was twice told incorrectly they were to be at 1500 ft. The captain pushed the control column quickly forward and started a steep descent. The plane could not recover from the dive and crashed. How much above the correct altitude did the captain think they were when he started his rapid descent and lost control? (It turns out that aircraft altitudes are given in feet throughout the world except in China, Mongolia, and the former Soviet states where meters are used.)

8. Paula swims across a river that is 10.2 m wide. She can swim at 0.833 m/s in still water, but the river flows with a speed of 1.43 m/s. If Paula swims in such a way that she crosses the river in as short a time as possible, how far downstream is she when she gets to the opposite shore?

9. Peter is collecting paving stones from a quarry. He harnesses two dogs, Sandy and Rufus, in tandem to the loaded cart. Sandy pulls with force $\vec{\mathbf{F}}$ at a 15° angle to the north of east; Rufus pulls with 1.5 times the force of Sandy and at an angle of 30.0° south of east. Use a ruler and protractor to draw the force vectors to scale (choose a simple scale, such as $2.0 \text{ cm} \leftrightarrow F$). Find the sum of the two force vectors graphically. Measure its length and find the magnitude of the sum from the scale used and the direction with the protractor. Will the cart stay on the road that runs directly west to east?

10. A tire swing hangs at a 12° angle to the vertical when a stiff breeze is blowing. In terms of the tire's weight W, (a) what is the magnitude of the horizontal force exerted on the tire by the wind? (b) What is the tension in the rope supporting the tire? Ignore the weight of the rope.

11. An astronaut of mass 60.0 kg and a small asteroid of mass 40.0 kg are initially at rest with respect to the space station. The astronaut pushes the asteroid with a constant force of magnitude 250 N for 0.35 s. Gravitational forces are negligible. (a) How far apart are the astronaut and the asteroid 5.00 s after the astronaut stops pushing? (b) What is their relative speed at this time?

12. In the fairy tale, Rapunzel, the beautiful maiden let her long golden hair hang down from the tower in which she was held prisoner so that her prince could use her hair as a climbing rope to climb the tower and rescue her. (a) Estimate how much force is required to pull a strand of hair out of your head. (b) There are about 10^5 hairs growing out of Rapunzel's head. If the prince has a mass of 60 kg, estimate the average force pulling on each strand of hair. Will Rapunzel be bald by the time the prince reaches the top of the 30-m tower?

13. Marie slides a paper plate with a slice of pizza across a horizontal table to her friend Jaden. The coefficient of friction between the table and plate is 0.32. If the pizza must travel 44 cm to get from Marie to Jaden, what initial speed should Marie give the plate of pizza so that it just stops when it gets to Jaden?

14. Two wooden crates with masses as shown are tied together by a horizontal cord. Another cord is tied to the first crate and it is pulled with a force of 195 N at an angle of 20.0°, as shown. Each crate has a coefficient of

kinetic friction of 0.550. Find the tension in the rope between the crates and the magnitude of the acceleration of the system.

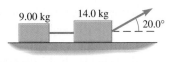

+15. A boy has stacked two blocks on the floor so that a 5.00-kg block is on top of a 2.00-kg block. (a) If the coefficient of static friction between the two blocks is 0.400 and the coefficient of static friction between the bottom block and the floor is 0.220, with what minimum force should the boy push horizontally on the upper block to make both blocks start to slide together along the floor? (b) If he pushes too hard, the top block starts to slide off the lower block. What is the maximum force with which he can push without that happening if the coefficient of kinetic friction between the bottom block and the floor is 0.200?

+16. A binary star consists of two stars of masses M_1 and $4.0M_1$ a distance d apart. Is there any point where the gravitational field due to the two stars is zero? If so, where is that point?

17. Two boys are trying to break a cord. Gerardo says they should each pull in opposite directions on the two ends; Stefan says they should tie the cord to a pole and both pull together on the opposite end. Which plan is more likely to work?

+18. Fish don't move as fast as you might think. A small trout has a top swimming speed of only about 2 m/s, which is about the speed of a brisk walk (for a human, not a fish!). It may seem to move faster because it is capable of large *accelerations*—it can dart about, changing its speed or direction very quickly. (a) If a trout starts from rest and accelerates to 2 m/s in 0.05 s, what is the trout's average acceleration? (b) During this acceleration, what is the average net force on the trout? Express your answer as a multiple of the trout's weight. (c) Explain how the trout gets the water to push it forward.

+19. A spotter plane sees a school of tuna swimming at a steady 5.00 km/h northwest. The pilot informs a fishing trawler, which is just then 100.0 km due south of the fish. The trawler sails along a straight-line course and intercepts the tuna after 4.0 h. How fast did the trawler move? [*Hint:* First find the velocity of the trawler relative to the tuna.]

20. Julia is delivering newspapers. Suppose she is driving at 15 m/s along a straight road and wants to drop a paper out the window from a height of 1.00 m so it slides along the shoulder and comes to rest in the customer's driveway. At what horizontal distance before the driveway should she drop the paper? The coefficient of kinetic friction between the newspaper and the ground is 0.40. Ignore air resistance and assume no bouncing or rolling.

21. Three rocks are thrown from a cliff with the same initial speeds but in different directions: one straight down, one straight up, and one horizontally. Ignore air resistance. (a) Compare the speeds of the three rocks just before they hit the flat ground at the bottom of the cliff. (b) Illustrate your answer by calculating the final speeds for three rocks thrown in the specified directions with initial speeds of 10.0 m/s from a cliff that is 15.00 m high. [*Hint:* Remember that the speed is the magnitude of the velocity vector.]

22. You are watching the Super Bowl where your favorite team is leading by a score of 21 to 20. The other team is lining up to try to kick the winning field goal. You watched their kicker warm up and you saw that he could kick the football with a velocity of 21 m/s. He lines up for a 45-yd kick. You watch as he kicks the ball at an angle of 35° above the horizontal. Assuming he kicks the ball straight and with the same speed as during the warmup, will the ball clear the 10-ft-high goal post, or will your favorite team win the Super Bowl?

23. A coin is placed on a turntable 13.0 cm from the center. The coefficient of static friction between the coin and the turntable is 0.110. Once the turntable is turned on, its angular acceleration is 1.20 rad/s². How long will it take until the coin begins to slide?

+24. Carlos and Shannon are sledding down a snow-covered slope that is angled at 12° below the horizontal. When sliding on snow, Carlos's sled has a coefficient of friction $\mu_k = 0.10$; Shannon has a "supersled" with $\mu_k = 0.010$. Carlos takes off down the slope starting from rest. When Carlos is 5.0 m from the starting point, Shannon starts down the slope from rest. (a) How far have they traveled when Shannon catches up to Carlos? (b) How fast is Shannon moving with respect to Carlos as she passes by?

25. A proposed "space elevator" consists of a cable going all the way from the ground to a space station in geo-synchronous orbit (always above the same point on Earth's surface). Elevator "cars" would climb the cable to transport cargo to outer space. Consider a cable connected between the equator and a space station at height H above the surface. Ignore the mass of the cable*. (a) Find the height H. (b) Suppose there is an elevator car of mass 100 kg sitting halfway up at height $H/2$. What tension T would be required in the cable to hold the car in place? Which part of the cable would be under tension (above the car or below it)?

26. Anthony is going to drive a flat-bed truck up a hill that makes an angle of 10° with respect to the horizontal direction. A 36.0-kg package sits in the back of the truck. The coefficient of static friction between the package

---

*More realistically, the mass of the cable is one of the primary engineering challenges of a space elevator. The cable is so long that it would have a very large mass and would have to withstand an enormous tension to support its own weight. The cable would need to be supported by a counterweight positioned beyond the geosynchronous orbit. Some believe *carbon nanotubes* hold the key to producing a cable with the required properties.

and the truck bed is 0.380. What is the maximum acceleration the truck can have without the package falling off the back?

27. A road with a radius of 75.0 m is banked so that a car can navigate the curve at a speed of 15.0 m/s without any friction. On a cold day when the street is icy, the coefficient of static friction between the tires and the road is 0.120. What is the *slowest* speed the car can go around this curve without sliding *down* the bank?

28. You want to lift a heavy box with a mass of 98.0 kg using the two-pulley system as shown. With what minimum force do you have to pull down on the rope in order to lift the box at a constant velocity? One pulley is attached to the ceiling and one to the box.

29. At time $t = 0$, block A of mass 0.225 kg and block B of mass 0.600 kg rest on a horizontal frictionless surface a distance 3.40 m apart, with block A located to the left of block B. A horizontal force of 2.00 N directed to the right is applied to block A for a time interval $\Delta t = 0.100$ s. During the same time interval, a 5.00-N horizontal force directed to the left is applied to block B. How far from B's initial position do the two blocks meet? How much time has elapsed from $t = 0$ until the blocks meet?

30. A hamster of mass 0.100 kg gets onto his 20.0-cm-diameter exercise wheel and runs along inside the wheel for 0.800 s until its frequency of rotation is 1.00 Hz. (a) What is the tangential acceleration of the wheel, assuming it is constant? (b) What is the normal force on the hamster just before he stops? The hamster is at the bottom of the wheel during the entire 0.800 s.

31. A pellet is fired from a toy cannon with a velocity of 12 m/s directed 60° above the horizontal. After 0.10 s, a second identical pellet is fired with the same initial velocity. After an additional 0.15 s have passed, what is the velocity of the first pellet with respect to the second? Ignore air resistance.

32. A crate is sliding down a frictionless ramp that is inclined at 35.0°. (a) If the crate is released from rest, how far does it travel down the incline in 2.50 s if it does not get to the bottom of the ramp before the time has elapsed? (b) How fast is the crate moving after 2.50 s of travel?

33. The invention of the cannon in the fourteenth century made the catapult unnecessary and ended the safety of castle walls. Stone walls were no match for balls shot from cannons. Suppose a cannonball of mass 5.00 kg is launched from a height of 1.10 m, at an angle of elevation of 30.0° with an initial velocity of 50.0 m/s, toward a castle wall of height 30 m and located 215 m away from the cannon. (a) The range of a projectile is defined as the horizontal distance traveled when the projectile returns to its

original height. Derive an equation for the range in terms of v_i, g, and angle of elevation θ. (b) What will be the range reached by the projectile, if it is not intercepted by the wall? (c) If the cannonball travels far enough to hit the wall, find the height at which it strikes.

34. Two blocks are connected by a lightweight, flexible cord that passes over a single frictionless pulley. If $m_1 \gg m_2$, find (a) the acceleration of each block and (b) the tension in the cord.

35. A runner runs three-quarters of the way around a circular track of radius 60.0 m, when she collides with another runner and trips. (a) How far had the runner traveled on the track before the collision? (b) What was the magnitude of the displacement of the runner from her starting position when the accident occurred?

36. A solar sailplane is going from Earth to Mars. Its sail is oriented to give a solar radiation force of 8.00×10^2 N. The gravitational force due to the Sun is 173 N and the gravitational force due to the Earth is 1.00×10^2 N. All forces are in the plane formed by Earth, Sun, and sailplane. The mass of the sailplane is 14 500 kg. (a) What is the net force (magnitude and direction) acting on the sailplane? (b) What is the acceleration of the sailplane?

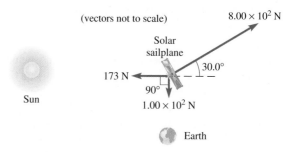

37. A star near the visible edge of a galaxy travels in a uniform circular orbit. It is 40 000 ly (light-years) from the galactic center and has a speed of 275 km/s. (a) Estimate the total mass of the galaxy based on the motion of the star? [*Hint:* For this estimate, assume the total mass to be concentrated at the galactic center and relate it to the gravitational force on the star.] (b) The total *visible* mass (i.e., matter we can detect via electromagnetic radiation) of the galaxy is 10^{11} solar masses. What fraction of the total mass of the galaxy is visible*, according to this estimate?

38. One of the tricky things about learning to sail is distinguishing the true wind from the apparent wind. When you are on a sailboat and you feel the wind on your face, you are experiencing the *apparent wind*—the motion of

*In many galaxies the stars appear to have roughly the *same orbital speed* over a large range of distances from the center. A popular hypothesis to explain such galaxy rotation velocities is the existence of *dark matter*—matter that we cannot detect via electromagnetic radiation. Dark matter is thought to account for the majority of the mass of some galaxies and nearly a fourth of the total mass of the universe.

the air relative to you. The true wind is the speed and direction of the air relative to the water while the apparent wind is the speed and direction of the air relative to the *sailboat*. The figure shows three different directions for the true wind along with one possible sail orientation as indicated by the position of the boom attached to the mast. (a) In each case, draw a vector diagram to establish the magnitude and direction of the apparent wind. (b) In which of the three cases is the apparent wind speed greater than the true wind speed? (Assume that the speed of the boat relative to the water is less than the true wind speed.) (c) In which of the three cases is the direction of the apparent wind direction forward of the true wind? ["Forward" means coming from a direction more nearly straight ahead. For example, (1) is forward of (2), which is forward of (3).]

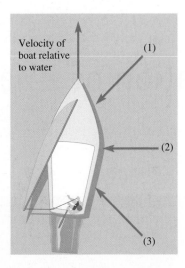

Velocity of boat relative to water

(1)

(2)

(3)

MCAT Review

Read the paragraph and then answer the following four questions:

The study of the flight of projectiles has many practical applications. The main forces acting on a projectile are air resistance and gravity. The path of a projectile is often approximated by ignoring the effects of air resistance. Gravity is then the only force acting on the projectile. When air resistance is included in the analysis, another force, $\mathbf{F_R}$, is introduced. F_R is proportional to the square of the velocity, v. The direction of the air resistance is exactly opposite the direction of motion. The equation for air resistance is $F_R = bv^2$, where b is a proportionality constant that depends on such factors as the density of the air and the shape of the projectile.

Air resistance was studied by launching a 0.5-kg projectile from a level surface. The projectile was launched with a speed of 30 m/s at a 40° angle to the surface. (Note: Assume air resistance is present unless otherwise specified.)

1. What is the magnitude of the vertical acceleration of the projectile immediately after it is launched? (Note: v_y = vertical velocity component.)

    A. $-g + (bvv_y)$

    B. $-g - (bvv_y)$

    C. $-g + (bvv_y)/(0.5 \text{ kg})$

    D. $-g - (bvv_y)/(0.5 \text{ kg})$

2. Approximately what horizontal distance does the projectile travel before returning to the elevation from which it was launched? (Note: Assume that the effects of air resistance are negligible.)

    A. 45 m   B. 60 m   C. 90 m   D. 120 m

3. What is the magnitude of the *horizontal component of air resistance* on the projectile at any point during flight? (Note: v_x = horizontal velocity component.)

    A. $(bvv_x) \cos 40°$

    B. $(bvv_x)/2$

    C. $(bvv_x) \sin 40°$

    D. bvv_x

4. How does the amount of time it takes a projectile to reach its maximum height compare to the time it takes to fall from its maximum height back to the ground? (Note: b is greater than zero.)

    A. The times are the same.

    B. The time to reach its maximum height is greater.

    C. The time to fall back to the ground is greater.

    D. Either can be greater depending on the magnitude of b.

Read the paragraph and then answer the following questions:

A raft is constructed from wood and used in a river that varies in depth, width, and current at several points along its length. The river at point A has a current of 2 m/s, a width of 200 m, and an average depth of 3 m.

5. Near point A, the raft is rowed at a constant velocity of 2 m/s relative to the river current and perpendicular to it. How far does the raft travel before it reaches the other side?

    A. 224 m

    B. 250 m

    C. 283 m

    D. 400 m

6. A rower at point A rows the raft at 3 m/s relative to the river current and wants to end up directly across the river from the point of origin. At what angle to the shore should the rower direct the raft?

    A. $\cos^{-1} \frac{5}{3}$

    B. $\cos^{-1} \frac{2}{5}$

    C. $\cos^{-1} \frac{3}{2}$

    D. $\cos^{-1} \frac{2}{3}$

7. A rock is dropped from a cliff that is 100 m above ground level. How long does it take the rock to reach the ground? (Note: Use $g = 10 \text{ m/s}^2$.)

    A. 4.5 s

    B. 10 s

    C. 14 s

    D. 20 s

6 Conservation of Energy

As a kangaroo hops along, the maximum height of each hop might be around 2.8 m. This height is only slightly higher than that achieved by an Olympic high jumper, but the kangaroo is able to achieve this height hop after hop as it travels with a horizontal velocity of 15 m/s or more. What features of kangaroo anatomy make this feat possible? It cannot simply be a matter of having more powerful leg muscles. If it were, the kangaroo would have to consume large amounts of energy-rich food to supply the muscles with enough chemical energy for each jump, but in reality a kangaroo's diet consists largely of grasses that are poor in energy content. (See p. 210 for the answer.)

- gravitational forces (Section 4.5)
- Newton's second law: force and acceleration (Sections 4.3–4.8)
- components of vectors (Section 3.2)
- circular orbits (Section 5.4)
- area under a graph (Sections 2.2 and 2.3)

Concepts & Skills to Review

6.1 THE LAW OF CONSERVATION OF ENERGY

Until now, we have relied on Newton's laws of motion to be the fundamental physical laws used to analyze the forces that act on objects and to predict the motion of objects. Now we introduce another physical principle: the conservation of energy. A **conservation law** is a physical principle that identifies some quantity that does not change with time. Conservation of energy means that every physical process leaves the total energy in the universe unchanged. Energy can be converted from one form to another, or transferred from one place to another. If we are careful to account for all the energy transformations, we find that the total energy remains the same.

Conservation law: a physical law that identifies a quantity that does not change with time.

The Law of Conservation of Energy

The total energy in the universe is unchanged by any physical process:

total energy before = total energy after.

"Turn down the thermostat—we're trying to conserve energy!" In ordinary language, *conserving energy* means trying not to waste useful energy resources. In the scientific meaning of *conservation*, energy is *always* conserved no matter what happens. When we "produce" or "generate" electric energy, for instance, we aren't creating any new energy; we're just converting energy from one form into another that's more useful to us.

Conservation of energy is one of the few universal principles of physics. No exceptions to the law of conservation of energy have been found. Conservation of energy is a powerful tool in the search to understand nature. It applies equally well to radioactive decay, the gravitational collapse of a star, a chemical reaction, a biological process such as respiration, and to the generation of electricity by a wind turbine (Fig. 6.1). Think about the energy conversions that make life possible. Green plants use photosynthesis to convert the energy they receive from the Sun into stored chemical energy. When animals eat the plants, that stored energy enables motion, growth, and maintenance of body temperature. Energy conservation governs every one of these processes.

Figure 6.1 At a California "wind farm," these wind turbines convert the energy of motion of the air into electric energy.

Choosing Between Alternative Solution Methods Some problems can be solved using either energy conservation *or* Newton's second law. Usually the energy method is easier. We often don't know the details of all the forces acting on an object, making a direct application of Newton's second law difficult. Using conservation of energy enables us to solve some of these problems more easily. When deciding which of these two approaches to use to solve a problem, try using energy conservation first. If the energy method does not lead to the solution, then try Newton's second law.

Historical Development of the Principle of Energy Conservation While many scientists contributed to the development of the law of conservation of energy, the law's first clear statement was made in 1842 by the German surgeon Julius Robert von Mayer (1814–1878). As a ship's physician on a voyage to what is now Indonesia, Mayer had noticed that the sailors' venous blood was a much deeper red in the tropics than it was in Europe. He concluded that less oxygen was being used because they didn't need to "burn" as much fuel to keep the body warm in the warmer climate.

Figure 6.2 The stored chemical energy in food enables a weightlifter to lift the barbell over her head.

Table 6.1	Some Common Forms of Energy
Form of Energy	**Brief Description**
Translational kinetic	Energy of translational motion (Chapter 6)
Elastic	Energy stored in a "springy" object or material when it is deformed (Chapter 6)*
Gravitational	Energy of gravitational interactions (Chapter 6)
Rotational kinetic	Energy of rotational motion (Chapter 8)*
Vibrational, acoustic, seismic	Energy of the oscillatory motions of atoms and molecules in a substance caused by a mechanical wave passing through it (Chapters 11 and 12)*
Internal	Energies of motion and interaction of atoms and molecules in solids, liquids, and gases, related to our sensation of temperature (Chapters 14 and 15)*
Electromagnetic	Energy of interaction of electric charges and currents; energy of electromagnetic fields, including electromagnetic waves such as light (Chapters 14, 17–22)
Rest	The total energy of a particle of mass m when it is at rest, given by Einstein's famous equation $E = mc^2$ (Chapters 26, 29, and 30)
Chemical	Energies of motion and interaction of electrons in atoms and molecules (Chapter 28)*
Nuclear	Energies of motion and interaction of protons and neutrons in atomic nuclei (Chapters 29 and 30)

*Not a *fundamental* form of energy; made up of microscopic kinetic or electromagnetic energy.

In 1843, the English physicist James Prescott Joule (1818–1889), whose "day job" was running the family brewery, performed precise experiments to show that gravitational potential energy could be converted into a previously unrecognized form of energy (internal energy). It had previously been thought that forces like friction "use up" energy. Thanks to Mayer, Joule, and others, we now know that friction converts mechanical forms of energy into internal energy and that total energy is always conserved.

Forms of Energy

Kinetic energy: energy of motion.

Potential energy: stored energy due to interaction.

Energy comes in many different forms (Fig. 6.2). Table 6.1 summarizes the main forms of energy discussed in this text and indicates the principal chapters that discuss each one. At the most fundamental level, there are only three kinds of energy: energy due to motion (**kinetic energy**), stored energy due to interaction (**potential energy**), and rest energy.

To apply the energy conservation principle, we need to learn how to calculate the amount of each form of energy. There isn't one formula that applies to all. Fortunately, we don't have to learn about all of them at once. This chapter focuses on three forms of macroscopic mechanical energy (kinetic energy, gravitational potential energy, and elastic potential energy). For now, we use energy conservation as a tool to understand the **translational** motion of objects, but we do not consider rotational motion or changes in the *internal* energy of an object. We assume that these moving objects are perfectly rigid, so every point on the object moves through the same displacement.

Translation: motion of an object in which any point of the object moves with the same velocity as any other point. (That is, the object does not rotate or change shape.)

6.2   WORK DONE BY A CONSTANT FORCE

To apply the principle of energy conservation, we need to learn how energy can be converted from one form to another. We begin with an example. Suppose the trunk in Fig. 6.3a weighs 220 N and must be lifted a height $h = 4.0$ m. To lift it at constant speed, Rosie must exert a force of 220 N on the rope, assuming an ideal pulley and rope. (We

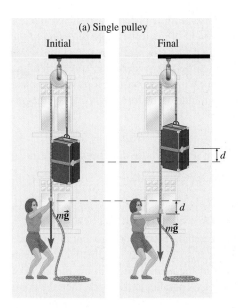
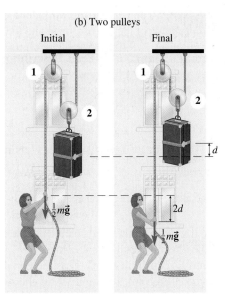

(a) Single pulley

Initial   Final

(b) Two pulleys

Initial   Final

Figure 6.3 (a) Rosie moves a trunk into her dorm room through the window. (b) The two-pulley system makes it easier for Rosie to lift the trunk: the *force* she must exert is halved. Is she getting something for nothing, or does she still have to do the same amount of *work* to lift the trunk?

ignore for now the brief initial time when she pulls with more than 220 N to accelerate the trunk from rest to its constant speed and the brief time she pulls with less than 220 N to let it come to rest.)

As discussed in Example 4.12, she would only have to exert half the force (110 N) if she were to use the two-pulley system of Fig. 6.3b. She doesn't get something for nothing, though. To lift the trunk 4.0 m, the sections of rope on *both* sides of pulley 2 must be shortened by 4.0 m, so Rosie must pull an 8.0-m length of rope. The two-pulley system enables her to pull with half the force, but now she must pull the rope through *twice the distance*.

Notice that the *product* of the magnitude of the force and the distance is the same in both cases:

$$220 \text{ N} \times 4.0 \text{ m} = 110 \text{ N} \times 8.0 \text{ m} = 880 \text{ N·m} = W$$

This product is called the **work** (W) done by Rosie on the rope. Work is a scalar quantity; it does not have a direction. The same symbol W is often used for the weight of an object. To avoid confusion, we write mg for weight and let W stand for work.

Don't be misled by the many different meanings the word *work* has in ordinary conversation. We talk about doing homework, or going to work, or having too much work to do. Not everything we call "work" in conversation is *work* as defined in physics.

The SI unit of work and energy is the newton-meter (N·m), which is given the name joule (symbol: J). Using either method, Rosie must do 880 J of work on the rope to lift the trunk. When we say that Rosie does 880 J of work, we mean that Rosie supplies 880 J of energy—the amount of energy required to lift the trunk 4.0 m. *Work is an energy transfer that occurs when a force acts on an object that moves.*

Rosie does no work on the rope while she holds it in one place because the displacement is zero. She can just as well fasten it and walk away (Fig. 6.4). *If there is no displacement, no work is done and no energy is transferred.* Why then does she get tired if she holds the rope in place for a long time? Although Rosie does no work *on the rope* when holding it in place, work *is* done inside her body by muscle fibers, which have to contract and expand continually to maintain tension in the muscle. This internal work converts chemical energy into internal energy—the muscle warms up—but no energy is transferred *to the trunk*.

Work Done by a Force not Parallel to the Displacement

The force that Rosie exerts on the rope is in the same direction as the displacement of that end of the rope. More generally, how much work is done by a constant force that is at some angle to the displacement? It turns out that only the *component* of the force *in the direction of the displacement*

Figure 6.4 While the trunk is held in place by tying the rope, no work is done and no energy transfers occur.

SI unit of work and energy is the joule: 1 J = 1 N·m.

Work: an energy transfer that occurs when a force acts on an object that moves.

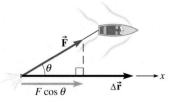

Figure 6.5 The work done by the force of the towrope on the water-skier during a displacement $\Delta\vec{\mathbf{r}}$ is $(F \cos \theta) \Delta r$, where $(F \cos \theta)$ is the component of $\vec{\mathbf{F}}$ in the direction of $\Delta\vec{\mathbf{r}}$.

The **scalar product** (or **dot product**) of two vectors is defined by the equation $\vec{\mathbf{A}} \cdot \vec{\mathbf{B}} = AB \cos \theta$, where θ is the angle between $\vec{\mathbf{A}}$ and $\vec{\mathbf{B}}$ when they are drawn starting at the same point. The special name and notation are used because this pattern occurs often in physics and mathematics. Work can be expressed using the scalar product: $W = \vec{\mathbf{F}} \cdot \Delta\vec{\mathbf{r}}$. See Appendix A.8 for more information on the scalar product.

does work. So, in general, the work done by a constant force is defined as the product of the magnitude of the displacement and the *component* of the force *in the direction of the displacement*. If θ represents the angle between the force and displacement vectors when they are drawn starting at the same point, then the force component in the direction of the displacement is $F \cos \theta$ (Fig. 6.5). Therefore, work done by a constant force on an object can be written $W = F \Delta r \cos \theta$, where F is the magnitude of the force and Δr is the magnitude of the displacement of the object.

Work done by a constant force $\vec{\mathbf{F}}$ acting on an object whose displacement is $\Delta\vec{\mathbf{r}}$:

$$W = F \Delta r \cos \theta \qquad (6\text{-}1)$$

(θ is the angle between $\vec{\mathbf{F}}$ and $\Delta\vec{\mathbf{r}}$)

If we choose the x-axis parallel to the displacement, then the component of the force in the direction of the displacement is $F_x = F \cos \theta$, so $W = F_x \Delta x$. Alternatively, we can identify $\Delta r \cos \theta$ in Eq. (6-1) as the component of the *displacement* in the direction of the *force* (Fig. 6.6). Therefore, if we choose the x-axis parallel to the *force*, then the component of the displacement in the direction of the force is Δx and $W = F_x \Delta x$, as before:

Work done by a constant force $\vec{\mathbf{F}}$ acting on an object whose displacement is $\Delta\vec{\mathbf{r}}$:

$$W = F_x \Delta x \qquad (6\text{-}2)$$

($\vec{\mathbf{F}}$ and/or $\Delta\vec{\mathbf{r}}$ parallel to the x-axis)

Work Can Be Positive, Negative, or Zero   When the angle between $\vec{\mathbf{F}}$ and $\Delta\vec{\mathbf{r}}$ is less than 90°, cos θ in Eq. (6-1) is positive, so the work done by the force is positive ($W > 0$). If the angle between $\vec{\mathbf{F}}$ and $\Delta\vec{\mathbf{r}}$ is greater than 90°, cos θ is negative and the work done by the force is negative ($W < 0$). Pay careful attention to the algebraic sign when calculating work. For example, the rope pulls Rosie's trunk in the direction of its displacement, so $\theta = 0$ and cos $\theta = 1$; the rope does positive work on the trunk. At the same time, gravity pulls downward in the direction opposite to the displacement, so $\theta = 180°$ and cos $\theta = -1$; gravity does *negative* work on the trunk.

If the force is perpendicular to the displacement, $\theta = 90°$ and cos 90° = 0, so the work done is zero. For example, the normal force exerted by a stationary surface on a sliding object does no work because it is perpendicular to the displacement of the object (Fig. 6.7a). Even if the surface is curved, at any instant the normal force is perpendicular to the velocity of the object. During a short time interval, then, the normal force is perpendicular to the displacement $\Delta\vec{\mathbf{r}} = \vec{\mathbf{v}} \Delta t$ (Fig. 6.7b), so the normal force still does zero work.

On the other hand, if the surface exerting the normal force is moving, then the normal force can do work. In Fig. 6.7c, the normal force exerted by the forklift on the pallet does positive work as it lifts the pallet.

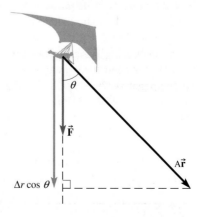

Figure 6.6 The work done by the force of gravity on the hang glider during a displacement $\Delta\vec{\mathbf{r}}$ is $F(\Delta r \cos \theta)$, which is F times the component of $\Delta\vec{\mathbf{r}}$ in the direction of $\vec{\mathbf{F}}$.

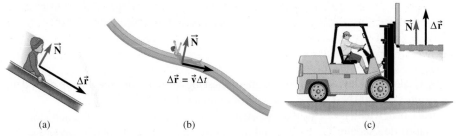

Figure 6.7 (a) The normal force does no work because it is perpendicular to the displacement. (b) Even while sliding on a curved surface, the direction of the normal force is always perpendicular to the displacement during a short Δt, so it does no work. (c) The normal force that the forklift exerts on the pallet does work; it is not perpendicular to the displacement.

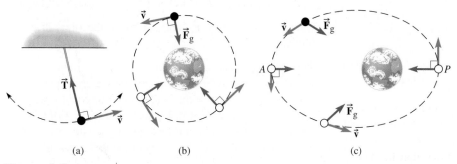

Figure 6.8 (a) The tension in the string of a pendulum is always perpendicular to the velocity of the pendulum bob, so the string does no work on the bob. (b) No matter where the satellite is in its circular orbit, it experiences a gravitational force directed toward the center of the Earth. This force is always perpendicular to the satellite's velocity; thus, gravity does no work on the satellite. (c) In an elliptical orbit, the gravitational force is *not* always perpendicular to the velocity. As the satellite moves counterclockwise in its orbit from point P to point A, gravity does negative work; from A to P, gravity does positive work.

No work is done by the tension in the string on a swinging pendulum bob because the tension is always perpendicular to the velocity of the bob (Fig. 6.8a). Similarly, no work is done by the Earth's gravitational force on a satellite in circular orbit (Fig. 6.8b). In a circular orbit, the gravitational force is always directed along a radius from the satellite to the center of the Earth. At every point in the orbit, the gravitational force is perpendicular to the velocity of the satellite (which is tangent to the circular orbit).

By contrast, gravity does work on a satellite in a noncircular orbit (Fig. 6.8c). Only at points A and P are the gravitational force and the satellite's velocity perpendicular. Wherever the angle between the gravitational force and the velocity is less than 90°, gravity is doing positive work, increasing the satellite's kinetic energy by making it move faster. Wherever the angle between the gravitational force and the velocity is greater than 90°, gravity is doing negative work, decreasing the satellite's kinetic energy by slowing it down.

Application of work: elliptical orbits

√ CHECKPOINT 6.2

A force is applied to a moving object, but no work is done. How is that possible?

Example 6.1

Antique Chest Delivery

A valuable antique chest, made in 1907 by Gustav Stickley, is to be moved into a truck. The weight of the chest is 1400 N. To get the chest from the ground onto the truck bed, which is 1.0 m higher, the movers must decide what to do. Should they lift it straight up, or should they push it up their 4.0-m-long ramp? Assume they push the chest on a wheeled dolly, which in a simplified model is equivalent to sliding it up a *frictionless* ramp.

(a) Find the work done by the movers on the chest if they lift it straight up 1.0 m at constant speed.
(b) Find the work done by the movers on the chest if they slide the chest up the 4.0-m-long *frictionless* ramp at constant speed by pushing parallel to the ramp.
(c) Find the work done by gravity on the chest in each case.
(d) Find the work done by the normal force of the ramp on the chest. Assume that all forces are constant.

Strategy To calculate work, we use either Eq. (6-1) or Eq. (6-2), whichever is easier. For (a) and (b), we must calculate the force exerted by the movers. Drawing the FBD helps us calculate the forces. The ramp is a simple machine—just as for Rosie's pulleys, the ramp cannot reduce the amount of work that must be done, so we expect the work done by the movers to be the same in both cases (ignoring friction). We expect the work done by gravity to be negative in both cases, since the chest is moving up while gravity pulls down. The normal force due to the ramp is perpendicular to the displacement, so it does zero work on the chest. Since more than one force does work on the chest, we use subscripts to clarify which work is being calculated.

Figure 6.9
FBD for the chest as the movers lift it straight up at constant speed.

Given: Weight of chest $mg = 1400$ N; length of ramp $d = 4.0$ m; height of ramp $h = 1.0$ m
To find: Work done by movers on the chest W_m and work done by gravity on the chest W_g in the two cases; work done by the normal force on the chest W_N.

Solution (a) The displacement is 1.0 m straight up. The movers must exert an upward force $\vec{\mathbf{F}}_m$ equal in magnitude to the weight of the chest to move it at constant speed (Fig. 6.9). The work done to lift it 1.0 m is

$$W_m = F_m \, \Delta r \cos \theta = 1400 \text{ N} \times 1.0 \text{ m} \times \cos 0 = +1400 \text{ J}$$

where $\theta = 0$ because $\vec{\mathbf{F}}_m$ and $\Delta\vec{\mathbf{r}}$ are in the same direction (upward).

(b) Figure 6.10 shows a sketch of the situation. We take the x-axis along the inclined ramp and the y-axis perpendicular to the ramp and resolve the gravitational force into its x- and y-components (Fig. 6.11a). Figure 6.11b is the FBD for the chest. Sliding along at constant speed, the chest's acceleration is zero, so the x-components of the forces add to zero.

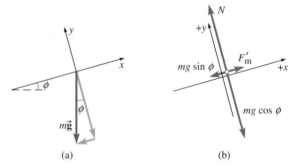

(a)   (b)

Figure 6.11
(a) Resolving $m\vec{\mathbf{g}}$ into x- and y-components; (b) FBD for the chest.

continued on next page

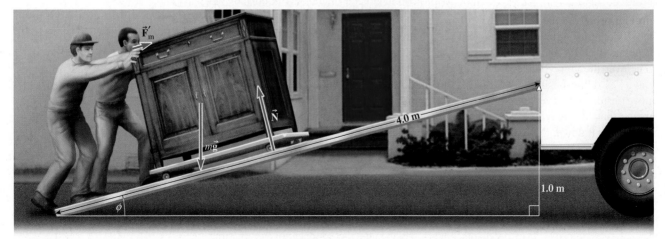

Figure 6.10 An antique chest is pushed up a ramp into a truck.

Example 6.1 continued

The *x*-component of the gravitational force acts in the −*x*-direction and the force exerted by the movers $\vec{F}'_m$ acts in the +*x*-direction. [The *prime* symbol indicates that the force exerted by the movers is different from what it was in part (a).]

$$\sum F_x = F'_m - mg \sin \phi = 0$$

From the right triangle formed by the ramp, the ground, and the truck bed in Fig. 6.12:

$$\sin \phi = \frac{\text{height of truck bed}}{\text{distance along ramp}} = \frac{h}{d}$$

We can now solve for F'_m:

$$F'_m = mg \sin \phi = \frac{mgh}{d}$$

The force and displacement are in the same direction, so $\theta = 0$:

$$W_m = F'_m \, d \cos 0 = \frac{mgh}{d} \times d \times 1 = mgh = +1400 \text{ J}$$

The work done by the movers is the same as in (a).

(c) In both cases, the force of gravity has magnitude *mg* and acts downward. Choosing the *y*-axis so it now points upward, $F_{gy} = -mg$. In both cases, the component of the displacement

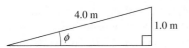

4.0 m

1.0 m

ϕ

Figure 6.12

Finding the angle of the incline.

along the *y*-axis is $\Delta y = h = 1.0$ m. The work done by gravity is the same for the two cases. Using Eq. (6-2),

$$W_g = F_{gy} \Delta y = -mg \, \Delta y$$

$$= -1400 \text{ N} \times 1.0 \text{ m} = -1400 \text{ J}$$

(d) The normal force of the ramp on the chest does no work because it acts in a direction perpendicular to the displacement of the chest.

$$W_N = N \, \Delta r \cos 90° = 0$$

Discussion   Since *d*, the length of the ramp, cancels when multiplying the force times the distance, the work done by the movers is the same for *any* length ramp (as long as the height is the same). Using the ramp, the movers apply one quarter the force over a displacement that is four times larger. With a *real* ramp, friction acts to oppose the motion of the chest, so the movers would have to do *more* than 1400 J of work to slide the chest up the ramp. There's no getting around it; if the movers want to get that chest into the truck, they're going to have to do *at least* 1400 J of work.

Practice Problem 6.1   Bicycling Uphill

A bicyclist climbs a 2.0-km-long hill that makes an angle of 7.0° with the horizontal. The total weight of the bike and the rider is 750 N. How much work is done on the bike and rider by gravity?

Total Work

When several forces act on an object, the total work is the sum of the work done by each force individually:

$$W_{total} = W_1 + W_2 + \cdots + W_N \qquad (6\text{-}3)$$

Total work is sometimes called *net* work because the work done by each force can be positive, negative, or zero, so the total work is often smaller than the work done by any one of the forces. Because we assume a rigid object with no rotational or internal motion, another way to calculate the total work is to find the work done by the *net* force as if there were a single force acting:

$$W_{total} = F_{net} \, \Delta r \cos \theta \qquad (6\text{-}4)$$

To show that these two methods give the same result, let's choose the *x*-axis in the direction of the displacement. Then the work done by each individual force is the *x*-component of the force times Δx. From Eq. (6-3),

$$W_{total} = F_{1x} \Delta x + F_{2x} \Delta x + \cdots + F_{Nx} \Delta x$$

Factoring out the Δx from each term,

$$W_{total} = (F_{1x} + F_{2x} + \cdots + F_{Nx}) \Delta x = \left(\sum F_x\right) \Delta x$$

$\sum F_x$ is the *x*-component of the net force. In Eq. (6-4), $F_{net} \cos \theta$ is the component of the net force in the direction of the displacement, which is the *x*-component of the net force. The two methods give the same total work.

Example 6.2

Fun on a Sled

Diane pulls a sled along a snowy path on level ground with her little brother Jasper riding on the sled (Fig. 6.13). The total mass of Jasper and the sled is 26 kg. The cord makes a 20.0° angle with the ground. As a simplified model, assume that the force of friction on the sled is determined by $\mu_k = 0.16$, even though the surfaces are not dry (some snow melts as the runners slide along it). Find (a) the work done by Diane and (b) the work done by the ground on the sled while the sled moves 120 m along the path at a constant 3 km/h. (c) What is the total work done on the sled?

Strategy   (a,b) To find the work done by a force on an object, we need to know the magnitudes and directions of the force and of the displacement of the object. The sled's acceleration is zero, so the vector sum of all the external forces (gravity, friction, rope tension, and the normal force) is zero. We draw the FBD and use Newton's second law to find the tension in the rope and the force of kinetic friction on the sled. Then we apply Eq. (6-1) or Eq. (6-2) to find the work done by each. (c) We have two methods to find the total work. We'll use Eq. (6-3) to calculate the total work and Eq. (6-4) as a check.

Solution   (a) The FBD is shown in Fig. 6.14. The x- and y-axes are parallel and perpendicular to the ground, respectively. After resolving the tension into its components (Fig. 6.15), Newton's second law with zero acceleration yields

$$\sum F_x = +T \cos \theta - f_k = 0 \qquad (1)$$

$$\sum F_y = +T \sin \theta - mg + N = 0 \qquad (2)$$

where T is the tension and $\theta = 20.0°$. The force of kinetic friction is

$$f_k = \mu_k N$$

Substituting this into Eq. (1)

$$T \cos \theta - \mu_k N = 0 \qquad (3)$$

To find the tension, we need to eliminate the unknown normal force N. Equation (2) also involves the normal force N. We multiply Eq. (2) by μ_k,

$$\mu_k T \sin \theta - \mu_k mg + \mu_k N = 0 \qquad (4)$$

Adding Eqs. (3) and (4) eliminates N. Then we solve for T.

$$T \cos \theta + \mu_k T \sin \theta - \mu_k mg = 0$$

$$T = \frac{\mu_k mg}{\mu_k \sin \theta + \cos \theta}$$

$$= \frac{0.16 \times 26 \text{ kg} \times 9.80 \text{ m/s}^2}{0.16 \times \sin 20.0° + \cos 20.0°} = 41 \text{ N}$$

Now that we know the tension, we find the work done by Diane. The component of the tension T acting parallel to the displacement is $T_x = T \cos \theta$ and the displacement is $\Delta x = 120$ m. The work done by Diane is

$$W_T = (T \cos \theta)\Delta x$$

$$= 41 \text{ N} \times \cos 20.0° \times 120 \text{ m} = +4600 \text{ J}$$

(b) The force on the sled due to the ground has two components: N and f_k. The normal force does no work since it is perpendicular to the displacement of the sled. Friction acts in a direction opposite to the displacement, so the angle between the force and displacement is 180°. The work done by friction is

$$W_f = f_k \, \Delta x \cos 180° = -f_k \, \Delta x$$

From Eq. (1),

$$f_k = T \cos \theta$$

Therefore, the work done by the ground—the work done by the frictional force—is

$$W_f = -f_k \, \Delta x = -(T \cos \theta) \, \Delta x$$

Except for the negative sign, W_f is the same as W_T: $W_f = -4600$ J.

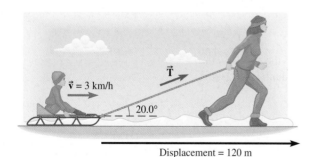

Figure 6.13

Jasper being pulled on a sled.

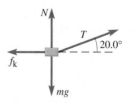

$\theta = 20.0°$

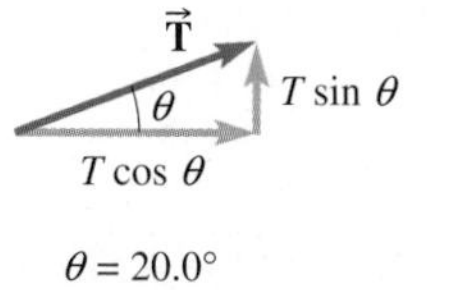

Figure 6.15

Resolving the tension into x- and y-components.

Figure 6.14

FBD.

continued on next page

Example 6.2 continued

(c) The tension and friction are the only forces that do work on the sled. The normal force and gravity are both perpendicular to the displacement, so they do zero work.

$$W_{\text{total}} = W_{\text{T}} + W_{\text{f}} = 4600 \text{ J} + (-4600 \text{ J}) = 0$$

Discussion   To check (c), note that the sled travels with constant velocity, so the net force acting on it is zero. $W_{\text{total}} = F_{\text{net}} \Delta r \cos \theta = 0$.

The speed (3 km/h) was not used in the solution. Assuming that the frictional force on the sled is independent of speed, Diane would exert the same force to pull the sled at *any* constant speed. Then the work she does is the same for a 120-m displacement. At a higher speed, though, she would have to do that amount of work in a shorter time interval.

Practice Problem 6.2   A Different Angle

Find the tension if Diane pulls at an angle $\theta = 30.0°$ instead of 20.0°, assuming the same coefficient of friction. What is the work done by Diane on the sled in this case for a 120-m displacement? Explain how the tension can be greater but the work done by Diane smaller.

Work Done by Dissipative Forces

The work done by kinetic friction was calculated in Example 6.2 according to a simplified model of friction. In this model, when friction truly does −4600 J of work on the sled, it transfers 4600 J of energy from the sled to the ground's internal energy—the ground warms up a bit. In reality, 4600 J of energy is converted into internal energy *shared* between the ground and the sled—both the ground and the sled warm up a little. So the 4600 J is not all transferred to the ground; some stays in the sled but is converted to a different form of energy.

Rather than saying friction does −4600 J of work, a more accurate statement is that friction *dissipates* 4600 J of energy. **Dissipation** is the conversion of energy from an organized form to a disorganized form such as the kinetic energy associated with the random motions of the atoms and molecules within an object, which is part of the object's internal energy. As a practical matter, we usually are not concerned with *where* the internal energy appears. When we can calculate the work done by friction using Eq. (6-1), we get the correct amount of energy dissipated; we just don't know how much of it is transferred to the stationary surface and how much remains in the sliding object. This is how we apply the term *work* to kinetic friction or to other dissipative forces such as air resistance. (In Chapters 14 and 15, when we study internal energy in detail, we will look at situations in which we *do* care where the internal energy appears.)

6.3   KINETIC ENERGY

Suppose a constant net force $\vec{F}_{\text{net}}$ acts on a rigid object of mass m during a displacement $\Delta\vec{r}$. Choosing the x-axis in the direction of the net force, the total work done on the object is

$$W_{\text{total}} = F_{\text{net}} \Delta x$$

where Δx is the x-component of the displacement. Newton's second law tells us that $\vec{F}_{\text{net}} = m\vec{a}$, so

$$W_{\text{total}} = ma_x \Delta x \qquad (6\text{-}5)$$

Since the acceleration is constant, we can use any of the equations for constant acceleration from Chapter 2. From Eq. (2-13), $v_{\text{fx}}^2 - v_{\text{ix}}^2 = 2a_x\Delta x$ or

$$a_x \Delta x = \tfrac{1}{2}(v_{\text{fx}}^2 - v_{\text{ix}}^2)$$

Substituting into Eq. (6-5) yields

$$W_{\text{total}} = \tfrac{1}{2}m(v_{\text{fx}}^2 - v_{\text{ix}}^2)$$

Since the net force is in the x-direction, a_y and a_z are both zero. Only the x-component of the velocity changes; v_y and v_z are constant. As a result,

$$v_f^2 - v_i^2 = (v_{fx}^2 + \cancel{v_{fy}^2} + \cancel{v_{fz}^2}) - (v_{ix}^2 + \cancel{v_{iy}^2} + \cancel{v_{iz}^2}) = v_{fx}^2 - v_{ix}^2$$

Therefore, the total work done is

$$W_{total} = \tfrac{1}{2}m(v_f^2 - v_i^2) = \tfrac{1}{2}mv_f^2 - \tfrac{1}{2}mv_i^2$$

The total work done is equal to the change in the quantity $\tfrac{1}{2}mv^2$, which is called the object's **translational kinetic energy** (symbol K). (Often we just say *kinetic energy* if it is understood that we mean translational kinetic energy.) Translational kinetic energy is the energy associated with motion of the object as a whole; it does not include the energy of rotational or internal motion.

Translational kinetic energy:

$$K = \tfrac{1}{2}mv^2 \tag{6-6}$$

Relation between total work and kinetic energy

Work-kinetic energy theorem:

$$W_{total} = \Delta K \tag{6-7}$$

Kinetic energy is a scalar quantity and is always positive if the object is moving or zero if it is at rest. Kinetic energy is never negative, although a *change* in kinetic energy can be negative. The kinetic energy of an object moving with speed v is equal to the work that must be done on the object to accelerate it to that speed starting from rest. When the total work done is positive, the object's speed increases, increasing the kinetic energy. When the total work done is negative, the object's speed decreases, decreasing the kinetic energy.

Conceptual Example 6.3

Collision Damage

Why is the damage caused by an automobile collision so much worse when the vehicles involved are moving at high speeds?

Strategy When a collision occurs, the kinetic energy of the automobiles gets converted into other forms of energy. We can use the kinetic energy as a rough measure of how much damage can be done in a collision.

Solution and Discussion Suppose we compare the kinetic energy of a car at two different speeds: 60.0 mi/h and 72.0 mi/h (which is 20.0% greater than 60.0 mi/h). If kinetic energy were proportional to speed, then a 20.0% increase in speed would mean a 20.0% increase in kinetic energy. However, since kinetic energy is proportional to the *square* of the speed, a 20.0% speed increase causes an increase in kinetic energy greater than 20.0%. Working by proportions, we can find the percent increase in kinetic energy:

$$\frac{K_2}{K_1} = \frac{\tfrac{1}{2}\cancel{m}v_2^2}{\tfrac{1}{2}\cancel{m}v_1^2} = \left(\frac{72.0 \text{ mi/h}}{60.0 \text{ mi/h}}\right)^2 = 1.44$$

Therefore, a 20.0% increase in speed causes a 44% increase in kinetic energy. What seems like a relatively modest difference in speed makes a lot of difference when a collision occurs.

Practice Problem 6.3   Two Different Cars Collide with a Stone Wall

Suppose a sports utility vehicle and a small electric car both collide with a stone wall and come to a dead stop. If the SUV mass is 2.5 times that of the small car and the speed of the SUV is 60.0 mph while that of the other car is 40.0 mph, what is the ratio of the kinetic energy changes for the two cars (SUV to small car)?

Example 6.4

Bungee Jumping

A bungee jumper makes a jump in the Gorge du Verdon in southern France. The jumping platform is 182 m above the bottom of the gorge. The jumper weighs 780 N. If the jumper falls to within 68 m of the bottom of the gorge, how much work is done by the bungee cord on the jumper during his descent? Ignore air resistance.

Strategy   Ignoring air resistance, only two forces act on the jumper during the descent: gravity and the tension in the cord. Since the jumper has zero kinetic energy at both the highest and lowest points of the jump, the change in kinetic energy for the descent is zero. Therefore, the total work done by the two forces on the jumper must equal zero.

Solution   Let W_g and W_c represent the work done on the jumper by gravity and by the cord. Then

$$W_{total} = W_g + W_c = \Delta K = 0$$

The work done by gravity is

$$W_g = F_y \, \Delta y = -mg \, \Delta y$$

where the weight of the jumper is $mg = 780$ N. With $y = 0$ at the bottom of the gorge, the vertical component of the displacement is

$$\Delta y = y_f - y_i = 68 \text{ m} - 182 \text{ m} = -114 \text{ m}$$

Then the work done by gravity is

$$W_g = -(780 \text{ N}) \times (-114 \text{ m}) = +89 \text{ kJ}$$

The work done by the cord is $W_c = W_{total} - W_g = -89$ kJ.

Discussion   The work done by gravity is positive, since the force and the displacement are in the same direction (downward). If not for the negative work done by the cord, the jumper would have a kinetic energy of 89 kJ after falling 114 m.

The length of the bungee cord is not given, but it does not affect the answer. At first the jumper is in free fall as the cord plays out to its full length; only then does the cord begin to stretch and exert a force on the jumper, ultimately bringing him to rest again. Regardless of the length of the cord, the total work done by gravity and by the cord must be zero since the change in the jumper's kinetic energy is zero.

Practice Problem 6.4   The Bungee Jumper's Speed

Suppose that during the jumper's descent, at a height of 111 m above the bottom of the gorge, the cord has done −21.7 kJ of work on the jumper. What is the jumper's speed at that point?

✓ CHECKPOINT 6.3

Kinetic energy and work are related. Can kinetic energy ever be negative? Can work ever be negative?

6.4   GRAVITATIONAL POTENTIAL ENERGY (1)

Gravitational Potential Energy When Gravitational Force Is Constant

Toss a stone up with initial speed v_i. Ignoring air resistance, how high does the stone go? We can solve this problem with Newton's second law, but let's use work and energy instead. The stone's initial kinetic energy is $K_i = \frac{1}{2}mv_i^2$. For an upward displacement Δy, gravity does negative work $W_{grav} = -mg \, \Delta y$. No other forces act, so this is the total work done on the stone. The stone is momentarily at rest at the top, so $K_f = 0$. Then

$$W_{grav} = K_f - K_i$$

$$-mg \, \Delta y = -\frac{1}{2} \, mv_i^2 \quad \Rightarrow \quad \Delta y = \frac{v_i^2}{2g}$$

From the standpoint of energy conservation, where did the stone's initial kinetic energy go? If total energy cannot change, it must be "stored" somewhere. Furthermore,

The symbol for potential energy is U.

the stone gets its kinetic energy back as it falls from its highest point to its initial position, so the energy is stored in a way that is easily recovered as kinetic energy. Stored energy due to the interaction of an object with something else (here, Earth's gravitational field) that can easily be recovered as kinetic energy is called **potential energy** (symbol U).

The change in gravitational potential energy when an object moves up or down is the *negative* of the work done by gravity:

Change in gravitational potential energy:

$$\Delta U_{grav} = -W_{grav} \tag{6-8}$$

If the gravitational field is uniform, the work done by gravity is

$$W_{grav} = F_y \, \Delta y = -mg \, \Delta y$$

where the y-axis points up. Therefore,

Change in gravitational potential energy:

$$\Delta U_{grav} = mg \, \Delta y \tag{6-9}$$

(uniform $\vec{g}$, y-axis up)

Equation (6-9) holds even if the object does not move in a straight-line path.

Significance of the Negative Sign in Eq. (6-8)   When the stone moves up, Δy is positive. The gravitational force and the displacement of the stone are in opposite directions, so the work done by gravity is negative, gravity is taking away kinetic energy and adding it to its stored potential energy, so the potential energy increases (Fig. 6.16a). If the stone moves down, Δy is negative. The work done by gravity is positive; gravity is giving back kinetic energy by depleting its storage of potential energy, so the potential energy decreases (Fig. 6.16b).

√ CHECKPOINT 6.4

A stone is tossed straight up in the air and is moving upward. (a) Does the gravitational potential energy increase, decrease, or stay the same? (b) What about the kinetic energy? (c) What force, if any, does work on the stone once it leaves the hand of the one who threw it?

Figure 6.16 (a) When the stone moves up, the gravitational potential energy increases. (b) When the stone moves down, the gravitational potential energy decreases.

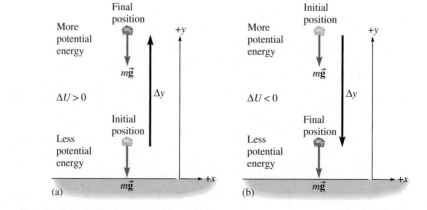

Other Forms of Potential Energy   In addition to gravitational potential energy, other kinds of potential energy include elastic potential energy (Section 6.7) and electric potential energy (Chapter 17). Forces that have potential energies associated with them are called **conservative forces,** for reasons we explain shortly. Not every force has an associated potential energy. For instance, there is no such thing as "frictional potential energy." When kinetic friction does work, it converts energy into a disorganized form that is not easily recoverable as kinetic energy.

Mechanical Energy   The total work done on an object can always be written as the sum of the work done by conservative forces (W_{cons}) plus the work done by nonconservative forces (W_{nc}). Since the total work is equal to the change in the object's kinetic energy [Eq. (6-7)],

$$W_{total} = W_{cons} + W_{nc} = \Delta K \quad \Rightarrow \quad W_{nc} = \Delta K - W_{cons} \qquad \text{(6-10)}$$

Following the same reasoning we used for gravity [see Eq. (6-8)], the change in the total potential energy is equal to the negative of the work done by the conservative forces:

$$\Delta U = -W_{cons} \qquad \text{(6-11)}$$

Combining Eqs. (6-10) and (6-11) yields

$$W_{nc} = \Delta K + \Delta U = \Delta E_{mech} \qquad \text{(6-12)}$$

or

$$(K_i + U_i) + W_{nc} = (K_f + U_f)$$

The sum of the kinetic and potential energies ($K + U$) is called the **mechanical energy** (E_{mech}). W_{nc} is equal to the change in mechanical energy. When finding the change in mechanical energy, do not include the work done by conservative forces. Conservative forces such as gravity do not change the mechanical energy; they just change one form of mechanical energy into another. Work done by conservative forces is already accounted for by the change in potential energy.

Mechanical energy: the sum of the kinetic and potential energies

The term *conservative force* comes from a time before the general law of conservation of energy was understood and when no forms of energy other than mechanical energy were recognized. Back then, it was thought that certain forces conserved energy and others did not. Now we believe that *total* energy is *always* conserved. Nonconservative forces do not conserve *mechanical* energy, but they do conserve *total* energy.

Conservation of Mechanical Energy

When nonconservative forces do no work, mechanical energy is conserved:
$$E_i = E_f$$

Example 6.5

Rock Climbing in Yosemite

A team of climbers is rappelling down steep terrain in the Yosemite valley (Fig. 6.17). Mei-Ling (mass 60.0 kg) slides down a line starting from rest 12.0 m above a horizontal shelf. If she lands on the shelf below with a speed of 2.0 m/s, calculate the energy dissipated by the kinetic frictional forces acting between her and the line. The local value of g is 9.78 N/kg. Ignore air resistance.

continued on next page

Example 6.5 continued

Figure 6.17

Mei-Ling rappelling downward from a position 12.0 m above a shelf.

Given: mass of climber, $m = 60.0$ kg; $\Delta y = -12.0$ m; $v_i = 0$ m/s and $v_f = 2.0$ m/s, just before stopping; local field strength $g = 9.78$ N/kg.

To find: change in mechanical energy ΔE.

Solution $W_{nc} = \Delta E_{mech} = \Delta K + \Delta U$, so we need to calculate the changes in kinetic and potential energy. Mei-Ling's kinetic energy is initially zero since she starts at rest. The change in her kinetic energy is

$$\Delta K = \tfrac{1}{2}mv_f^2 - \tfrac{1}{2}mv_i^2 = \tfrac{1}{2}mv_f^2 - 0 = \tfrac{1}{2}(60.0 \text{ kg}) \times (2.0 \text{ m/s})^2$$
$$= +120 \text{ J}$$

The change in her potential energy is

$$\Delta U = mg\,\Delta y = 60.0 \text{ kg} \times 9.78 \text{ m/s}^2 \times (0 - 12.0 \text{ m}) = -7040 \text{ J}$$

The work done by friction is

$$\Delta E_{mech} = \Delta K + \Delta U = 120 \text{ J} + (-7040 \text{ J}) = -6920 \text{ J}$$

The amount of energy dissipated by friction (converted from mechanical energy into internal energy) is 6920 J. Fortunately, Mei-Ling is wearing gloves, so her hands don't get burned.

Discussion   If the line had broken when Mei-Ling was at the top, her final kinetic energy would have been +7040 J—disastrously large since it corresponds to a final speed of

$$v = \sqrt{\frac{K}{\tfrac{1}{2}m}} = \sqrt{\frac{7040 \text{ J}}{30.0 \text{ kg}}} = 15.3 \text{ m/s}$$

Instead, kinetic friction reduces her final kinetic energy to a manageable +120 J (which corresponds to a final speed of 2.0 m/s). Mei-Ling can absorb this much kinetic energy safely by landing on the shelf while bending her knees.

Practice Problem 6.5   Energy Dissipated by Air Resistance

A ball thrown straight up at an initial speed of 14.0 m/s reaches a maximum height of 7.6 m. What fraction of the ball's initial kinetic energy is dissipated by air resistance as the ball moves upward?

Strategy   The forces acting on Mei-Ling are gravity and kinetic friction (Fig. 6.18). The only force whose work is not included in the change in potential energy is the work done by kinetic friction. Therefore, the change in the mechanical energy, $\Delta K + \Delta U$, is equal to the work done by friction. Since we know Mei-Ling's initial and final speeds as well as her mass, we can calculate the change in her kinetic energy. From the change in height, we can calculate the change in potential energy.

Figure 6.18

FBD for Mei-Ling.

Choosing Where the Potential Energy Is Zero

Notice that when we apply Eq. (6-12), only the *change* in potential energy enters the calculation. Therefore, we can always assign the value of the potential energy for any *one* position. Most often, we choose some convenient position and assign it to have zero potential energy. Once that choice is made, the potential energy of every other configuration is determined by Eq. (6-11).

For gravitational potential energy in a uniform gravitational field, we usually choose the potential energy to be zero at some convenient place: on the floor, on a table, or at the top of a ladder. After assigning $y = 0$ to that place, the potential energy at any other place is $U = mgy$.

> **Gravitational potential energy:**
> $$U_{\text{grav}} = mgy \qquad (6\text{-}13)$$
> (uniform $\vec{\mathbf{g}}$, y-axis up, assign $U = 0$ to $y = 0$)

Potential energy is then positive above $y = 0$ and negative below it. There is no special significance to the sign of the potential energy. What matters is the sign of the potential energy *change*.

Example 6.6

A Quick Descent

A ski trail makes a vertical descent of 78 m. A novice skier, unable to control his speed, skis down this trail and is lucky enough not to hit any trees. What is his speed at the bottom of the trail, ignoring friction and air resistance?

Strategy When nonconservative forces do no work, $W_{\text{nc}} = \Delta E_{\text{mech}} = 0$ and mechanical energy does not change. A skilled skier can control his speed by, in effect, controlling how much work the frictional force does on the skis. Here we assume *no* friction or air resistance. Then the only forces acting on the skier are the normal force and gravity (Fig. 6.19). The normal force does no work, since it is always perpendicular to the skier's velocity, so $W_{\text{nc}} = 0$.

continued on next page

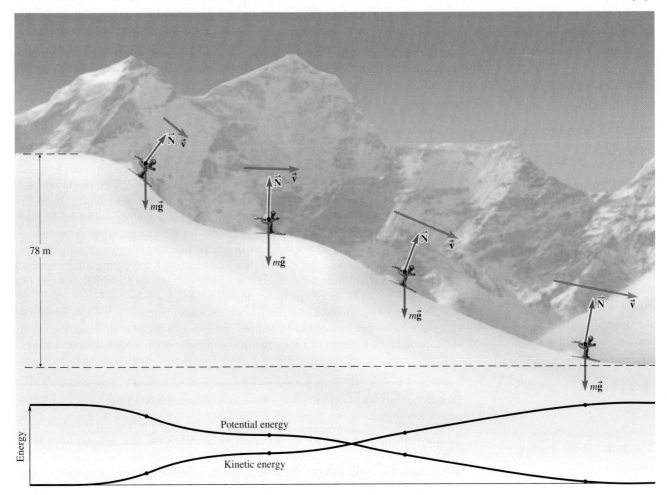

Figure 6.19 The final speed of the skier depends only on the initial and final altitudes if no friction acts.

Example 6.6 continued

Solution   Because $W_{nc} = 0$, the mechanical energy does not change:

$$K_i + U_i = K_f + U_f$$

If we choose the y-axis up and $y = 0$ at the bottom of the hill, $y_i = 78$ m and $y_f = 0$. Then

$$U_i = mgy_i \quad \text{and} \quad U_f = 0$$

If the skier starts with zero kinetic energy, then $K_i = 0$ and $K_f = \frac{1}{2}mv_f^2$. Setting the mechanical energies equal,

$$0 + mgy_i = \frac{1}{2}mv_f^2 + 0$$

Solving for the final speed v_f,

$$v_f = \sqrt{2gy_i} = \sqrt{2 \times 9.80 \text{ m/s}^2 \times 78 \text{ m}} = 39 \text{ m/s}$$

Discussion   Notice that the solution did not depend on the detailed shape of the path. If the slope were constant (Fig. 6.20), we could use Newton's second law to find the skier's acceleration and then the change in velocity:

$$\sum F_x = mg \sin \theta = ma_x \implies a_x = g \sin \theta$$

From Eq. (2-13),

$$\Delta x = \frac{v_{fx}^2 - \cancel{v_{ix}^2}}{2a_x} = \frac{v_{fx}^2}{2g \sin \theta} = \frac{h}{\sin \theta} \implies v_{fx} = \sqrt{2gh}$$

where $h = 78$ m.

This method shows that the final speed does not depend on the angle of the slope, but the energy method shows that the final speed is the same for *any* shape path, not just for constant slopes. On the other hand, the *time* that it takes the skier to reach the bottom *does* depend on the length and contour of the trail.

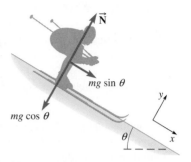

Figure 6.20
FBD for the skier on a constant slope.

A final speed of 39 m/s (87 mi/h) is dangerously fast. In reality, friction and air resistance would do negative work on the skier, so the final speed would be smaller.

Practice Problem 6.6   Speeding Roller Coaster

A roller coaster is hauled to the top of the first hill of the ride by a motorized chain drive. After that, the train of cars is released and no more energy is supplied by an external motor. The cars are moving at 4.0 m/s at the top of the first hill, 35.0 m above the ground. How fast are they moving at the top of the second hill, 22.0 m above the ground? Ignore friction and air resistance.

Recognizing a Conservative Force

In Example 6.6, the final speed doesn't depend on the shape of the trail: it could have been a steep descent, or a long gradual one, or have a complicated profile with varying slope. It could even be a vertical descent—the final speed is the same for free fall off a 78-m-high building. Any time the work done by a force is *independent of path*—that is, the work depends only on the initial and final positions—the force is conservative. We depend on the path-independence of the work done to define the potential energy in Eq. (6-11).

Energy stored as potential energy by a conservative force during a displacement from point A to point B can be recovered as kinetic energy. We can simply reverse displacement to get all of the energy back: $\Delta U_{B \to A} = -\Delta U_{A \to B}$.

The work done by friction, air resistance, and other contact forces *does* depend on path, so these forces cannot have potential energies associated with them. We cannot use friction to store energy in a form that is completely recoverable as kinetic energy.

6.5   GRAVITATIONAL POTENTIAL ENERGY (2)

The expressions for gravitational potential energy developed in Section 6.4 apply when the gravitational force is *constant* (or nearly constant). If the gravitational force is not constant, such as when a satellite is placed into orbit around the Earth, Eqs. (6-9) and (6-13) cannot be used. Instead, we need to use an expression for gravitational potential energy that corresponds to Newton's law of universal gravitation. Recall that the magnitude of the gravitational force that one body exerts on another is

$$F = \frac{Gm_1 m_2}{r^2} \qquad (2\text{-}6)$$

where r is the distance between the centers of the bodies. The corresponding expression for gravitational potential energy in terms of the distance between two bodies is

Gravitational potential energy:

$$U = -\frac{Gm_1 m_2}{r} \qquad (6\text{-}14)$$

(assign $U = 0$ when $r = \infty$)

A graph showing the gravitational potential energy as a function of r is shown in Fig. 6.21. Note that we have assigned the potential energy to be zero at infinite separation ($U = 0$ when $r = \infty$). Why this choice? Simply put, any other choice would mean adding a constant term to the expression for U. This constant term would *always subtract out* of our equations, which involve only *changes* in potential energy.

This choice ($U = 0$ when $r = \infty$) means that the gravitational potential energy is *negative* for any finite value of r, because potential energy decreases as the bodies get closer together and increases as they get farther apart.

Does Eq. (6-14) Contradict Eq. (6-9)?   Calculus is used to derive Eq. (6-14), but we can *verify* that it is consistent with Eq. (6-9) without using calculus. For a *very small* displacement from r_i to $r_f = r_i + \Delta y$ (Fig. 6.22), the potential energy change given by Eq. (6-14) must reduce to the constant-force case:

$$\Delta U = U_f - U_i = \left(-\frac{GM_E m}{r_i + \Delta y}\right) - \left(-\frac{GM_E m}{r_i}\right)$$

Rearranging and factoring out the common factors $GM_E m$ and then rewriting with a common denominator,

$$\Delta U = GM_E m \left(\frac{1}{r_i} - \frac{1}{r_i + \Delta y}\right) = GM_E m \frac{\cancel{r_i} + \Delta y - \cancel{r_i}}{r_i(r_i + \Delta y)} \qquad (6\text{-}15)$$

For values of Δy that are small compared with r_i, $r_i + \Delta y \approx r_i$. Making that approximation in the denominator of Eq. (6-15),

$$\Delta U = m\left(\frac{GM_E}{r_i^2}\right) \Delta y \quad (\Delta y \ll r_i) \qquad (6\text{-}16)$$

The quantity in the parentheses in Eq. (6-16) is the gravitational field strength g, the gravitational force on the object divided by its mass m. Then, $\Delta U = mg\,\Delta y$, in agreement with Eq. (6-9).

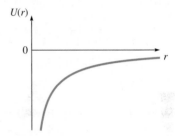

Figure 6.21 Gravitational potential energy as a function of r, the distance between the centers of the two bodies. The potential energy increases as the distance increases.

Figure 6.22 An object at a distance r from Earth's center moves up a small distance Δy (greatly exaggerated in the figure).

✓ CHECKPOINT 6.5

As Mercury travels in its elliptical orbit about the Sun, how does its mechanical energy at its nearest point (*perihelion*) to the Sun compare with that at its farthest point (*aphelion*) from the Sun? How does its potential energy compare at the same two points?

Example 6.7

Orbital Speed of Mercury

The orbit of the planet Mercury around the Sun is an ellipse. At its perihelion (4.60×10^7 km), its orbital speed is 59 km/s. What is its orbital speed at aphelion (6.98×10^7 km)?

Strategy   Ignoring the small gravitational forces exerted by other planets, the only force acting on Mercury is the gravitational force due to the Sun. Gravity is a conservative force, so the mechanical energy is constant. Figure 6.23 is a sketch of the orbit. At aphelion, Mercury is farther from the Sun than at perihelion, so the potential energy is greater. Then the kinetic energy must be smaller, so the answer must be less than 59 km/s.

Given: $v_p = 5.9 \times 10^4$ m/s, $r_p = 4.60 \times 10^{10}$ m, $r_a = 6.98 \times 10^{10}$ m.
To find: v_a.

Solution   Mechanical energy is constant:

$$K_p + U_p = K_a + U_a$$

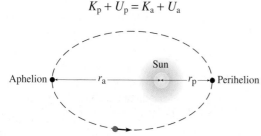

Figure 6.23
Sketch of Mercury's orbit.

The kinetic energy of Mercury at perihelion is $K_p = \frac{1}{2}mv_p^2$, where m is the mass of Mercury; the kinetic energy at aphelion is $K_a = \frac{1}{2}mv_a^2$. The potential energies at perihelion and at aphelion are

$$U_p = -\frac{GM_S m}{r_p} \quad \text{and} \quad U_a = -\frac{GM_S m}{r_a}$$

respectively, where $M_S = 1.99 \times 10^{30}$ kg is the mass of the Sun. From conservation of energy:

$$\frac{1}{2}m v_p^2 + \left(-\frac{GM_S m}{r_p}\right) = \frac{1}{2}m v_a^2 + \left(-\frac{GM_S m}{r_a}\right)$$

The mass of Mercury cancels out. Solving for v_a,

$$\frac{1}{2}v_a^2 = \frac{1}{2}v_p^2 + \left(-\frac{GM_S}{r_p}\right) - \left(-\frac{GM_S}{r_a}\right)$$

$$v_a = \sqrt{v_p^2 + 2GM_S\left(\frac{1}{r_a} - \frac{1}{r_p}\right)}$$

Substituting numerical values yields $v_a = 39$ km/s.

Discussion   The speed at aphelion is less than the speed at perihelion, as expected.

Practice Problem 6.7   Speed at a Different Distance

What is Mercury's orbital speed when its distance from the Sun is 5.80×10^7 km?

Example 6.8

Escape Speed

Ignoring air resistance, what is the minimum initial speed a projectile must have at Earth's surface if the projectile is to escape Earth's gravitational pull?

Strategy   What does "escape Earth's gravitational pull" mean? The gravitational force on the projectile due to Earth *approaches* zero at large distances, but never *reaches* zero. We are looking for the initial speed so that, even though Earth's gravity keeps pulling the projectile back, the projectile can keep moving away from Earth. The gravitational force is not constant, and the trajectory of the projectile may be complicated, so using $\Sigma \vec{F} = m\vec{a}$ is impractical. We try an energy approach.

The only force acting on the projectile is gravity, so the mechanical energy is constant. To escape, the projectile must have enough initial kinetic energy so that it can reach an unlimited distance from Earth.

Solution   The mechanical energy is constant:

$$K_i + U_i = K_f + U_f$$

Initially the projectile is at a distance R_E from Earth's center and is moving at initial speed v_i. At some later time, the projectile has speed v_f at distance r_f from Earth. Then

$$\frac{1}{2}mv_i^2 + \left(-\frac{GM_E m}{R_E}\right) = K_f + U_f$$

continued on next page

Example 6.8 continued

To escape, the projectile must be able to reach any value of r_f, no matter how large. As r_f gets larger and larger, the potential energy approaches its maximum value, which is zero. (Mathematically, as $r_f \to \infty$, $U_f \to 0$.) The *minimum* value of v_i gives the projectile *just enough* energy. So we assume that the projectile can reach its maximum potential energy without any kinetic energy left over ($K_f = 0$):

$$\frac{1}{2}mv_i^2 + \left(-\frac{GM_E m}{R_E}\right) = 0 + 0$$

Solving for v_i,

$$\frac{1}{2}\cancel{m}v_i^2 = \frac{GM_E \cancel{m}}{R_E} \quad \Rightarrow \quad v_i = \sqrt{\frac{2GM_E}{R_E}} = 11.2 \text{ km/s}$$

Discussion This speed is called the **escape speed** of Earth. Note that the escape speed is independent of the mass of the projectile because both the kinetic energy and the potential energy are proportional to the projectile's mass.

The concept of escape speed helps explain why there is little hydrogen gas (H_2) or helium gas (He) in Earth's atmosphere. We will see in Chapter 13 that the molecules in a gas have an average kinetic energy determined by the temperature of the gas. In a mixture of gases, the molecules with the smallest mass have the highest average speeds. The average speeds of hydrogen and helium in our atmosphere are large enough that they can escape the atmosphere. A negligible fraction of the nitrogen, oxygen, or water molecules have speeds greater than the escape speed, so they persist in the atmosphere.

Practice Problem 6.8   Protons Streaming Away from the Sun

Particles such as protons and electrons are continually streaming away from the Sun in all directions. They carry off some of the energy released in the thermonuclear reactions occurring in the Sun. How fast must a proton be moving at a distance of 7.00×10^9 m from the center of the Sun for it to escape the Sun's gravitational pull and leave the solar system?

6.6   WORK DONE BY VARIABLE FORCES: HOOKE'S LAW

So far we have considered only constant forces when calculating work. The advantage of using energy methods really shines in problems dealing with variable forces, where it's difficult to use Newton's second law. How can we calculate the work done by a variable force? Consider an archer drawing back a compound bow (Fig. 6.24). The compound bow is designed to make it easier to draw the string back and hold it back because, at a certain point, the force required to draw the string farther stops increasing. A convenient way to describe how the force varies with string position is to plot a graph. Figure 6.25 shows the force that must be applied to hold the string back as a function of distance. How can we calculate the work done by the archer as he draws the string back 40 cm?

We've asked analogous questions in previous chapters. Recall how we find the displacement Δx when the velocity v_x is not constant (Section 2.2). We divide the time interval into a series of *short* time intervals and sum up the displacements that occur during each one.

To approximate the work done by a variable force F_x, we divide the overall displacement into a series of small displacements Δx. During each small displacement, the work done is

$$\Delta W = F_x \Delta x \qquad (6\text{-}17)$$

On a graph of $F_x(x)$, each ΔW is the area of a rectangle of height F_x and width Δx (Fig. 6.26). The total work done is the sum of the areas of these rectangles. This approximation gets better as we make the rectangles thinner and thinner, so *the total work done is the area under the graph of $F_x(x)$ from x_i to x_f.*

In Fig. 6.25, the "area" of each rectangle represents (0.050 m $\times 20.0$ N) $= 1.0$ J of work. There are approximately 36 rectangles under the graph between $x = 0$ and $x = 40$ cm, so the work done by the archer is $+36$ J.

Figure 6.24 Application of work done by a variable force: drawing a compound bow.

CONNECTION

See Sections 2.2 and 2.3 to review how we found that the area under a graph of $v_x(t)$ is Δx and that the area under a graph of $a_x(t)$ is Δv_x.

W = the area under a graph of $F_x(x)$

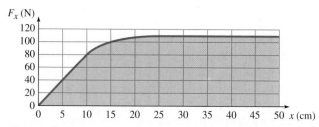

Figure 6.25 The force to draw back the compound bow depends on how far it is drawn. In this graph, the "area" represented by each rectangle is 0.050 m × 20.0 N = 1.0 J.

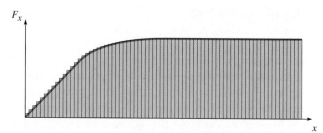

Figure 6.26 Each rectangle's area approximates the work done during a small displacement. The total area of the rectangles approximates the total work done.

Example 6.9

Archery Practice

To draw back a *simple* bow, the force the archer exerts on the string continues to increase as the displacement of the string increases and the bow bends slightly. The force-versus-position graph of Fig. 6.27 describes such a bow. Calculate the work done by the archer on the string as he draws the string back 40.0 cm.

Strategy The work done by the archer is the area under the force-versus-position graph. This time, instead of counting rectangles, we can calculate the triangular area formed by the force-versus-position graph.

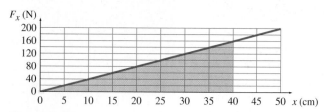

Figure 6.27
A simple bow requires a force proportional to the displacement of the string.

Solution We want to find the work done by the archer to draw the string back 40.0 cm, so the base of the triangle is 40.0 cm. The altitude of the triangle is the force at 40.0 cm: 160 N. The area of a triangle is $\frac{1}{2}$(base × altitude), so

$$W = \tfrac{1}{2}(0.400 \text{ m} \times 160 \text{ N}) = +32 \text{ J}$$

Discussion To check, we can count the number of rectangles (including the half rectangles) that lie under the graph. There are 32 rectangles and each represents 20 N × 0.05 m = 1 J of work, so the answer is correct.

By doing 32 J of work on the bowstring, the archer stores this much energy in the bow. When the arrow is released, the bowstring does 32 J of work on the arrow, giving the arrow a kinetic energy of 32 J.

Practice Problem 6.9   A Gentle Pull

How much work would you do to draw the string of the compound bow (Fig. 6.25) back 10.0 cm instead of 40.0 cm?

Hooke's Law and Ideal Springs

In Example 6.9, the displacement of the bowstring is proportional to the force exerted by the archer. Robert Hooke (1635–1703) observed that, for many objects, the deformation—change in size or shape—of the object is proportional to the magnitude of the force that causes the deformation. This observation, called **Hooke's law**, is an approximation and is valid only within limits. For example, the compound bow of Fig. 6.25 is described by Hooke's law only for an applied force less than 80 N.

Many springs are described by Hooke's law as long as they are not stretched or compressed too far. That is, the extension or compression—the increase or decrease in length from the relaxed length—is proportional to the force applied to the ends of the spring. When we refer to an **ideal spring**, we mean a spring that is described by Hooke's law and is also massless.

Hooke's law: the deformation is proportional to the deforming force.

Hooke's law for an ideal spring:

$$F = k \, \Delta L \qquad \text{(6-18)}$$

In Eq. (6-18), F is the *magnitude* of the force exerted *on each end* of the spring and ΔL is the distance that the spring is stretched or compressed from its relaxed length.

The constant k is called the **spring constant** for a particular spring. The SI unit of force is the newton and the SI unit of length is the meter, so the SI units of a spring constant are N/m. The spring constant is a measure of how hard it is to stretch or compress a spring. A stiffer spring has a larger spring constant because larger forces must be exerted on the ends of the spring to stretch or compress it. Example 1.10 describes an experiment to measure the spring constant of a real spring and shows a graph of length of the spring as a function of the forces applied to its ends (Fig. 1.5).

In many situations, we are more interested in the forces exerted *by* the spring than in the forces exerted *on* it. From Newton's third law, the forces exerted *by* the spring on whatever is attached to its ends are equal in magnitude and opposite in direction to the forces exerted *by* those objects *on* the ends of the spring. Suppose that an ideal spring is aligned with the x-axis. One end is fixed in place and the other end can move along the x-axis (Fig. 6.28). Choose the origin so the moveable end is at $x = 0$ when the spring is relaxed. Then the force exerted by the moveable end of the spring on whatever is attached to it is

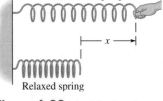

Figure 6.28 An ideal spring is stretched a distance x beyond its relaxed length.

Force exerted *by* an ideal spring (Hooke's law):

$$F_x = -kx \qquad (6\text{-}19)$$

(F_x is the force exerted *by* the moveable end when its position is x; the spring is relaxed at $x = 0$.)

The negative sign in Eq. (6-19) indicates the direction of the force. The moveable end of the spring always pushes or pulls toward its relaxed position. If it is displaced in the $+x$-direction, the force it exerts is in the $-x$-direction (back toward $x = 0$). If it is displaced in the $-x$-direction, the force it exerts is in the $+x$-direction (again, back toward $x = 0$).

Example 6.10

Getting Down to Nuts and Bolts

In many hardware stores, bulk nuts and bolts are sold by weight. A spring scale in the store stretches 4.8 cm when 24.0 N of bolts are weighed. On the scale, what is the distance in centimeters between calibration marks that are marked in increments of 1 N? Assume an ideal spring.

Strategy The bolts are in equilibrium, so the spring scale is pulling upward on them with a force of 24.0 N (see Fig. 6.29). Using Hooke's law and the data given, we can find the spring constant k. Then we can use Hooke's law again to find out how much the spring stretches when the applied force is increased by 1 N.

Solution Let the x-axis point up. When the pan of the scale is at $x = -4.8$ cm, it exerts a force $F_x = +24.0$ N on the bolts. From Hooke's law, $F_x = -kx$ and the spring constant is

$$k = -\frac{F_x}{x} = \frac{-24.0 \text{ N}}{-4.8 \text{ cm}} = 5.0 \text{ N/cm}$$

Now let $F_x = 1.00$ N and solve for x:

$$x = -\frac{F_x}{k} = -\frac{1.00 \text{ N}}{5.0 \text{ N/cm}} = -0.20 \text{ cm}$$

Since the relation between F and x is linear, the spring stretches an additional 0.20 cm for each additional newton of force. Therefore, the 1-N marks should be 0.20 cm apart.

Discussion A variation on the solution is to look back at the question and notice that we are asked how many centimeters the spring stretches for each newton of force, which is the *reciprocal* of the spring constant. The reciprocal of the spring constant is

$$\frac{1}{k} = -\frac{x}{F} = -\frac{-4.8 \text{ cm}}{24.0 \text{ N}} = 0.20 \text{ cm/N}$$

The answer is reasonable: since it takes 5 N to make the spring stretch 1 cm, 1 N makes the spring stretch $\frac{1}{5}$ cm.

Scale pulling up

Weight of bolts

Figure 6.29

FBD for the pan of the scale.

Practice Problem 6.10 Stretching a Spring

16.0 N of nuts are placed in the pan of the scale of Example 6.10. How far does the spring stretch?

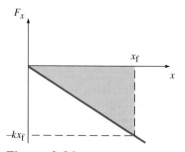

Figure 6.30 The work done by the spring is the (negative) area under the $F_x(x)$ graph.

Work Done by an Ideal Spring

To find the work done by an ideal spring, first we draw the $F_x(x)$ graph (Fig. 6.30). The unstretched position of the moveable end is $x = 0$. The work done by the spring as its moveable end moves from equilibrium ($x_i = 0$) to the final position x_f is the area of the shaded right triangle whose base is x and altitude is $-kx$:

$$W = \tfrac{1}{2}(\text{base} \times \text{altitude}) = -\tfrac{1}{2}kx^2 \qquad (6\text{-}20)$$

The area is negative because the graph is underneath the x-axis. Think of $-\tfrac{1}{2}kx^2$ as the average force ($-\tfrac{1}{2}kx$) times the displacement (x).

More generally, if the moveable end starts at position x_i, not necessarily at the equilibrium point, the work done by the spring is

$$W_{\text{spring}} = (-\tfrac{1}{2}kx_f^2) - (-\tfrac{1}{2}kx_i^2) = -\tfrac{1}{2}kx_f^2 + \tfrac{1}{2}kx_i^2 \qquad (6\text{-}21)$$

Imagine the spring starting at equilibrium and ultimately ending up at a displacement x_f after passing through x_i. The total work done by the spring is $-\tfrac{1}{2}kx_f^2$; then we subtract the work that was done to get the spring to position x_i from equilibrium ($-\tfrac{1}{2}kx_i^2$) to get the work done from x_i to x_f.

6.7   ELASTIC POTENTIAL ENERGY

The work done by an ideal spring [Eq. (6-21)] depends on the initial and final positions of the moveable end, but *not* on the path that was taken. Therefore, the force exerted by an ideal spring is *conservative* and we can associate a potential energy with it. The kind of potential energy stored in a spring is called **elastic potential energy**.

Just as for gravity [see Eqs. (6-8) and (6-11)], the change in elastic potential energy is the *negative* of the work done by the spring:

$$\Delta U_{\text{elastic}} = -W_{\text{spring}} \qquad (6\text{-}22)$$

> **CONNECTION:**
>
> The change in potential energy is always equal to the negative of the work done by the associated force. See Eq. (6-11).

For example, if you increase the elastic energy stored in a spring by compressing it, the spring does *negative* work because the force its end exerts on your hand is in the direction opposite to its displacement. This stored elastic energy can be recovered as kinetic energy by, say, using the spring to shoot a stone. As the spring expands back to its original length, it does positive work on the stone to increase the stone's kinetic energy and the stored elastic energy decreases.

From Eqs. (6-21) and (6-22),

$$\Delta U_{\text{elastic}} = \tfrac{1}{2}kx_f^2 - \tfrac{1}{2}kx_i^2 \qquad (6\text{-}23)$$

Remember that only changes in potential energy enter our calculations, so we can assign $U = 0$ to any convenient position. The most convenient choice is to assign $U = 0$ when the spring is relaxed ($x = 0$):

> **Elastic potential energy stored in an ideal spring:**
>
> $$U_{\text{elastic}} = \tfrac{1}{2}kx^2 \qquad (6\text{-}24)$$
>
> $U = 0$ when $x = 0$ (relaxed spring)

Conservation of Energy with More than One Form of Potential Energy When applying conservation of energy using $W_{\text{nc}} = \Delta K + \Delta U$ [Eq. (6-12)], ΔU must include the change in all forms of potential energy. For now, with two forms of potential energy,

$$\Delta U = \Delta U_{\text{grav}} + \Delta U_{\text{elastic}} \qquad (6\text{-}25)$$

W_{nc} is the work done by all forces *other than* those included in the potential energy. When $W_{nc} = 0$, the mechanical energy $K + U$ is constant.

CHECKPOINT 6.7

If a spring is compressed horizontally on a table and then released so it expands to its original relaxed position, where does the spring have the greatest elastic potential energy?

Example 6.11

The Dart Gun

In a dart gun (Fig. 6.31), a spring with $k = 400.0$ N/m is compressed 8.0 cm when the dart (mass $m = 20.0$ g) is loaded (Fig. 6.31a). What is the muzzle speed of the dart when the spring is released (Fig. 6.31b)? Ignore friction.

Strategy The elastic energy initially stored in the spring is converted into the kinetic energy of the dart as the spring expands. There is no change in gravitational potential energy since the motion of the dart is horizontal. The vertical normal forces do no work because they are perpendicular to the displacement of the dart. The spring pushes the dart to the right until it reaches its relaxed length. Assuming the spring can't pull the dart to the left (as it would if they stick together), the dart loses contact with the spring when the spring is at its relaxed length. We choose the origin at the relaxed position of the spring; therefore, $x_f = 0$. Using the x-axis in Fig. 6.31, $x_i = -8.0$ cm. The dart starts from rest, so $v_i = 0$. To find: v_f.

Solution Since we ignore friction, no work is done by nonconservative forces. Therefore, the mechanical energy is constant:

$$K_i + U_i = K_f + U_f$$

We can ignore the gravitational potential energy because it does not change. Using Eq. (6-24) for the elastic potential energy in the spring,

$$\tfrac{1}{2}mv_i^2 + \tfrac{1}{2}kx_i^2 = \tfrac{1}{2}mv_f^2 + \tfrac{1}{2}kx_f^2$$

After setting $x_f = 0$ and $v_i = 0$,

$$0 + \tfrac{1}{2}kx_i^2 = \tfrac{1}{2}mv_f^2 + 0$$

Solving for v_f,

$$v_f = \sqrt{\frac{k}{m}}\, x_i = \sqrt{\frac{400.0 \text{ N/m}}{0.0200 \text{ kg}}} \times 0.080 \text{ m} = 11 \text{ m/s}$$

Discussion Checking the units,

$$\sqrt{\frac{\text{N/m}}{\text{kg}}} \times \text{m} = \sqrt{\frac{(\text{kg} \cdot \text{m/s}^2)/\text{m}}{\text{kg}}} \times \text{m} = \frac{\text{m}}{\text{s}}$$

Notice that the muzzle speed is proportional to the distance the spring is compressed when the gun is cocked. If the spring is compressed halfway, it stores only *one quarter* as much elastic energy. The dart then acquires one quarter the kinetic energy, which means its speed is half as much. A more massive dart fired from the same gun would have a smaller muzzle speed, but the *same* kinetic energy.

Practice Problem 6.11   A Misfire

The same dart gun is cocked by compressing the spring the same distance (8.0 cm). This time the spring gets caught inside the gun, stopping at the point where it is still compressed by 4.0 cm. The dart is not caught inside the gun, but is released. Find the muzzle speed of the dart. [*Hint:* What is x_f in this case?]

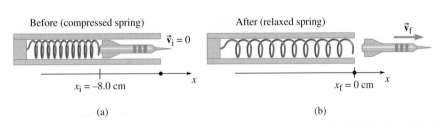

Before (compressed spring)   After (relaxed spring)

$\vec{v}_i = 0$ $\vec{v}_f$

$x_i = -8.0$ cm $x_f = 0$ cm

(a)   (b)

Figure 6.31
Dart gun (a) before and (b) after firing. The spring was compressed by 8.0 cm when the gun was cocked.

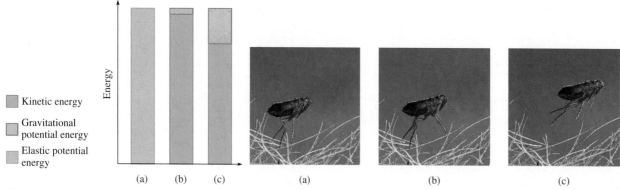

Figure 6.32 Energy transformations in the jump of a flea.

Kinetic energy

Gravitational potential energy

Elastic potential energy

Application of Energy Conversion: Jumping

How does the kangaroo keep jumping?

When a human jumps, the muscles supply the energy to propel the body upward. Try jumping as high as you can from a standing start. You no doubt start by crouching down. Then you accelerate upward, straightening your legs and your body; your muscles convert chemical energy into the mechanical energy of your jump. If you are very athletic, you might be able to jump about 1 m above the floor.

The kangaroo uses a different mechanism. It has long, elastic tendons and small muscles in its hind legs, in contrast to the relatively large muscles and short, stiffer tendons found in humans. The kangaroo folds its legs before a jump, using its muscles to stretch the tendons and converting chemical energy into elastic potential energy. The kangaroo then quickly extends its legs, relaxing the tendons like a released spring. The elastic energy stored in the tendons supplies much of the energy needed for the jump; the rest is supplied by the kangaroo's leg muscles, which convert some more chemical energy into mechanical energy.

When the kangaroo lands on the ground, the tendons are stretched again as its legs bend. Thus, rather than dissipating all of the energy from the previous jump, a large fraction of it is recaptured as elastic energy in the tendons and then released to assist the next jump. This process reduces the amount of energy the muscles must supply for subsequent jumps and makes the kangaroo one of the most energy-efficient travelers among animals. The human body also stores some elastic energy in stretched tendons and in flexed foot bones when we run or jump, but not to the extent that its specialized anatomy enables the kangaroo to do.

Some insects jump using a catapult technique. The knee joint of a flea contains an elastic material called resilin (a rubber-like protein). The flea slowly bends its knee, stretching out the resilin and storing elastic energy, and then locks its knee in place (Fig. 6.32a). When the flea is ready to jump, the knee is unlocked and the resilin quickly contracts with a sudden conversion of the stored elastic energy into kinetic energy (Fig. 6.32b). Some of this kinetic energy is then converted into gravitational potential energy as the flea moves higher and higher (Fig. 6.32c). Ignoring air resistance and other dissipative forces, the total mechanical energy (kinetic energy + gravitational potential energy + elastic potential energy) does not change during the jump.

Example 6.12

The Hopping Kangaroo

Suppose the height h of a kangaroo's hop (Fig. 6.33) after it stretches its tendons a distance x_1 (beyond their unstretched length) is 2.0 m. How high would the hop be after it stretched the tendons 10% more than before (that is, a distance $1.10x_1$

beyond their unstretched length)? In a simplified model, we assume that all the energy for a kangaroo's hop comes from the elastic energy stored in the tendons, which behave as ideal springs. Ignore air resistance and other energy dissipation.

continued on next page

Example 6.12 continued

Strategy   Ignoring dissipation, the mechanical energy does not change. We have to include both gravitational and elastic potential energies in the mechanical energy. At first we consider a kangaroo jumping straight up. Then we try to generalize to more typical hopping with forward motion as well as upward motion.

Solution   The mechanical energy does not change:

$$K_i + U_{i,grav} + U_{i,elastic} = K_f + U_{f,grav} + U_{f,elastic}$$

Initially, when the kangaroo is crouched before the jump, it has zero kinetic energy. For convenience, we choose the initial gravitational potential energy to be zero. Thinking of the elastic potential energy as being stored in a single ideal spring with spring constant k, the initial mechanical energy is

$$K_i + U_{i,grav} + U_{i,elastic} = 0 + 0 + \tfrac{1}{2}kx_i^2$$

where x_i represents the initial stretch of the tendons. With the kangaroo at the high point of the jump, the kinetic energy is again zero if it jumped straight up. The tendons are no longer stretched, so the elastic potential energy is zero. But now there is gravitational potential energy. At a height h above the initial point, the final mechanical energy is

$$K_f + U_{f,grav} + U_{f,elastic} = 0 + mgh + 0$$

where m is the kangaroo's mass. Setting the mechanical energies equal,

$$\tfrac{1}{2}kx_i^2 = mgh \quad \Rightarrow \quad h = \frac{kx_i^2}{2mg}$$

We don't know all of the constants (mass, spring constant, initial amount of stretch), so we set up a ratio:

$$\frac{h_2}{h_1} = \frac{\cancel{k}x_2^2/(2\cancel{mg})}{\cancel{k}x_1^2/(2\cancel{mg})} = \frac{x_2^2}{x_1^2}$$

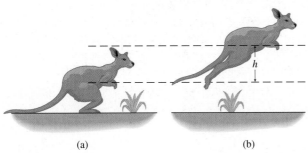

(a)   (b)

Figure 6.33

(a) Kangaroo crouched and ready to hop. (b) Kangaroo at the highest point in its hop.

For a 10% increase in stretch, $x_2 = 1.10x_1$ and

$$h_2 = \left(\frac{x_2}{x_1}\right)^2 h_1 = (1.10)^2 h_1 = 1.21 \times 2.0 \text{ m} = 2.4 \text{ m}$$

Using a 10% increase in the stretch of the tendon, the kangaroo jumps about 21% higher.

When the kangaroo is hopping along, it does not jump straight up. Will the kangaroo's jump still be 21% higher when jumping at another angle? Imagine the kangaroo hopping along so that it leaves the ground at a 45° angle, which gives the maximum horizontal range per hop in the absence of air resistance. The elastic energy in the tendon is first converted to kinetic energy. This time, not all of the kinetic energy is converted to gravitational potential energy. The kinetic energy at the highest point of the jump is *not* zero because the kangaroo is still moving forward. The initial velocity can be resolved into components:

$$v^2 = v_x^2 + v_y^2 = 2v_x^2 \quad \text{(since } v_x = v_y \text{ for a 45° angle)}$$

At the highest point of the jump, the kinetic energy is $\tfrac{1}{2}mv_x^2$, which is half of the initial kinetic energy. Overall, *half* of the elastic energy of the tendon is converted to gravitational potential energy:

$$\tfrac{1}{2} \times \left(\tfrac{1}{2}kx_i^2\right) = mgh$$

Since h is still proportional to x_i^2, the height of the jump still increases by 21% if the stretch of the tendon is increased by 10%.

Discussion   The storage of elastic energy in the tendon is a clever way for the kangaroo to get more "miles per gallon." Without such an energy storage system, most of the kangaroo's mechanical energy would be converted to an unrecoverable form of energy at the end of each hop. The tendons store some of the energy that would otherwise be lost and then release it to help the next jump. Since less mechanical energy is "lost" on each landing, the energy supplied by the kangaroo's muscles is less than it would otherwise be. Humans use a similar energy-saving mechanism when running (see Problem 103).

Practice Problem 6.12   Jumping with Joey

Suppose the kangaroo has a baby kangaroo (a *joey*) riding in her pouch. If the joey has grown to be one sixth the mass of its mother, how high can the kangaroo jump with the additional load? Assume that, without the joey, she can jump 2.8 m.

6.8   POWER

Sometimes the *rate* of energy conversion is important. When shopping for a sports car, you wouldn't ask the salesman how much work the engine can do. A tiny economy car like the Toyota Prius does more work than a Ferrari if the Prius is used for daily commuting while the Ferrari sits in the garage most of the time. But the Ferrari can do work *at a much faster rate* than the Prius can. In other words, it can change chemical energy in the gasoline into mechanical energy of the car at a much faster rate—it has a larger maximum power output. The higher power output enables the Ferrari to accelerate to high speeds much faster than the Prius. We give the name **power** (symbol *P*) to the rate of energy transfer. The average power is the amount of energy converted (ΔE) divided by the time the transfer takes (Δt):

Power: the *rate* of energy conversion

Average power:

$$P_{av} = \frac{\Delta E}{\Delta t} \qquad (6\text{-}26)$$

The SI unit of power, the joule per second, is given the name watt (1 W = 1 J/s), after James Watt (1736–1819), a Scottish inventor who greatly improved the efficiency of steam engines. Remember that the unit symbol W stands for *watt*, not *work*.

In the United States, the maximum power output of an electric motor or automobile engine is usually specified in horsepower, which is a non-SI unit of power (1 hp = 746 W).

The *kilowatt-hour* (kW·h) is a unit of energy, *not* a unit of power. One kilowatt-hour is the amount of energy transferred at a constant rate of 1 kW during a time interval of 1 h. The kilowatt-hour is commonly used by utility companies to measure the amount of electric energy used by consumers.

The work done by a force during a small time interval Δt is

$$W = F\,\Delta r \cos\theta \qquad (6\text{-}1)$$

The magnitude of the displacement is

$$\Delta r = v\,\Delta t$$

Hence, the power—the rate at which the force does work—can be found from the force and the velocity.

$$P = \frac{W}{\Delta t} = \frac{F\Delta r \cos\theta}{\Delta t} = F\frac{\Delta r}{\Delta t}\cos\theta = Fv\cos\theta$$

Instantaneous power (rate at which work is done):

$$P = Fv\cos\theta \qquad (6\text{-}27)$$

(θ is the angle between $\vec{\mathbf{F}}$ and $\vec{\mathbf{v}}$)

Example 6.13

Air Resistance on a Hill-Climbing Car

A 1000.0-kg car climbs a hill with a 4.0° incline at a constant 12.0 m/s (Fig. 6.34). (a) At what rate is the gravitational potential energy increasing? (b) If the mechanical power output of the engine is 20.0 kW, find the force of air resistance on the car. (Assume that air resistance is responsible for all of the energy dissipation.)

Strategy   (a) We can find the rate of gravitational potential energy increase in two ways. One is to find the potential energy change during a time interval Δt and divide it by the time interval, which is equivalent to using the definition of average power [Eq. (6-26)]. The other possibility is to use Eq. (6-27) to find the rate at which the gravitational force does work.

continued on next page

Example 6.13 continued

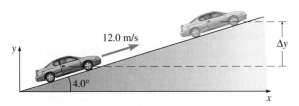

Figure 6.34
Car climbing a hill at constant speed.

(b) The car moves at constant speed, so its kinetic energy is not changing. Therefore, during any time interval, the work done by the engine (W_e) plus the (negative) work done by air resistance (W_a) is equal to the increase in the gravitational potential energy.

Given: car mass = 1000.0 kg; v = 12.0 m/s; 4.0° incline.
To find: (a) rate of potential energy change, $\Delta U/\Delta t$;
(b) force due to air resistance, $\vec{F}_a$.

Solution (a) For a small change in elevation Δy, the change in potential energy is

$$\Delta U = mg\,\Delta y$$

The *rate* of potential energy change is

$$\frac{\Delta U}{\Delta t} = \frac{mg\,\Delta y}{\Delta t} = mg\,\frac{\Delta y}{\Delta t} = mgv_y$$

where $v_y = \Delta y/\Delta t$ is the y-component of the velocity. From Fig. 6.35, $v_y = v \sin \phi$, where $\phi = 4.0°$. Then,

$$\frac{\Delta U}{\Delta t} = mgv \sin \phi = 1000.0 \text{ kg} \times 9.80 \text{ m/s}^2 \times 12.0 \text{ m/s} \times \sin 4.0°$$
$$= 8200 \text{ W}$$

(b) During any time interval Δt, the (positive) work done by the engine plus the (negative) work done by air resistance must equal the increase in the gravitational potential energy:

$$W_{\text{total}} = W_e + W_a = \Delta U$$

Dividing each term by Δt, we find

$$\frac{W_e}{\Delta t} + \frac{W_a}{\Delta t} = \frac{\Delta U}{\Delta t} \quad \Rightarrow \quad P_e + P_a = \frac{\Delta U}{\Delta t}$$

where P_e and P_a represent the power output of the engine and the rate at which air resistance does (negative) work on the car, respectively. Then,

$$P_a = \frac{\Delta U}{\Delta t} - P_e = 8.2 \text{ kW} - 20.0 \text{ kW} = -11.8 \text{ kW}$$

So, of the 20.0 kJ of mechanical work that the engine does each second, 8.2 kJ goes into gravitational potential energy

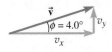

Figure 6.35
Resolving the velocity into x- and y-components.

and 11.8 kJ goes into pushing air out of the way and stirring it up in the process.

The direction of the force of air resistance $\vec{F}_a$ on the car is opposite to the car's velocity, so

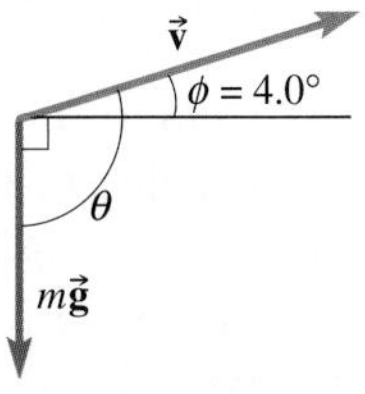

$$P_a = F_a v \cos 180° = -F_a v$$

Solving for F_a,

$$F_a = -\frac{P_a}{v} = -\frac{-11\,800 \text{ W}}{12.0 \text{ m/s}} = 983 \text{ N}$$

Discussion We can check (a) by using Eq. (6-27) to find the rate at which the gravitational force does work: $P = Fv \cos \theta$, where $F = mg$. The angle θ is *not* the same as ϕ. In Eq. (6-27), θ is the angle between the force and velocity vectors, which is 94.0° (Fig. 6.36). Then,

$$P = mgv \cos 94.0°$$
$$= 1000.0 \text{ kg} \times 9.80 \text{ m/s}^2 \times 12.0 \text{ m/s} \times \cos 94.0°$$
$$= -8200 \text{ W}$$

Gravity does work on the car at a rate of −8200 W, which means the potential energy is *increasing* at a rate of +8200 W.

We can also figure out what mechanical power the engine must supply to go 12.0 m/s on level ground. With no change in potential energy, all of the mechanical power output of the engine goes into stirring up the air, so $P_e + P_a = 0$. The magnitude of the force of air resistance is the same (983 N) since the speed is the same. Then air resistance dissipates energy at the same rate as before:

$$P_a = -F_a v = -983 \text{ N} \times 12.0 \text{ m/s} = -11.8 \text{ kW}$$

Therefore, $P_e = 11.8$ kW. On level ground, the gravitational potential energy isn't increasing, so the engine only needs to do enough work to counteract the tendency of air resistance to slow down the car.

In this example, we have assumed that all of the mechanical power output of the engine is delivered to the wheels to propel the car forward. In reality, some of the engine's power output is used to run auxiliary devices such as headlights, radios, and windshield wipers. Friction (in the moving parts of the engine, transmission, and drivetrain) also reduces the amount of power that is actually delivered to the wheels.

Figure 6.36
The angle between the force and the velocity is $\theta = 94.0°$. (The angle is exaggerated for clarity.)

Practice Problem 6.13 Mechanical Power Output on Flat Ground or Going Downhill

What mechanical power must the engine supply (a) to drive on level ground at 12.0 m/s and (b) to go down a 4.0° incline at 12.0 m/s? (Since this is the same speed as in Example 6.13, the force of air resistance is the same.)

Master the Concepts

- Conservation law: a physical law phrased in terms of a quantity that does not change with time.
- The law of conservation of energy: the total energy of the universe is unchanged by any physical process.
- Work is an energy transfer due to the application of a force. The work done by a constant force $\vec{F}$ acting on an object during a displacement $\Delta\vec{r}$ is

$$W = F\,\Delta r\cos\theta \qquad (6\text{-}1)$$

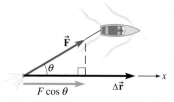

where θ is the angle between $\vec{F}$ and $\Delta\vec{r}$. If $\vec{F}$ or $\Delta\vec{r}$ is parallel to the x-axis,

$$W = F_x\,\Delta x \qquad (6\text{-}2)$$

- When several forces act on an object, the total work is the sum of the work done by each force individually.
- Translational kinetic energy is the energy associated with motion of the object as a whole. The translational kinetic energy of an object of mass m moving with speed v is

$$K = \tfrac{1}{2}mv^2 \qquad (6\text{-}6)$$

- The gravitational potential energy for an object of mass m in a *uniform* gravitational field is

$$U_{\text{grav}} = mgy \qquad (6\text{-}13)$$

where the $+y$-axis points up and we assign $U = 0$ to $y = 0$.

- The gravitational potential energy for two bodies of masses m_1 and m_2 whose centers are separated by a distance r is

$$U = -\frac{Gm_1 m_2}{r} \qquad (6\text{-}14)$$

where we assign $U = 0$ to infinite separation ($r = \infty$).

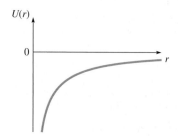

- There is no special significance to the sign of the potential energy. What matters is the sign of the potential energy *change*. Only *changes* in potential energy enter our calculations.

- The work done by a variable force directed along the x-axis during a displacement Δx is the area under the $F_x(x)$ graph from x_i to x_f.

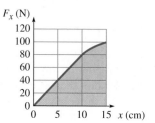

- Hooke's law: for many objects, the deformation is proportional to the magnitude of the force that causes the deformation. An ideal spring is massless and follows Hooke's law. The force exerted *by* the moveable end of an ideal spring when it is at position x is

$$F_x = -kx \qquad (6\text{-}19)$$

where the origin is chosen so the spring is relaxed at $x = 0$ and k is called the spring constant.

- If we assign $U = 0$ to the relaxed spring ($x = 0$), the elastic potential energy stored in an ideal spring of spring constant k is

$$U_{\text{elastic}} = \tfrac{1}{2}kx^2 \qquad (6\text{-}24)$$

- Mechanical energy is the sum of the kinetic and potential energies. The change in potential energy accounts for the work done by the forces associated with the potential energy. The work done by nonconservative forces is equal to the change in the mechanical energy:

$$W_{\text{nc}} = \Delta K + \Delta U = \Delta E_{\text{mech}} \qquad (6\text{-}12)$$

When nonconservative forces do no net work, the mechanical energy does not change.

$$\text{If } W_{\text{nc}} = 0,\ \Delta K + \Delta U = 0$$

- Average power is the average rate of energy conversion.

$$P_{\text{av}} = \frac{\Delta E}{\Delta t} \qquad (6\text{-}26)$$

The instantaneous rate at which a force $\vec{F}$ does work when the object it acts on moves with velocity $\vec{v}$ is

$$P = Fv\cos\theta \qquad (6\text{-}27)$$

where θ is the angle between $\vec{F}$ and $\vec{v}$.

- The SI unit of work and energy is the joule. $1\,\text{J} = 1\,\text{N·m}$. The SI unit of power is the watt. $1\,\text{W} = 1\,\text{J/s}$.

Conceptual Questions

1. An object moves in a circle. Is the total work done on the object by external forces necessarily zero? Explain.

2. You are walking to class with a backpack full of books. As you walk at constant speed on flat ground, does the force exerted on the backpack by your back and shoulders do any work? If so, is it positive or negative? Answer the same questions in two other situations: (1) you are walking down some steps at constant speed; (2) you start to run faster and faster on a level sidewalk to catch a bus.

3. Why do roads leading to the top of a mountain wind back and forth? [*Hint:* Think of the road as an inclined plane.]

4. A mango falls to the ground. During the fall, does the Earth's gravitational field do positive or negative work W_m on the mango? Does the mango's gravitational field do positive or negative work W_E on the Earth? Compare the signs and the magnitudes of W_m and W_E.

5. Can static friction do work? If so, give an example. [*Hint:* Static friction acts to prevent *relative* motion along the contact surface.]

6. In the design of a roller coaster, is it possible for any hill of the ride to be higher than the first one? If so, how?

7. When a ball is dropped to the floor from a height h, it strikes the ground and briefly undergoes a change of shape before rebounding to a maximum height less than h. Explain why it does not return to the same height h.

8. A gymnast is swinging in a vertical circle about a crossbar. In terms of energy conservation, explain why the speed of the gymnast's body is slowest at the top of the circle and fastest at the bottom.

9. A bicycle rider notices that he is approaching a steep hill. Explain, in terms of energy, why the bicyclist pedals hard to gain as much speed as possible on level road before reaching the hill.

10. You need to move a heavy crate by sliding it across a smooth floor. The coefficient of sliding friction is 0.2. You can either push the crate horizontally or pull the crate using an attached rope. When you pull on the rope, it makes a 30° angle with the floor. Which way should you choose to move the crate so that you do the least amount of work? How can you answer this question without knowing the weight of the crate or the displacement of the crate?

11. The main effort of running is the work done by the muscles to accelerate and decelerate the legs. When a foot strikes the ground, it is momentarily brought to rest while the remainder of the animal's body continues to move forward. When the foot is picked up, it is accelerated forward by one set of muscles in order to move ahead of the rest of the body. Then the foot is slowed down by a second set of muscles until it is brought to rest on the ground again. The muscles expend energy both when accelerating and when decelerating the leg. How are thoroughbred horses, deer, and greyhounds adapted so that they can run at great speed?

12. Explain why an ideal spring *must* exert forces of equal magnitude on the objects attached to each end, even if the spring itself has a nonzero acceleration. [*Hint:* Use one of Newton's laws of motion and remember that an ideal spring has zero mass.] Is the amount of work done by the spring on the two objects necessarily the same? Explain. If the answer is no, give an example to illustrate.

13. Zorba and Boris are at a water park. There are two water slides with straight slopes that start at the same height and end at the same height. Slide A has a more gradual slope than slide B. Boris says he likes slide B better because you reach a faster speed and he notes that he got to the bottom level in less time on slide B as measured with his stop watch. His brother Zorba says you reach the same speed with either slide. Who is correct and why? Both slides have negligible friction.

Multiple-Choice Questions

1. After getting on the Santa Monica Freeway, a sports car accelerates from 30 mi/h to 90 mi/h. Its kinetic energy
   (a) increases by a factor of $\sqrt{3}$.
   (b) increases by a factor of 3.
   (c) increases by a factor of 9.
   (d) increases by a factor that depends on the car's mass.

2. If a kangaroo on Earth can jump from a standing start so that its feet reach a height h above the surface, approximately how high can the same kangaroo jump from a standing start on the Moon's surface? $g_{\text{Moon}} \approx \frac{1}{6} g_{\text{Earth}}$. (Assume the kangaroo has an oxygen tank and pressure suit with negligible mass.)
   (a) h (b) $6h$ (c) $\frac{1}{6}h$
   (d) $36h$ (e) $\frac{1}{36}h$ (f) $\sqrt{6}h$

Questions 3–5. The orbit of Pluto is much more eccentric than the orbits of the planets. That is, instead of being nearly circular, the orbit is noticeably elliptical. The point in the orbit nearest the Sun is called the *perihelion* and the point farthest from the Sun is called the *aphelion*.

Answer choices for Questions 3–5:

   (a) its maximum value   (b) its minimum value

   (c) the same value as at every other point in the orbit

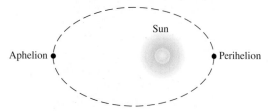

Multiple-Choice Questions 3–5

3. At perihelion, the gravitational potential energy of Pluto's orbit has

4. At perihelion, the kinetic energy of Pluto has

5. At perihelion, the mechanical energy of Pluto's orbit has

6. As Pluto moves from the perihelion to the aphelion, the work done by gravity on Pluto is

   (a) zero.   (b) positive.   (c) negative.

7. Two balls are thrown from the roof of a building with the same initial speed. One is thrown horizontally while the other is thrown at an angle of 20° above the horizontal. Which hits the ground with the greatest speed? Ignore air resistance.

   (a) The one thrown horizontally

   (b) The one thrown at 20°

   (c) They hit the ground with the same speed.

   (d) The answer cannot be determined with the given information.

8. A hiker descends from the South Rim of the Grand Canyon to the Colorado River. During this hike, the work done by gravity on the hiker is

   (a) positive and depends on the path taken.

   (b) negative and depends on the path taken.

   (c) positive and independent of the path taken.

   (d) negative and independent of the path taken.

   (e) zero.

Questions 9 and 10. A simple catapult, consisting of a leather pouch attached to rubber bands tied to two forks of a wooden Y, has a spring constant k and is used to shoot a pebble horizontally. When the catapult is stretched by a distance d, it gives a pebble of mass m a launch speed v. Answer choices for Questions 9 and 10:

   (a) $\sqrt{3}v$ (b) $3v$ (c) $3\sqrt{3}v$ (d) $9v$ (e) $27v$

9. What speed does the catapult give a pebble of mass m when stretched to a distance $3d$?

10. What speed does the catapult give a pebble of mass $m/3$ when stretched to a distance d?

11. A projectile is launched at an angle θ above the horizontal. Ignoring air resistance, what fraction of its initial kinetic energy does the projectile have at the top of its trajectory?

   (a) $\cos\theta$ (b) $\sin\theta$ (c) $\tan\theta$ (d) $\dfrac{1}{\tan\theta}$ (e) $\dfrac{1}{2}$

   (f) $\cos^2\theta$ (g) $\sin^2\theta$ (h) 0   (i) 1

Problems

   🄒   Combination conceptual/quantitative problem

   ⚕   Biological or medical application

   ✦   Challenging problem

  Blue #   Detailed solution in the Student Solutions Manual

  ①②   Problems paired by concept

  Ⓦ   Text website interactive or tutorial

Section 6.2 Work Done by a Constant Force

1. How much work must Denise do to drag her basket of laundry of mass 5.0 kg a distance of 5.0 m along a floor, if the force she exerts is a constant 30.0 N at an angle of 60.0° with the horizontal?

2. A sled is dragged along a horizontal path at a constant speed of 1.5 m/s by a rope that is inclined at an angle of 30.0° with respect to the horizontal. The total weight of the sled is 470 N. The tension in the rope is 240 N. How much work is done by the rope on the sled in a time interval of 10.0 s?

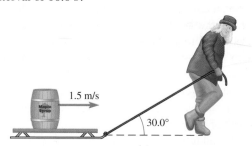

3. Hilda holds a gardening book of weight 10 N at a height of 1.0 m above her patio for 50 s. How much work does she do *on the book* during that 50 s?

4. The tension in the horizontal towrope pulling a water-skier is 240 N while the skier moves due west a distance of 54 m. How much work does the towrope do on the water-skier?

5. A barge of mass 5.0×10^4 kg is pulled along the Erie Canal by two mules, walking along towpaths parallel to the canal on either side of it. The ropes harnessed to the mules make angles of 45° to the canal. Each mule is pulling on its rope with a force of 1.0 kN. How much work is done on the barge by both of these mules together as they pull the barge 150 m along the canal?

6. A 402-kg pile driver is raised 12 m above ground. (a) How much work must be done to raise the pile driver? (b) How much work

does gravity do on the driver as it is raised? (c) The driver is now dropped. How much work does gravity do on the driver as it falls?

7. Jennifer lifts a 2.5-kg carton of cat litter from the floor to a height of 0.75 m. (a) How much *total* work is done on the carton during this operation? Jennifer then pours 1.2 kg of the litter into the cat's litter box on the floor. (b) How much work is done by gravity on the 1.2 kg of litter as it falls into the litter box?

8. Dirk pushes on a packing box with a horizontal force of 66.0 N as he slides it along the floor. The average friction force acting on the box is 4.80 N. How much *total* work is done on the box in moving it 2.50 m along the floor?

9. Juana slides a crate along the floor of the moving van. The coefficient of kinetic friction between the crate and the van floor is 0.120. The crate has a mass of 56.8 kg and Juana pushes with a horizontal force of 124 N. If 74.4 J of total work are done on the crate, how far along the van floor does it move?

Section 6.3 Kinetic Energy

10. An automobile with a mass of 1600 kg has a speed of 30.0 m/s. What is its kinetic energy?

11. A record company executive is on his way to a TV interview and is carrying a promotional CD in his briefcase. The mass of the briefcase and its contents is 5.00 kg. The executive realizes that he is going to be late. Starting from rest, he starts to run, reaching a speed of 2.50 m/s. What is the work done by the executive on the briefcase during this time? Ignore air resistance.

12. In 1899, Charles M. "Mile a Minute" Murphy set a record for speed on a bicycle by pedaling for a mile at an average of 62.3 mph (27.8 m/s) on a 3-mi track of plywood planks set over railroad ties in the draft of a Long Island Railroad train. In 1985, a record was set for this type of "motor pacing" by Olympic cyclist John Howard who raced at 152.2 mph (68.04 m/s) in the wake of a race car at Bonneville Salt Flats. The race car had a modified tail assembly designed to reduce the air drag on the cyclist. What was the kinetic energy of the bicycle plus rider in each of these feats? Assume that the mass of bicycle plus rider is 70.5 kg in each case.

13. Sam pushes a 10.0-kg sack of bread flour on a frictionless horizontal surface with a constant horizontal force of 2.0 N starting from rest. (a) What is the kinetic energy of the sack after Sam has pushed it a distance of 35 cm? (b) What is the speed of the sack after Sam has pushed it a distance of 35 cm?

14. Josie and Charlotte push a 12-kg bag of playground sand for a sandbox on a frictionless, horizontal, wet polyvinyl surface with a constant, horizontal force for a distance of 8.0 m, starting from rest. If the final speed of the sand bag is 0.40 m/s, what is the magnitude of the force with which they pushed?

15. A ball of mass 0.10 kg moving with speed of 2.0 m/s hits a wall and bounces back with the same speed in the opposite direction. What is the change in the ball's kinetic energy?

16. Jim rides his skateboard down a ramp that is in the shape of a quarter circle with a radius of 5.00 m. At the bottom of the ramp, Jim is moving at 9.00 m/s. Jim and his skateboard have a mass of 65.0 kg. How much work is done by friction as the skateboard goes down the ramp? (🕸 tutorial: energy, parts (a) and (b))

17. A 69.0-kg short-track ice skater is racing at a speed of 11.0 m/s when he falls down and slides across the ice into a padded wall that brings him to rest. Assuming that he doesn't lose any speed during the fall or while sliding across the ice, how much work is done by the wall while stopping the ice skater?

18. A plane weighing 220 kN (25 tons) lands on an aircraft carrier. The plane is moving horizontally at 67 m/s (150 mi/h) when its tailhook grabs hold of the arresting cables. The cables bring the plane to a stop in a distance of 84 m. (a) How much work is done on the plane by the arresting cables? (b) What is the force (assumed constant) exerted on the plane by the cables? (Both answers will be *underestimates*, since the plane lands with the engines full throttle forward; in case the tailhook fails to grab hold of the cables, the pilot must be ready for immediate takeoff.)

19. A shooting star is a meteoroid that burns up when it reaches Earth's atmosphere. Many of these meteoroids are quite small. Calculate the kinetic energy of a meteoroid of mass 5.0 g moving at a speed of 48 km/s and compare it to the kinetic energy of a 1100-kg car moving at 29 m/s (65 mi/h).

Section 6.4 Gravitational Potential Energy (1)

20. Sean climbs a tower that is 82.3 m high to make a jump with a parachute. The mass of Sean plus the parachute is 68.0 kg. If $U = 0$ at ground level, what is the potential energy of Sean and the parachute at the top of the tower? (🕸 tutorial: energy, parts (c) and (d))

21. Justin moves a desk 5.0 m across a level floor by pushing on it with a constant horizontal force of 340 N. (It slides for a negligibly small distance before coming to a stop when the force is removed.) Then, changing his mind, he moves it back to its starting point, again by pushing with a constant force of 340 N. (a) What is the change in the desk's gravitational potential energy during the round-trip? (b) How much work has Justin done on the desk? (c) If the work done by Justin is not equal to the change in gravitational potential energy of the desk, then where has the energy gone?

22. An airline executive decides to economize by reducing the amount of fuel required for long-distance flights. He orders the ground crew to remove the paint from the outer surface of each plane. The paint removed from a single plane has a mass of approximately 100 kg. (a) If the airplane cruises at an altitude of 12000 m, how much energy is saved in not having to lift the paint to that altitude? (b) How much energy is saved by not having to move that amount of paint from rest to a cruising speed of 250 m/s?

23. Emil is tossing an orange of mass 0.30 kg into the air. (a) Emil throws the orange straight up and then catches it, throwing and catching it at the same point in space. What is the change in the potential energy of the orange during its trajectory? Ignore air resistance. (b) Emil throws the orange straight up, starting 1.0 m above the ground. He fails to catch it. What is the change in the potential energy of the orange during this flight?

24. A brick of mass 1.0 kg slides down an icy roof inclined at 30.0° with respect to the horizontal. (a) If the brick starts from rest, how fast is it moving when it reaches the edge of the roof 2.00 m away? Ignore friction. (b) Redo part (a) if the coefficient of kinetic friction is 0.10. (![interactive icon] interactive: sliding brick)

25. An arrangement of two pulleys, as shown in the figure, is used to lift a 48.0-kg mass a distance of 4.00 m above the starting point. Assume the pulleys and rope are ideal and that all rope sections are essentially vertical. (a) What is the mechanical advantage of this system? (In other words, by what factor is the force you exert to lift the weight multiplied by the pulley system?) (b) What is the change in the potential energy of the weight when it is lifted a distance of 4.00 m? (c) How much work must be done to lift the 48.0-kg mass a distance of 4.00 m? (d) What length of rope must be pulled by the person lifting the weight 4.00 m higher in the air? (![tutorial icon] tutorial: block-and-tackle)

48.0 kg

$\vec{F}$

26. In Example 6.1, find the work done by the movers as they slide the chest up the ramp if the coefficient of friction between the chest and the ramp is 0.20. (![tutorial icon] tutorial: ramp)

27. A cart moving to the *right* passes point 1 at a speed of 20.0 m/s. Let $g = 9.81$ m/s². (a) What is the speed of the cart as it passes point 3? (b) Will the cart reach position 4? Ignore friction.

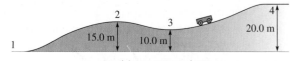

2

3

4

20.0 m

15.0 m

10.0 m

1

Problems 27 and 28

28. A cart starts from position 4 with a velocity of 15 m/s to the left. Find the speed with which the cart reaches positions 1, 2, and 3. Ignore friction.

29. Bruce stands on a bank beside a pond, grasps the end of a 20.0-m-long rope attached to a nearby tree and swings out to drop into the water. If the rope starts at an angle of 35.0° with the vertical, what is Bruce's speed at the bottom of the swing?

30. The maximum speed of a child on a swing is 4.9 m/s. The child's height above the ground is 0.70 m at the lowest point in his motion. How high above the ground is he at his highest point?

31. If the skier of Example 6.6 is moving at 12 m/s at the bottom of the trail, calculate the total work done by friction and air resistance during the run. The skier's mass is 75 kg.

32. A 750-kg automobile is moving at 20.0 m/s at a height of 5.0 m above the bottom of a hill when it runs out of gasoline. The car coasts down the hill and then continues coasting up the other side until it comes to rest. Ignoring frictional forces and air resistance, what is the value of h, the highest position the car reaches above the bottom of the hill?

750 kg

20.0 m/s

5.0 m

h = ?

33. Rachel is on the roof of a building, h meters above ground. She throws a heavy ball into the air with a speed v, at an angle θ with respect to the horizontal. Ignore air resistance. (a) Find the speed of the ball when it hits the ground in terms of h, v, θ, and g. (b) For what value(s) of θ is the speed of the ball greatest when it hits the ground?

34. A crate of mass m_1 on a frictionless inclined plane is attached to another crate of mass m_2 by a massless rope. The rope passes over an ideal pulley so the mass m_2 is suspended in air. The plane is inclined at an angle $\theta = 36.9°$. Use conservation of energy to find how fast crate m_2 is moving after m_1 has traveled a distance of 1.4 m along the incline, starting from rest. The mass of m_1 is 12.4 kg and the mass of m_2 is 16.3 kg.

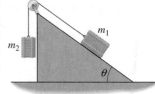

m_2 m_1 θ

35. The forces required to extend a spring to various lengths are measured. The results are shown in the following table. Using the data in the table, plot a graph that helps you to answer the following two questions: (a) What is the spring constant? (b) What is the relaxed length of the spring?

Force (N)	1.00	2.00	3.00	4.00	5.00
Spring length (cm)	14.5	18.0	21.5	25.0	28.5

Section 6.5 Gravitational Potential Energy (2)

36. A 75.0-kg skier starts from rest and slides down a 32.0-m frictionless slope that is inclined at an angle of 15.0°

with the horizontal. Ignore air resistance. (a) Calculate the work done by gravity on the skier and the work done by the normal force on the skier. (b) If the slope is not frictionless so that the skier has a final velocity of 10.0 m/s, calculate the work done by gravity, the work done by the normal force, the work done by friction, the force of friction (assuming it is constant), and the coefficient of kinetic friction. (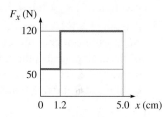 tutorial: water slide)

37. You are on the Moon and would like to send a probe into space so that it does not fall back to the surface of the Moon. What launch speed do you need?

38. A planet with a radius of 6.00×10^7 m has a gravitational field of magnitude 30.0 m/s^2 at the surface. What is the escape speed from the planet?

39. The escape speed from the surface of Planet Zoroaster is 12.0 km/s. The planet has no atmosphere. A meteor far away from the planet moves at speed 5.0 km/s on a collision course with Zoroaster. How fast is the meteor going when it hits the surface of the planet?

40. The escape speed from the surface of the Earth is 11.2 km/s. What would be the escape speed from another planet of the same density (mass per unit volume) as Earth but with a radius twice that of Earth?

41. A satellite is placed in a noncircular orbit about the Earth. The farthest point of its orbit (apogee) is 4 Earth radii from the center of the Earth, while its nearest point (perigee) is 2 Earth radii from the Earth's center. If we define the gravitational potential energy U to be zero for an infinite separation of Earth and satellite, find the ratio $U_{\text{perigee}}/U_{\text{apogee}}$.

42. What is the minimum speed with which a meteor strikes the top of the Earth's stratosphere (about 40 km above Earth's surface), assuming that the meteor begins as a bit of interplanetary debris far from Earth? Assume the drag force is negligible until the meteor reaches the stratosphere.

43. A projectile with mass of 500 kg is launched straight up from the Earth's surface with an initial speed v_i. What magnitude of v_i enables the projectile to just reach a maximum height of $5R_E$, measured from the *center* of the Earth? Ignore air friction as the projectile goes through the Earth's atmosphere.

44. The orbit of Halley's comet around the Sun is a long thin ellipse. At its aphelion (point farthest from the Sun), the comet is 5.3×10^{12} m from the Sun and moves with a speed of 10.0 km/s. What is the comet's speed at its perihelion (closest approach to the Sun) where its distance from the Sun is 8.9×10^{10} m?

45. Suppose a satellite is in a circular orbit 3.0 Earth radii above the surface of the Earth (4.0 Earth radii from the center of the Earth). By how much does it have to increase its speed in order to be able to escape Earth? [Hint: You need to calculate the orbital speed and the escape speed.]

46. An asteroid hits the Moon and ejects a large rock from its surface. The rock has enough speed to travel to a point between the Earth and the Moon where the gravitational forces on it from the Earth and the Moon are equal in magnitude and opposite in direction. At that point the rock has a very small velocity toward Earth. What is the speed of the rock when it encounters Earth's atmosphere at an altitude of 700 km above the surface?

Section 6.6 Work Done by Variable Forces: Hooke's Law

47. How much work is done on the bowstring of Example 6.9 to draw it back by 20.0 cm? [Hint: Rather than recalculate from scratch, use proportional reasoning.]

48. An ideal spring has a spring constant $k = 20.0$ N/m. What is the amount of work that must be done to stretch the spring 0.40 m from its relaxed length?

49. The force that must be exerted to drive a nail into a wall is roughly as shown in the graph. The first 1.2 cm are through soft drywall; then the nail enters the solid wooden stud. How much work must be done to hammer the nail a horizontal distance of 5.0 cm into the wall?

50. (a) If the length of the Achilles tendon increases 0.50 cm when the force exerted on it by the muscle increases from 3200 N to 4800 N, what is the "spring constant" of the tendon? (b) How much work is done by the muscle in stretching the tendon 0.50 cm as the force increases from 3200 N to 4800 N?

51. (a) If forces of magnitude 5.0 N applied to each end of a spring cause the spring to stretch 3.5 cm from its relaxed length, how far do forces of magnitude 7.0 N cause the same spring to stretch? (b) What is the spring constant of this spring? (c) How much work is done by the applied forces in stretching the spring 3.5 cm from its relaxed length? (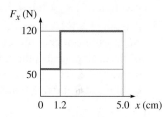 tutorial: spring)

52. A block of wood is compressed 2.0 nm when inward forces of magnitude 120 N are applied to it on two opposite sides. (a) Assuming Hooke's law holds, what is the effective spring constant of the block? (b) Assuming Hooke's law still holds, how much is the same block compressed by inward forces of magnitude 480 N? (c) How much work is done by the applied forces during the compression of part (b)?

53. The length of a spring increases by 7.2 cm from its relaxed length when a mass of 1.4 kg is hanging in equilibrium from the spring. (a) What is the spring constant? (b) How much elastic potential energy is stored in the spring? (c) A different mass is suspended and the spring length increases by 12.2 cm from its relaxed length to its new equilibrium position. What is the second mass?

54. A spring fixed at one end is compressed from its relaxed position by a distance of 0.20 m. See the graph of the applied external force, F_x, versus the position, x, of the spring. (a) Find the work done by the external force in compressing the spring 0.20 m starting from its relaxed position. (b) Find the work done by the external force to compress the spring from 0.10 m to 0.20 m.

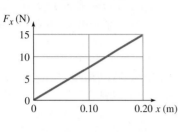

55. Rhonda keeps a 2.0-kg model airplane moving at constant speed in a horizontal circle at the end of a string of length 1.0 m. The tension in the string is 18 N. How much work does the string do on the plane during each revolution? (tutorial: circular motion)

56. The graph shows the force exerted on an object versus the position of that object along the x-axis. The force has no components other than along the x-axis. What is the work done by the force on the object as the object is displaced from 0 to 3.0 m?

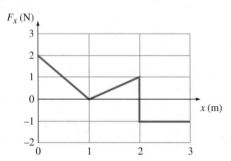

Section 6.7 Elastic Potential Energy

57. A kangaroo decides to see how high it can hop on *one leg*. Assuming the elastic energy stored in the tendon is the same as for Example 6.12, how high can it jump using a single leg?

58. When the spring on a toy gun is compressed by a distance x, it will shoot a rubber ball straight up to a height of h. Ignoring air resistance, how high will the gun shoot the same rubber ball if the spring is compressed by an amount $2x$? Assume $x \ll h$.

59. You shoot a 51-g pebble straight up with a catapult whose spring constant is 320 N/m. The catapult is initially stretched by 0.20 m. How high above the starting point does the pebble fly? Ignore air resistance.

60. A block (mass m) hangs from a spring (spring constant k). The block is released from rest a distance d above its *equilibrium* position. (a) What is the speed of the block as it passes through the equilibrium point? (b) What is the maximum distance below the equilibrium point that the block will reach?

61. A gymnast of mass 52 kg is jumping on a trampoline. She jumps so that her feet reach a maximum height of 2.5 m above the trampoline and, when she lands, her feet stretch the trampoline down 75 cm. How far does the trampoline stretch when she stands on it at rest? [*Hint:* Assume the trampoline obeys Hooke's law when it is stretched.]

62. Jorge is going to bungee jump from a bridge that is 55.0 m over the river below. The bungee cord has an unstretched length of 27.0 m. To be safe, the bungee cord should stop Jorge's fall when he is at least 2.00 m above the river. If Jorge has a mass of 75.0 kg, what is the minimum spring constant of the bungee cord? (tutorial: spring scale)

63. A 2.0-kg block is released from rest and allowed to slide down a frictionless surface and into a spring. The far end of the spring is attached to a wall, as shown. The initial height of the block is 0.50 m above the lowest part of the slide and the spring constant is 450 N/m. (a) What is the block's speed when it is at a height of 0.25 m above the base of the slide? (b) How far is the spring compressed? (c) The spring sends the block back to the left. How high does the block rise?

Problems 63 and 105

Section 6.8 Power

64. Lars, of mass 82.4 kg, has been working out and can do work for about 2.0 min at the rate of 1.0 hp (746 W). How long will it take him to climb three flights of stairs, a vertical height of 12.0 m?

65. Show that 1 kilowatt-hour (kW·h) is equal to 3.6 MJ.

66. If a man has an average useful power output of 40.0 W, what minimum time would it take him to lift fifty 10.0-kg boxes to a height of 2.00 m?

67. In Section 6.2, Rosie lifts a trunk weighing 220 N up 4.0 m. If it take her 40 s to lift the trunk, at what average rate does she do work?

68. A bicycle and its rider together has a mass of 75 kg. What power output of the rider is required to maintain a constant speed of 4.0 m/s (about 9 mph) up a 5.0% grade (a road that rises 5.0 m for every 100 m along the pavement)? Assume that frictional losses of energy are negligible.

69. The power output of a cyclist moving at a constant speed of 6.0 m/s on a level road is 120 W. (a) What is the force exerted on the cyclist and the bicycle by the air? (b) By bending low over the handlebars, the cyclist reduces the air resistance to 18 N. If she maintains a power output of 120 W, what will her speed be?

70. A car with mass of 1000.0 kg accelerates from 0 m/s to 40.0 m/s in 10.0 s. Ignore air resistance. The engine has a 22% efficiency, which means that 22% of the energy released by the burning gasoline is converted into mechanical energy. (a) What is the average mechanical power output of the engine? (b) What volume of gasoline is consumed? Assume that the burning of 1.0 L of gasoline releases 46 MJ of energy.

71. A motorist driving a 1200-kg car on level ground accelerates from 20.0 m/s to 30.0 m/s in a time of 5.0 s. Neglecting friction and air resistance, determine the *average* mechanical power in watts the engine must supply during this time interval.

72. A 62-kg woman takes 6.0 s to run up a flight of stairs. The landing at the top of the stairs is 5.0 m above her starting place. (a) What is the woman's average power output while she is running? (b) Would that be equal to her average power *input*—the rate at which chemical energy in food or stored fat is used? Why or why not?

73. How many grams of carbohydrate does a person of mass 74 kg need to metabolize to climb five flights of stairs (15 m height increase)? Each gram of carbohydrate provides 17.6 kJ of energy. Assume 10.0% efficiency—that is, 10.0% of the available chemical energy in the carbohydrate is converted to mechanical energy. What happens to the other 90% of the energy?

74. An object moves in the positive x-direction under the influence of a force F_x. A graph of F_x versus v_x is shown. (a) What is the instantaneous power on the object when its velocity is 11 m/s? (b) What is the instantaneous power on the object when its velocity is 16 m/s?

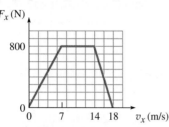

75. A top fuel drag racer with a mass of 500.0 kg completes a quarter-mile (402 m) drag race in a time of 4.2 s starting from rest. The car's final speed is 125 m/s. What is the engine's average power output? Ignore friction and air resistance.

76. (a) Calculate the change in potential energy of 1 kg of water as it passes over Niagara Falls (a vertical descent of 50 m). (b) At what rate is gravitational potential energy lost by the water of the Niagara River? The rate of flow is 5.5×10^6 kg/s. (c) If 10% of this energy can be converted into electric energy, how many households would the electricity supply? (An average household uses an average electrical power of about 1 kW.)

Comprehensive Problems

77. If a high jumper needs to make his center of gravity rise 1.2 m, how fast must he be able to sprint? Assume all of his kinetic energy can be transformed into potential energy. For an extended object, the gravitational potential energy is $U = mgh$, where h is the height of the center of gravity.

78. A pole-vaulter converts the kinetic energy of running to elastic potential energy in the pole, which is then converted to gravitational potential energy. If a pole-vaulter's center of gravity is 1.0 m above the ground while he sprints at 10.0 m/s, what is the maximum height of his center of gravity during the vault? For an extended object, the gravitational potential energy is $U = mgh$, where h is the height of the center of gravity. (In 1988, Sergei Bubka was the first pole-vaulter ever to clear 6 m.)

79. A hang glider moving at speed 9.5 m/s dives to an altitude 8.2 m lower. Ignoring drag, how fast is it then moving?

80. A car moving at 30 mi/h is stopped by jamming on the brakes and locking the wheels. The car skids 50 ft before coming to rest. How far would the car skid if it were initially moving at 60 mi/h? [*Hint:* You will not have to do any unit conversions if you set up the problem as a proportion.]

81. Prove that $U = -2K$ for any gravitational circular orbit. [*Hint:* Use Newton's second law to relate the gravitational force to the acceleration required to maintain uniform circular motion.]

82. A spring gun ($k = 28$ N/m) is used to shoot a 56-g ball horizontally. Initially the spring is compressed by 18 cm. The ball loses contact with the spring and leaves the gun when the spring is still compressed by 12 cm. What is the speed of the ball when it hits the ground, 1.4 m below the spring gun?

83. Two springs with equal spring constants k are connected first in series (one after the other) and then in parallel (side by side) with a weight hanging from the bottom of the combination. What is the effective spring constant of the two different arrangements? In other words, what would be the spring constant of a single spring that would behave exactly as (a) the series combination and (b) the parallel combination? Ignore the weight of the springs. [*Hint* for (a): *each* spring stretches an amount $x = F/k$, but only one spring exerts a force on the hanging object. *Hint* for (b): *each* spring exerts a force $F = kx$.]

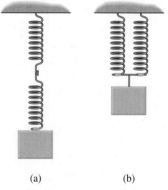

(a)   (b)

84. A roller coaster car (mass = 988 kg including passengers) is about to roll down a track. The diameter of the circular loop is 20.0 m and the car starts out from rest 40.0 m above the lowest point of the track. Ignore friction and air resistance. (a) At what speed does the car reach the top of the loop?

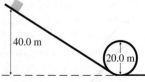

(b) What is the force exerted on the car by the track at the top of the loop? (c) From what minimum height above the bottom of the loop can the car be released so that it does not lose contact with the track at the top of the loop?

◆85. A 4.0-kg block is released from rest at the top of a frictionless plane of length 8.0 m that is inclined at an angle of 15° to the horizontal. A cord is attached to the block and trails along behind it. When the block reaches a point 5.0 m along the incline from the top, someone grasps the cord and exerts a constant tension parallel to the incline. The tension is such that the block just comes to rest when it reaches the bottom of the incline. (The person's force is a nonconservative force.) What is this constant tension? Solve the problem twice, once using work and energy and again using Newton's laws and the equations for constant acceleration. Which method do you prefer?

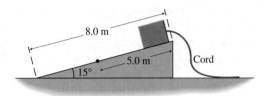

◆86. The bungee jumper of Example 6.4 made a jump into the Gorge du Verdon in southern France from a platform 182 m above the bottom of the gorge. The jumper weighed 780 N and came within 68 m of the bottom of the gorge. The cord's unstretched length is 30.0 m. (a) Assuming that the bungee cord follows Hooke's law when it stretches, find its spring constant. [*Hint:* The cord does not begin to stretch until the jumper has fallen 30.0 m.] (b) At what speed is the jumper falling when he reaches a height of 92 m above the bottom of the gorge?

87. A spring with $k = 40.0$ N/m is at the base of a frictionless 30.0° inclined plane. A 0.50-kg object is pressed against the spring, compressing it 0.20 m from its equilibrium position. The object is then released. If the object is not attached to the spring, how far up the incline does it travel before coming to rest and then sliding back down?

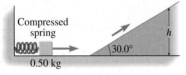

88. In an adventure movie, a 62.5-kg stunt woman falls 8.10 m and lands in a huge air bag. Her speed just before she hit the air bag was 10.5 m/s. (a) What is the total work done on the stunt woman during the fall? (b) How much work is done by gravity on the stunt woman? (c) How much work is done by air resistance on the stunt woman? (d) Estimate the magnitude of the average force of air resistance by assuming it is constant throughout the fall.

◆89. When a 0.20-kg mass is suspended from a vertically hanging spring, it stretches the spring from its original length of 5.0 cm to a total length of 6.0 cm. The spring with the same mass attached is then placed on a horizontal frictionless surface. The mass is pulled so that the spring stretches to a *total* length of 10.0 cm; then the mass is released and it oscillates back and forth. What is the maximum speed of the mass as it oscillates?

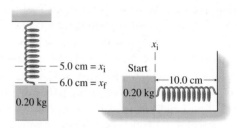

90. Yosemite Falls in California is about 740 m high. (a) What average power would it take for a 70-kg person to hike up to the top of Yosemite Falls in 1.5 h? (b) The human body is about 25% efficient at converting chemical energy to mechanical energy. How much chemical energy is used in this hike? (c) One food Calorie is equal to 4.186×10^3 J. How many Calories of food energy would a person use in this hike?

◆91. A 1500-kg car coasts in neutral down a 2.0° hill. The car attains a terminal speed of 20.0 m/s. (a) How much power must the engine deliver to drive the car on a *level* road at 20.0 m/s? (b) If the maximum useful power that can be delivered by the engine is 40.0 kW, what is the steepest hill the car can climb at 20.0 m/s?

◆92. A spring used in an introductory physics laboratory stores 10.0 J of elastic potential energy when it is compressed 0.20 m. Suppose the spring is cut in half. When one of the halves is compressed by 0.20 m, how much potential energy is stored in it? [*Hint:* Does the half spring have the same k as the original uncut spring?]

◆93. An elevator can carry a maximum load of 1202 kg (including the mass of the elevator car). The elevator has an 801-kg counterweight that always moves with the same speed but in the *opposite direction* to the car. (a) What is the average power that must be delivered by the motor to carry the maximum load up 40.0 m in 60.0 s? (b) How would your answer be different if there were no counterweight?

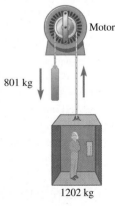

94. (a) How much work does a major-league pitcher do on the baseball when he throws a 90.0 mi/h (40.2 m/s) fastball? The mass of a baseball is 153 g. (b) How many fastballs would a pitcher have to throw to "burn off" a 1520-Calorie meal? (1 Calorie = 1000 cal = 1 kcal.)

Assume that 80.0% of the chemical energy in the food is converted to thermal energy and only 20.0% becomes the kinetic energy of the fastballs.

95. The number of kilocalories per day required by a person resting under standard conditions is called the basal metabolic rate (BMR). (a) To generate 1 kcal, Jermaine's body needs approximately 0.010 mol of oxygen. If Jermaine's net intake of oxygen through breathing is 0.015 mol/min while he is resting, what is his BMR in kcal/day? (b) If Jermaine fasts for 24 h, how many pounds of fat does he lose? Assume that only fat is consumed. Each gram of fat consumed generates 9.3 kcal.

96. Tarzan is running along the ground and approaching a deep gully. A tree branch with a vine hangs over the gully. Tarzan must grab the vine and swing across the gully to the other side, where the ground surface is 1.7 m higher than the ground surface from which Tarzan starts. How fast does Tarzan have to be running to accomplish this feat?

97. Jane is running from the ivory hunters in the jungle. Cheetah throws a 7.0-m-long vine toward her. Jane leaps onto the vine with a speed of 4.0 m/s. When she catches the vine, it makes an angle of 20° with respect to the vertical. (a) When Jane is at her lowest point, she has moved downward a distance h from the height where she originally caught the vine. Show that h is given by $h = L - L \cos 20°$, where L is the length of the vine. (b) How fast is Jane moving when she is at the lowest point in her swing? (c) How high can Jane swing above the lowest point in her swing?

98. The escape speed from Earth is 11.2 km/s, but that is only the minimum speed needed to escape *Earth's* gravitational pull; it does not give the object enough energy to leave the solar system. What is the minimum speed for an object near the Earth's surface so that the object escapes both the Earth's and the Sun's gravitational pulls? Ignore drag due to the atmosphere and the gravitational forces due to the Moon and the other planets. Also ignore the rotation and the orbital motion of the Earth.

99. A skier starts from rest at the top of a frictionless slope of ice in the shape of a hemispherical dome with radius R and slides down the slope. At a certain height h, the normal force becomes zero and the skier leaves the surface of the ice. What is h in terms of R?

100. Two springs with spring constants k_1 and k_2 are connected in series. (a) What is the effective spring constant of the combination? (b) If a hanging object attached to the combination is displaced by 4.0 cm from the relaxed position, what is the potential energy stored in the spring for $k_1 = 5.0$ N/cm and $k_2 = 3.0$ N/cm? [See Problem 83(a).]

101. Two springs with spring constants k_1 and k_2 are connected in parallel. (a) What is the effective spring constant of the combination? (b) If a hanging object attached to the combination is displaced by 2.0 cm from the relaxed position, what is the potential energy stored in the spring for $k_1 = 5.0$ N/cm and $k_2 = 3.0$ N/cm? [See Problem 83(b).]

102. A pendulum, consisting of a bob of mass M on a cord of length L, is interrupted in its swing by a peg a distance d below its point of suspension. (a) If the bob is to travel in a full circle of radius $(L - d)$ around the peg, what is the minimum possible speed it can have at the lowest point in its motion, just before it starts to go around? Ignore any decrease in the length of the string due to the peg's circumference. (b) From what minimum angle θ must the pendulum be released so that the bob attains the speed calculated in (a)?

103. Human feet and legs store elastic energy when walking or running. They are not nearly as efficient at doing so as kangaroo legs, but the effect is significant nonetheless. If not for the storage of elastic energy, a 70-kg man running at 4 m/s would lose about 100 J of mechanical energy each time he sets down a foot. Some of this energy is stored as elastic energy in the Achilles tendon and in the arch of the foot; the elastic energy is then converted back into the kinetic and gravitational potential energy of the leg, reducing the expenditure of metabolic energy. If the maximum tension in the Achilles tendon when the foot is set down is 4.7 kN and the tendon's spring constant is 350 kN/m, calculate how far the tendon stretches and how much elastic energy is stored in it.

104. The graph shows the tension in a rubber band as it is first stretched and then allowed to contract. As you stretch a rubber band, the tension force at a particular length (on the way to a maximum stretch) is larger than the force at that same length as you let the rubber band contract. That is why the graph shows two separate lines, one for stretching and one for contracting; the lines are not superimposed as you might have thought they would be. (a) Make a rough estimate of the total work done by the external force applied to the rubber band for the entire process. (b) If the rubber band obeyed Hooke's law, what would the answer to (a) have to be? (c) While the rubber band is stretched, is all of the work done on it accounted for by the increase in elastic potential energy? If not, what happens to the rest of it? [*Hint:* Take a rubber band and stretch it rapidly several times. Then hold it against your wrist or your lip.]

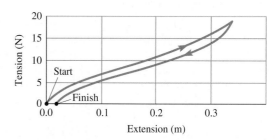

105. A 0.50-kg block, starting at rest, slides down a 30.0° incline with kinetic friction coefficient 0.25 (see the

figure with Problem 63). After sliding 85 cm down the incline, it slides across a frictionless horizontal surface and encounters a spring ($k = 35$ N/m). (a) What is the maximum compression of the spring? (b) After the compression of part (a), the spring rebounds and shoots the block back up the incline. How far along the incline does the block travel before coming to rest?

✦106. A wind turbine converts some of the kinetic energy of the wind into electric energy. Suppose that the blades of a small wind turbine have length $L = 4.0$ m. (a) When a 10 m/s (22 mi/h) wind blows head-on, what volume of air (in m³) passes through the circular area swept out by the blades in 1.0 s? (b) What is the mass of this much air? Each cubic meter of air has a mass of 1.2 kg. (c) What is the translational kinetic energy of this mass of air? (d) If the turbine can convert 40% of this kinetic energy into electric energy, what is its electric power output? (e) What happens to the power output if the wind speed decreases to $\frac{1}{2}$ of its initial value? What can you conclude about electric power production by wind turbines?

✦107. Use dimensional analysis to show that the electric power output of a wind turbine is proportional to the *cube* of the wind speed. The relevant quantities on which the power can depend are the length L of the rotor blades, the density ρ of air (SI units kg/m³), and the wind speed v.

✦108. Use this method to find how the speed with which animals of similar shape can run up a hill depends on the size of the animal. Let L represent some characteristic length, such as the height or diameter of the animal. Assume that the maximum rate at which the animal can do work is proportional to the animal's surface area: $P_{max} \propto L^2$. Set the maximum power output equal to the rate of increase of gravitational potential energy and determine how the speed v depends on L.

109. The potential energy of a particle constrained to move along the x-axis is shown in the graph. At $x = 0$, the particle is moving in the +x-direction with a kinetic energy of 200 J. Can this particle get into the region 3 cm $< x <$ 8 cm? Explain. If it can, what is its kinetic energy in that region? If it can't, what happens to it?

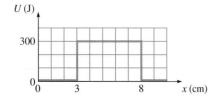

110. The potential energy of a particle constrained to move along the x-axis is shown in the graph. At $x = 0$, the particle is moving in the +x-direction with a kinetic energy of 400 J. Can this particle get into the region 3 cm $< x <$ 8 cm? Explain. If it can, what is its kinetic energy in that region? If it can't, what happens to it?

Answers to Practice Problems

6.1 −180 kJ

6.2 43 N; 4500 J; she pulls with a greater force but its component in the direction of the displacement is smaller.

6.3 $(2.5 \, m)(1.50v)^2/(mv^2) = 5.6$

6.4 29 m/s

6.5 0.24

6.6 16.5 m/s

6.7 48 km/s

6.8 195 km/s

6.9 4.0 J

6.10 3.2 cm

6.11 9.8 m/s

6.12 2.4 m

6.13 (a) 11.8 kW (b) 3.6 kW

Answers to Checkpoints

6.2 The force is perpendicular to the displacement.

6.3 Kinetic energy is never negative. Work can be positive, negative, or zero, because kinetic energy can increase, decrease, or stay the same.

6.4 (a) The gravitational potential energy increases until it reaches its maximum value when the stone reaches its highest point above the ground. (b) The kinetic energy decreases as the potential energy increases. It is zero at the highest point. (c) The force of gravity does work on the stone throughout its motion.

6.5 The mechanical energy is the same throughout Mercury's orbit. The kinetic energy is greatest at the perihelion because the potential energy is smallest there.

6.7 The greatest elastic potential energy is at the maximum compression.

Linear Momentum

After a collision, an accident investigator measures the lengths of skid marks on the road. How can the investigator use this information to figure out the velocities of the vehicles immediately *before* the collision? (See p. 246 for the answer.)

Concepts & Skills to Review

- conservation laws (Section 6.1)
- Newton's third law of motion (Section 4.4)
- Newton's second law of motion (Section 4.3)
- velocity (Section 2.2)
- components of vectors (Section 3.2)
- vector subtraction (Section 3.1)
- kinetic energy (Section 6.3)

7.1   A CONSERVATION LAW FOR A VECTOR QUANTITY

In Chapter 4 we learned how to determine the acceleration of an object by finding the net force acting on it and applying Newton's second law of motion. If the forces happen to be constant, then the resulting constant acceleration enables us to calculate changes in velocity and position. Calculating velocity and position changes when the forces are not constant is much more difficult. In many cases, the forces cannot even be easily determined. Conservation of energy is one tool that enables us to draw conclusions about motion without knowing all the details of the forces acting. Recall, for example, how easily we can calculate the escape speed of a projectile using conservation of energy, without even knowing the path the object takes. Now imagine how difficult the same calculation would be using Newton's second law, with a gravitational force that changes magnitude and direction depending on the path taken.

In this chapter we develop another conservation law. Conservation laws are powerful tools. If a quantity is conserved, then no matter how complicated the situation, we can set the value of the conserved quantity at one time equal to its value at a later time. The "before-and-after" aspect of a conservation law enables us to draw conclusions about the results of a complicated set of interactions without knowing all of the details.

The new conserved quantity, *momentum*, is a vector quantity, in contrast to energy, which is a scalar. When momentum is conserved, both the magnitude and the direction of the momentum must be constant. Equivalently, the *x*- and *y*-components of momentum are constant. When we find the total momentum of more than one object, we must add the momentum vectors according to the procedure by which vectors are always added.

CONNECTION:

Conservation laws can involve scalars, such as energy, or vectors, such as momentum.

7.2   MOMENTUM

The word *momentum* is often heard in broadcasts of sporting events. A sports broadcaster might say, "The home team has won five consecutive games; they have the momentum in their favor." The team with "momentum" is hard to stop; they are moving forward on a winning streak. A football player, running for the goal line with a football tucked under his arm, has momentum; he is hard to stop. This use of the word *momentum* is closer to the physics usage. In physics we would agree that the runner has momentum, but we have a precise definition in mind.

In everyday use, momentum has something to do with mass as well as with velocity. Would you rather have a running child bump into you, or a football player running with the same velocity? The child has much less momentum than the football player, even though their velocities are the same.

Could a quantity combining mass and velocity be useful in physics? Imagine a collision between two spaceships (Fig. 7.1). Let the spaceships be so far from planets and stars that we can ignore gravitational interactions with celestial bodies. The spaceships exert forces on one another while they are in contact. According to Newton's third law, these forces are equal and opposite. The force on ship 2 exerted by ship 1 is equal and opposite to the force exerted on ship 1 by ship 2:

$$\vec{F}_{21} = -\vec{F}_{12}$$

$\vec{F}_{21}$ is the force exerted *on* object 2 *by* object 1.

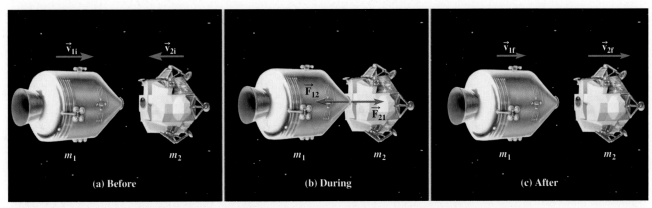

Figure 7.1 (a) Two spaceships about to collide. (b) During the collision, the spaceships exert forces on one another that are equal in magnitude and opposite in direction. (c) The velocities of the spaceships after the collision.

The changes in *velocities* of the two spaceships are *not* equal and opposite if the masses are different. Suppose a large spaceship (mass m_1) collides with a much smaller ship (mass $m_2 \ll m_1$). Assume for now that the forces are constant during the time interval Δt that the spaceships are in contact. Although the forces have the same magnitude, the magnitudes of the accelerations of the two ships are different because their masses are different. The ship with the larger mass has the smaller acceleration.

The acceleration of either spaceship causes its velocity to change by

$$\Delta \vec{v} = \vec{a}\, \Delta t = \frac{\vec{F}}{m} \Delta t$$

The time interval Δt is the duration of the interaction between the two ships, so it must be the same for both ships.

Since the changes in velocity are inversely proportional to the masses, the changes in the *products* of mass and velocity are equal and opposite for the two bodies involved in the interaction:

$$m_1 \Delta \vec{v}_1 = \vec{F}_{12} \Delta t$$

$$m_2 \Delta \vec{v}_2 = \vec{F}_{21} \Delta t = (-\vec{F}_{12})\Delta t = -(m_1 \Delta \vec{v}_1)$$

This is a useful insight, so we give the product of mass and velocity a name and symbol: **linear momentum** (symbol $\vec{p}$, SI unit kg·m/s). Linear momentum (or just *momentum*) is a vector quantity having the same direction as the velocity.

Definition of linear momentum:

$$\vec{p} = m\vec{v} \tag{7-1}$$

The collision of the two spaceships causes changes in their momenta that are equal in magnitude and opposite in direction:

$$\Delta \vec{p}_2 = -\Delta \vec{p}_1$$

In any interaction between two objects, momentum can be transferred from one object to the other. The momentum changes of the two objects are always equal and opposite, so the total momentum of the two objects is unchanged by the interaction. (By *total momentum* we mean the vector sum of the individual momenta of the objects.)

Example 7.1 gives some practice in finding the change in momentum of an object whose velocity changes. Remember that momentum is a vector quantity, so changes in momentum must be found by subtracting momentum vectors, not by subtracting the magnitudes of the momenta.

CONNECTION:

Newton's third law implies that during an interaction momentum is transferred from one body to another.

Example 7.1

Change of Momentum of a Moving Car

A car weighing 12 kN is driving due north at 30.0 m/s. After driving around a sharp curve, the car is moving east at 13.6 m/s. What is the change in momentum of the car?

Strategy   The definition of momentum is $\vec{\mathbf{p}} = m\vec{\mathbf{v}}$. We can start by finding the car's mass. There are two potential pitfalls:

1. momentum depends not on weight but on mass, and
2. momentum is a vector, so we must take its direction into consideration as well as its magnitude. To find the change in momentum, we need to do a *vector subtraction*.

Solution   The car's mass is

$$m = \frac{W}{g} = \frac{1.2 \times 10^4\,\text{N}}{9.8\,\text{m/s}^2} = 1220\,\text{kg}$$

The car's initial velocity is

$$\vec{\mathbf{v}}_i = 30.0\,\text{m/s, north}$$

The car's initial momentum is then

$$\vec{\mathbf{p}}_i = m\vec{\mathbf{v}}_i = 1220\,\text{kg} \times 30.0\,\text{m/s north}$$
$$= 3.66 \times 10^4\,\text{kg·m/s north}$$

After the curve, the final velocity is

$$\vec{\mathbf{v}}_f = 13.6\,\text{m/s, east}$$

The final momentum is

$$\vec{\mathbf{p}}_f = m\vec{\mathbf{v}}_f = 1220\,\text{kg} \times 13.6\,\text{m/s east}$$
$$= 1.66 \times 10^4\,\text{kg·m/s east}$$

Momentum vectors are added and subtracted according to the same methods used for other vectors. To find the change in the momentum, we draw vector arrows representing the addition of $\vec{\mathbf{p}}_f$ and $-\vec{\mathbf{p}}_i$ (Fig. 7.2). Since *in this case* the three vectors in Fig. 7.2 form a right triangle, the magnitude of $\Delta\vec{\mathbf{p}}$ can be found from the Pythagorean theorem

$$|\Delta\vec{\mathbf{p}}| = \sqrt{p_i^2 + p_f^2}$$
$$= \sqrt{(3.66 \times 10^4\,\text{kg·m/s})^2 + (1.66 \times 10^4\,\text{kg·m/s})^2}$$
$$= 4.02 \times 10^4\,\text{kg·m/s}$$

From the vector diagram, $\Delta\vec{\mathbf{p}}$ is directed at an angle θ east of south. Using trigonometry,

$$\tan\theta = \frac{\text{opposite}}{\text{adjacent}} = \frac{p_f}{p_i} = \frac{1.66 \times 10^4\,\text{kg·m/s}}{3.66 \times 10^4\,\text{kg·m/s}} = 0.454$$

$$\theta = \tan^{-1} 0.454 = 24.4°$$

Since the weight is given with two significant figures, we report the change in momentum of the car as 4.0×10^4 kg·m/s directed 24° east of south.

Discussion   As with displacements, velocities, accelerations, and forces, it is crucial to remember that momentum is a vector. When finding changes in momentum, we must find the difference between final and initial momentum *vectors*. If the initial and final momenta had not been perpendicular, we would have had to resolve the vectors into *x*- and *y*-components in order to subtract them.

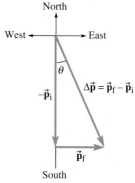

Figure 7.2
Vector subtraction to find the change in momentum.

Practice Problem 7.1   Falling Apple

(a) What is the momentum of an apple weighing 1.0 N just before it hits the ground, if it falls out of a tree from a height of 3.0 m? (b) The apple falls because of the gravitational interaction between the apple and the Earth. How much does this interaction change *Earth's* momentum? How much does it change Earth's velocity?

✓ CHECKPOINT 7.2

In Example 7.1, if the speed of the car had remained constant, would $\Delta\vec{\mathbf{p}}$ have been zero?

7.3   THE IMPULSE-MOMENTUM THEOREM

We found that the change in momentum of an object when a single force acts on it is equal to the product of the force acting on the object and the time interval during which the force acts:

$$\Delta\vec{\mathbf{p}} = \vec{\mathbf{F}}\,\Delta t$$

The product $\vec{\mathbf{F}}\,\Delta t$ is given the name **impulse**. Since the impulse is the product of a vector (the force) and a positive scalar (the time), impulse is a vector quantity having the same direction as that of the force. In words, $\Delta\vec{\mathbf{p}} = \vec{\mathbf{F}}\,\Delta t$ can be read as *the change in momentum equals the impulse.* The SI units of impulse are newton-seconds (N·s) and those of momentum are kilogram-meters per second (kg·m/s). These are equivalent units, as can be demonstrated using the definition of the newton (Problem 3).

If an object is involved in more than one interaction, then its change in momentum during any time interval is equal to the *total* impulse during that time interval. The total impulse is the vector sum of the impulses due to each force. The total impulse is also equal to the net force times the time interval:

$$\text{total impulse} = \vec{\mathbf{F}}_1\,\Delta t + \vec{\mathbf{F}}_2\,\Delta t + \cdots$$
$$= (\vec{\mathbf{F}}_1 + \vec{\mathbf{F}}_2 + \cdots)\,\Delta t = \sum\vec{\mathbf{F}}\,\Delta t$$

The total impulse on an object is equal to the change in the object's momentum during the same time interval. This relationship between total impulse and momentum change is called the impulse-momentum theorem and is especially useful in solving problems that involve collisions and impacts.

Impulse = $\vec{\mathbf{F}}\,\Delta t$

CONNECTION

Impulse is a momentum transfer due to a force; work is an energy transfer due to a force.

	Impulse	Work
Definition	$\vec{\mathbf{F}}\,\Delta t$	$\vec{\mathbf{F}}\cdot\Delta\vec{\mathbf{r}}$
Vector or Scalar?	Vector	Scalar*
Physical meaning	Momentum transfer	Energy transfer

*The scalar or dot product of two vectors is introduced in Section 6.2.

Impulse-Momentum Theorem

$$\Delta\vec{\mathbf{p}} = \sum\vec{\mathbf{F}}\,\Delta t \qquad (7\text{-}2)$$

Impulse When Forces Are Not Constant Our discussion so far has assumed that the forces acting are constant or that Δt is very small so the change in $\vec{\mathbf{F}}$ is negligible. That is a rather unusual situation; the concept of momentum would be of limited use if it were applicable only when forces are constant. However, everything we have said still applies to situations where the forces are not constant, as long as we use the *average* force to calculate the impulse.

When the force is not constant, the impulse can be found using the average force.

$$\text{impulse} = \vec{\mathbf{F}}_{\text{av}}\,\Delta t \qquad (7\text{-}3)$$

Conceptual Example 7.2

Big Force–Short Time Versus Small Force–Long Time

Which causes the larger change in momentum of an object, an average force of 5 N acting for 4 s or an average force of 2 N acting for 10 s? How might this principle be used when designing products to protect the human body from injury? Give an example.

Solution and Discussion The change in momentum is equal to the impulse. The product of the force and the time interval gives the momentum change of the object. Over a period of 4 s, the 5-N force causes a momentum change of magnitude (5 N × 4 s) = 20 N·s, and the 2-N force acting for 10 s also causes a momentum change of magnitude (2 N × 10 s) = 20 N·s. The smaller force causes the same change in momentum because it acts for a longer time interval.

When designing products to protect the human body, one goal is to lengthen the time period during which a velocity change occurs. For example, when a movie stuntman falls from a great height, he lands on a large air bag (Fig. 7.3),

Figure 7.3

A stuntman lands safely in an air bag to break his fall. The air bag reduces the risk of injury in two ways. It changes the stuntman's momentum more gradually, so that forces of smaller magnitude act on his body. It also spreads these forces over a larger area so they are less likely to cause serious injury.

continued on next page

Example 7.2 continued

which changes his momentum much more gradually than if he were to fall onto concrete. The average force exerted by the air bag on the stuntman is much smaller than the average force exerted by concrete would be. Nets used under circus acrobats serve the same purpose. The net gives and dips downward when the acrobat falls into it, gradually reducing the speed of the fall over a longer time interval than if she fell directly onto the ground.

Practice Problem 7.2   Pole-Vaulter Landing on a Padded Surface

A pole-vaulter vaults over the bar and falls onto thick padding. He lands with a speed of 9.8 m/s; the padding then brings him to a stop in a time of 0.40 s. What is the average force on his body *due to the padding* during that time interval? Express your answer as a fraction or multiple of his weight. [*Hint:* The force due to the padding is not the only force acting on the vaulter during the 0.40-s interval.]

Application of momentum conservation to automotive design.

Designing a Safer Automobile   One automotive design change implemented to minimize injury on collision is the foam padding built into automobile dashboards (Fig. 7.4). Automobile bumpers have shock absorbers built in to lessen damage to the car body in small collisions. The structure of the car itself is often a single piece of metal with reinforced supports (*unibody* construction) so that the entire body can crumple and absorb the change in momentum more slowly than it would if it were made of separate sections of metal that would slide into or over each other or fall off the car. The safety glass in a windshield has two advantages. One is that it does not shatter and send sharp shards of glass into human tissue, but the other is that it distorts when struck by solid objects like human bones or a human head. The glass doesn't give much, but in a crash every little bit helps.

The use of seat belts plus the air bag is better than either alone. Without a seat belt, the body continues moving with the same speed the car had before the crash. The rapidly inflating air bag moves toward the body and the effective velocity is then the sum of the two velocities (air bag velocity + body velocity) when the two collide. The body flying into the air bag can be injured more than a restrained body making more gradual contact with the air bag. An adult should sit at least 12 in. from the air bag container to avoid injury from the deploying air bag itself. Small children should always be placed in the back seat, in proper car seats for their size, to ensure their safety.

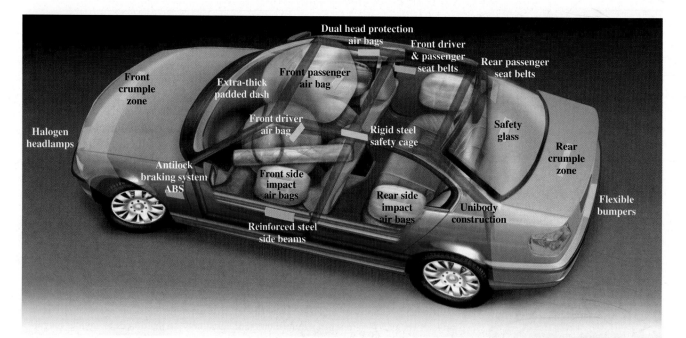

Figure 7.4 Some safety features of the modern automobile. Many of these features serve to lengthen the time interval during which a momentum change occurs in a crash, thereby lessening the forces acting on the passengers.

Example 7.3

Collision Between an Automobile and a Tree

A car moving at 20.0 m/s (44.7 mi/h) crashes into a tree. Find the magnitude of the average force acting on a passenger of mass 65 kg in each of the following cases. (a) The passenger is not wearing a seat belt. He is brought to rest by a collision with the windshield and dashboard that lasts 3.0 ms. (b) The car is equipped with a passenger-side air bag. The force due to the air bag acts for 30 ms, bringing the passenger to rest.

Strategy From the impulse-momentum theorem, $\Delta \vec{p} = \vec{F}_{av} \Delta t$, where $\vec{F}_{av}$ is the average force acting on the passenger and Δt is the time interval during which the force acts. The change in the passenger's momentum is the same in the two cases. What differs is the time interval during which the change occurs. It takes a larger force to change the momentum in a shorter time interval.

Solution The magnitude of the passenger's initial momentum is

$$|\vec{p}_i| = |m\vec{v}_i| = 65 \text{ kg} \times 20.0 \text{ m/s} = 1300 \text{ kg·m/s}$$

His final momentum is zero, so the magnitude of the momentum change is

$$|\Delta \vec{p}| = 1300 \text{ kg·m/s}$$

This momentum change divided by the time interval gives the magnitude of the average force in each case.

(a) No seat belt: $|\vec{F}_{av}| = \dfrac{|\Delta \vec{p}|}{\Delta t} = \dfrac{1300 \text{ kg·m/s}}{0.0030 \text{ s}} = 4.3 \times 10^5 \text{ N}$

(b) Air bag: $|\vec{F}_{av}| = \dfrac{|\Delta \vec{p}|}{\Delta t} = \dfrac{1300 \text{ kg·m/s}}{0.030 \text{ s}} = 4.3 \times 10^4 \text{ N}$

Discussion The average forces required to bring the passenger to rest are inversely proportional to the time interval over which those forces act. It is a far happier situation to have the momentum change over as long a period as possible to make the forces smaller. Automotive safety engineers design cars to minimize the average forces on the passengers during sudden stops and collisions.

The air bag also spreads the force over a much larger area than impact with a hard surface like the windshield, further reducing the risk of injury.

Practice Problem 7.3 Catching a Fastball

A baseball catcher is catching a fastball that is thrown at 43 m/s (96 mi/h) by the pitcher. If the mass of the ball is 0.15 kg and if the catcher moves his mitt backward toward his body by 8.0 cm as the ball lands in the glove, what is the magnitude of the average force acting on the catcher's mitt? Estimate the time interval required for the catcher to move his hands.

PHYSICS AT HOME

Try playing catch with a friend [on the lawn] while using a raw egg or a water balloon as a ball. How do you move your hands to minimize the chance of breaking the egg or balloon when you catch it? What is likely to happen if you forget that the "ball" is an egg or balloon and catch it as you would a ball?

Graphical Calculation of Impulse

When a force is changing, how can we find the impulse? We've asked similar questions in previous chapters. For simplicity we consider components along the x-axis. Recall:

- displacement = $\Delta x = v_{av,x} \Delta t$ = area under $v_x(t)$ graph
- change in velocity = $\Delta v_x = a_{av,x} \Delta t$ = area under $a_x(t)$ graph

In both cases, the mathematical relationship is that of a rate of change. Velocity is the rate of change of position with time and acceleration is the rate of change of velocity with time. Now we have force as the rate of change of momentum with time. By analogy:

- impulse = $F_{av,x} \Delta t$ = area under $F_x(t)$ graph

So to find the impulse for a variable force, we find the area under the $F_x(t)$ graph. Then, if we wish to know the average force, we can divide the impulse by the time interval during which the force is applied.

CONNECTION:

See Sections 2.2, 2.3, and 6.6 to review how we used the area under a graph to find displacement, change in velocity, and work done by a force.

A "graph of $v_x(t)$" means the quantity v_x is plotted as a function of the variable t with v_x on the vertical axis and t on the horizontal axis.

Figure 7.5 (a) The area under the $F_x(t)$ graph for a variable force is the impulse. (b) The average force for a given time interval is the constant force that would produce the same impulse.

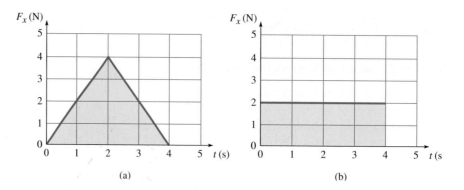

(a)   (b)

The variable force of Fig. 7.5a increases linearly from 0 to 4 N in a time of 2 s; then it decreases from 4 N to 0 N in 2 s. The area under the $F_x(t)$ graph is found from the triangular area

$$\text{area} = \tfrac{1}{2}\,\text{base} \times \text{height} = 2\;\text{s} \times 4\;\text{N} = 8\;\text{N·s} = \text{impulse}$$

The average force during the 4-s time interval is

$$\text{average force} = \frac{\text{impulse}}{\text{time interval}} = \frac{8\;\text{N·s}}{4\;\text{s}} = 2\;\text{N}$$

Figure 7.5b shows the average force over the 4-s time interval; the area under the curve (the impulse) is the same as in Fig. 7.5a.

Example 7.4

Hitting the Wall

An experimental robotic car of mass 10.2 kg moving at 1.2 m/s in the +x-direction crashes into a brick wall and rebounds. A force sensor on the car's bumper records the force that the wall exerts on the car as a function of time. These data are shown in graphical form in Fig. 7.6. (a) What is the maximum magnitude of the force exerted on the car? (b) What is the average force on the car during the collision? (c) At what speed does the car rebound from the wall?

Strategy The maximum force can be read directly from the graph. To solve parts (b) and (c) of this problem, we must

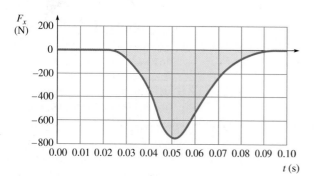

Figure 7.6

Force versus time for a car colliding with a wall.

find the impulse exerted on the car. Since impulse is the area under the $F_x(t)$ curve, we'll make an estimate of the area. The impulse is then equal to the average force times the time interval and also to the car's change in momentum. Once we find the change in momentum, we use it to find the car's final speed.

Given: $m = 10.2$ kg; $v_{ix} = 1.2$ m/s; graph of $F_x(t)$

To find: (a) F_{max}; (b) $F_{av,x}$; (c) v_{fx}

Solution (a) From Fig. 7.6, the maximum force is approximately 750 N in magnitude.

(b) Each division on the horizontal axis represents 0.01 s, and each vertical division represents 200 N. Then the area of each grid box represents $(200\;\text{N} \times 0.01\;\text{s}) = 2$ N·s. Counting the number of grid boxes between the $F_x(t)$ curve and the time axis, estimating fractions of boxes, yields about 10 boxes. Then the magnitude of the impulse is approximately

$$10\;\text{boxes} \times 2\;\text{N·s/box} = 20\;\text{N·s}$$

The collision is underway when the force is nonzero. So the collision begins at about $t = 0.025$ s and ends at about $t = 0.095$ s. The duration of the collision is

$$\Delta t = 0.07\;\text{s}$$

continued on next page

Example 7.4 continued

The magnitude of the average force is approximately

$$|F_{av,x}| = \frac{|\text{impulse}|}{\Delta t} = \frac{20 \text{ N·s}}{0.07 \text{ s}} = 300 \text{ N}$$

(c) The impulse gives us the momentum change. The force exerted by the wall is in the $-x$-direction. Thus, the x-component of the impulse is negative. In the graph of F_x versus t, the area lies under the time axis and so is counted as negative. So, working with x-components,

$$\Delta p_x = mv_{fx} - mv_{ix} = F_{av,x} \Delta t = -20 \text{ N·s}$$

Solving for v_{fx},

$$v_{fx} = \frac{\Delta p_x + mv_{ix}}{m} = \frac{\Delta p_x}{m} + v_{ix}$$

Substituting numerical values in this expression yields

$$v_{fx} = \frac{-20 \text{ N·s}}{10.2 \text{ kg}} + 1.2 \text{ m/s} = -0.8 \text{ m/s}$$

The car rebounds at a speed of 0.8 m/s.

Discussion   As a check, we compare the average force with the maximum force. The average force is a bit less than half of the maximum force. If the force were a linear function of time, the average would be exactly half the maximum. Here,

the average force is less than that because more time is spent at smaller values of force than at the larger values.

Practice Problem 7.4   Car-Van Collision

A car weighing 13.6 kN is moving at 10.0 m/s in the $+x$-direction when it collides head-on with a van weighing 33.0 kN. The horizontal force exerted on the car before, during, and after the collision is shown in Fig. 7.7. What is the car's velocity just after the collision?

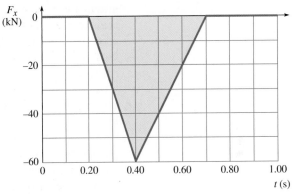

Figure 7.7

Varying force on a car during a car-van collision.

A Restatement of Newton's Second Law

We can use the relationship between impulse and momentum to find a new way to understand Newton's second law. Let's rewrite the impulse-momentum theorem this way:

$$\sum \vec{F}_{av} = \frac{\Delta \vec{p}}{\Delta t}$$

What happens if we let the time interval Δt get smaller and smaller, approaching zero? Then the average force is taken over a smaller and smaller time interval, approaching the instantaneous force:

$$\vec{F} = \lim_{\Delta t \to 0} \frac{\Delta \vec{p}}{\Delta t}$$

If more than one force acts, we must replace $\vec{F}$ with $\sum \vec{F}$. Then our restatement of Newton's second law becomes

Newton's Second Law

$$\sum \vec{F} = \lim_{\Delta t \to 0} \frac{\Delta \vec{p}}{\Delta t} \tag{7-4}$$

CONNECTION:

Equation (7-4) is closer to Newton's original statement of his second law and is more general than $\sum \vec{F} = m\vec{a}$.

Figure 7.8 The Space Shuttle is propelled upward as hot gases are exhausted downward at high speeds.

In words, *the net force is the rate of change of momentum.*

Equation (7-4) is *more general* than $\sum \vec{F} = m\vec{a}$, the form of Newton's second law used in Chapters 4 through 6, which holds only when mass is constant. One situation in which mass is not constant is the rocket engine. In a rocket engine, fuel combustion produces hot gases that are then expelled at high speeds (Fig. 7.8). The rocket's mass decreases as the exhaust gases are expelled.

When mass is constant, then it can be factored out:

$$\Sigma\vec{F} = \lim_{\Delta t \to 0} \frac{\Delta \vec{p}}{\Delta t} = \lim_{\Delta t \to 0} \frac{\Delta(m\vec{v})}{\Delta t} = m \lim_{\Delta t \to 0} \frac{\Delta \vec{v}}{\Delta t} = m\vec{a}$$

Thus, Eq. (7-4) reduces to the familiar form of Newton's second law from Chapters 4–6 when mass is constant.

7.4   CONSERVATION OF MOMENTUM

Consider two pucks that bump into one another after sliding along a frictionless table. Figure 7.9 shows what happens to the two pucks before, during, and after their interaction. If we think of the two pucks as constituting a single system, then the gravitational interactions with the Earth and the contact interactions with the table are *external* interactions—interactions with objects external to the system. The force of gravity on each object is balanced by the normal force on the same object and, thus, there is no net impulse up or down. Together, these forces produce a net external force of zero, so they leave the system's momentum unchanged. Since these two always cancel, we can ignore these external interactions and just focus on the interaction between the pucks. Therefore, we omit the normal and gravitational forces in Fig. 7.9.

Until contact is made, there is no interaction between the pucks (ignoring the small gravitational interaction between the two). During the collision, the pucks exert forces on each other. Force $\vec{F}_{12}$ is the contact force acting on mass m_1 and force $\vec{F}_{21}$ is the contact force acting on mass m_2. If we continue to regard the two pucks as parts of a single interacting system, then those forces are *internal* forces of this system. When they collide, some momentum is transferred from one puck to the other. The changes in momentum of the two are equal and opposite:

$$\Delta\vec{p}_1 = -\Delta\vec{p}_2$$

Since the change in momentum is the final momentum minus the initial momentum, we write:

$$\vec{p}_{1f} - \vec{p}_{1i} = -(\vec{p}_{2f} - \vec{p}_{2i})$$

Moving the initial momenta to the right side and the final momenta to the left:

$$\vec{p}_{1f} + \vec{p}_{2f} = \vec{p}_{1i} + \vec{p}_{2i} \qquad (7\text{-}5)$$

Equation (7-5) says the sum of the momenta of the pucks after the interaction is equal to the sum of the momenta before the interaction; or, more simply, the total momentum of the objects is unchanged by the collision. This isn't surprising since, if some momentum is just transferred from one to the other, the total hasn't changed. We say that momentum is *conserved* for this collision. The interaction between the pucks changes the momentum of each puck, but the total momentum of the system is unchanged.

Figure 7.9 Two sliding pucks with different masses before, during, and after collision.

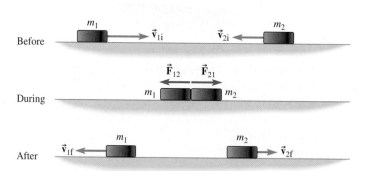

In a system composed of more than two objects, interactions between objects inside the system do not change the total momentum of the system—they just transfer some momentum from one part of the system to another. Only external interactions can change the total momentum of the system. To summarize:

- The total momentum of a system is the vector sum of the momenta of each object in the system.
- External interactions can change the total momentum of a system.
- Internal interactions do not change the total momentum of a system.

In the absence of external interactions, momentum is conserved:

Law of Conservation of Linear Momentum

If the net external force acting on a system is zero, then the momentum of the system is conserved.

$$\text{If } \sum \vec{F}_{ext} = 0, \quad \vec{p}_i = \vec{p}_f \tag{7-6}$$

By definition, an *isolated*, or closed, system is subject to no external interactions; thus, *linear momentum is always conserved for an isolated system.* Remember that momentum is a vector quantity, so both the magnitude *and the direction* of the momentum at the beginning and end of the interaction must be the same. In component form, both p_x and p_y are unchanged by the interaction.

✓ CHECKPOINT 7.4

When is the momentum of a system *not* conserved?

Example 7.5

Adrift on a Raft

Diana is standing on a raft of mass 100.0 kg that is floating on a still lake. She decides to walk the length of the raft (Fig. 7.10). If Diana's mass is 55 kg and she walks with a velocity of 0.91 m/s *with respect to the shore*, how fast and in what direction does the raft move while Diana is walking?

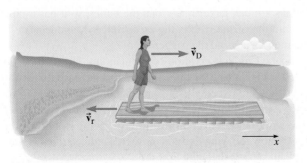

Figure 7.10

Diana walking along a raft. Velocities $\vec{v}_D$ and $\vec{v}_r$ are measured with respect to the shore.

Assume the raft is stationary with respect to the shore before Diana starts walking.

Strategy   Diana and the raft can be considered to be a single *isolated system*: as long as frictional forces on the raft due to the water and air are small enough to ignore, the net external force on the system is zero. Then the momentum of this system (raft + Diana) is conserved. We let the subscripts D stand for Diana and r for the raft and set the change in momentum of the system equal to zero.

Solution   To walk forward, Diana must exert a backward force on the raft: the static frictional force between her feet and the raft. This is an internal interaction within the isolated system, so it cannot change the total momentum of the system. Only something acting from outside the system could do that. As Diana walks in one direction, she acquires some momentum. The rest of the system (the raft) must acquire an equal and opposite momentum, because the momentum of the isolated system (Diana + raft) is conserved, which means that the change in momentum of the system is zero.

continued on next page

Example 7.5 continued

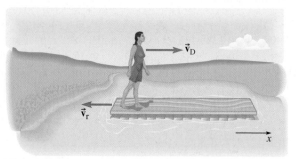

Figure 7.10
Diana walking along a raft. Velocities $\vec{v}_D$ and $\vec{v}_r$ are measured with respect to the shore.

First we set the change in momentum of the system equal to zero:

$$\Delta\vec{p} = 0 = \Delta\vec{p}_D + \Delta\vec{p}_r$$

or

$$\Delta\vec{p}_D = -\Delta\vec{p}_r$$

This means that the momentum changes of Diana and of the raft are equal and opposite. Since momentum is the product of mass and velocity and the masses of the raft and Diana do not change,

$$m_D\,\Delta\vec{v}_D = -m_r\,\Delta\vec{v}_r$$

Solving for the change in velocity of the raft gives

$$\Delta\vec{v}_r = -\frac{m_D}{m_r}\,\Delta\vec{v}_D$$

Finally we substitute numerical values from the given information in the statement of the problem. Let Diana walk in the +x-direction.

$$\Delta\vec{v}_r = \frac{-55\ \text{kg} \times 0.91\ \text{m/s (in the } +x\text{-direction)}}{100.0\ \text{kg}}$$

$$= 0.50\ \text{m/s in the } -x\text{-direction}$$

The negative sign reverses the direction: $\Delta\vec{v}_r$ is in the $-x$-direction and has a magnitude of 0.50 m/s.

The raft moves in a direction opposite to Diana's motion to keep the momentum unchanged and thus conserved. Since the raft was originally stationary, this is the new velocity of the raft.

Discussion   In any momentum conservation problem there are two equivalent ways to proceed. In this example we set the momentum change of the system equal to zero. We could just as well write an equation that sets the initial total momentum equal to the final momentum of the system. The raft and Diana are initially at rest, so the initial momentum is zero:

$$0 = m_D\vec{v}_D + m_r\vec{v}_r$$

where $\vec{v}_D$ and $\vec{v}_r$ are the final velocities of Diana and the raft.

Practice Problem 7.5   Skaters Pushing Apart

Two skaters on in-line skates, Lisa and Bart, are initially at rest. They push apart and start moving in opposite directions. If Lisa's speed just after they push apart is 2.0 m/s and her mass is 85% of Bart's mass, how fast is Bart moving at that time?

Application of momentum conservation: recoil of a rifle

When a bullet is fired from a rifle, the system of rifle plus bullet must conserve momentum. Suppose the rifle is at rest before the bullet is fired. The momentum of the system is zero. When the bullet is fired, part of the system's mass breaks away and travels in one direction with a certain momentum. The rifle, which is the remaining mass of the system, moves in the exact opposite direction such that the total momentum of the system is still zero. The rifle has a much larger mass than the bullet, so it has a much smaller speed. The backward motion of the rifle is the *recoil* felt by anyone who has held a rifle against her shoulder and squeezed the trigger.

Application of momentum conservation: jets, rockets, and airplane wings

Jet engines and rockets operate by conservation of momentum. Hot combustion gases are forced out of nozzles at high speed by the engines. The increased backward momentum of the hot gases as they are expelled is accompanied by an increased forward momentum of the engines. Airplane wings generate lift by conservation of momentum. The main purpose of the wing is to deflect air downward, giving it a downward momentum component. (Exactly how the wing does this is the complicated part.) Since the wing pushes air downward, air pushes the wing upward.

Conceptual Example 7.6

Escape on Slippery Ice

A pilot parachutes from his disabled aircraft and lands on the frozen surface of a lake. There is no breeze blowing and the lake surface is too slippery to walk on. What can the pilot do to reach the shore?

Strategy and Solution Since the person in jeopardy is a pilot, he begins to think about how hot gases forced backward from a jet engine cause the plane to move forward. That gives him an idea: he bundles the parachute into a package and pushes it as hard as possible in a direction away from the nearest point of the shore. If friction is negligible, the net external force on the system of pilot plus parachute is zero and the total momentum of the system cannot change. The momentum of the parachute plus the momentum of the pilot must still equal zero. By conservation of momentum, the pilot begins sliding in the opposite direction and glides toward the shore.

Discussion If friction brings the pilot to rest before he reaches the shore, he can search his pockets and belt loops for other items to throw away. Once he reaches shore, he can tie one end of a rope to a tree and, holding onto the other end, venture back out onto the ice to retrieve any essential items. The rope provides him with an external force so he can get back to shore.

Practice Problem 7.6   Recoil of a Rifle

During an afternoon of target practice, you fire a Winchester .308 rifle of mass 3.8 kg. The bullets have a mass of 9.72 g and leave the rifle at a muzzle velocity of 860 m/s. If you are sloppy and fire a round when the butt of the rifle is not firmly up against your shoulder, at what speed does the rifle butt smash into your shoulder? (Ouch!)

PHYSICS AT HOME

In case you and a friend ever end up stuck in the middle of the ice, practice the technique the pilot used to escape to the lakeshore. Bring a heavy medicine ball out to the middle of the ice rink and face each other with your skates aligned parallel. Toss the ball to your friend. What happens to you? What happens to your friend when he catches the ball? Can you both be "saved" by tossing the ball back and forth? (The same technique works using in-line skates.)

7.5   CENTER OF MASS

We have seen that the momentum of an isolated system is conserved even though parts of the system may interact with other parts; internal interactions transfer momentum between parts of the system but do not change the total momentum of the system. We can define a point called the **center of mass** (CM) that serves as an average location of the system. In Section 7.6, we prove that the center of mass of an isolated system must move with constant velocity, regardless of how complicated the motions of parts of the system may be. Then we can treat the mass of the system as if it were all concentrated at the CM, like a point particle. The CM of an object is not necessarily located within the object; for some objects, such as a boomerang, the center of mass is located outside of the object itself (Fig. 7.11a).

What if a system is not isolated, but has external interactions? Again imagine all of the mass of the system concentrated into a single point particle located at the CM. The motion of this fictitious point particle is determined by Newton's second law, where the net force is the sum of all of the external forces acting on *any part* of the system. In the case of a complex system composed of many parts interacting with each other, the motion of the CM is considerably simpler than the motion of an arbitrary particle of the system (Fig. 7.11b,c).

Figure 7.11 (a) The center of mass of a boomerang is a point outside of the boomerang. (b) The path followed by the center of mass when a hammer is tossed through the air. (c) British pole-vaulter Ben Challenger's center of mass actually passes *beneath* the bar as his body passes over the bar.

Location of Center of Mass   For a system composed of two particles, the center of mass lies somewhere on a line between the two particles. In Fig. 7.12, particles of masses m_1 and m_2 are located at positions x_1 and x_2, respectively. We define the location of the CM for these two particles as

$$x_{CM} = \frac{m_1 x_1 + m_2 x_2}{m_1 + m_2} \tag{7-7}$$

The CM is a *weighted average* of the positions of the two particles. Here we use the word *weighted* in its statistical sense. The position of a particle with more mass counts more—carries more *statistical* weight—than does the position of a particle with a smaller mass. We can rewrite Eq. (7-7) as a weighted average:

$$x_{CM} = \frac{m_1}{M} x_1 + \frac{m_2}{M} x_2 \tag{7-8}$$

Here $M = m_1 + m_2$ represents the total mass of the system. The statistical weight used for the location of each particle is the mass of that particle as a fraction of the total mass of the system.

Suppose masses m_1 and m_2 are equal. Then we expect the CM to be located midway between the two particles (Fig. 7.12a). If $m_1 = 2m_2$, as in Fig. 7.12b, then the CM is closer to the particle of mass m_1. Figure 7.12b shows that, in this case, the CM is twice as far from m_2 as from m_1.

For a system of N particles, at arbitrary locations in three-dimensional space, the definition of the CM is a generalization of Eq. (7-7).

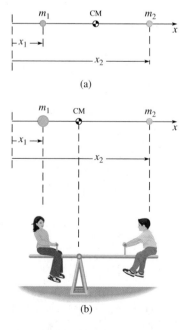

Figure 7.12 (a) Two particles of equal mass located at positions x_1 and x_2 from the origin. The CM is midway between the two. (b) Two particles of unequal mass. The CM is closer to the more massive particle. For two children balanced on a see-saw, the CM is at the fulcrum.

Definition of center of mass:

Vector form:
$$\vec{r}_{CM} = \frac{\sum m_i \vec{r}_i}{M} \qquad (7\text{-}9)$$

Component form: $x_{CM} = \dfrac{\sum m_i x_i}{M} \qquad y_{CM} = \dfrac{\sum m_i y_i}{M} \qquad z_{CM} = \dfrac{\sum m_i z_i}{M}$

where $i = 1, 2, 3, \ldots, N$ and $M = \sum m_i$

Remember that the symbol $\sum$ stands for *sum*. The shorthand notation $\sum m_i x_i$ is interpreted as

$$\sum m_i x_i = m_1 x_1 + m_2 x_2 + \cdots + m_N x_N$$

For particles in two-dimensional space, we use only two of these equations for the *x-y* plane and find the *x-* and *y*-components of the CM.

Example 7.7

Center of Mass of a Binary Star System

Due to the gravitational interaction between the two stars in a binary star system, each moves in a circular orbit around their CM. One star has a mass of 15.0×10^{30} kg; its center is located at $x = 1.0$ AU and $y = 5.0$ AU. The other has a mass of 3.0×10^{30} kg; its center is at $x = 4.0$ AU and $y = 2.0$ AU. Find the CM of the system composed of the two stars. (AU stands for *astronomical unit*. 1 AU = the average distance between the Earth and the Sun = 1.5×10^8 km.)

Strategy We treat the stars as point particles located at their centers. Since we are given *x-* and *y*-coordinates, the easiest way to proceed is to find the *x-* and *y*-coordinates of the CM. There is no particular advantage here in finding the position vector of the CM in terms of its length and direction.

Given: $m_1 = 15.0 \times 10^{30}$ kg $\quad x_1 = 1.0$ AU $\quad y_1 = 5.0$ AU
$\qquad\quad m_2 = 3.0 \times 10^{30}$ kg $\quad x_2 = 4.0$ AU $\quad y_2 = 2.0$ AU

To find: x_{CM}; y_{CM}

Solution The total mass of the system is the sum of the individual masses:

$$M = m_1 + m_2 = 15.0 \times 10^{30} \text{ kg} + 3.0 \times 10^{30} \text{ kg} = 18.0 \times 10^{30} \text{ kg}$$

For the *x*-position, we find

$$x_{CM} = \frac{m_1}{M} x_1 + \frac{m_2}{M} x_2$$
$$= \frac{15.0 \times 10^{30} \text{ kg}}{18.0 \times 10^{30} \text{ kg}} \times 1.0 \text{ AU} + \frac{3.0 \times 10^{30} \text{ kg}}{18.0 \times 10^{30} \text{ kg}} \times 4.0 \text{ AU}$$
$$= 1.5 \text{ AU}$$

and for the *y*-position, we find

$$y_{CM} = \frac{m_1}{M} y_1 + \frac{m_2}{M} y_2$$
$$= \frac{15.0}{18.0} \times 5.0 \text{ AU} + \frac{3.0}{18.0} \times 2.0 \text{ AU} = 4.5 \text{ AU}$$

Discussion In Fig. 7.13, we mark the position of the CM. As we expect for the case of two particles, it is located closer to the larger mass and on a line connecting the two. Once the CM position is found in a problem, check to be sure its location is reasonable. Suppose we had made an error in this example and found the CM to be at $x = 1.5$ AU and $y = 1.7$ AU. This is not a reasonable location for the CM since it is not along the line connecting the two and is closer to the less massive star; we then would go back to look for the error.

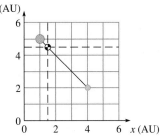

Figure 7.13
Finding the CM for the system of two stars.

Practice Problem 7.7 **Three Balls with Unequal Masses**

Three spherical objects are shown in Fig. 7.14. Their masses are $m_1 = m_3 = 1.0$ kg and $m_2 = 4.0$ kg. Find the location of the CM for the three objects.

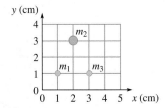

Figure 7.14
Three spheres located at *x, y* positions (1.0 cm, 1.0 cm), (2.0 cm, 3.0 cm), and (3.0 cm, 1.0 cm).

Using Symmetry to Locate the Center of Mass   Most objects we deal with in real life are not composed of a small set of point particles or spherically symmetrical objects. In Example 7.7, we use the location of the center of each star to find the CM. Due to spherical symmetry, the CM of either star (by itself) is at its geometric center. The same technique can be applied to other shapes with symmetry. A standard 2 by 4, which is an 8-ft-long uniform piece of lumber used in building 1.5 in. deep × 3.5 in. high, has its center of mass at its geometric center. By contrast, a "loaded" die does *not* have its CM at its geometric center, since a small metal plug has been inserted near one face to make the distribution of mass in the die asymmetrical. The definition of the CM [Eq. (7-9)] still holds as long as (x_i, y_i, z_i) are the coordinates of the CM of a part of the system with mass m_i.

7.6   MOTION OF THE CENTER OF MASS

Now that we know how to find the position of the CM of a system, we turn our attention to the motion of the CM. How is the velocity of the CM related to the velocities of the various parts of the system?

During a short time interval Δt, the displacement of the i-th particle is $\Delta \vec{r}_i = \vec{v}_i \, \Delta t$ and the displacement of the center of mass is $\Delta \vec{r}_{CM} = \vec{v}_{CM} \, \Delta t$. From the definition of the CM [Eq. (7-9)], the displacements must be related as follows:

$$\Delta \vec{r}_{CM} = \frac{\sum m_i \Delta \vec{r}_i}{M} \quad \Rightarrow \quad \vec{v}_{CM} \, \Delta t = \frac{\sum m_i \vec{v}_i \, \Delta t}{M}$$

Dividing both sides by Δt and multiplying by M yields

$$M\vec{v}_{CM} = \sum m_i \vec{v}_i \tag{7-10}$$

The right side of Eq. (7-10) is the sum of the momenta of the particles that constitute the system—the total momentum of the system $\vec{p}$. Therefore,

$$\vec{p} = M\vec{v}_{CM} \tag{7-11}$$

For two-dimensional motion, Eq. (7-11) is equivalent to two component equations

$$p_x = Mv_{CM,x} \quad \text{and} \quad p_y = Mv_{CM,y} \tag{7-12}$$

In Section 7.4, we showed that, for an isolated system, the total linear momentum is conserved. In such a system, Eq. (7-11) implies that the CM must move with constant velocity regardless of the motions of the individual particles. On the other hand, what if the system is not isolated? If a net external force acts on a system, the CM does not move with constant velocity. Instead, it moves as if all the mass were concentrated there into a fictitious point particle with all the external forces acting on that point. The motion of the CM obeys the following statement of Newton's second law:

$$\sum \vec{F}_{ext} = M\vec{a}_{CM} \tag{7-13}$$

where M is the total mass of the system, $\sum \vec{F}_{ext}$ is the net external force, and $\vec{a}_{CM}$ is the acceleration of the CM. [Eq. (7-13) is proved in Problem 38.]

✓ CHECKPOINT 7.6

Turn back to Fig. 7.11b. Why does the CM of the hammer move along a parabolic path?

Example 7.8

An Exploding Rocket

A model rocket is fired from the ground in a parabolic trajectory. At the top of the trajectory, a horizontal distance of 260 m from the launch point, an explosion occurs within the rocket, breaking it into two fragments. One fragment, having one third of the mass of the rocket, falls straight down to Earth as if it had been dropped from rest at that point. At what horizontal distance from the launch point does the other fragment land? Ignore air resistance. [*Hint:* The two fragments land simultaneously.]

Strategy Two different strategies can be used to solve this problem.

Strategy 1: We apply conservation of momentum to the explosion. The momentum of the rocket *just before* the explosion is equal to the total momentum of the two fragments *just after* the explosion. Why can momentum conservation be assumed here? There is an external force—gravity—acting on the system. External forces change momentum. However, the explosion takes place in a *very short time interval*. From the impulse-momentum theorem [Eq. (7-2)], the momentum change of the system is the force of gravity multiplied by the time interval. As long as the time interval considered is sufficiently short, the momentum change of the system can be ignored.

Strategy 2: The explosion is caused by an *internal* interaction between two parts of the rocket. The motion of the CM of the system is unaffected by internal interactions, so it continues in the same parabolic path. Just before the explosion, the rocket is at the top of its trajectory, so it has $p_y = 0$ (with the y-axis pointing up). Just after the explosion, one fragment is at rest. Then the other fragment must have $p_y = 0$; otherwise, conservation of momentum would be violated. Then both fragments have $v_y = 0$ just after the explosion. Ignoring air resistance, they land simultaneously. At that same instant, the CM also reaches the ground.

Solution 1 First we make a sketch of the situation (Fig. 7.15). At the top of the trajectory, where the explosion occurs, $v_y = 0$; the rocket is moving in the x-direction. The initial momentum just before the explosion is entirely in the x-direction. If M is the mass of the rocket, then

$$p_{ix} = Mv_{ix}$$

Just after the explosion, one-third of the mass of the rocket is at rest; it then drops straight down under the influence of the gravitational force. This piece has zero momentum just after the explosion. To conserve momentum, the other two thirds of the rocket must have a momentum equal to the momentum just before the explosion.

$$p_{ix} = p_{1x} + p_{2x}$$
$$Mv_{ix} = 0 + (\tfrac{2}{3}M)\, v_{2x}$$

Solving for v_{2x}, we find

$$v_{2x} = \tfrac{3}{2} v_{ix}$$

The y-component of momentum must also be conserved:

$$p_{iy} = p_{1y} + p_{2y}$$

We know that both p_{iy} and p_{1y} are zero; therefore, p_{2y} is zero as well. Just after the explosion, both parts of the rocket have zero vertical components of velocity. Then both parts take the same time to fall to the ground as if the rocket had not exploded. With a horizontal velocity larger by a factor of $\tfrac{3}{2}$, the second piece of the rocket travels a horizontal distance from the explosion a factor of $\tfrac{3}{2}$ larger than 260 m (see Fig. 7.15). The distance from the launch point where this piece lands is

$$\Delta x = 260 \text{ m} + \tfrac{3}{2} \times 260 \text{ m} = 650 \text{ m}$$

Figure 7.15
Rocket motion after explosion.

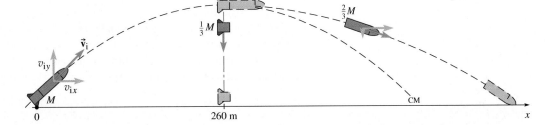

continued on next page

Example 7.8 continued

Solution 2   The piece with mass $\frac{1}{3}M$ falls straight down and lands 260 m from the launch point. After the explosion, the CM continues to travel just as the rocket itself would have done if it had not broken apart. From the symmetry of the parabola, the CM touches the ground at a distance of 2×260 m $= 520$ m from the launch point. Since we know the location of the CM and that of one of the pieces, we can find where the second piece lands:

$$Mx_{CM} = \tfrac{1}{3}Mx_1 + \tfrac{2}{3}Mx_2$$

After canceling the common factor of M,

$$x_{CM} = \tfrac{1}{3}x_1 + \tfrac{2}{3}x_2$$

Solving for x_2 yields

$$x_2 = \frac{3x_{CM} - x_1}{2} = \frac{3 \times 520 \text{ m} - 260 \text{ m}}{2} = 650 \text{ m}$$

which is the same answer that we found in Solution 1.

Discussion   The insight that the motion of the CM is unaffected by internal interactions can be of enormous help.

Note, however, that solution 2 would not be so simple if the two fragments did not land simultaneously. As soon as one fragment (fragment 1) hits the ground, the external force on the system is no longer due exclusively to gravity, so the CM doesn't continue to follow the same parabolic path. The normal and frictional forces acting on fragment 1 affect its subsequent motion and the subsequent motion of the CM even though the motion of fragment 2 is unaffected.

Practice Problem 7.8   Diana and the Raft Revisited

In Example 7.5, Diana (mass 55 kg) walks at 0.91 m/s (relative to the water) on a raft of mass 100.0 kg. The raft moves in the opposite direction at 0.50 m/s. Suppose it takes her 3.0 s to walk from one end of the raft to the other. (a) How far does Diana walk (relative to the water)? (b) How far does the raft move while Diana is walking? (c) How far does the CM of Diana and the raft move during the 3.0 s?

7.7   COLLISIONS IN ONE DIMENSION

What Is a Collision?   In the macroscopic world, a moving body bumps into another body that may be at rest or in motion. The two bodies exert forces on one another while they are in contact; as a result, their velocities change. In the microscopic and submicroscopic world, our picture of a collision is different. When atoms collide, they don't "touch" each other: the atom doesn't have a definite spatial boundary, so there are no surfaces to make "contact." However, the collision model is still useful for atoms and subatomic particles whenever there is an interaction in which the forces are strong over a short time interval, so that there is a clear "before collision" and a clear "after collision."

Analyzing Collisions Using Momentum Conservation   We can often use conservation of momentum to analyze collisions even when external forces act on the colliding objects. If the net external force is small compared with the internal forces the colliding objects exert on each other during the collision, then the change in the total momentum of the two objects is small compared with the transfer of momentum from one object to the other. Then the total momentum after the collision is *approximately* the same as it was before the collision.

   The same techniques that are used for collisions in the macroscopic world (car crashes, billiard ball collisions, baseball bats hitting balls) are also used in collisions in the microscopic world (gas molecules colliding with each other and with surfaces, radioactive decays of nuclei). First, we study collisions limited to motion along a line; later, we consider collisions limited to motion in a plane (in two dimensions).

Example 7.9

Collision in the Air

A krypton atom (mass 83.9 u) moving with a velocity of 0.80 km/s to the right and a water molecule (mass 18.0 u) moving with a velocity of 0.40 km/s to the left collide head-on. The water molecule has a velocity of 0.60 km/s to the right after

continued on next page

Example 7.9 continued

the collision. What is the velocity of the krypton atom after the collision? (The symbol "u" stands for the atomic mass unit.)

Strategy Since we know both initial velocities and one of the final velocities, we can find the second final velocity by applying momentum conservation. Let the subscript "1" refer to the krypton atom and let the subscript "2" refer to the water molecule. Let the x-axis point to the right. Figure 7.16 shows before and after pictures of the collision.

Solution Momentum conservation requires that the final momentum be equal to the initial momentum:

$$\vec{p}_{1f} + \vec{p}_{2f} = \vec{p}_{1i} + \vec{p}_{2i}$$

Now we substitute $\vec{p} = m\vec{v}$ for each momentum. It is easiest to work in terms of components. For simplicity we drop the "x" subscripts, remembering that all quantities refer to x-components:

$$m_1 v_{1f} + m_2 v_{2f} = m_1 v_{1i} + m_2 v_{2i}$$

Since $m_1/m_2 = 83.9/18.0 = 4.661$, we can substitute $m_1 = 4.661 m_2$:

$$4.661\, m_2 v_{1f} + m_2 v_{2f} = 4.661\, m_2 v_{1i} + m_2 v_{2i}$$

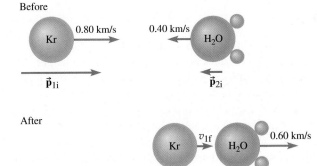

Before

After

Figure 7.16
Before and after snapshots of a collision.

The common factor m_2 cancels out. Solving for v_{1f},

$$v_{1f} = \frac{4.661 v_{1i} + v_{2i} - v_{2f}}{4.661}$$

$$= \frac{4.661 \times 0.80\ \text{km/s} + (-0.40\ \text{km/s}) - 0.60\ \text{km/s}}{4.661}$$

$$= 0.59\ \text{km/s}$$

After the collision, the krypton atom moves to the right with a speed of 0.59 km/s.

Discussion To check this result, we calculate the total momentum (x-component) before and after the collision:

$$m_1 v_{1i} + m_2 v_{2i} = (83.9\ \text{u})(0.80\ \text{km/s}) + (18.0\ \text{u})(-0.40\ \text{km/s})$$
$$= 60\ \text{u·km/s}$$

$$m_1 v_{1f} + m_2 v_{2f} = (83.9\ \text{u})(0.59\ \text{km/s}) + (18.0\ \text{u})(0.60\ \text{km/s})$$
$$= 60\ \text{u·km/s}$$

Momentum is conserved. There is no need to convert u to kg since we only need to compare these two values.

If we made the mistake of thinking of momentum as a scalar, we would get the wrong answer. The sum of the *magnitudes* of the momenta before the collision is *not* equal to the sum of the *magnitudes* of the momenta after the collision. Conservation of energy is perhaps easier to understand intuitively since energy is a scalar quantity. Converting kinetic energy to potential energy is analogous to moving money from a checking account to a savings account; the total amount of money is the same before and after. This sort of analogy does *not* work with momentum!

Practice Problem 7.9   Head-On Collision

A 5.0-kg ball is at rest when it is struck head-on by a 2.0-kg ball moving along a track at 10.0 m/s. If the 2.0-kg ball is at rest after the collision, what is the speed of the 5.0-kg ball after the collision?

Suppose we observe a bumper car traveling at speed v_i toward a second car that is at rest. The masses of the two cars are equal. When the first car hits the second, what happens?

Based on momentum considerations *alone*, there are many possible outcomes. One possibility is that the first car stops moving and the second car moves off with the same velocity that the first one had to begin with (Fig. 7.17a). This possibility satisfies conservation of momentum because the total momentum is the same before and after.

Another possibility is that the two cars stick together, moving away together (Fig. 7.17b). With what speed do they move after the collision? If the momentum is to be the same with twice as much mass moving, the speed must be half the initial speed of the first car. There are many other possibilities. Conservation of momentum doesn't tell us which of these outcomes actually happens, but if we know one car's velocity after the collision, we can use momentum conservation to determine the other car's velocity.

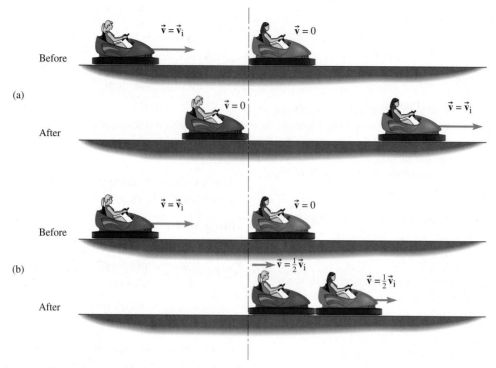

Figure 7.17 Two of the many possible outcomes of a collision between bumper cars of equal mass with one of them initially at rest.

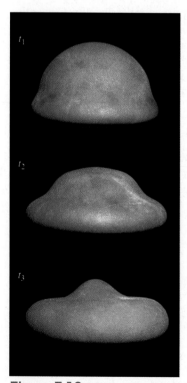

Figure 7.18 Three successive photos of a racquetball during its collision with the floor ($t_1 < t_2 < t_3$).

Elastic and Inelastic Collisions

Collisions are often classified based on what happens to the kinetic energy of the colliding objects. A ball dropped from a height h does not rebound to the same height. The kinetic energy of the ball just after the collision with the floor or ground is less than it was just before the collision; the amount of the kinetic energy decrease depends on the makeup of the ball and the ground. A racquetball dropped onto a hard wooden floor may rebound nearly to its original height, but a watermelon rebounds very little or not at all. Why do some objects rebound much better than others?

Imagine a racquetball colliding with the floor (Fig. 7.18). The bottom of the ball is flattened. What makes the ball rebound from the floor? The forces holding the ball together are like springs; the kinetic energy of the ball has been transformed largely into potential energy stored in these springs. When the ball bounces back up, this energy is transformed back into kinetic energy. Then why does the watermelon not rebound? The watermelon, too, is deformed when it collides with the floor, but this deformation is not reversible. The kinetic energy of the watermelon is changed mostly into thermal energy rather than into potential energy.

A collision in which the *total* kinetic energy is the same before and after is called **elastic**. When the final kinetic energy is less than the initial kinetic energy, the collision is said to be **inelastic**. Collisions between macroscopic objects are generally inelastic to some degree, but sometimes the change in kinetic energy is so small that we treat them as elastic. When a collision results in two objects sticking together, the collision is **perfectly inelastic**. The decrease of kinetic energy in a perfectly inelastic collision is as large as *possible* (consistent with the conservation of momentum). Now that we have defined elastic and inelastic collisions, we can put together a problem-solving strategy for collision problems.

Problem-Solving Strategy for Collisions Involving Two Objects

1. Draw before and after diagrams of the collision.
2. Collect and organize information on the masses and velocities of the two objects before and after the collision. Express the velocities in component form (with correct algebraic signs).
3. Set the sum of the momenta of the two before the collision equal to the sum of the momenta after the collision. Write one equation for each direction:

$$m_1 v_{1ix} + m_2 v_{2ix} = m_1 v_{1fx} + m_2 v_{2fx}$$

$$m_1 v_{1iy} + m_2 v_{2iy} = m_1 v_{1fy} + m_2 v_{2fy}$$

4. If the collision is known to be perfectly inelastic, set the final velocities equal:

$$v_{1fx} = v_{2fx} \quad \text{and} \quad v_{1fy} = v_{2fy}$$

5. If the collision is known to be perfectly elastic, then set the final kinetic energy equal to the initial kinetic energy:

$$\tfrac{1}{2}m_1 v_{1i}^2 + \tfrac{1}{2}m_2 v_{2i}^2 = \tfrac{1}{2}m_1 v_{1f}^2 + \tfrac{1}{2}m_2 v_{2f}^2$$

6. Solve for the unknown quantities.

There is no *conservation law* for kinetic energy by itself. Total energy is always conserved, but that does not preclude some kinetic energy being transformed into another type of energy. The elastic collision is just a special kind of collision in which no kinetic energy is changed into other forms of energy. Momentum is conserved regardless of whether a collision is elastic or inelastic.

It can be proved (see Problem 56) that for *any* elastic collision between two objects, the relative speed is the same before and after the collision. (This fact is most useful in one-dimensional collisions; in two-dimensional collisions the *direction* of the relative velocity changes due to the collision.) Since the relative velocity is in the opposite direction after a one-dimensional collision—first the objects move together, then they move apart—we can write:

$$v_{2ix} - v_{1ix} = -(v_{2fx} - v_{1fx}) \tag{7-14}$$

assuming the objects move along the *x*-axis. For a one-dimensional elastic collision, Eq. (7-14) is a useful alternative to setting the final kinetic energy equal to the initial kinetic energy.

CHECKPOINT 7.7

Is momentum conserved in a perfectly inelastic collision?

Example 7.10

Collision at the Highway Entry Ramp

At a Route 3 highway on-ramp, a car of mass 1.50×10^3 kg is stopped at a stop sign, waiting for a break in traffic before merging with the cars on the highway. A pickup of mass 2.00×10^3 kg comes up from behind and hits the stopped car.

Assuming the collision is elastic, how fast was the pickup going just before the collision if the car is pushed straight ahead onto the highway at 20.0 m/s just after the collision?

continued on next page

Example 7.10 continued

Strategy   Conservation of momentum provides one equation relating the initial and final velocities. That the collision is elastic provides another equation. With two unknown velocities, these two equations enable us to solve for both. Let "1" refer to the car stopped at the stop sign and "2" refer to the pickup. All motions are in one direction, which we call the x-axis. To simplify the notation, we drop the x subscripts and let all p's and v's refer to x-components. Figure 7.19 shows a before and after diagram for the collision.

Given: $m_1 = 1.50 \times 10^3$ kg; $m_2 = 2.00 \times 10^3$ kg; before the collision, $v_{1i} = 0$; after the collision, $v_{1f} = 20.0$ m/s

To find: v_{2i} (speed of the pickup just before the collision)

Solution   From conservation of momentum,

$$m_1 v_{1i} + m_2 v_{2i} = m_1 v_{1f} + m_2 v_{2f} \qquad (1)$$

where we cross out the first term because $v_{1i} = 0$. The collision is elastic, so the relative velocity after the collision is equal and opposite to the relative velocity before the collision [Eq. (7-14)]:

$$v_{2i} - v_{1i} = -(v_{2f} - v_{1f}) \qquad (2)$$

We want to solve these two equations for v_{2i}, so we can eliminate v_{2f}. Multiplying Eq. (2) through by m_2 and rearranging yields

$$m_2 v_{2i} = m_2 v_{1f} - m_2 v_{2f} \qquad (3)$$

Adding Eqs. (1) and (3) gives

$$2m_2 v_{2i} = (m_1 + m_2) v_{1f} \qquad (4)$$

Finally, we solve Eq. (4) for v_{2i}:

$$v_{2i} = \frac{m_1 + m_2}{2m_2} v_{1f} = \frac{1500 \text{ kg} + 2000 \text{ kg}}{4000 \text{ kg}} \times 20.0 \text{ m/s} = 17.5 \text{ m/s}$$

Discussion   To check this answer, first solve for v_{2f}. Then you can verify that momentum is conserved [Eq. (1)] and that the relative velocity changes sign [Eq. (2)]. You can also calculate the total kinetic energy before and after the collision and show they are equal, as they must be for an elastic collision. We leave these checks to you for practice.

The road exerts frictional forces on the vehicles, so the net external force on the vehicles was *not* zero during the collision. We still use conservation of momentum because during the short time interval of the collision, friction doesn't have time to change the system's momentum significantly.

Practice Problem 7.10   **Perfectly Inelastic Collision Between the Cars**

Instead of colliding elastically, suppose the two vehicles lock bumpers when they collide. With the same initial conditions ($v_{1i} = 0$ and $v_{2i} = 17.5$ m/s), find the speed at which the car would be pushed out onto the highway.

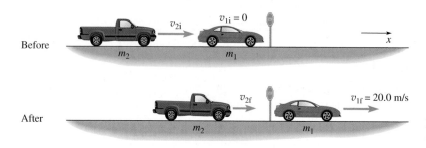

Before

v_{2i} $v_{1i} = 0$ x

m_2 m_1

After

v_{2f} $v_{1f} = 20.0$ m/s

m_2 m_1

Figure 7.19

Before and after diagrams of the collision (side view).

How are skid marks used to find car velocities just before the collision?

Suppose in Example 7.10 that the entry ramp speed limit is 20 mi/h (8.94 m/s). By measuring the length of the skid marks from the stop sign and estimating the coefficient of friction, the accident investigator can determine that the car was pushed onto the highway at a speed of 20.0 m/s. Witnesses confirm that the car was stopped before the collision. Then the investigator calculates the speed of the pickup just before the collision using conservation of momentum. The duration Δt of the collision is so short that we can ignore momentum changes due to external forces and treat the two vehicles as an isolated system. Finding that the pickup exceeded the speed limit, the investigator adds speeding to the charges against the driver of the pickup.

7.8 COLLISIONS IN TWO DIMENSIONS

Most collisions are not limited to motion in one dimension in the absence of a track or other device to constrain motion to a single line. In a two-dimensional collision, we use the same techniques we used for one-dimensional collisions, as long as we remember that momentum is a vector. (⬤ interactive: the virtual pool table.) To apply conservation of momentum, it is usually easiest to work with x- and y-components.

CONNECTION:

See the Problem-Solving Strategy in Section 7.7. The same strategy applies to collisions in two or three dimensions.

Example 7.11

Colliding Pucks on an Air Table

A small puck (mass $m_1 = 0.10$ kg) is sliding to the right with an initial speed of 8.0 m/s on an air table (Fig. 7.20a). An air table has many tiny holes through which air is blown; the resulting air cushion allows objects to slide with very little friction. The puck collides with a larger puck (mass $m_2 = 0.40$ kg), which is initially at rest. Figure 7.20b shows the outcome of the collision: the pucks move off at angles $\phi_1 = 60.0°$ above and $\phi_2 = 30.0°$ below the initial direction of motion of the small puck. (a) What are the final speeds of the pucks? (b) Is this an elastic collision or an inelastic collision? (c) If inelastic, what fraction of the initial kinetic energy is converted to other forms of energy in the collision?

Strategy The system of two pucks is an isolated system because the net external force is zero. Therefore, we can apply conservation of momentum. Since motions in two dimensions are involved, we treat the horizontal and vertical components of momentum separately.

Figure 7.20 shows the pucks before and after the collision. Now we collect information on the known quantities, writing velocities in component form.

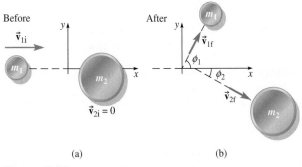

Before   After

(a)   (b)

Figure 7.20

Snapshots in time, (a) before and (b) after a collision.

Masses: $m_1 = 0.10$ kg; $m_2 = 0.40$ kg

Before collision: $v_{1ix} = 8.0$ m/s; $v_{1iy} = v_{2ix} = v_{2iy} = 0$

After collision: $v_{1fx} = v_{1f} \cos \phi_1$; $v_{1fy} = v_{1f} \sin \phi_1$;
$v_{2fx} = v_{2f} \cos \phi_2$; $v_{2fy} = -v_{2f} \sin \phi_2$
($\phi_1 = 60.0°$ and $\phi_2 = 30.0°$)

To find: v_{1f} and v_{2f}; total kinetic energy before and after the collision

Solution (a) Working with components means that we set the total x-component of momentum before the collision equal to the total x-component of momentum after the collision. We treat the y-components in the same way. The initial momentum is in the x-direction only. Thus, the total momentum y-component after the collision must be zero.

First we set the x-component of the total momentum after the collision equal to the x-component of the total momentum before the collision:

$$p_{1fx} + p_{2fx} = p_{1ix} + p_{2ix}$$

Each momentum component is now rewritten using $p_x = mv_x$:

$$m_1 v_{1f} \cos \phi_1 + m_2 v_{2f} \cos \phi_2 = m_1 v_{1ix} + 0$$

Since $m_2 = 4m_1$,

$$m_1 v_{1f} \cos 60.0° + 4m_1 v_{2f} \cos 30.0° = m_1 v_{1ix}$$

After canceling the common factor m_1 and substituting numerical values for cos 60.0° and cos 30.0°, this reduces to

$$0.500v_{1f} + 3.46v_{2f} = 8.0 \text{ m/s} \qquad (1)$$

For conservation of the y-component of the momentum:

$$p_{1fy} + p_{2fy} = p_{1iy} + p_{2iy} = 0$$

continued on next page

Example 7.11 continued

The y-component of $\vec{\mathbf{p}}_{2f}$ is negative because the y-component of $\vec{\mathbf{v}}_{2f}$ is negative.

$$m_1 v_{1f} \sin \phi_1 + (-4m_1 v_{2f} \sin \phi_2) = 0$$

$$v_{1f} \sin 60.0° - 4v_{2f} \sin 30.0° = 0$$

We solve for v_{2f} in terms of v_{1f}:

$$v_{2f} = \frac{\sin 60.0°}{4 \sin 30.0°} v_{1f} = 0.433v_{1f} \qquad (2)$$

Equations (1) and (2) contain two unknowns. To eliminate one unknown, we substitute $0.433v_{1f}$ for v_{2f} in Eq. (1):

$$0.500v_{1f} + 3.46(0.433v_{1f}) = 2.00v_{1f} = 8.0 \text{ m/s}$$

Solving this equation gives the value of v_{1f}:

$$v_{1f} = 4.0 \text{ m/s}$$

Then by substitution into Eq. (2), we find the value of v_{2f}:

$$v_{2f} = 0.433v_{1f} = 1.73 \text{ m/s} \rightarrow 1.7 \text{ m/s}$$

(b) Now that we have the final speeds, we can compare the initial and final kinetic energies.

$$K_i = \tfrac{1}{2}m_1 v_{1i}^2$$

$$K_i = \tfrac{1}{2}(0.10 \text{ kg}) \times (8.0 \text{ m/s})^2 = 3.2 \text{ J}$$

and

$$K_f = \tfrac{1}{2}m_1 v_{1f}^2 + \tfrac{1}{2}m_2 v_{2f}^2$$
$$= \tfrac{1}{2}(0.10 \text{ kg}) \times (4.0 \text{ m/s})^2 + \tfrac{1}{2}(0.40 \text{ kg}) \times (1.73 \text{ m/s})^2$$
$$= 0.80 \text{ J} + 0.60 \text{ J} = 1.40 \text{ J}$$

The final kinetic energy is less than the initial kinetic energy, so the collision is inelastic.

(c) The amount of kinetic energy converted to other forms of energy (primarily internal energy of the pucks) is

$$3.2 \text{ J} - 1.40 \text{ J} = 1.8 \text{ J}$$

We divide by the initial kinetic energy to find the fraction of the initial kinetic energy converted to other forms:

$$\frac{1.8 \text{ J}}{3.2 \text{ J}} = 0.56$$

Less than half of the kinetic energy of the incident puck therefore survives the collision as the kinetic energies of the two pucks.

Discussion   Although a two-dimensional collision problem tends to require more complicated algebra than a one-dimensional problem, the physical principles are the same. As long as the net external force on the system is zero (or negligibly small), the total vector momentum must be conserved.

Practice Problem 7.11   Colliding Balls

A ball of mass m_1 moves at speed v_i along the $+x$-axis toward a second ball of mass m_2, which is initially at rest. The second ball has five times the mass of the first ball. After the collision between these two objects, m_1 moves along the $+y$-axis at a speed v_1, and m_2 moves at a speed $v_2 = \tfrac{1}{4}v_i$ at an angle of 36.9° below the $+x$-axis. Find v_1 in terms of v_i.

Conceptual Example 7.12

Eric at the Pool Table

Playing a game of pool, Eric is trying to decide whether to attempt a shot to sink the 4-ball in the pocket at corner B without *scratching* (sinking the cue ball "C" in corner A). He notices that the lines from the 4-ball to the two corner pockets happen to make a right angle (Fig. 7.21). The collision of the balls is nearly elastic. Assume Eric is an amateur player and does not know how to do fancy things, like putting sidespin on a ball. Should he attempt the shot?

Strategy   We assume a perfectly elastic collision between the balls. They have the same mass. The cue ball moves with an initial velocity $\vec{\mathbf{v}}_i$ and strikes the 4-ball, which is initially at rest. The 4-ball falls in pocket B if its velocity after the collision, $\vec{\mathbf{v}}_4$, points toward the pocket. Assuming it does, we use conservation of momentum and kinetic energy to find the direction of the cue ball velocity, $\vec{\mathbf{v}}_c$, after the collision.

Solution   Conservation of momentum requires that

$$m\vec{\mathbf{v}}_i = m\vec{\mathbf{v}}_c + m\vec{\mathbf{v}}_4$$

or

$$\vec{\mathbf{v}}_i = \vec{\mathbf{v}}_c + \vec{\mathbf{v}}_4$$

This vector addition is shown graphically in Fig. 7.22a. Since the collision is elastic, the total kinetic energy doesn't change:

$$\tfrac{1}{2}mv_i^2 = \tfrac{1}{2}mv_c^2 + \tfrac{1}{2}mv_4^2$$

continued on next page

Conceptual Example 7.12 continued

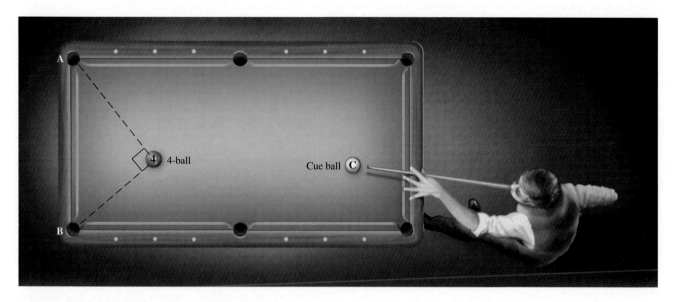

Figure 7.21 Should Eric try to sink the 4-ball?

or

$$v_i^2 = v_c^2 + v_4^2$$

Since v_i, v_c, and v_4 are the sides of a triangle, this is a statement of the Pythagorean theorem—the triangle must be a *right* triangle with v_i as the hypotenuse (Fig. 7.22b). Therefore, the velocities of the 4-ball and the cue ball after the collision are perpendicular to each other.

Figure 7.22
(a) Graphical addition of velocity vectors as required by the conservation of momentum. (b) Since $v_i^2 = v_c^2 + v_4^2$, the three velocities form a right triangle.

If Eric sinks the 4-ball, the cue ball falls into pocket A. He shouldn't attempt this shot until he learns how to put some spin on the ball.

Discussion Note that we did not resolve the velocities into *x*- and *y*-components. Doing so would have made the solution longer in this case.

We found that the two balls move at right angles after the collision. This result is true for any two-dimensional elastic collision between two objects of equal masses if one of them is initially at rest. In Example 7.11 the two pucks move at right angles after the collision, but the collision is inelastic—the masses are unequal.

Practice Problem 7.12 Finding the Speed Ratio

Suppose that the cue ball initially moves in the −*x*-direction. After the collision, the cue ball moves at 52.0° above the −*x*-axis and the 4-ball moves at 38.0° below the −*x*-axis. Find the ratio of the balls' speeds v_c/v_4 after the collision.

Master the Concepts

- Definition of linear momentum:

$$\vec{p} = m\vec{v} \qquad (7\text{-}1)$$

- During an interaction, momentum is transferred from one body to another, but the total momentum of the two is unchanged.

$$\Delta\vec{p}_2 = -\Delta\vec{p}_1$$

- Impulse is the average force times the time interval.
- The total impulse equals the change in momentum:

$$\Delta\vec{p} = \sum\vec{F}\,\Delta t \qquad (7\text{-}2)$$

continued on next page

Master the Concepts continued

- A conserved quantity is one that remains unchanged as time passes.
- Impulse is the area under a graph of force versus time.

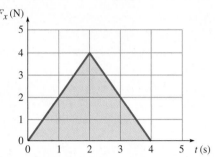

- The net force is the rate of change of momentum.

$$\sum \vec{\mathbf{F}} = \lim_{\Delta t \to 0} \frac{\Delta \vec{\mathbf{p}}}{\Delta t} \qquad (7\text{-}4)$$

- External interactions may change the total momentum of a system.
- Internal interactions do not change the total momentum of a system.
- Conservation of linear momentum: if the net external force acting on a system is zero, then the momentum of the system is conserved.
- The position of the CM of a system of N particles is

$$x_{\text{CM}} = \frac{m_1 x_1 + m_2 x_2 + \cdots + m_N x_N}{M}$$

and

$$y_{\text{CM}} = \frac{m_1 y_1 + m_2 y_2 + \cdots + m_N y_N}{M} \qquad (7\text{-}9)$$

where M is the total mass of the particles:

$$M = m_1 + m_2 + \cdots + m_N$$

- The total momentum of a system is equal to the total mass times the velocity of the center of mass:

$$\vec{\mathbf{p}} = \vec{\mathbf{p}}_1 + \vec{\mathbf{p}}_2 + \cdots + \vec{\mathbf{p}}_N = M \vec{\mathbf{v}}_{\text{CM}} \qquad (7\text{-}11)$$

- No matter how complicated a system is, the CM moves as if all the mass of the system were concentrated to a point particle with all the external forces acting on it:

$$\sum \vec{\mathbf{F}}_{\text{ext}} = M \vec{\mathbf{a}}_{\text{CM}} \qquad (7\text{-}13)$$

- The CM of an isolated system moves at constant velocity.
- Conservation of momentum is used to solve problems involving collisions, explosions, and the like. Even when external forces are acting, the momentum of the system just before a collision is nearly equal to the momentum just after if the collision interaction is brief. The impulse, and, therefore, the change in momentum of the system, is small since the time interval is small.

Conceptual Questions

1. You are trapped on the second floor of a burning building. The stairway is impassable, but there is a balcony outside your window. Describe what might happen in the following situations. (a) You jump from the second-story balcony to the pavement below, landing stiff-legged on your feet. (b) You jump into a privet hedge, landing on your back and rolling to your feet. (c) You jump into a firefighters' net, landing on your back. What happens to the net as you land in it? What do the firefighters do to cushion your fall even more?

2. A force of 30 N is applied for 5 s to each of two bodies of different masses. (a) Which one has the greatest momentum change? (b) The greatest velocity change? (c) The greatest acceleration?

3. If you take a rifle and saw off part of the barrel, the muzzle speed (the speed at which bullets emerge from the barrel) will be smaller. Why?

4. A firecracker at rest explodes, sending fragments off in all directions. Initially the firecracker has zero momentum, but after the explosion the fragments flying off each have quite a lot of momentum. Hasn't momentum been created? If not, explain why not.

5. An astronaut in deep space is taking a space walk when the tether connecting him to his spaceship breaks. How can he get back to the ship? He doesn't have a rocket propulsion backpack, unfortunately, but he is carrying a big wrench.

6. An astronaut hits a golf ball on the surface of the Moon. Is the momentum of the ball conserved while it is in flight? Is there a *component* of its momentum that is conserved?

7. Which would be more effective: a hammer that collides *elastically* with a nail, or one that collides perfectly *inelastically*? Assume that the mass of the hammer is much larger than that of the nail.

8. Squid are the fastest swimmers among invertebrates. A cavity within the squid is filled with water. The *mantle*, a powerful muscle, squeezes the cavity and expels the water through a narrow opening (the *siphon*) at high speed. Using momentum conservation, explain how this propels the squid forward. How is the squid's swimming mechanism like a rocket engine?

9. In your own words, phrase each of Newton's three laws of motion as a statement about momentum.

10. Two objects with different masses have the same kinetic energy. Which has the larger magnitude of momentum?

11. A woman is 1.60 m tall. When standing straight, is her CM necessarily 0.80 m above the floor? Explain.

12. The momentum of a system can only be changed by an external force. What is the external force that changes the momentum of a bicycle (with its rider) as it speeds up, slows down, or changes direction? Is it true that changes in the bicycle's kinetic energy must come from an external force? Explain.

13. In an egg toss, two people try to toss a raw egg back and forth without breaking it as they move farther and farther apart. Discuss a strategy in terms of impulse and momentum for catching the egg without breaking it.

14. In the "executive toy," two balls are pulled back and then released. After the collision, two balls move away on the opposite side. Why do we never see three balls move away following this action, although with a lower velocity so that linear momentum is still conserved?

15. A baseball batting coach emphasizes the importance of "follow-through" when a batter is trying for a home run. The coach explains that the follow-through keeps the bat in contact with the ball for a longer time so the ball will travel a greater distance. Explain the reasoning behind this statement in terms of the impulse-momentum theorem.

16. Micah is standing on his frictionless skateboard facing a concrete wall. He wants to project himself backward by throwing small balls at the wall. His friend Jeremy says that Micah need not throw the balls against the wall, he just needs to throw the balls away from himself, but Micah says the balls need something to push against if they are to propel him backward. Who is right and why?

17. Mary and Daryl are new to the sport of rock climbing. Mary says she wants a stiff rope because a stiff rope is a strong rope. Daryl insists that a good climbing rope must have some stretch. Who is correct, and why?

Multiple-Choice Questions

1. A ball of mass m with initial speed v collides with another ball of mass M, initially at rest. After the collision the two balls stick together, moving with speed V. The ratio of the final speed V to the initial speed v is $V/v =$

   (a) $\dfrac{M}{M+m}$ (b) $\dfrac{M+m}{M}$

   (c) $\dfrac{m}{M+m}$ (d) $\dfrac{M+m}{m}$

   (e) $\sqrt{\dfrac{M}{M+m}}$ (f) $\sqrt{\dfrac{m}{M+m}}$

2. Two particles A and B of equal mass are located at some distance from each other. Particle A is at rest while B moves away from A at speed v. What happens to the center of mass of the system of two particles?

   (a) It does not move.

   (b) It moves with a speed v away from A.

   (c) It moves with a speed v toward A.

   (d) It moves with a speed $\frac{1}{2}v$ away from A.

   (e) It moves with a speed $\frac{1}{2}v$ toward A.

3. Two uniform spheres with equal mass per unit volume are in contact with one another. The mass of sphere A is five times that of sphere B. The center of mass of the system is

   (a) at the point where A and B touch.

   (b) inside sphere B somewhere on the line joining the centers of A and B.

   (c) inside sphere A somewhere on the line joining the centers.

   (d) at the center of sphere A.

   (e) outside of both spheres.

4. An object at rest suddenly explodes into three parts of equal mass. Two of the parts move away at right angles to each other and with equal speeds v. What is the velocity of the third part just after the explosion?

   (a) Direction of vector 1 and magnitude $2v$

   (b) Direction of vector 2 and magnitude $\sqrt{2}v$

   (c) Direction of vector 3 and magnitude $\dfrac{1}{\sqrt{2}}v$

   (d) Direction of vector 2 and magnitude $\dfrac{1}{\sqrt{2}}v$

   (e) Direction of vector 1 and magnitude $\frac{1}{2}v$

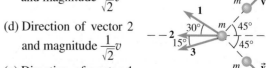

5. A 3.0-kg object is initially at rest. It then receives an impulse of magnitude 15 N·s. After the impulse, the object has
   (a) a speed of 45 m/s.
   (b) a momentum of magnitude 5.0 kg·m/s.
   (c) a speed of 7.5 m/s.
   (d) a momentum of magnitude 15 kg·m/s.

6. An object of mass m drops from rest a little above the Earth's surface for a time t. Ignore air resistance. After time t the magnitude of its momentum is
   (a) mgt^2
   (b) mgt
   (c) $mg\sqrt{t}$
   (d) $\sqrt{mgt}$
   (e) $\dfrac{mgt^2}{2}$

Multiple-Choice Questions 7–12 refer to a situation in which a golf ball is projected straight upward in the $+y$-direction. Ignore air resistance. The answer choices are found in the figure.

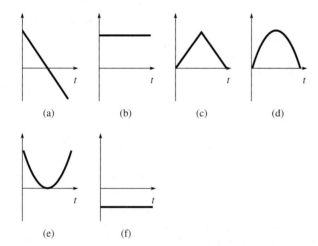

(a)   (b)   (c)   (d)

(e)   (f)

7. Which graph shows the acceleration a_y of the ball as a function of time?

8. Which graph shows the vertical position y of the ball as a function of time?

9. Which graph shows the momentum p_y of the ball as a function of time?

10. Which graph shows the kinetic energy of the ball as a function of time?

11. Which graph shows the potential energy of the ball as a function of time?

12. Which graph shows the total energy of the ball as a function of time?

Problems

⊙ Combination conceptual/quantitative problem
♆ Biological or medical application
✦ Challenging problem

7.2 Momentum; 7.3 The Impulse-Momentum Theorem

1. Two cars, each of mass 1300 kg, are approaching each other on a head-on collision course. Each speedometer reads 19 m/s. What is the magnitude of the total momentum of the system?

2. What is the momentum of an automobile (weight = 9800 N) when it is moving at 35 m/s to the south?

3. Verify that the SI unit of impulse is the same as the SI unit of momentum.

4. A cue stick hits a cue ball with an average force of 24 N for a duration of 0.028 s. If the mass of the ball is 0.16 kg, how fast is it moving after being struck?

5. A system consists of three particles with these masses and velocities: mass 3.0 kg, moving north at 3.0 m/s; mass 4.0 kg, moving south at 5.0 m/s; and mass 7.0 kg, moving north at 2.0 m/s. What is the total momentum of the system?

6. A sports car traveling along a straight line increases its speed from 20.0 mi/h to 60.0 mi/h. (a) What is the ratio of the final to the initial magnitude of its momentum? (b) What is the ratio of the final to the initial kinetic energy?

7. A ball of mass 5.0 kg moving with a speed of 2.0 m/s in the $+x$-direction hits a wall and bounces back with the same speed in the $-x$-direction. What is the change of momentum of the ball?

8. An object of mass 3.0 kg is projected into the air at a 55° angle. It hits the ground 3.4 s later. What is its change in momentum while it is in the air? Ignore air resistance.

9. An object of mass 3.0 kg is allowed to fall from rest under the force of gravity for 3.4 s. What is the change in its momentum? Ignore air resistance.

10. What average force is necessary to bring a 50.0-kg sled from rest to a speed of 3.0 m/s in a period of 20.0 s? Assume frictionless ice.

11. For a safe re-entry into the Earth's atmosphere, the pilots of a space capsule must reduce their speed from 2.6×10^4 m/s to 1.1×10^4 m/s. The rocket engine produces a backward force on the capsule of 1.8×10^5 N. The mass of the capsule is 3800 kg. For how long must they fire their engine? [*Hint:* Ignore the change in mass of the capsule due to the expulsion of exhaust gases.]

12. A 0.15-kg baseball traveling in a horizontal direction with a speed of 20 m/s hits a bat and is popped straight up with a speed of 15 m/s. (a) What is the change in momentum (magnitude and direction) of the baseball? (b) If the bat was in contact with the ball for 50 ms, what was the average force of the bat on the ball?

13. An automobile traveling at a speed of 30.0 m/s applies its brakes and comes to a stop in 5.0 s. If the automobile has a mass of 1.0×10^3 kg, what is the average horizontal force exerted on it during braking? Assume the road is level.

14. A 3.0-kg body is initially moving northward at 15 m/s. Then a force of 15 N, toward the east, acts on it for 4.0 s. (a) At the end of the 4.0 s, what is the body's final velocity? (b) What is the change in momentum during the 4.0 s?

✦15. A boy of mass 60.0 kg is rescued from a hotel fire by leaping into a firefighters' net. The window from which he leapt was 8.0 m above the net. The firefighters lower their arms as he lands in the net so that he is brought to a complete stop in a time of 0.40 s. (a) What is his change in momentum during the 0.40-s interval? (b) What is the impulse on the net due to the boy during the interval? [*Hint:* Do not ignore gravity.] (c) What is the average force on the net due to the boy during the interval?

16. A 115-g ball is traveling to the left with a speed of 30 m/s when it is struck by a racket. The force on the ball, directed to the right and applied over 21 ms of contact time, is shown in the graph. What is the speed of the ball immediately after it leaves the racket? (🐾 tutorial: impulse)

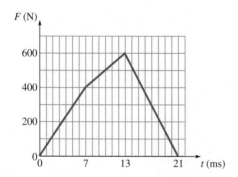

✦17. A pole-vaulter of mass 60.0 kg vaults to a height of 6.0 m before dropping to thick padding placed below to cushion her fall. (a) Find the speed with which she lands. (b) If the padding brings her to a stop in a time of 0.50 s, what is the average force on her body due to the padding during that time interval?

7.4 Conservation of Momentum

18. A rifle has a mass of 4.5 kg and it fires a bullet of mass 10.0 g at a muzzle speed of 820 m/s. What is the recoil speed of the rifle as the bullet leaves the gun barrel?

19. A 0.030-kg bullet is fired vertically at 200 m/s into a 0.15-kg baseball that is initially at rest. The bullet lodges in the baseball and, after the collision, the baseball/bullet rise to a height of 37 m. (a) What was the speed of the baseball/bullet right after the collision? (b) What

was the average force of air resistance while the baseball/bullet was rising?

20. A submarine of mass 2.5×10^6 kg and initially at rest fires a torpedo of mass 250 kg. The torpedo has an initial speed of 100.0 m/s. What is the initial recoil speed of the submarine? Neglect the drag force of the water.

21. A uranium nucleus (mass 238 u), initially at rest, undergoes radioactive decay. After an alpha particle (mass 4.0 u) is emitted, the remaining nucleus is thorium (mass 234 u). If the alpha particle is moving at 0.050 times the speed of light, what is the recoil speed of the thorium nucleus? (Note: "u" is a unit of mass; it is *not* necessary to convert it to kg.)

22. Dash is standing on his frictionless skateboard with three balls, each with a mass of 100 g, in his hands. The combined mass of Dash and his skateboard is 60 kg. How fast should dash throw the balls forward if he wants to move backward with a speed of 0.50 m/s? Do you think Dash can succeed? Explain.

23. A 58-kg astronaut is in space, far from any objects that would exert a significant gravitational force on him. He would like to move toward his spaceship, but his jet pack is not functioning. He throws a 720-g socket wrench with a velocity of 5.0 m/s in a direction away from the ship. After 0.50 s, he throws a 800-g spanner in the same direction with a speed of 8.0 m/s. After another 9.90 s, he throws a mallet with a speed of 6.0 m/s in the same direction. The mallet has a mass of 1200 g. How fast is the astronaut moving after he throws the mallet?

24. A man with a mass of 65 kg skis down a frictionless hill that is 5.0 m high. At the bottom of the hill the terrain levels out. As the man reaches the horizontal section, he grabs a 20-kg backpack and skis off a 2.0-m-high ledge. At what horizontal distance from the edge of the ledge does the man land?

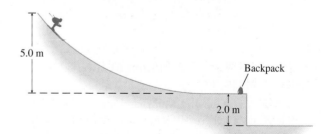

✦25. A cannon on a railroad car is facing in a direction parallel to the tracks. It fires a 98-kg shell at a speed of 105 m/s (relative to the ground) at an angle of 60.0° above the horizontal. If the cannon plus car have a mass of 5.0×10^4 kg, what is the recoil speed of the car if it was at rest before the cannon was fired? [*Hint:* A *component*

of a system's momentum along an axis is conserved if the net external force acting on the system has no component along that axis.]

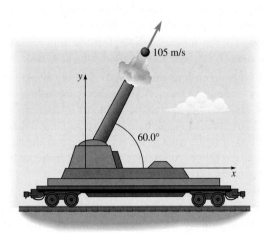

◆26. A marksman standing on a motionless railroad car fires a gun into the air at an angle of 30.0° from the horizontal. The bullet has a speed of 173 m/s (relative to the ground) and a mass of 0.010 kg. The man and car move to the left at a speed of 1.0×10^{-3} m/s after he shoots. What is the mass of the man and car? (See the hint in Problem 25.)

7.5 Center of Mass; 7.6 Motion of the Center of Mass

27. Particle A is at the origin and has a mass of 30.0 g. Particle B has a mass of 10.0 g. Where must particle B be located if the coordinates of the CM are $(x, y) = (2.0$ cm, 5.0 cm)?

28. Particle A has a mass of 5.0 g and particle B has a mass of 1.0 g. Particle A is located at the origin and particle B is at the point $(x, y) = (25$ cm, 0). What is the location of the CM? (ⓦ tutorial: center of mass)

29. The three bodies in the figure each have the same mass. If one of the bodies is moved 12 cm in the positive x-direction, by how much does the CM move?

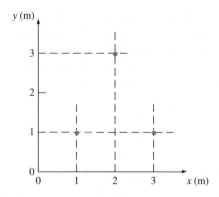

30. The positions of three particles, written as (x, y) coordinates, are: particle 1 (mass 4.0 kg) at (4.0 m, 0 m); particle 2 (mass 6.0 kg) at (2.0 m, 4.0 m); particle 3 (mass 3.0 kg) at (−1.0 m, −2.0 m). What is the location of the CM?

◆31. Belinda needs to find the CM of a sculpture she has made so that it will hang in a gallery correctly. The sculpture is all in one plane and consists of various shaped uniform objects with masses and sizes as shown. Where is the CM of this sculpture? Assume the thin rods connecting the larger pieces have no mass and place the reference frame origin at the top left corner of the sculpture.

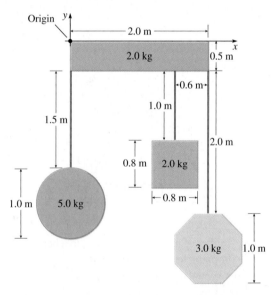

32. Jane is sitting on a chair with her lower leg at a 30.0° angle with respect to the vertical, as shown. You need to develop a computer model of her leg to assist in some medical research. If you assume that her leg can be modeled as two uniform cylinders, one with mass $M = 20$ kg and length $L = 35$ cm and one with mass $m = 10$ kg and length $l = 40$ cm, where is the CM of her leg?

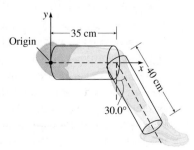

33. Find the *x*-coordinate of the CM of the composite object shown in the figure. The sphere, cylinder, and rectangular solid all have a uniform composition. Their masses and dimensions are: sphere: 200 g, diameter = 10 cm; cylinder: 450 g, length = 17 cm, radius = 5.0 cm; rectangular solid: 325 g, length in *x*-direction = 16 cm, height = 10 cm, depth = 12 cm.

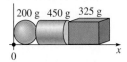

34. Consider two falling bodies. Their masses are 3.0 kg and 4.0 kg. At time *t* = 0, the two are released from rest. What is the velocity of their CM at *t* = 10.0 s? Ignore air resistance.

35. Body A of mass 3 kg is moving in the +*x*-direction with a speed of 14 m/s. Body B of mass 4 kg is moving in the −*y*-direction with a speed of 7 m/s. What are the *x*- and *y*-components of the velocity of the CM of the two bodies?

◆36. If a particle of mass 5.0 kg is moving east at 10 m/s and a particle of mass 15 kg is moving west at 10 m/s, what is the velocity of the CM of the pair?

37. An object located at the origin and having mass *M* explodes into three pieces having masses *M*/4, *M*/3, and 5*M*/12. The pieces scatter on a horizontal frictionless *xy*-plane. The piece with mass *M*/4 flies away with velocity 5.0 m/s at 37° above the *x*-axis. The piece with mass *M*/3 has velocity 4.0 m/s directed at an angle of 45° above the −*x*-axis. (a) What are the velocity components of the third piece? (b) Describe the motion of the CM of the system after the explosion.

38. Prove Eq. (7-13) $\sum \vec{F}_{ext} = M\vec{a}_{CM}$. [*Hint:* Start with $\sum \vec{F}_{ext} = \lim_{\Delta t \to 0} (\Delta \vec{p}/\Delta t)$, where $\sum \vec{F}_{ext}$ is the net external force acting on a system and $\vec{p}$ is the total momentum of the system.]

7.7 Collisions in One Dimension

39. A helium atom (mass 4.00 u) moving at 618 m/s to the right collides with an oxygen molecule (mass 32.0 u) moving in the same direction at 412 m/s. After the collision, the oxygen molecule moves at 456 m/s to the right. What is the velocity of the helium atom after the collision?

40. A toy car with a mass of 120 g moves to the right with a speed of 0.75 m/s. A small child drops a 30.0-g piece of clay onto the car. The clay sticks to the car and the car continues to the right. What is the change in speed of the car? Consider the frictional force between the car and the ground to be negligible.

41. In the railroad freight yard, an empty freight car of mass *m* rolls along a straight level track at 1.0 m/s and collides with an initially stationary, fully loaded boxcar of mass 4.0*m*. The two cars couple together on collision. (a) What is the speed of the two cars after the collision? (b) Suppose instead that the two cars are at rest after the collision. With what speed was the loaded boxcar moving before the collision if the empty one was moving at 1.0 m/s? (tutorial: sticking collision)

42. A 0.020-kg bullet traveling at 200.0 m/s east hits a motionless 2.0-kg block and bounces off it, retracing its original path with a velocity of 100.0 m/s west. What is the final velocity of the block? Assume the block rests on a perfectly frictionless horizontal surface.

43. A block of wood of mass 0.95 kg is initially at rest. A bullet of mass 0.050 kg traveling at 100.0 m/s strikes the block and becomes embedded in it. With what speed do the block of wood and the bullet move just after the collision?

44. A 0.020-kg bullet is shot horizontally and collides with a 2.00-kg block of wood. The bullet embeds in the block and the block slides along a horizontal surface for 1.50 m. If the coefficient of kinetic friction between the block and surface is 0.400, what was the original speed of the bullet?

45. A 2.0-kg block is moving to the right at 1.0 m/s just before it strikes and sticks to a 1.0-kg block initially at rest. What is the total momentum of the two blocks after the collision?

46. A 75-kg man is at rest on ice skates. A 0.20-kg ball is thrown to him. The ball is moving horizontally at 25 m/s just before the man catches it. How fast is the man moving just after he catches the ball?

47. A BMW of mass 2.0×10^3 kg is traveling at 42 m/s. It approaches a 1.0×10^3 kg Volkswagen going 25 m/s in the same direction and strikes it in the rear. Neither driver applies the brakes. Neglect the relatively small frictional forces on the cars due to the road and due to air resistance. (a) If the collision slows the BMW down to 33 m/s, what is the speed of the VW after the collision? (b) During the collision, which car exerts a larger force on the other, or are the forces equal in magnitude? Explain.

48. A 100-g ball collides elastically with a 300-g ball that is at rest. If the 100-g ball was traveling in the positive *x*-direction at 5.00 m/s before the collision, what are the velocities of the two balls after the collision? (tutorial: elastic collision)

49. A projectile of 1.0-kg mass approaches a stationary body of 5.0 kg at 10.0 m/s and, after colliding, rebounds in the reverse direction along the same line with a speed of 5.0 m/s. What is the speed of the 5.0-kg body after the collision?

50. A 2.0-kg object is at rest on a perfectly frictionless surface when it is hit by a 3.0-kg object moving at 8.0 m/s. If the two objects are stuck together after the collision, what is the speed of the combination?

51. A spring of negligible mass is compressed between two blocks, A and B, which are at rest on a frictionless horizontal surface at a distance of 1.0 m from a wall on the left and 3.0 m from a wall on the right. The sizes of the blocks and spring are small. When the spring is released, block A moves toward the left wall and strikes it at the same instant that block B strikes the right wall. The mass of A is 0.60 kg. What is the mass of B?

52. Two identical gliders on an air track are held together by a piece of string, compressing a spring between the gliders. While they are moving to the right at a common speed of 0.50 m/s, someone holds a match under the string and burns it, letting the spring force the gliders apart. One glider is then observed to be moving to the right at 1.30 m/s. (a) What velocity does the other glider have? (b) Is the total kinetic energy of the two gliders after the collision greater than, less than, or equal to the total kinetic energy before the collision? If greater, where did the extra energy come from? If less, where did the "lost" energy go?

53. A 0.010-kg bullet traveling horizontally at 400.0 m/s strikes a 4.0-kg block of wood sitting at the edge of a table. The bullet is lodged into the wood. If the table height is 1.2 m, how far from the table does the block hit the floor?

54. Two objects with masses m_1 and m_2 approach each other with equal and opposite momenta so that the total momentum is zero. Show that, if the collision is elastic, the final *speed* of each object must be the same as its initial speed. (The final *velocity* of each object is *not* the same as its initial velocity, however.)

55. A 6.0-kg object is at rest on a perfectly frictionless surface when it is struck head-on by a 2.0-kg object moving at 10 m/s. If the collision is perfectly elastic, what is the speed of the 6.0-kg object after the collision? [*Hint:* You will need two equations.]

56. Use the result of Problem 54 to show that in *any* elastic collision between two objects, the relative speed of the two is the same before and after the collision. [*Hints:* Look at the collision in its CM *frame*—the reference frame in which the CM is at rest. The *relative* speed of two objects is the same in any inertial reference frame.]

7.8 Collisions in Two Dimensions

57. A firecracker is tossed straight up into the air. It explodes into three pieces of equal mass just as it reaches the highest point. Two pieces move off at 120 m/s at right angles to each other. How fast is the third piece moving?

58. Body A of mass M has an original velocity of 6.0 m/s in the +x-direction toward a stationary body (body B) of the same mass. After the collision, body A has velocity components of 1.0 m/s in the +x-direction and 2.0 m/s in the +y-direction. What is the magnitude of body B's velocity after the collision?

59. (a) With reference to Practice Problem 7.11, find the momentum change of the ball of mass m_1 during the collision. Give your answer in x- and y-component form; express the components in terms of m_1 and v_i. (b) Repeat for the ball of mass m_2. How are the momentum changes related?

60. A hockey puck moving at 0.45 m/s collides elastically with another puck that was at rest. The pucks have equal mass. The first puck is deflected 37° to the right and moves off at 0.36 m/s. Find the speed and direction of the second puck after the collision.

61. Puck 1 sliding along the x-axis strikes stationary puck 2 of the same mass. After the elastic collision, puck 1 moves off at speed v_{1f} in the direction 60.0° above the x-axis; puck 2 moves off at speed v_{2f} in the direction 30.0° below the x-axis. Find v_{2f} in terms of v_{1f}.

62. Block A, with a mass of 220 g, is traveling north on a frictionless surface with a speed of 5.0 m/s. Block B, with a mass of 300 g travels west on the same surface until it collides with A. After the collision, the blocks move off together with a velocity of 3.13 m/s at an angle of 42.5° to the north of west. What was B's speed just before the collision?

63. A projectile of mass 2.0 kg approaches a stationary target body at 5.0 m/s. The projectile is deflected through an angle of 60.0° and its speed after the collision is 3.0 m/s. What is the magnitude of the momentum of the target body after the collision?

64. A 1500-kg car moving east at 17 m/s collides with a 1800-kg car moving south at 15 m/s and the two cars stick together. (a) What is the velocity of the cars right after the collision? (b) How much kinetic energy was converted to another form during the collision?

65. A car with a mass of 1700 kg is traveling directly northeast (45° between north and east) at a speed of 14 m/s (31 mph), and collides with a smaller car with a mass of 1300 kg that is traveling directly south at a speed of 18 m/s (40 mph). The two cars stick together during the collision. With what speed and direction does the tangled mess of metal move right after the collision?

66. In a nuclear reactor, a neutron moving at speed v_i in the positive x-direction strikes a deuteron, which is at rest. The neutron is deflected by 90.0° and moves off with speed $v_i/\sqrt{3}$ in the positive y-direction. Find the x- and y-components of the deuteron's velocity after the collision. (The mass of the deuteron is twice the mass of the neutron.)

67. Two identical pucks are on an air table. Puck A has an initial velocity of 2.0 m/s in the +x-direction. Puck B is at rest. Puck A collides elastically with puck B and A moves off at 1.0 m/s at an angle of 60° above the x-axis. What is the speed and direction of puck B after the collision?

68. In a circus trapeze act, two acrobats actually fly through the air and grab on to one another, then together grab a

swinging bar. One acrobat, with a mass of 60 kg, is moving at 3.0 m/s at an angle of 10° above the horizontal and the other, with a mass of 80 kg, is approaching her with a speed of 2.0 m/s at an angle of 20° above the horizontal. What is the direction and speed of the acrobats right after they grab on to each other? Let the positive x-axis be in the horizontal direction and assume the first acrobat has positive velocity components in the positive x- and y-directions.

✦69. Two African swallows fly toward one another, carrying coconuts. The first swallow is flying north horizontally with a speed of 20 m/s. The second swallow is flying at the same height as the first and in the opposite direction with a speed of 15 m/s. The mass of the first swallow is 0.270 kg and the mass of his coconut is 0.80 kg. The second swallow's mass is 0.220 kg and her coconut's mass is 0.70 kg. The swallows collide and lose their coconuts. Immediately after the collision, the 0.80-kg coconut travels 10° west of south with a speed of 13 m/s, and the 0.70-kg coconut moves 30° east of north with a speed of 14 m/s. The two birds are tangled up with one another and stop flapping their wings as they travel off together. What is the velocity of the birds immediately after the collision?

Comprehensive Problems

70. A sled of mass 5.0 kg is coasting along on a frictionless ice-covered lake at a constant speed of 1.0 m/s. A 1.0-kg book is dropped vertically onto the sled. At what speed does the sled move once the book is on it?

71. An automobile weighing 13.6 kN is moving at 17.0 m/s when it collides with a stopped car weighing 9.0 kN. If they lock bumpers and move off together, what is their speed just after the collision?

72. For a system of three particles moving along a line, an observer in a laboratory measures the following masses and velocities. What is the velocity of the CM of the system?

Mass (kg)	v_x (m/s)
3.0	+290
5.0	−120
2.0	+52

73. An intergalactic spaceship is traveling through space far from any planets or stars, where no human has gone before. The ship carries a crew of 30 people (of total mass 2.0×10^3 kg). If the speed of the spaceship is 1.0×10^5 m/s and its mass (excluding the crew) is 4.8×10^4 kg, what is the magnitude of the total momentum of the ship and the crew?

74. A baseball player pitches a fastball toward home plate at a speed of 41 m/s. The batter swings, connects with the ball of mass 145 g, and hits it so that the ball leaves the bat with a speed of 37 m/s. Assume that the ball is moving horizontally just before and just after the collision with the bat. (a) What is the magnitude of the change in momentum of the ball? (b) What is the impulse delivered to the ball by the bat? (c) If the bat and ball are in contact for 3.0 ms, what is the magnitude of the average force exerted on the ball by the bat?

✦75. A tennis ball of mass 0.060 kg is served. It strikes the ground with a velocity of 54 m/s (120 mi/h) at an angle of 22° below the horizontal. Just after the bounce it is moving at 53 m/s at an angle of 18° above the horizontal. If the interaction with the ground lasts 0.065 s, what average force did the ground exert on the ball?

76. A uniform rod of length 30.0 cm is bent into the shape of an inverted U. Each of the three sides is of length 10.0 cm. Find the location, in x- and y-coordinates, of the CM as measured from the origin.

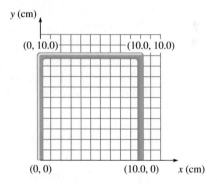

77. A child places 12 wooden blocks together, as shown in the figure. If each block has the same mass and density, where is the CM of these blocks? Each block is a cube with sides of 1.0 inch length. The origin of the coordinate system is at the center of the farthest block to the left.

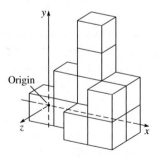

78. To contain some unruly demonstrators, the riot squad approaches with fire hoses. Suppose that the rate of flow of water through a fire hose is 24 kg/s and the stream of water from the hose moves at 17 m/s. What force is exerted by such a stream on a person in the crowd? Assume that the water comes to a dead stop against the demonstrator's chest.

79. An inexperienced catcher catches a 130 km/h fastball of mass 140 g within 1 ms, whereas an experienced catcher slightly retracts his hand during the catch, extending the

stopping time to 10 ms. What are the average forces imparted to the two gloved hands during the catches?

◆80. A stationary 0.1-g fly encounters the windshield of a 1000-kg automobile traveling at 100 km/h. (a) What is the change in momentum of the car due to the fly? (b) What is the change of momentum of the fly due to the car? (c) Approximately how many flies does it take to reduce the car's speed by 1 km/h?

81. A 0.15-kg baseball is pitched with a speed of 35 m/s (78 mph). When the ball hits the catcher's glove, the glove moves back by 5.0 cm (2 in.) as it stops the ball. (a) What was the change in momentum of the baseball? (b) What impulse was applied to the baseball? (c) Assuming a constant acceleration of the ball, what was the average force applied by the catcher's glove?

◆82. A projectile of mass 2.0 kg approaches a stationary target body at 8.0 m/s. The projectile is deflected through an angle of 90.0° and its speed after the collision is 6.0 m/s. What is the speed of the target body after the collision if the collision is perfectly elastic?

83. A radioactive nucleus is at rest when it spontaneously decays by emitting an electron and neutrino. The momentum of the electron is 8.20×10^{-19} kg·m/s and it is directed at right angles to that of the neutrino. The neutrino's momentum has magnitude 5.00×10^{-19} kg·m/s. (a) In what direction does the newly formed (daughter) nucleus recoil? (b) What is its momentum?

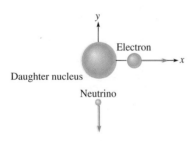

◆84. A 60.0-kg woman stands at one end of a 120-kg raft that is 6.0 m long. The other end of the raft is 0.50 m from a pier. (a) The woman walks toward the pier until she gets to the other end of the raft and stops there. Now what is the distance between the raft and the pier? (b) In (a), how far did the woman walk (relative to the pier)?

85. A police officer is investigating the scene of an accident where two cars collided at an intersection. One car with a mass of 1100 kg moving west had collided with a 1300-kg car moving north. The two cars, stuck together, skid at an angle of 30° north of west for a distance of 17 m. The coefficient of kinetic friction between the tires and the road is 0.80. The speed limit for each car was 70 km/h. Was either car speeding?

◆86. A jet plane is flying at 130 m/s relative to the ground. There is no wind. The engines take in 81 kg of air per second. Hot gas (burned fuel and air) is expelled from the engines at high speed. The engines provide a forward force on the plane of magnitude 6.0×10^4 N. At what speed relative to the ground is the gas being expelled? [*Hint:* Look at the momentum change of the air taken in by the engines during a time interval Δt.] This calculation is approximate since we are ignoring the 3.0 kg of fuel consumed and expelled with the air each second.

◆87. Within cells, small organelles containing newly synthesized proteins are transported along microtubules by tiny molecular motors called kinesins. What force does a kinesin molecule need to deliver in order to accelerate an organelle with mass 0.01 pg (10^{-17} kg) from 0 to 1 µm/s within a time of 10 µs?

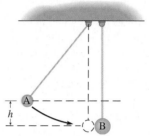

Problems 88 and 89.

◆88. The pendulum bobs in the figure are made of soft clay so that they stick together after impact. The mass of bob A is half that of bob B. Bob B is initially at rest. What is the ratio of the kinetic energy of the combined bobs, just after impact, to the kinetic energy of bob A just before impact?

◆89. The pendulum bobs in the figure are made of soft clay so that they stick together after impact. The mass of bob A is half that of bob B. Bob B is initially at rest. If bob A is released from a height h above its lowest point, what is the maximum height attained by bobs A and B after the collision?

◆90. A flat, circular metal disk of uniform thickness has a radius of 3.0 cm. A hole is drilled in the disk that is 1.5 cm in radius. The hole is tangent to one side of the disk. Where is the CM of the disk now that the hole has been drilled? [*Hint:* The original disk (before the hole is drilled) can be thought of as having two pieces—the disk with the hole plus the smaller disk of metal drilled out. Write an equation that expresses x_{CM} of the original disk in terms of the x_{CM}'s of the two pieces. Since the thickness is uniform, the mass of any piece is proportional to its area.]

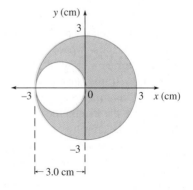

91. Two pendulum bobs have equal masses and lengths (5.1 m). Bob A is initially held horizontally while bob B hangs vertically at rest. Bob A is released and collides elastically with bob B. How fast is bob B moving immediately after the collision?

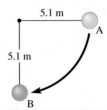

92. Two identical gliders, each with elastic bumpers and mass 0.10 kg, are on a horizontal air track. Friction is negligible. Glider 2 is stationary. Glider 1 moves toward glider 2 from the left with a speed of 0.20 m/s. They collide. After the collision, what are the velocities of glider 1 and glider 2?

93. A radium nucleus (mass 226 u) at rest decays into a radon nucleus (symbol Rn, mass 222 u) and an alpha particle (symbol α, mass 4 u). (a) Find the ratio of the speeds v_α/v_{Rn} after the decay. (b) Find the ratio of the magnitudes of the momenta p_α/p_{Rn}. (c) Find the ratio of the kinetic energies K_α/K_{Rn}. (Note: "u" is a unit of mass; it is *not* necessary to convert it to kg.)

Answers to Practice Problems

7.1 (a) 0.78 kg·m/s downward; (b) 0.78 kg·m/s toward the apple; 1.3×10^{-25} m/s

7.2 3.5 times his weight

7.3 1700 N; 0.0037 s

7.4 0.8 m/s in the $-x$-direction

7.5 1.7 m/s

7.6 2.2 m/s

7.7 (2.0 cm, 2.3 cm)

7.8 (a) 2.7 m; (b) 1.5 m in the other direction; (c) the CM does not move

7.9 4.0 m/s

7.10 10.0 m/s

7.11 0.751 v_i

7.12 0.781

Answers to Checkpoints

7.2 No, because the *direction* of the car's momentum would have changed.

7.4 When external forces act on a system, the momentum of the system is not conserved.

7.6 Despite the fact that the hammer is rotating, it is in free fall and its CM follows the same trajectory as a point particle in free fall.

7.7 Yes. Momentum is conserved in both elastic and inelastic collisions. In an inelastic collision, the initial and final *kinetic energies* are not equal.

Torque and Angular Momentum

In gymnastics, the iron cross is a notoriously difficult feat requiring incredible strength. Why does it require such great strength? (See p. 282 for the answer.)

Concepts & Skills to Review

- translational equilibrium (Section 4.2)
- uniform circular motion and circular orbits (Sections 5.1 and 5.4)
- angular acceleration (Section 5.6)
- conservation of energy (Section 6.1)
- center of mass and its motion (Sections 7.5 and 7.6)
- rolling without slipping (Section 5.1)

8.1   ROTATIONAL KINETIC ENERGY AND ROTATIONAL INERTIA

When a rigid object is rotating about a fixed axis, it has kinetic energy because each particle other than those on the axis of rotation is moving in a circle around the axis. In principle, we can calculate the kinetic energy of rotation by summing the kinetic energy of each particle. To say the least, that sounds like a laborious task. We need a simpler way to express the rotational kinetic energy of such an object so that we don't have to calculate this sum over and over. Our simpler expression exploits the fact that the speed of each particle is proportional to the angular speed of rotation ω.

If a rigid object consists of N particles, the sum of the kinetic energies of the particles can be written mathematically using a subscript to label the mass and speed of each particle:

$$K_{\text{rot}} = \tfrac{1}{2}m_1 v_1^2 + \tfrac{1}{2}m_2 v_2^2 + \cdots + \tfrac{1}{2}m_N v_N^2 = \sum_{i=1}^{N} \tfrac{1}{2}m_i v_i^2$$

The speed of each particle is related to its distance from the axis of rotation. Particles that are farther from the axis move faster. In Section 5.1, we found that the speed of a particle moving in a circle is

$$v = r\omega \tag{5-7}$$

where ω is the angular speed and r is the distance between the rotation axis and the particle (Fig. 8.1). By substitution, the rotational kinetic energy can be written

$$K_{\text{rot}} = \sum_{i=1}^{N} \tfrac{1}{2}m_i r_i^2 \omega^2$$

The entire object rotates at the same angular velocity ω, so the constants $\tfrac{1}{2}$ and ω^2 can be factored out of each term of the sum:

$$K_{\text{rot}} = \tfrac{1}{2}\left(\sum_{i=1}^{N} m_i r_i^2 \right) \omega^2$$

The quantity in the parentheses *cannot change* since the distance between each particle and the rotation axis stays the same if the object is rigid and doesn't change shape. However difficult it may be to compute the sum in the parentheses, we only need to do it *once* for any given mass distribution and axis of rotation.

Let's give the quantity in the brackets the symbol I. In Chapter 5, we found it useful to draw analogies between translational variables and their rotational equivalents. By using the symbol I, we can see that translational and rotational kinetic energy have similar forms: translational kinetic energy is

$$K_{\text{tr}} = \tfrac{1}{2}mv^2$$

and **rotational kinetic energy** is

Rotational kinetic energy:

$$K_{\text{rot}} = \tfrac{1}{2}I\omega^2 \tag{8-1}$$

The symbol Σ stands for a sum. $\displaystyle\sum_{i=1}^{N}$ means the sum for $i = 1, 2, \ldots, N$.

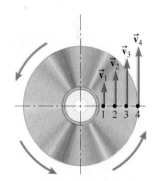

Figure 8.1 Four points on a spinning CD. Points at greater distances from the center are moving faster than points closer to the center.

Since $v = r\omega$ was used to derive Eq. (8-1), ω must be expressed in radians per unit time (normally rad/s).

The quantity *I* is called the **rotational inertia**:

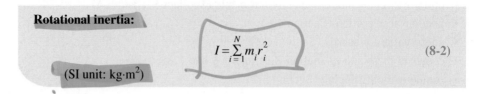

Rotational inertia:

$$I = \sum_{i=1}^{N} m_i r_i^2$$

(8-2)

(SI unit: kg·m²)

CONNECTION:

Rotational and translational kinetic energies have the same form: $\frac{1}{2}$ inertia × speed².

Comparing the expressions for translational and rotational kinetic energies, we see that angular speed ω takes the place of speed v and rotational inertia I takes the place of mass m. Mass is a measure of the inertia of an object, or, in other words, how difficult it is to change the object's velocity. Similarly, for a rigid rotating object, I is a measure of its rotational inertia—how hard it is to change its angular velocity. That is why the quantity I is called the rotational inertia; it is also sometimes called the **moment of inertia**.

When a problem requires you to find a rotational inertia, there are three principles to follow.

Finding the Rotational Inertia

1. If the object consists of a *small* number of particles, calculate the sum $I = \sum_{i=1}^{N} m_i r_i^2$ directly.
2. For symmetrical objects with simple geometric shapes, calculus can be used to perform the sum in Eq. (8-2). Table 8.1 lists the results of these calculations for the shapes most commonly encountered.
3. Since the rotational inertia is a sum, you can always mentally decompose the object into several parts, find the rotational inertia of each part, and then add them. This is an example of the *divide-and-conquer* problem-solving technique.

See the text website for more information on rotational inertia, including the parallel- and perpendicular-axis theorems.

Keep in mind that the rotational inertia of an object depends on the location of the rotation axis. For instance, imagine taking the hinges off the side of a door and putting them on the top so that the door swings about a horizontal axis like a cat flap door (Fig. 8.2b). The door now has a considerably larger rotational inertia than before the hinges were moved because the door's height is greater than its width. The door has the same mass as before, but its mass now lies on average much farther from the axis of rotation than that

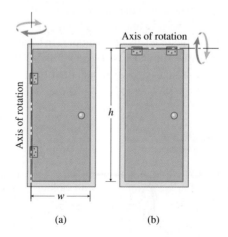

Figure 8.2 The rotational inertia of a door depends on the rotation axis. (a) The door with hinges at the side has a smaller rotational inertia, $I = \frac{1}{3}Mw^2$, than (b) the rotational inertia, $I = \frac{1}{3}Mh^2$, of the same door with hinges at the top, because the door is taller than it is wide.

Table 8.1   Rotational Inertia for Uniform Objects with Various Geometrical Shapes

Shape	Axis of Rotation	Rotational Inertia	Shape	Axis of Rotation	Rotational Inertia
Thin hollow cylindrical shell (or hoop)	Central axis of cylinder	MR^2	Solid sphere	Through center	$\frac{2}{5}MR^2$
Solid cylinder (or disk)	Central axis of cylinder	$\frac{1}{2}MR^2$	Thin hollow spherical shell	Through center	$\frac{2}{3}MR^2$
Hollow cylindrical shell or disk	Central axis of cylinder	$\frac{1}{2}M(a^2 + b^2)$	Thin rod (or rectangular plate)	Perpendicular to rod through end (or along edge of plate)	$\frac{1}{3}ML^2$
Rectangular plate	Perpendicular to plate through center	$\frac{1}{12}M(a^2 + b^2)$	Thin rod (or rectangular plate)	Perpendicular to rod through center (or parallel to edge of plate through center)	$\frac{1}{12}ML^2$

of the door in Fig. 8.2a. In applying Eq. (8-2) to find the rotational inertia of the door, the values of r_i range from 0 to the height of the door (h), whereas with the hinges in the normal position the values of r_i range from 0 only to the width of the door (w).

PHYSICS AT HOME

The change in rotational inertia of a rod as the rotation axis changes can be easily felt. Hold a baseball bat in the usual way, with your hands gripping the bottom of the bat. Swing the bat a few times. Now "choke up" on the bat—move your hands up the bat—and swing a few times. The bat is easier to swing because it now has a smaller rotational inertia. Children often choke up on a bat that is too massive for them. Even major league baseball players occasionally choke up on the bat when they want more control over their swing to place a hit in a certain spot (Fig. 8.3). On the other hand, choking up on the bat makes it impossible to hit a home run. To hit a long fly ball, you want the pitched baseball to encounter a bat that is swinging with a lot of rotational inertia.

Figure 8.3 Hank Aaron choking up on the bat.

✓ CHECKPOINT 8.1

According to Table 8.1, the rotational inertia of a uniform cylinder or disk about its central axis depends only on the mass and radius. Why does it not depend on the height of the cylinder (or thickness of the disk)?

Example 8.1

Rotational Inertia of a Barbell

A barbell consists of two plates, each a uniform disk of mass 20 kg and radius 15 cm, attached 20 cm from each end of a uniform rod of mass 10 kg, radius 1.25 cm, and length 2.20 m (Fig. 8.4). Find the rotational inertia of the barbell about two different axes of rotation: (a) axis a, the central axis of the bar, and (b) axis b, perpendicular to the bar and through its midpoint. Ignore the thickness of the disks and the holes in the disks.

Strategy   The rotational inertia of this composite object is the sum of the rotational inertias of the three parts (two disks and rod). Table 8.1 gives formulas for the rotational inertias of disks and rods, but *only for certain axes of rotation*. In particular, for axis b we have two disks rotating about an axis external to the disks, so none of the formulas in Table 8.1 apply; instead we'll return to the basic definition of rotational inertia [Eq. (8-2)] and make an approximation. Based on the distances between parts of the barbell and the two axes, we expect a smaller rotational inertia about axis a than about axis b. Let M and R be the mass and radius of each disk, and m, r, and L the mass, radius, and length of the rod, respectively.

Solution   (a) Each of the three component parts, the two disks and the rod, are solid cylinders rotating about their central axes. (The two formulas in Table 8.1 for thin rods are for axes *perpendicular* to the rod, so they are not useful here.) From Table 8.1,

$$I = \tfrac{1}{2}MR^2 + \tfrac{1}{2}MR^2 + \tfrac{1}{2}mr^2$$
$$= 2 \times [\tfrac{1}{2} \times 20 \text{ kg} \times (0.15 \text{ m})^2] + \tfrac{1}{2} \times 10 \text{ kg} \times (0.0125 \text{ m})^2$$
$$= 2 \times 0.225 \text{ kg·m}^2 + 0.00078 \text{ kg·m}^2 = 0.45 \text{ kg·m}^2$$

(b) Table 8.1 gives the rotational inertia of the rod about axis b as $\tfrac{1}{12}mL^2$. The center of each disk (assumed to have negligible thickness) is a distance $d = \tfrac{1}{2}(2.20 \text{ m} - 0.40 \text{ m}) = 0.90 \text{ m}$ from the midpoint of the rod. If we think of breaking a disk

into tiny pieces and applying Eq. (8-2), each of the distances r_i is at least $d = 0.90$ m (to the center) but no more than $\sqrt{d^2 + R^2} \approx 0.91$ m (to the edge). Therefore, to a good approximation, we can assume each disk to be a point mass at a distance d from the axis. Then

$$I = Md^2 + Md^2 + \tfrac{1}{12}mL^2$$
$$= 2 \times [20 \text{ kg} \times (0.90 \text{ m})^2] + \tfrac{1}{12} \times 10 \text{ kg} \times (2.20 \text{ m})^2$$
$$= 2 \times 16.2 \text{ kg·m}^2 + 4.03 \text{ kg·m}^2 = 36 \text{ kg·m}^2$$

As expected, the rotational inertia is much smaller about axis a than about axis b.

Discussion   The rod makes only a slight contribution to the rotational inertia about axis a because the radius of the rod is so much smaller than the radii of the disks, so its mass is on average much closer to the axis of rotation. The rod makes a more significant contribution to the rotational inertia about axis b because now the length, not radius, of the rod is relevant—its mass is distributed at distances from 0 to 1.10 m from the axis of rotation. Even if we account for the thickness of the disks, as long as their thicknesses are small relative to d, our estimate Md^2 of the contribution to I from each disk about axis b is still valid.

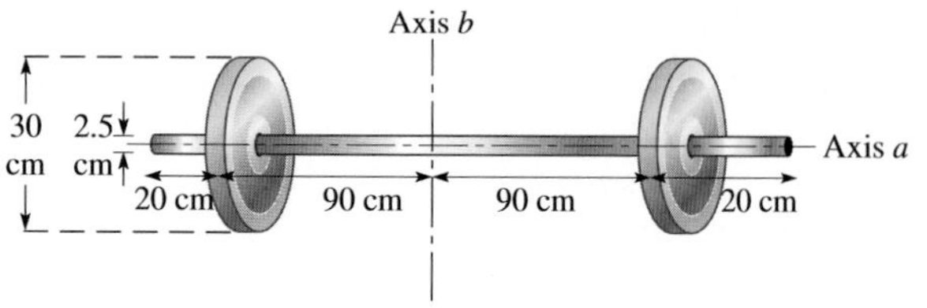

Practice Problem 8.1   Playground Merry-Go-Round

A playground merry-go-round is essentially a uniform disk that rotates about a vertical axis through its center (Fig. 8.5). Suppose the disk has a radius of 2.0 m and a mass of 160 kg; a child of mass 18.4 kg sits at the edge of the merry-go-round. What is the merry-go-round's rotational inertia, including the contribution due to the child? [*Hint:* Treat the child as a point mass at the edge of the disk.]

Figure 8.4
A barbell with two different rotation axes.

Axis b

30 cm   2.5 cm   20 cm   90 cm   90 cm   20 cm   — Axis a

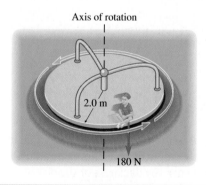

Axis of rotation

2.0 m

180 N

Figure 8.5
Child on a merry-go-round.

When applying conservation of energy to objects that rotate, the rotational kinetic energy is included in the mechanical energy. In Eq. (6-12),

$$W_{nc} = \Delta K + \Delta U = \Delta E_{mech} \qquad \text{(6-12)}$$

just as U stands for the sum of the elastic and gravitational potential energies, K stands for the sum of the translational and rotational kinetic energies:

$$K = K_{tr} + K_{rot}$$

Example 8.2

Atwood's Machine

Atwood's machine consists of a cord around a pulley of rotational inertia I, radius R, and mass M, with two blocks (masses m_1 and m_2) hanging from the ends of the cord as in Fig. 8.6. (Note that in Example 3.11 we analyzed Atwood's machine for the special case of a massless pulley; for a massless pulley $I = 0$.) Assume that the pulley is free to turn without friction and that the cord does not slip. Ignore air resistance. If the masses are released from rest, find how fast they are moving after they have moved a distance h (one up, the other down).

Strategy   Ignoring both air resistance and friction means that no nonconservative forces act on the system; therefore, its mechanical energy is conserved:

$$\Delta U + \Delta K = 0$$

Gravitational potential energy is converted into the translational kinetic energies of the two blocks and the rotational kinetic energy of the pulley.

Solution   For our convenience, we assume that $m_1 > m_2$. Mass m_1, therefore, moves down and m_2 moves up. After the masses have each moved a distance h, the changes in gravitational potential energy are

$$\Delta U_1 = -m_1 g h$$

$$\Delta U_2 = +m_2 g h$$

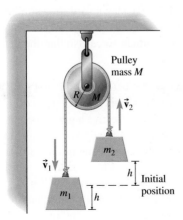

Pulley
mass M

R M

$\vec{v}_2$

m_2

$\vec{v}_1$

m_1 h

h Initial
position

Figure 8.6

Atwood's machine.

The mechanical energy of the system includes the kinetic energies of three objects: the two masses and the pulley. All start with zero kinetic energy, so

$$\Delta K = \tfrac{1}{2}(m_1 + m_2)v^2 + \tfrac{1}{2}I\omega^2$$

The speed v of the masses is the same since the cord's length is fixed. The speed v and the angular speed of the pulley ω are related if the cord does not slip: the tangential speed of the pulley must equal the speed at which the cord moves. The tangential speed of the pulley is its angular speed times its radius:

$$v = \omega R$$

After substituting v/R for ω, the energy conservation equation becomes

$$\Delta U + \Delta K = [-m_1 g h + m_2 g h] + \left[\tfrac{1}{2}(m_1 + m_2)v^2 + \tfrac{1}{2}I\left(\tfrac{v}{R}\right)^2\right] = 0$$

or

$$\left[\tfrac{1}{2}(m_1 + m_2) + \tfrac{1}{2}\tfrac{I}{R^2}\right]v^2 = (m_1 - m_2)gh$$

Solving this equation for v yields

$$v = \sqrt{\frac{2(m_1 - m_2)gh}{m_1 + m_2 + I/R^2}}$$

Discussion   This answer is rich in information, in the sense that we can ask many "What if?" questions. Not only do these questions provide checks as to whether the answer is reasonable, they also enable us to perform thought experiments, which could then be checked by constructing an Atwood's machine and comparing the results.

For instance: What if m_1 is only slightly greater than m_2? Then the final speed v is small—as m_2 approaches m_1, v approaches 0. This makes intuitive sense: a small imbalance in weights produces a small acceleration. You should practice this kind of reasoning by making other such checks.

It is also enlightening to look at terms in an algebraic solution and connect them with physical interpretations. The quantity $(m_1 - m_2)g$ is the imbalance in the gravitational

continued on next page

Example 8.2 continued

forces pulling on the two sides. The denominator $(m_1 + m_2 + I/R^2)$ is a measure of the total inertia of the system—the sum of the two masses plus an inertial contribution due to the pulley. The pulley's contribution is *not* simply equal to its mass. If, for example, the pulley is a uniform disk with $I = \frac{1}{2}MR^2$, the term I/R^2 would be equal to *half* the mass of the pulley.

The same principles used to analyze Atwood's machine have many applications in the real world. One such application is in elevators, where one of the hanging masses is the elevator and the other is the counterweight. Of course, the elevator and counterweight are not allowed to hang freely from a pulley—we must also consider the energy supplied by the motor.

Practice Problem 8.2   Modified Atwood's Machine

Figure 8.7 shows a modified form of Atwood's machine where one of the blocks slides on a table instead of hanging from the pulley. The blocks are released from rest. Find the speed of the blocks after they have moved a distance h in terms of m_1, m_2, I, R, and h. Ignore friction.

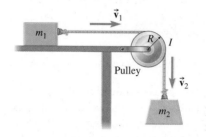

Figure 8.7
Modified Atwood's machine.

8.2   TORQUE

Suppose you place a bicycle upside down to repair it. First, you give one of the wheels a spin. If everything is working as it should, the wheel spins for quite a while; its angular acceleration is small. If the wheel doesn't spin for very long, then its angular velocity changes rapidly and the angular acceleration is large in magnitude; there must be excessive friction somewhere. Perhaps the brakes are rubbing on the rim or the bearings need to be repacked.

If we could eliminate *all* the frictional forces acting on the wheel, including air resistance, then we would expect the wheel to keep spinning without diminishing angular speed. In that case, its angular acceleration would be zero. The situation is reminiscent of Newton's first law: a body with no external interactions, or at least no net force acting on it, moves with constant velocity. We can state a "Newton's first law for rotation": a rotating body with no external interactions, and whose rotational inertia doesn't change, keeps rotating at constant angular velocity.

Of course, the hypothetical frictionless bicycle wheel does have external interactions. The Earth's gravitational field exerts a downward force and the axle exerts an upward force to keep the wheel from falling. Then is it true that, as long as there is no net external force, the angular acceleration is zero? No; it is easy to give the wheel an angular acceleration while keeping the net force zero. Imagine bringing the wheel to rest by pressing two hands against the tire on opposite sides. On one side, the motion of the rim of the tire is downward and the kinetic frictional force is upward (Fig. 8.8). On the other side, the tire moves upward and the frictional force is downward. In a similar way, we could apply equal and opposite forces to the opposite sides of a wheel at rest to make it start spinning. In either case, we exert equal magnitude forces, so that the net force is zero, and still give the wheel an angular acceleration.

Figure 8.8 A spinning bicycle wheel slowed to a stop by friction. Each hand exerts a normal force and a frictional force on the tire. The two normal forces add to zero and the two frictional forces add to zero.

The *radial* direction is directly toward or away from the axis of rotation. The *perpendicular* or *tangential* direction is perpendicular to both the radial direction and the axis of rotation; it is tangent to the circular path followed by a point as the object rotates.

Torque   A quantity related to force, called **torque**, plays the role in rotation that force itself plays in translation. A torque is not separate from a force; it is impossible to exert a torque without exerting a force. Torque is a measure of how effective a given force is at twisting or turning something. For something rotating about a fixed axis such as the bicycle wheel, a torque can *change* the rotational motion either by making it rotate faster or by slowing it down.

When stopping the bicycle wheel with two equal and opposite forces, as in Fig. 8.8, the net applied force is zero and, thus, the wheel is in translational equilibrium; but the net torque is not zero, so it is not in rotational equilibrium. Both forces tend to give the wheel the same sign of angular acceleration; they are both making the wheel slow down. The two torques are in fact equal, with the same sign.

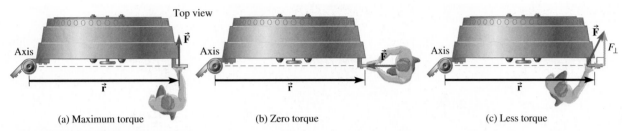

(a) Maximum torque   (b) Zero torque   (c) Less torque

Figure 8.9 Torque on a bank vault door depends on the direction of the applied force. (a) Pushing perpendicularly gives the maximum torque. (b) Pushing radially inward with the same magnitude force gives zero torque. (c) The torque is proportional to the perpendicular component of the force ($F_\perp$).

Relationship Between Force and Torque   What determines the torque produced by a particular force? Imagine trying to push open a massive bank vault door. Certainly you would push as hard as you can; the torque is proportional to the magnitude of the force. It also matters where and in what direction the force is applied. For maximum effectiveness, you would push perpendicularly to the door (Fig. 8.9a). If you pushed radially, straight in toward the axis of rotation that passes through the hinges, the door wouldn't rotate, no matter how hard you push (Fig. 8.9b). A force acting in any other direction could be decomposed into radial and perpendicular components, with the radial component contributing nothing to the torque (Fig. 8.9c). Only the perpendicular component of the force ($F_\perp$) produces a torque.

The symbol $\perp$ stands for *perpendicular;* $\parallel$ stands for *parallel.*

Furthermore, *where* you apply the force is critical (Fig. 8.10). Instinctively, you would push at the outer edge, as far from the rotation axis as possible. If you pushed close to the axis, it would be difficult to open the door. Torque is proportional to the distance between the rotation axis and the **point of application** of the force (the point at which the force is applied).

To satisfy the requirements of the previous paragraphs, we define the magnitude of the torque as the product of the distance between the rotation axis and the point of application of the force (r) with the perpendicular component of the force ($F_\perp$):

Definition of torque:

$$\tau = \pm r F_\perp \tag{8-3}$$

where r is the shortest distance between the rotation axis and the point of application of the force and $F_\perp$ is the perpendicular component of the force.

The symbol for torque is τ, the Greek letter tau. The SI unit of torque is the N·m. The SI unit of *energy*, the joule, is equivalent to N·m, but we do not write torque in joules. Even though both energy and torque can be written using the same SI base units, the two quantities have different meanings; torque is not a form of energy. To help maintain the distinction, the joule is used for energy but *not* for torque.

✓ **CHECKPOINT 8.2**

You are trying to loosen a nut, without success. Why might it help to switch to a wrench with a longer handle?

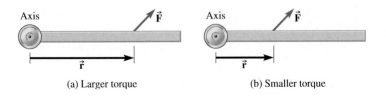

(a) Larger torque   (b) Smaller torque

Figure 8.10 Torques; the same force at different distances from the axis.

Figure 8.11 (a) When the cyclist climbs a hill, the top half of the chain exerts a large force $\vec{\mathbf{F}}$ on the sprocket attached to the rear wheel. As viewed here, the torque about the axis of rotation (the axle) due to this force is clockwise. By convention, we call this a negative torque. (b) When the brakes are applied, the brake pads are pressed onto the rim, giving rise to frictional forces on the rim. As viewed here, the frictional force $\vec{\mathbf{f}}$ causes a counterclockwise (positive) torque on the wheel about the axle.

In a more general treatment of torque, torque is a vector quantity defined as the cross product $\vec{\boldsymbol{\tau}} = \vec{\mathbf{r}} \times \vec{\mathbf{F}}$. See Appendix A.8 for the definition of the cross product. For an object rotating about a fixed axis, Eq. (8-3) gives the component of $\vec{\boldsymbol{\tau}}$ along the axis of rotation.

Sign Convention for Torque   The sign of the torque indicates the direction of the angular acceleration that torque would cause *by itself.* Recall from Section 5.1 that by convention a positive angular velocity ω means counterclockwise (CCW) rotation and a negative angular velocity ω means clockwise (CW) rotation. A positive angular acceleration α either increases the rate of CCW rotation (increases the magnitude of a positive ω) or decreases the rate of CW rotation (decreases the magnitude of a negative ω).

We use the same sign convention for torque. A force whose perpendicular component tends to cause rotation in the CCW direction gives rise to a positive torque; if it is the only torque acting, it would cause a positive angular acceleration α (Fig. 8.11). A force whose perpendicular component tends to cause rotation in the CW direction gives rise to a negative torque. The symbol $\pm$ in Eq. (8-3) reminds us to assign the appropriate algebraic sign each time we calculate a torque.

The sign of the torque is *not* determined by the sign of the angular velocity (in other words, whether the wheel is spinning CCW or CW); rather, it is determined by the sign of the angular *acceleration* the torque would cause if acting alone. To determine the sign of a torque, imagine which way the torque would make the object begin to spin if it is initially not rotating.

Example 8.3

A Spinning Bicycle Wheel

To stop a spinning bicycle wheel, suppose you push radially inward on opposite sides of the wheel, as shown in Fig. 8.8, with equal forces of magnitude 10.0 N. The radius of the wheel is 32 cm and the coefficient of kinetic friction between the tire and your hand is 0.75. The wheel is spinning in the CW sense. What is the net torque on the wheel?

continued on next page

Example 8.3 continued

Strategy   The 10.0-N forces are directed radially toward the rotation axis, so they produce no torques themselves; only perpendicular components of forces give rise to torques. The forces of kinetic friction between the hands and the tire are tangent to the tire, so they do produce torques. The normal force applied to the tire is 10.0 N on each side; using the coefficient of friction, we can find the frictional forces.

Solution   The frictional force exerted by each hand on the tire has magnitude

$$f = \mu_k N = 0.75 \times 10.0 \text{ N} = 7.5 \text{ N}$$

The frictional force is tangent to the wheel, so $f_\perp = f$. Then the magnitude of each torque is

$$|\tau| = rf_\perp = 0.32 \text{ m} \times 7.5 \text{ N} = 2.4 \text{ N·m}$$

The two torques have the same sign, since they are both tending to slow down the rotation of the wheel. Is the torque positive or negative? The angular velocity of the wheel is negative since it rotates CW. The angular acceleration has

the opposite sign because the angular speed is decreasing. Since $\alpha > 0$, the net torque is also positive. Therefore,

$$\sum \tau = +4.8 \text{ N·m}$$

Discussion   The trickiest part of calculating torques is determining the sign. To check, look at the frictional forces in Fig. 8.8. Imagine which way the forces would make the wheel begin to rotate if the wheel were not originally rotating. The frictional forces point in a direction that would tend to cause a CCW rotation, so the torques are positive.

Practice Problem 8.3   Disc Brakes

In the disc brakes that slow down a car, a pair of brake pads squeeze a spinning rotor; friction between the pads and the rotor provides the torque that slows down the car. If the normal force that each pad exerts on a rotor is 85 N and the coefficient of friction is 0.62, what is the frictional force on the rotor due to each of the pads? If this force acts 8.0 cm from the rotation axis, what is the magnitude of the torque on the rotor due to the pair of brake pads?

Lever Arms

There is another, completely equivalent, way to calculate torques that is often more convenient than finding the perpendicular component of the force. Figure 8.12 shows a force $\vec{F}$ acting at a distance r from an axis. The distance r is the length of a line perpendicular to the axis that runs from the axis to the force's point of application. The force makes an angle θ with that line. The torque is then

$$\tau = \pm rF_\perp = \pm r(F \sin \theta)$$

The factor $\sin \theta$ could be grouped with r instead of with F. Then $\tau = \pm(r \sin \theta)F$, or

$$\tau = \pm r_\perp F \qquad (8\text{-}4)$$

The distance $r_\perp$ is called the **lever arm** (or **moment arm**). The magnitude of the torque is, therefore, the magnitude of the force times the lever arm.

Finding Torques Using the Lever Arm

1. Draw a line parallel to the force through the force's point of application; this line (dashed in Fig. 8.12) is called the force's **line of action**.

2. Draw a line from the rotation axis to the line of action. This line must be perpendicular to both the axis and the line of action. The distance from the axis to the line of action along this perpendicular line is the lever arm ($r_\perp$). If the line of action of the force goes through the rotation axis, the lever arm and the torque are both zero (see Fig. 8.9b).

3. The magnitude of the torque is the magnitude of the force times the lever arm:

$$\tau = \pm r_\perp F$$

4. Determine the algebraic sign of the torque as before.

Figure 8.12 Finding the magnitude of a torque using the lever arm.

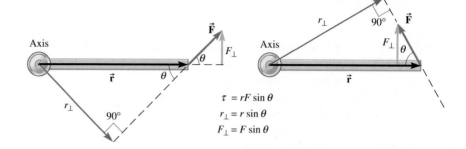

$$\tau = rF \sin \theta$$
$$r_\perp = r \sin \theta$$
$$F_\perp = F \sin \theta$$

Example 8.4

Screen Door Closer

An automatic screen door closer attaches to a door 47 cm away from the hinges and pulls on the door with a force of 25 N, making an angle of 15° with the door (Fig. 8.13). Find the magnitude of the torque exerted on the door due to this force about the rotation axis through the hinges using (a) the perpendicular component of the force and (b) the lever arm. (c) What is the sign of this torque as viewed from above?

Strategy For method (a), we must find the component of the 25-N force perpendicular to the radial direction. Then this component is multiplied by the length of the radial line. For method (b), we draw in the line of action of the force. Then the lever arm is the perpendicular distance from the line of action to the rotation axis. The torque is the magnitude of the force times the lever arm. We must be careful not to combine the two methods: the torque is *not* equal to the perpendicular force component times the lever arm. For (c), we determine whether this torque would tend to make the door rotate CCW or CW.

Solution (a) As shown in Fig. 8.14a, the radial component of the force ($F_\parallel$) passes through the rotation axis. The angle labeled 15° would actually be a bit larger than 15°, but since the thickness of the door is much less than 47 cm, we approximate it as 15°. The perpendicular component is

$$F_\perp = F \sin 15° \qquad {}^,$$

The magnitude of the torque is

$$|\tau| = rF_\perp = 0.47 \text{ m} \times 25 \text{ N} \times \sin 15° = 3.0 \text{ N·m}$$

(b) Figure 8.14b shows the line of action of the force, drawn parallel to the force and passing through the point of application. The lever arm is the perpendicular distance between the rotation axis and the line of action. The distance r is approximately 47 cm (again neglecting the thickness of the door). Then the lever arm is

$$r_\perp = r \sin 15°$$

and the magnitude of the torque is

$$|\tau| = r_\perp F = 0.47 \text{ m} \times \sin 15° \times 25 \text{ N} = 3.0 \text{ N·m}$$

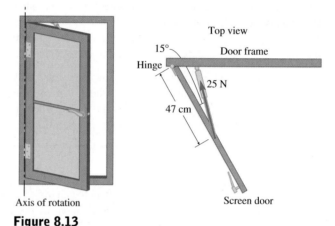

Figure 8.13

Screen door with automatic closing mechanism.

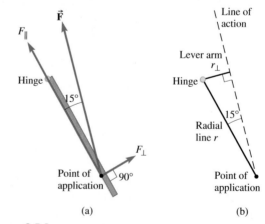

Figure 8.14

(a) Finding the perpendicular component of the force.
(b) Finding the lever arm.

(c) Using the top view of Fig. 8.13, the torque tends to close the door by making it rotate counterclockwise (assuming the door is initially at rest and no other torques act). The torque is therefore positive as viewed from above.

continued on next page

Example 8.4 continued

Discussion The most common mistake to make in either solution method would be to use cosine instead of sine (or, equivalently, to use the complementary angle 75° instead of 15°). A check is a good idea. If the automatic closer were more nearly parallel to the door, the angle would be less than 15°. The torque would be smaller because the force is more nearly pulling straight in toward the axis. Since the sine function gets smaller for angles closer to zero, the expression checks out correctly.

It might seem silly for a door closer to pull at such an angle that the perpendicular component is relatively small. The reason it's done that way is so the door closer does not get in the way. A closer that pulled in a perpendicular direction would stick straight out from the door. As discussed in Section 8.5, the situation is much the same in our bodies. In order to not inhibit the motion of our limbs, our tendons and muscles are nearly parallel to the bones. As a result, the forces they exert must be much larger than we might expect.

Practice Problem 8.4   Exercise Is Good for You

A person is lying on an exercise mat and lifts one leg at an angle of 30.0° from the horizontal with an 89-N (20-lb) weight attached to the ankle (Fig. 8.15). The distance between the ankle weight and the hip joint (which is the rotation axis for the leg) is 84 cm. What is the torque due to the ankle weight on the leg?

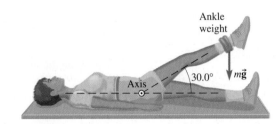

Figure 8.15
Exercise leg lifts.

Center of Gravity

We have seen that the torque produced by a force depends on the point of application of the force. What about gravity? The gravitational force on a body is not exerted at a single point, but is distributed throughout the volume of the body. When we talk of "the" force of gravity on something, we really mean the total force of gravity acting on each particle making up the system.

Fortunately, when we need to find the total torque due to the forces of gravity acting on an object, the total force of gravity can be considered to act at a single point. This point is called the **center of gravity**. The torque found this way is the same as finding all the torques due to the forces of gravity acting at every point in the body and adding them together. As you can verify in Problem 95, if the gravitational field is uniform in magnitude and direction, then the center of gravity of an object is located at the object's center of mass.

When calculating the torque due to gravity, consider the entire gravitational force to act at the center of gravity.

8.3   CALCULATING WORK DONE FROM THE TORQUE

Torques can do work, as anyone who has started a lawnmower with a pull cord can verify. Actually, it is the force that does the work, but in rotational problems it is often simpler to calculate the work done from the torque. Just as the work done by a constant force is the product of force and the parallel component of displacement, work done by a constant torque can also be calculated as the torque times the *angular* displacement.

Imagine a torque acting on a wheel that spins through an angular displacement $\Delta\theta$ while the torque is applied. The work done by the force that gives rise to the torque is the product of the perpendicular component of the force ($F_\perp$) with the arc length s through which the point of application of the force moves (Fig. 8.16). We use the perpendicular force component because that is the component parallel to the *displacement*, which is instantaneously tangent to the arc of the circle. Thus,

$$W = F_\perp s \qquad (8\text{-}5)$$

CONNECTION:

We're not introducing a different kind of work, just a different way to calculate work.

Figure 8.16 The work done by a torque is the product of the perpendicular force component $F_\perp$ and the arc length s.

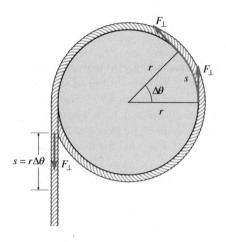

To write the work in terms of torque, note that $\tau = rF_\perp$ and $s = r\Delta\theta$; then

$$W = F_\perp s = \frac{\tau}{r} \times r\Delta\theta = \tau\Delta\theta$$

$$W = \tau\Delta\theta \quad (\Delta\theta \text{ in radians}) \qquad (8\text{-}6)$$

Work is indeed the product of torque and the angular displacement. If τ and $\Delta\theta$ have the same sign, the work done is positive; if they have opposite signs, the work done is negative. The *power* due to a constant torque—the rate at which work is done—is

$$P = \tau\omega \qquad (8\text{-}7)$$

Example 8.5

Work Done on a Potter's Wheel

A potter's wheel is a heavy stone disk on which the pottery is shaped. Potter's wheels were once driven by the potter pushing on a foot treadle; today most potter's wheels are driven by electric motors. (a) If the potter's wheel is a uniform disk of mass 40.0 kg and diameter 0.50 m, how much work must be done by the motor to bring the wheel from rest to 80.0 rpm? (b) If the motor delivers a constant torque of 8.2 N·m during this time, through how many revolutions does the wheel turn in coming up to speed?

Strategy   Work is an energy transfer. In this case, the motor is increasing the rotational kinetic energy of the potter's wheel. Thus, the work done by the motor is equal to the change in rotational kinetic energy of the wheel, ignoring frictional losses. In the expression for rotational kinetic energy, we must express ω in rad/s; we cannot substitute 80.0 rpm for ω. Once we know the work done, we use the torque to find the angular displacement.

Solution   (a) The change in rotational kinetic energy of the wheel is

$$\Delta K = \tfrac{1}{2}I(\omega_f^2 - \omega_i^2) = \tfrac{1}{2}I\omega_f^2$$

Initially the wheel is at rest, so the initial angular velocity ω_i is zero. From Table 8.1, the rotational inertia of a uniform disk is

$$I = \tfrac{1}{2}MR^2$$

Substituting this for I,

$$\Delta K = \tfrac{1}{4}MR^2\omega_f^2$$

Before substituting numerical values, we convert 80.0 rpm to rad/s:

$$\omega_f = 80.0 \,\frac{\text{rev}}{\text{min}} \times 2\pi\,\frac{\text{rad}}{\text{rev}} \times \frac{1}{60}\,\frac{\text{min}}{\text{s}} = 8.38 \text{ rad/s}$$

Substituting the known values for mass and radius,

$$\Delta K = \tfrac{1}{4} \times 40.0 \text{ kg} \times \left(\tfrac{0.50}{2} \text{ m}\right)^2 \times (8.38 \text{ rad/s})^2 = 43.9 \text{ J}$$

Therefore, the work done by the motor, rounded to two significant figures, is 44 J.

(b) The work done by a constant torque is

$$W = \tau\Delta\theta$$

Solving for the angular displacement $\Delta\theta$ gives

$$\Delta\theta = \frac{W}{\tau} = \frac{43.9 \text{ J}}{8.2 \text{ N·m}} = 5.35 \text{ rad}$$

Since 2π rad = 1 revolution,

$$\Delta\theta = 5.35 \text{ rad} \times \frac{1 \text{ rev}}{2\pi \text{ rad}} = 0.85 \text{ rev}$$

continued on next page

Example 8.5 continued

Discussion As always, work is an energy transfer. In this problem, the work done by the motor is the means by which the potter's wheel acquires its rotational kinetic energy. But work done by a torque does not *always* appear as a change in rotational kinetic energy. For instance, when you wind up a mechanical clock or a windup toy, the work done by the

torque you apply is stored as elastic potential energy in some sort of spring.

Practice Problem 8.5   Work Done on an Air Conditioner

A belt wraps around a pulley of radius 7.3 cm that drives the compressor of an automobile air conditioner. The tension in the belt on one side of the pulley is 45 N and on the other side of the pulley it is 27 N (Fig. 8.17). How much work is done by the belt on the compressor during one revolution of the pulley?

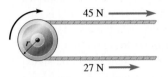

45 N

27 N

Figure 8.17
Air conditioner belt and pulley.

8.4   ROTATIONAL EQUILIBRIUM

In Chapter 4, we said that an object is in translational equilibrium when the net force acting on it is zero. It is quite possible for the net force acting to be zero, while the net torque is nonzero; the object would then have a nonzero angular acceleration. When designing a bridge or a new house, it would be unacceptable for any of the parts to have nonzero angular acceleration! Zero net force is sufficient to ensure *translational* equilibrium; if an object is also in *rotational* equilibrium, then the net torque acting on it must also be zero.

Conditions for equilibrium (both translational and rotational):

$$\sum \vec{F} = 0 \quad \text{and} \quad \sum \tau = 0 \qquad (8\text{-}8)$$

Choosing an Axis of Rotation in Equilibrium Problems   Before tackling equilibrium problems, we must resolve a conundrum: if something is not rotating, then where is the axis of rotation? How can we calculate torques without knowing where the axis of rotation is? In some cases, perhaps involving axles or hinges, there may be a clear axis about which the object would rotate if the balance of forces and torques is disturbed. In many cases, though, it is not clear what the rotation axis would be, and in general it depends on exactly how the equilibrium is upset. Fortunately, the axis can be chosen *arbitrarily* when calculating torques *in equilibrium problems.*

In equilibrium, the net torque about *any* rotation axis must be zero. Does that mean that we have to write down an infinite number of torque equations, one for each possible axis of rotation? Fortunately, no. Although the proof is complicated, it can be shown that if the net force acting on an object is zero and the net torque about one rotation axis is zero, then the net torque about every other axis parallel to that axis must also be zero. Therefore, one torque equation is all we need.

Since the torque can be calculated about any desired axis, a judicious choice can greatly simplify the solution of the problem. The best place to choose the axis is usually at the point of application of an unknown force so that the unknown force does not appear in the torque equation.

√ CHECKPOINT 8.4

Is it possible for the net torque on an object to be zero and the net force nonzero?
Is it possible for the net force to be zero and the net torque nonzero?

> ## Problem-Solving Steps in Equilibrium Problems
>
> - Identify an object or system in equilibrium. Draw a diagram showing all the forces acting on that object, each drawn at its point of application. Use the center of gravity as the point of application of any gravitational forces.
> - To apply the force condition $\Sigma \vec{F} = 0$, choose a convenient coordinate system and resolve each force into its x- and y-components.
> - To apply the torque condition $\Sigma \tau = 0$, choose a convenient rotation axis—generally one that passes through the point of application of an unknown force. Then find the torque due to each force. Use whichever method is easier: either the lever arm times the magnitude of the force or the distance times the perpendicular component of the force. Determine the direction of each torque; then either set the sum of all the torques (with their algebraic signs) equal to zero or set the magnitude of the CW torques equal to the magnitude of the CCW torques.
> - Not all problems require all three equations (two force component equations and one torque equation). Sometimes it is easier to use more than one torque equation, with a different axis. Before diving in and writing down all the equations, think about which approach is the easiest and most direct.

Example 8.6

Carrying a 6 × 6 Beam

Two carpenters are carrying a uniform 6×6 beam. The beam is 8.00 ft (2.44 m) long and weighs 425 N (95.5 lb). One of the carpenters, being a bit stronger than the other, agrees to carry the beam 1.00 m in from the end; the other carries the beam at its opposite end. What is the upward force exerted on the beam by each carpenter?

Strategy   The conditions for equilibrium are that the net external force equal zero and the net external torque equal zero:

$$\Sigma \vec{F} = 0 \quad \text{and} \quad \Sigma \tau = 0$$

Should we start with forces or with torques? In this problem, it is easiest to start with torques. If we choose the axis of rotation where one of the unknown forces acts, then that force has a lever arm of zero and its torque is zero. The torque equation can be solved for the other unknown force. Then with only one force still unknown, we set the sum of the y-components of the forces equal to zero.

Solution   The first step is to draw a force diagram (Fig. 8.18). Each force is drawn at the point where it acts. Known distances are labeled.

We choose a rotation axis perpendicular to the xy-plane and passing through the point of application of $\vec{F}_2$. The simplest way to find the torques for this example is to multiply each force by its lever arm. The lever arm for $\vec{F}_1$ is

$$2.44 \text{ m} - 1.00 \text{ m} = 1.44 \text{ m}$$

and the magnitude of the torque due to this force is

$$|\tau| = Fr_{\perp} = F_1 \times 1.44 \text{ m}$$

Since the beam is uniform, its center of gravity is at its midpoint. We imagine the entire gravitational force to act at this point. Then the lever arm for the gravitational force is

$$\tfrac{1}{2} \times 2.44 \text{ m} = 1.22 \text{ m}$$

and the torque due to gravity has magnitude

$$|\tau| = Fr_{\perp} = 425 \text{ N} \times 1.22 \text{ m} = 518.5 \text{ N·m}$$

The torque due to $\vec{F}_1$ is negative since, if it were the only torque, it would make the beam start to rotate clockwise about our chosen axis of rotation. The torque due to gravity is positive since, if it were the only torque, it would make the beam start to rotate counterclockwise. Therefore,

$$\Sigma \tau = -F_1 \times 1.44 \text{ m} + 518.5 \text{ N·m} = 0$$

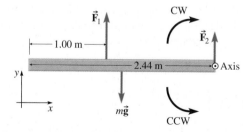

Figure 8.18 Diagram of the beam with rotation axis, forces, and distances shown.

continued on next page

Example 8.6 continued

Solving for F_1,

$$F_1 = \frac{518.5 \text{ N·m}}{1.44 \text{ m}} = 360 \text{ N}$$

Since another condition for equilibrium is that the net force be zero,

$$\sum F_y = F_1 + F_2 - mg = 0$$

Solving for F_2,

$$F_2 = 425 \text{ N} - 360 \text{ N} = 65 \text{ N}$$

Discussion A good way to check this result is to make sure that the net torque about a *different axis* is zero—for an object in equilibrium, the net torque about any axis must be zero. Suppose we choose an axis through the point of application of $\vec{F}_1$. Then the lever arm for $m\vec{g}$ is $1.22 \text{ m} - 1.00 \text{ m} = 0.22 \text{ m}$ and the lever arm for $\vec{F}_2$ is $2.44 \text{ m} - 1.00 \text{ m} = 1.44 \text{ m}$. Setting the net torque equal to zero:

$$\sum \tau = -425 \text{ N} \times 0.22 \text{ m} + F_2 \times 1.44 \text{ m} = 0$$

Solving for F_2 gives

$$F_2 = \frac{425 \text{ N} \times 0.22 \text{ m}}{1.44 \text{ m}} = 65 \text{ N}$$

which agrees with the value calculated before. We could have used this second torque equation to find F_2 instead of setting $\sum F_y$ equal to zero.

Practice Problem 8.6 A Diving Board

A uniform diving board of length 5.0 m is supported at two points; one support is located 3.4 m from the end of the board and the second is at 4.6 m from the end (Fig. 8.19). The supports exert vertical forces on the diving board. A diver stands at the end of the board over the water. Determine the directions of the support forces. (tutorial: plank) [*Hint:* In this problem, consider torques about different rotation axes.]

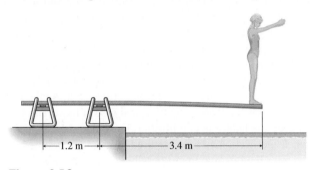

Figure 8.19
Diving board.

The Cantilever A diving board is an example of a cantilever—a beam or pole that extends beyond its support. The forces exerted by the supports on a diving board are considerably larger than if the same board were supported at both ends (see Problem 32). The advantage is that the far end of the board is left free to vibrate; as it does, the support forces adjust themselves to keep the board from tipping over. The architect Frank Lloyd Wright was fond of using cantilever construction to open up the sides and corners of a building, allowing corner windows that give buildings a lighter and more spacious feel (Fig. 8.20).

Application of Rotational Equilibrium: cantilever building construction

Figure 8.20 The cantilevered master bedroom in the north wing of Wingspread by Frank Lloyd Wright juts well out over its brick foundation. The cypress trellis extending even farther beyond the bedroom balcony filters the natural light and serves to emphasize the free-floating nature of the structure with views of the landscape below.

Example 8.7

The Slipping Ladder

A 15.0-kg uniform ladder leans against a wall in the atrium of a large hotel (Fig. 8.21a). The ladder is 8.00 m long; it makes an angle $\theta = 60.0°$ with the floor. The coefficient of static friction between the floor and the ladder is $\mu_s = 0.45$. How far along the ladder can a 60.0-kg person climb before the ladder starts to slip? Assume that the wall is frictionless. (interactive: ladder and tutorial: ladder)

Strategy Consider the ladder and the climber as a single system. Until the ladder starts to slip, this system is in equilibrium. Therefore, the net external force and the net external torque acting on the system are both equal to zero. Normal forces act on the ladder due to the wall ($\vec{N}_w$) and the floor ($\vec{N}_f$). A frictional force acts on the base of the ladder due to the floor ($\vec{f}$), but no frictional force acts on the top of the ladder since the wall is frictionless. Gravitational forces act on the ladder and on the person climbing it. As the person ascends the ladder, the frictional force $\vec{f}$ has to increase to keep the ladder in equilibrium. The ladder begins to slip when the frictional force required to maintain equilibrium is larger than its maximum possible value $\mu_s N_f$. The ladder is about to slip when $f = \mu_s N_f$.

$$\sum F_x = 0, \quad \sum F_y = 0, \quad \text{and} \sum \tau = 0$$

Solution The first step is to make a careful drawing of the ladder and label all distances and forces (Fig. 8.21b). Instead of cluttering the diagram with numerical values, we use

L (= 8.00 m) for the length of the ladder, d for the unknown distance from the bottom of the ladder to the point where the person stands, and M (= 60.0 kg) and m (= 15.0 kg) for the masses of the person and ladder, respectively. The weight of the ladder acts at the ladder's center of gravity, which is the ladder's midpoint since it is uniform.

Now we apply the conditions for equilibrium. Starting with $\sum F_x = 0$, we find

$$N_w - f = 0$$

where, if the climber is at the highest point possible, the frictional force must have its maximum possible magnitude:

$$f = \mu_s N_f$$

Combining these two equations, we obtain a relationship between the magnitudes of the two normal forces:

$$N_w = \mu_s N_f$$

Next we use the condition $\sum F_y = 0$, which gives

$$N_f - Mg - mg = 0$$

The only unknown quantity in this equation is N_f, so we can solve for it:

$$N_f = Mg + mg = (M + m)g$$

Now we can find the other normal force, N_w:

$$N_w = \mu_s N_f = \mu_s (M + m)g$$

(a)

(b)

Figure 8.21

(a) A ladder and (b) forces acting on the ladder.

continued on next page

Example 8.7 continued

At this point, we have expressions for the magnitudes of all the forces. We do not know the distance d, which is the goal of the problem. To find d we must set the net torque equal to zero.

First we choose a rotation axis. The most convenient choice is an axis perpendicular to the plane of Fig. 8.21 and passing through the bottom of the ladder. Since two of the five forces ($\vec{N}_f$ and $\vec{f}$) act at the bottom of the ladder, these two forces have zero lever arms and, thus, produce zero torque. Another reason why this is a convenient choice of axis is that the distance d is measured from the bottom of the ladder.

In this situation, with the forces either vertical or horizontal, it is probably easiest to use lever arms to find the torques. In three diagrams (Fig. 8.22), we first draw the line of action for each force; then the lever arm is the perpendicular distance between the axis and the line of action.

Using the usual convention that CCW torques are positive, the torque due to $\vec{N}_w$ is negative and the torques due to gravity are positive. The magnitude of each torque is the magnitude of the force times its lever arm:

$$\tau = Fr_\perp$$

Setting the net torque equal to zero yields

$$-N_w L \sin\theta + mg\left(\tfrac{1}{2}L\cos\theta\right) + Mgd\cos\theta = 0$$

Now we solve for d algebraically.

$$\frac{-N_w L \sin\theta}{Mg\cos\theta} + \frac{\tfrac{1}{2}mgL\cos\theta}{Mg\cos\theta} + d = 0$$

$$d = \frac{N_w L \tan\theta}{Mg} - \frac{mL}{2M}$$

Substituting for N_w, we have

$$d = L\left(\frac{\mu_s(M+m)\tan\theta}{M} - \frac{m}{2M}\right)$$

$$= 8.00\ \text{m} \times \left(\frac{0.45 \times 75.0\ \text{kg} \times \tan 60.0°}{60.0\ \text{kg}} - \frac{15.0\ \text{kg}}{2 \times 60.0\ \text{kg}}\right)$$

$$= 6.8\ \text{m}$$

The person can climb 6.8 m up the ladder without having it slip. This is the distance *along the ladder*, not the height above the ground, which is

$$h = 6.8\ \text{m} \times \sin 60.0° = 5.9\ \text{m}$$

Discussion If the person goes any higher, then his weight produces a larger CCW torque about our chosen rotation axis. To stay in equilibrium, the total CW torque would have to get larger. The only force providing a CW torque is the normal force due to the wall, which pushes to the right. However, if this force were to get larger, the frictional force would have to get larger to keep the net horizontal force equal to zero. Since friction already has its maximum magnitude, there is no way for the ladder to be in equilibrium if the person climbs any higher.

Practice Problem 8.7 Another Ladder Leaning on a Wall

A uniform ladder of mass 10.0 kg and length 3.2 m leans against a frictionless wall with its base located 1.5 m from the wall. If the ladder is not to slip, what must be the minimum coefficient of static friction between the bottom of the ladder and the ground? Assume the wall is frictionless.

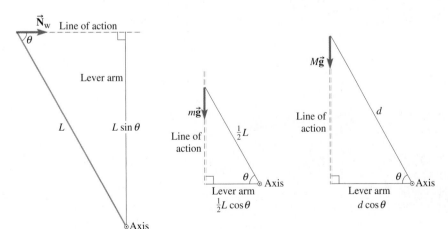

Figure 8.22

Finding the lever arm for each force.

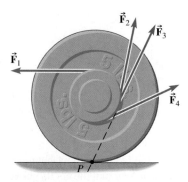

Figure 8.23 Forces $\vec{F}_1$ and $\vec{F}_2$ make the dumbbell roll to the left; $\vec{F}_4$ makes it roll to the right; $\vec{F}_3$ does not make it roll.

PHYSICS AT HOME

Take a dumbbell and wrap some string around the center of its axle. (An alternative: slide two spools of thread onto a pencil near its center with a small gap between the spools. Wrap some thread around the pencil between the two spools.) Place the dumbbell on a table (or on the floor). Unwind a short length of string and try pulling perpendicularly to the axle at different angles to the horizontal (Fig. 8.23). Depending on the direction of your pull, the dumbbell can roll in either direction. Try to find the angle at which the rolling changes direction; at this angle the dumbbell does not roll at all. (If using the pencil and spools of thread, pull gently and try to find the angle at which the whole thing *slides* along the table without any rotation.)

What is special about this angle? Since the dumbbell is in equilibrium when pulling at this angle, we can analyze the torques using any rotation axis we choose. A convenient choice is the axis that passes through point P, the point of contact with the table. Then the contact force between the table and the dumbbell acts at the rotation axis and its torque is zero. The torque due to gravity is also zero, since the line of action passes through point P. The dumbbell can only be in equilibrium if the torque due to the remaining force (the tension in the string) is zero. This torque is zero if the lever arm is zero, which means the line of action passes through point P.

Example 8.8

The Sign and the Breaking Cord

A uniform beam of weight 196 N and of length 1.00 m is attached to a hinge on the outside wall of a restaurant. A cord is attached at the center of the beam and is attached to the wall, making an angle of 30.0° with the beam (Fig. 8.24a). The cord keeps the beam perpendicular to the wall. If the breaking tension of the cord is 620 N, how large can the mass of the sign be without breaking the cord?

Strategy   The beam is in equilibrium; both the net force and the net torque acting on it must be zero. To find the maximum weight of the sign, we let the tension in the cord have its maximum value of 620 N. We do not know the force exerted by the hinge on the beam, so we choose an axis of rotation through

the hinge. Then the force exerted by the hinge on the beam has a zero lever arm and does not enter the torque equation.

Before doing anything else, we draw a diagram showing each force acting on the beam and the chosen rotation axis. The FBD in previous chapters often placed all the force vectors starting from a single point. Now we draw each force vector starting at its point of application so that we can find the torque—either by finding the lever arm or by finding the perpendicular force component and the distance from the axis to the point of application.

Solution   Figure 8.24b shows the forces acting on the beam; three of these contribute to the torque. The gravitational

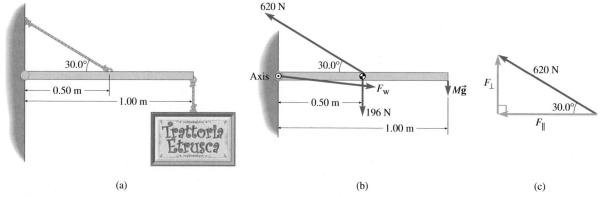

Figure 8.24

(a) A sign outside a restaurant. (b) Forces acting on the beam. (c) Finding the components of the tension in the cord.

continued on next page

Example 8.8 continued

force on the beam can be taken to act at the midpoint of the beam since it is uniform. The force due to the cord has a perpendicular component (Fig. 8.24c) of

$$F_\perp = 620 \text{ N} \times \sin 30.0° = 310 \text{ N}$$

The two gravitational forces tend to rotate the beam CW, while the tension in the cord tends to rotate it CCW. The net torque must be equal to zero:

$$-0.50 \text{ m} \times 196 \text{ N} - 1.00 \text{ m} \times Mg + 0.50 \text{ m} \times 310 \text{ N} = 0$$

or

$$1.00 \text{ m} \times Mg = 0.50 \text{ m} \times (310 \text{ N} - 196 \text{ N})$$

Now we solve for the unknown mass M:

$$M = \frac{0.50 \text{ m} \times (310 \text{ N} - 196 \text{ N})}{1.00 \text{ m} \times 9.80 \text{ N/kg}} = 5.8 \text{ kg}$$

Discussion   In this problem, we did not have to set the net force equal to zero. By placing the axis of rotation at the hinge, we eliminated two of the three unknowns from the torque equation: the horizontal and vertical components of the hinge force (or, equivalently, its magnitude and direction). If we wanted to find the hinge force as well, setting the net force equal to zero would be necessary.

Practice Problem 8.8   Hinge Forces

Find the vertical component of the force exerted by the hinge in two different ways: (a) setting the net force equal to zero and (b) using a torque equation about a different axis.

Distributed Forces

Gravity is not the only force that is distributed rather than acting at a point. Contact forces, including both the normal component and friction, are spread over the contact surface. Just as for gravity, we can consider the contact force to act at a single point, but the location of that point is often not at all obvious. For a book sitting on a horizontal table, it seems reasonable that the normal force effectively acts at the geometric center of the book cover that touches the table. It is less clear where that effective point is if the book is on an incline or is sliding. As Example 8.9 shows, when something is about to topple over, contact is about to be lost everywhere except at the corner around which the toppling object is about to rotate. That corner then must be the location of the contact forces.

Example 8.9

The Toppling File Cabinet

A file cabinet of height a and width b is on a ramp at angle θ (Fig. 8.25a). The file cabinet is filled with papers in such a way that its center of gravity is at its geometric center. Find the largest θ for which the file cabinet does not tip over. Assume the coefficient of static friction is large enough to prevent sliding. (🌐 tutorial: file cabinet)

Strategy   Until the file cabinet begins to tip over, it is in equilibrium; the net force acting on it must be zero and the total torque about any axis must also be zero. We first draw a force diagram showing the three forces (gravity, normal, friction) acting on the file cabinet. The point of application of the two contact forces (normal, friction) must be at the lower edge of the file cabinet if it is on the steepest possible incline, just about to tip over. In that case, contact has been lost over the rest of the bottom surface of the file cabinet so that only the lower edge makes good contact with the ramp.

As in all equilibrium problems, a good choice of rotation axis makes the problem easier to solve. We know that, at the maximum angle, the contact forces act at the bottom edge of

the file cabinet. A good choice of rotation axis is along the bottom edge of the file cabinet, because then the normal and frictional forces have zero lever arm.

Solution   Figure 8.25b shows the forces acting on the file cabinet at the maximum angle θ. The gravitational force is drawn at the center of gravity. Instead of drawing a single

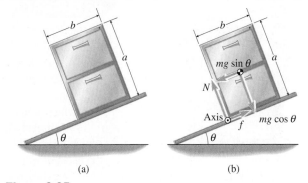

Figure 8.25

(a) File cabinet on an incline. (b) Forces acting on the file cabinet.

continued on next page

Example 8.9 continued

vector arrow for the gravitational force, we represent the gravitational force by its components parallel and perpendicular to the ramp. Then we find the lever arm for each of the components. The lever arm for the parallel component of the weight ($mg \sin \theta$) is $\frac{1}{2}a$ and the lever arm for the perpendicular component ($mg \cos \theta$) is $\frac{1}{2}b$. Setting the net torque equal to zero:

$$\sum \tau = -mg \cos \theta \times \tfrac{1}{2}b + mg \sin \theta \times \tfrac{1}{2}a = 0$$

After dividing out the common factors of $\frac{1}{2}mg$,

$$\cos \theta \times b = \sin \theta \times a$$

Solving for θ,

$$\theta = \tan^{-1} \frac{b}{a}$$

Discussion   As a check, we can regard the normal and friction forces as two components of a single contact force. We can think of that contact force as acting at a single point—a "center of contact" analogous to the center of gravity. As the file cabinet is put on steeper and steeper surfaces, the effective point of application of the contact force moves toward the lower edge of the file cabinet (Fig. 8.26). If we take the rotation axis through the center of gravity so there is no gravitational torque, then the torque due to the contact force must be zero. The only

Figure 8.27
Yuri Chechi of Italy holds the pike position on the rings at the World Gymnastic Championships in Sabae, Japan.

way that can happen is if its lever arm is zero, which means that the contact force must point directly toward the center of gravity. If the angle θ has its maximum value, the contact force acts at the lower edge and $\tan \theta = b/a$. The file cabinet is about to tip when its center of gravity is directly above the lower edge. Any object supported only by contact forces can be in equilibrium only if the point of application of the total contact force is directly below the object's center of gravity.

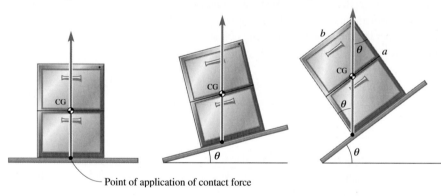

Point of application of contact force

Conceptual Practice Problem 8.9   Gymnast Holding a Pike Position

Figure 8.27 shows a gymnast holding a pike position. What can you say about the location of the gymnast's center of gravity?

Figure 8.26

Contact force for various incline angles.

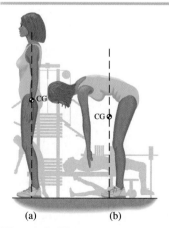

(a)   (b)

Figure 8.28 Location of the center of gravity when (a) standing and (b) reaching for the floor.

PHYSICS AT HOME

When a person stands up straight, the body's center of gravity lies directly above a point between the feet, about 3 cm in front of the ankle joint (see Fig. 8.28a). When a person bends over to touch her toes, the center of gravity lies outside the body (Fig. 8.28b). Note that the lower half of the body must move backward to keep the center of gravity from moving out in front of the toes, which would cause the person to fall over.

An interesting experiment can be done that illustrates what happens to your balance when you shift your center of gravity. Stand against a wall with the heels of your feet touching the wall and your back pressed against the wall. Then carefully try to bend over as if to touch your toes, without bending your knees. Can you do this without falling over? Explain.

8.5 EQUILIBRIUM IN THE HUMAN BODY

We can use the concepts of torque and equilibrium to understand some of how the musculoskeletal system of the human body works. A muscle has tendons at each end that connect it to two different bones across a joint (the flexible connection between the bones). When the muscle contracts, it pulls the tendons, which in turn pull on the bones. Thus, the muscle produces a pair of forces of equal magnitude, one acting on each of the two bones. The biceps muscle (Fig. 8.29) in the upper arm attaches the scapula to the forearm (radius) across the inside of the elbow joint. When the biceps contracts, the forearm is pulled toward the upper arm. The biceps is a *flexor* muscle; it moves one bone closer to another.

A muscle can pull but not push, so a flexor muscle such as the biceps cannot reverse its action to push the forearm away from the upper arm. The *extensor* muscles make bones move apart from each other. In the upper arm (Fig. 8.29), an extensor muscle—the triceps—connects the scapula and humerus to the ulna (a bone in the forearm parallel to the radius) across the outside of the elbow. Since the biceps and triceps connect to the forearm on opposite sides of the elbow joint, they tend to cause rotation about the joint in opposite directions. When the triceps contracts it pulls the forearm away from the upper arm. Using flexor and extensor muscles on opposite sides of the joint, the body can produce both positive and negative torques, although both muscles pull in the same direction.

Suppose the arm is held in a horizontal position. The deltoid muscle (the muscle shown in Fig. 8.30) exerts a force $\vec{\mathbf{F}}_m$ on the humerus at an angle of about 15° above the horizontal. This force has to do two things. The vertical component (magnitude $F_m \sin 15° \approx 0.26 F_m$) supports the weight of the arm, while the horizontal component (magnitude $F_m \cos 15° \approx 0.97 F_m$) stabilizes the joint by pulling the humerus in against the shoulder (scapula). In Example 8.10, we estimate the magnitude of $\vec{\mathbf{F}}_m$.

Application of Conditions for Equilibrium to the Human Body

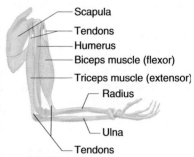

Figure 8.29 The biceps is a flexor muscle; the triceps is an extensor muscle.

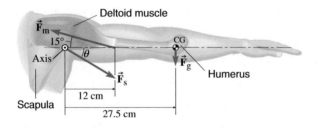

Figure 8.30 Forces exerted on an outstretched arm by the deltoid muscle ($\vec{\mathbf{F}}_m$), the scapula ($\vec{\mathbf{F}}_s$), and gravity ($\vec{\mathbf{F}}_g$).

Example 8.10

Force to Hold Arm Horizontal

A person is standing with his arm outstretched in a horizontal position. The weight of the arm is 30.0 N and its center of gravity is at the elbow joint, 27.5 cm from the shoulder joint (Fig. 8.30). The deltoid pulls on the upper arm at an angle of 15° above the horizontal and at a distance of 12 cm from the joint. What is the magnitude of the force exerted by the deltoid muscle on the arm?

Strategy The arm is in equilibrium, so we can apply the conditions for equilibrium: $\Sigma\vec{\mathbf{F}} = 0$ and $\Sigma\tau = 0$. When calculating torques, we choose the rotation axis at the shoulder joint because then the unknown force $\vec{\mathbf{F}}_s$, which acts on the arm at the joint, has a zero lever arm and produces zero torque. With only one unknown in the torque equation, we can solve immediately for F_m. We do not need to apply the condition $\Sigma\vec{\mathbf{F}} = 0$ unless we want to find $\vec{\mathbf{F}}_s$.

Solution The gravitational force is perpendicular to the line between its point of application and the rotation axis. Gravity produces a CW torque of magnitude

$$|\tau| = Fr = 30.0 \text{ N} \times 0.275 \text{ m} = 8.25 \text{ N·m}$$

For the torque due to $\vec{\mathbf{F}}_m$, we find the component of $\vec{\mathbf{F}}_m$ that is perpendicular to the line between its point of application and the rotation axis. Since this line is horizontal, we need the vertical component of $\vec{\mathbf{F}}_m$, which is $F_m \sin 15°$. Then the magnitude of the CCW torque due to $\vec{\mathbf{F}}_m$ is

$$|\tau| = F_\perp r = F_m \sin 15° \times 0.12 \text{ m}$$

The sum of these torques is zero. With the usual sign convention. (CCW is +),

$$F_m \sin 15° \times 0.12 \text{ m} - 8.25 \text{ N·m} = 0$$

continued on next page

Example 8.10 continued

Solving for F_m,

$$F_m = \frac{8.25 \text{ N·m}}{\sin 15° \times 0.12 \text{ m}} = 270 \text{ N}$$

Discussion   The force exerted by the muscle is much larger than the 30.0-N weight of the arm. The muscle must exert a larger force because the lever arm is small; the point of application is less than half as far from the joint as the center of gravity [0.12 m/(0.275 m) ≈ 4/9]. Also, the muscle cannot pull straight up on the arm; the vertical component of the muscle force is only about $\frac{1}{4}$ of the magnitude of the force. These two factors together make the weight supported (30.0 N) only $\frac{4}{9} \times \frac{1}{4} = \frac{1}{9}$ as large as the force exerted by the muscle.

Practice Problem 8.10   Holding a Juice Carton

Find the force exerted by the same person's deltoid muscle when holding a 1-L juice carton (weight 9.9 N) with the arm outstretched and parallel to the floor (as in Fig. 8.30). Assume that the juice carton is 60.0 cm from the shoulder.

Why does the iron cross require great strength?

The Iron Cross   When a gymnast does the iron cross (Fig. 8.31a), the primary muscles involved are the latissimus dorsi ("lats") and pectoralis major ("pecs"). Since the rings are supporting the gymnast's weight, they exert an upward force on the gymnast's arms. Thus, the task for the muscles is not to hold the arm up, but to pull it down. The lats pull on the humerus about 3.5 cm from the shoulder joint (Fig. 8.31b). The pecs pull on the humerus about 5.5 cm from the joint (Fig. 8.31c). The other ends of these two muscles connect to bone in many places, widely distributed over the back (lats) and chest (pecs). As a reasonable simplification, we can assume that these muscles pull at a 45° angle below the horizontal in the iron cross maneuver. We also assume that the two muscles exert equal forces, so we can replace the two with a single force acting at 4.5 cm from the joint.

To determine the force exerted, we look at the entire arm as a system in equilibrium. This time we can ignore the weight of the arm itself since the force exerted on the arm by the ring is much larger—half the gymnast's weight is supported by each ring. The ring exerts an upward force that acts on the hand about 60 cm from the shoulder joint (see Fig. 8.31d). Taking torques about the shoulder, in equilibrium we have

$$|\text{CW torque}| = |\text{CCW torque}|$$

$$F_m \times 0.045 \text{ m} \times \sin 45° = \tfrac{1}{2}W \times 0.60 \text{ m}$$

$$F_m = \frac{\tfrac{1}{2}W \times 0.60 \text{ m}}{0.045 \text{ m} \times \sin 45°} = 9.4W$$

Thus, the force exerted by the lats and pecs *on one side* of the gymnast's body is more than nine times his weight.

The structure of the human body makes large muscular forces necessary. Are there advantages to the structure? Due to the small lever arms, the muscle forces are much larger than they would otherwise be, but the human body has traded this for a wide range of movement of the bones. The biceps and triceps muscles can move the lower arms through almost 180° while they change their lengths by only a few centimeters. The muscles also remain nearly parallel to the bones. If the biceps and triceps muscles were attached to the lower arm much farther from the elbow, there would have to be a large flap of skin to allow them to move so far away from the bones. The arrangement of our bones and muscles favors a wide range of movement.

Another advantage of the body structure is that it tends to minimize the rotational inertia of our limbs. For example, the muscles that control the motion of the lower arm are contained mostly within the *upper* arm. This keeps the rotational inertia of the lower arms about the elbow smaller. It also keeps the rotational inertia of the entire arm about the shoulder smaller. Smaller rotational inertia means that the energy we have to expend to move our limbs around is smaller.

The biceps muscle with its tendons is almost parallel to the humerus. One interesting observation is that the tendon connects to the radius at different points in different people. In one person this point may be 5.0 cm from the elbow joint, while in another person whose arm is the same length it may be 5.5 cm from the elbow. Thus,

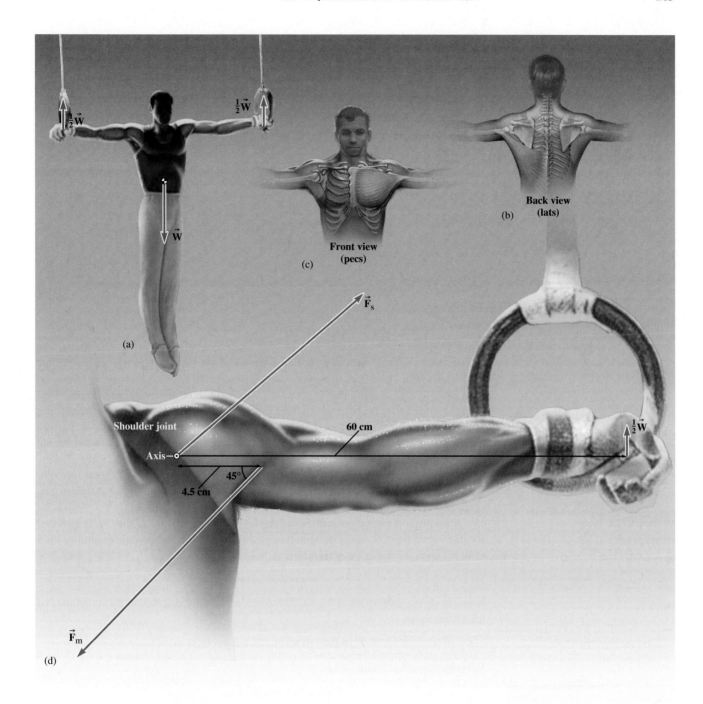

Figure 8.31 (a) Gymnast doing the iron cross. The principal muscles involved are (b) the "lats" and (c) the "pecs." (d) Simplified model of the forces acting on the arm of the gymnast.

some people are naturally stronger than others because of their internal structure. Chimpanzees have an advantage over humans because their biceps muscle has a longer lever arm. Do not make the mistake of arm wrestling with an adult chimp; challenge the chimp to a game of chess instead.

Application of Equilibrium Conditions: Heavy Lifting

When lifting an object from the floor, our first instinct is to bend over and pick it up. This is not a good way to lift something heavy. The spine is an ineffective lever and is susceptible to damage when a heavy object is lifted with bent waist. It is much better to squat down and use the powerful leg muscles to do the lifting instead of using our back muscles. Analyzing torques in a simplified model of the back can illustrate why.

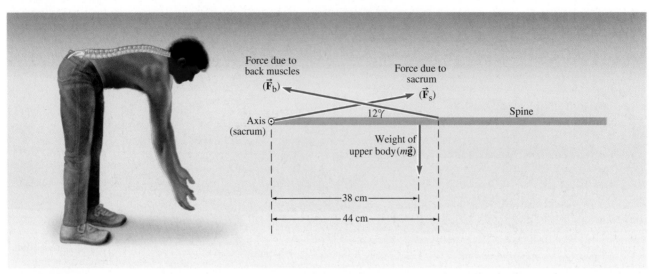

Figure 8.32 A simplified model of the human back when bent over.

The spine can be modeled as a rod with an axis at the tailbone (the sacrum). The sacrum exerts a force, marked $\vec{F}_s$ in Fig. 8.32, when a person bends at the waist with the back horizontal. The forces due to the complicated set of back muscles can be replaced with a single equivalent force $\vec{F}_b$ as shown. This equivalent force makes an angle of 12° with the spine and acts about 44 cm from the sacrum. The weight of the upper body, $m\vec{g}$ in Fig. 8.32, is about 65% of total body weight; its center of gravity is about 38 cm from the sacrum. By placing an axis at the sacrum we can ignore the force $\vec{F}_s$ in our torque equation. Since the vertical component of $\vec{F}_b$ is $F_b \sin 12° \approx 0.21F_b$, only about $\frac{1}{5}$ the magnitude of the forces exerted by the back muscles is supporting the body weight. The much larger horizontal component is pressing the rod representing the spine into the sacrum.

If we put some numbers into this example, we can get an idea of the forces required for just supporting the upper body in this position. If the person's total weight is 710 N (160 lb), then the upper body weight is

$$mg = 0.65 \times 710 \text{ N}$$

Now we set the magnitude of the CCW torques about the axis equal to the magnitude of the CW torques:

$$F_b \times 0.44 \text{ m} \times \sin 12° = mg \times 0.38 \text{ m}$$

Substituting and solving,

$$F_b = \frac{0.65 \times 710 \text{ N} \times 0.38 \text{ m}}{0.44 \text{ m} \times \sin 12°} = 1920 \text{ N}$$

The muscular force that compresses the spine is the horizontal component of $\vec{F}_b$:

$$F_b \cos 12° = 1900 \text{ N}$$

or about 430 lb. This is over four times the weight of the upper body.

Now if the person tries to lift something with his arms in this position, the lever arm for the weight of the load is even longer than for the weight of the upper body. The back muscles must supply a much larger force. The spine is now compressed with a dangerously large force. A cushioning disk called the lumbosacral disk, at the bottom of the spine, separates the last vertebra from the sacrum. This disk can be ruptured or deformed, causing great pain when the back is misused in such a fashion.

If, instead of bending over, we bend our knees and lower our body, keeping it vertically aligned as much as possible while lifting a load, the centers of gravity of the body and load are positioned more closely in a line above the sacrum, as in Fig. 8.33. Then

Figure 8.33 A safer way to lift a heavy object.

the lever arms of these forces with respect to an axis through the sacrum are relatively small and the force on the lumbosacral disk is roughly equal to the upper body weight plus the weight being lifted.

8.6   ROTATIONAL FORM OF NEWTON'S SECOND LAW

The concepts of torque and rotational inertia can be used to formulate a "Newton's second law for rotation"—a law that fills the role of $\sum \vec{\mathbf{F}} = m\vec{\mathbf{a}}$ for rotation about a fixed axis:

Rotational form of Newton's second law:

$$\sum \tau = I\alpha \qquad (8\text{-}9)$$

When calculating the net torque, remember to assign the correct algebraic sign to each torque before adding them.

The sum of the torques due to internal forces acting on a rigid object is always zero. Therefore, only *external* torques need be included in Eq. (8-9).

Thus, the angular acceleration of a rigid body is proportional to the net torque (more torque causes a larger α) and is inversely proportional to the rotational inertia (more inertia causes a smaller α). In rotational equilibrium, the angular acceleration must be zero; Eq. (8-9) then requires that the net torque be zero. We used $\sum \tau = 0$ as the condition for rotational equilibrium in Sections 8.4 and 8.5.

Equation (8-9) is proved in Problem 57. It is subject to an important restriction. Just as $\sum \vec{\mathbf{F}} = m\vec{\mathbf{a}}$ is valid only if the mass of the object is constant, $\sum \tau = I\alpha$ is valid only if the rotational inertia of the object is constant. For a *rigid* object rotating about a *fixed axis*, I cannot change, so Eq. (8-9) is always applicable.

Newton's second law for rotation explains why a tightrope walker carries a long pole to help maintain balance. Suppose the acrobat is about to topple over sideways. The pole has a large rotational inertia due to its length, so the angular acceleration of the system (acrobat plus pole) due to a small gravitational torque is much smaller than it would be without the pole. The smaller angular acceleration gives the acrobat more time to adjust his position and keep from falling.

Example 8.11

The Grinding Wheel

A grinding wheel is a solid, uniform disk of mass 2.50 kg and radius 9.00 cm. Starting from rest, what constant torque must a motor supply so that the wheel attains a rotational speed of 126 rev/s in a time of 6.00 s?

Strategy   Since the grinding wheel is a uniform disk, we can find its rotational inertia using Table 8.1. After converting the revolutions per second to radians per second, we can find the angular acceleration from the change in angular velocity over the given time interval. Once we have I and α, we can find the net torque from Newton's second law for rotation.

Solution   The grinding wheel is a uniform disk, so its rotational inertia is

$$I = \tfrac{1}{2}mr^2$$

$$\tfrac{1}{2} \times 2.50 \text{ kg} \times (0.0900 \text{ m})^2 = 0.010125 \text{ kg·m}^2$$

A single rotation of the wheel is equivalent to 2π radians, so

$$\omega = 126 \, \frac{\text{rev}}{\text{s}} \times 2\pi \frac{\text{rad}}{\text{rev}}$$

The angular acceleration is

$$\alpha = \frac{\Delta \omega}{\Delta t}$$

continued on next page

Example 8.11 continued

Then the torque required is

$$\sum \tau = I\alpha = I\frac{\Delta\omega}{\Delta t}$$

$$= 0.010125 \text{ kg·m}^2 \times \frac{126 \text{ rev/s} \times 2\pi \text{ rad/rev}}{6.00 \text{ s}}$$

$$= 1.34 \text{ N·m}$$

If there are no other torques on the wheel, the motor must supply a constant torque of 1.34 N·m.

Discussion   We assumed that no other torques are exerted on the wheel. There is certain to be at least a small frictional

torque on the wheel with a sign opposite to the sign of the motor's torque. Then the motor would have to supply a torque larger than 1.34 N·m. The *net* torque would still be 1.34 N·m.

Practice Problem 8.11   Another Approach

Verify the answer to Example 8.11 by: (a) finding the angular displacement of the wheel using equations for constant α; (b) finding the change in rotational kinetic energy of the wheel; and (c) finding the torque from $W = \tau\Delta\theta$.

8.7   THE MOTION OF ROLLING OBJECTS

A rolling object combines translational motion of the center of mass with rotation about an axis that passes through the center of mass (see Section 5.1). For an object that is rolling without slipping, $v_{CM} = \omega R$. As a result, there is a specific relationship between the rolling object's translational and rotational kinetic energies. The total kinetic energy of a rolling object is the sum of its translational and rotational kinetic energies.

A wheel with mass M and radius R has a rotational inertia that is some pure number times MR^2; it couldn't be anything else and still have the right units. We can write the rotational inertia about an axis through the CM as $I_{CM} = \beta MR^2$ where β is a pure number that measures how far from the axis of rotation the mass is distributed. Larger β means the mass is, on average, farther from the axis. From Table 8.1, a hoop has $\beta = 1$; a disk, $\beta = \frac{1}{2}$; and a solid sphere, $\beta = \frac{2}{5}$.

Using $I_{CM} = \beta MR^2$ and $v_{CM} = \omega R$, the rotational kinetic energy for a rolling object can be written

$$K_{rot} = \frac{1}{2}I_{CM}\omega^2 = \frac{1}{2} \times \beta MR^2 \times \left(\frac{v_{CM}}{R}\right)^2 = \beta \times \frac{1}{2}Mv_{CM}^2$$

Since $\frac{1}{2}Mv_{CM}^2$ is the translational kinetic energy,

$$K_{rot} = \beta K_{tr} \tag{8-10}$$

This is convenient since β depends only on the shape, not on the mass or radius of the object. For a given shape rolling without slipping, the ratio of its rotational to translational kinetic energy is always the same (β).

The total kinetic energy can be written

$$K = K_{tr} + K_{rot}$$

$$K = \frac{1}{2}Mv_{CM}^2 + \frac{1}{2}I_{CM}\omega^2 \tag{8-11}$$

or in terms of β,

$$K = (1 + \beta) K_{tr}$$

$$K = (1 + \beta) \frac{1}{2}Mv_{CM}^2 \tag{8-12}$$

Thus, two objects of the same mass rolling at the same translational speed do *not* necessarily have the same kinetic energy. The object with the larger value of β has more rotational kinetic energy.

Conceptual Example 8.12

Hollow and Solid Rolling Balls

Starting from rest, two balls roll down a hill as in Fig. 8.34. One is solid, the other hollow. Which one is moving faster when it reaches the bottom of the hill? (tutorial: rolling)

Strategy and Solution Energy conservation is the best way to approach this problem. As a ball rolls down the hill, its gravitational potential energy decreases as its kinetic energy increases by the same amount. The total kinetic energy is the sum of the translational and rotational contributions.

We do not know the mass or the radius of either ball and we cannot assume they are the same. Since both kinetic and potential energies are proportional to mass, mass does not affect the final speed. Also, the total kinetic energy does not depend on the radius of the ball [see Eq. (8-12)]. The final speeds of the two balls differ because different *fractions* of their total kinetic energies are translational.

One ball is a solid sphere and the other is approximately a spherical shell. The mass of a spherical shell is all concentrated on the surface of a sphere, while a solid sphere has its mass distributed throughout the sphere's volume. Therefore, the shell has a larger β than the solid sphere. When the shell rolls, it converts a bigger fraction of the lost potential energy into rotational kinetic energy; therefore, a smaller fraction becomes translational kinetic energy. The final speed of the solid sphere is larger since it puts a larger fraction of its kinetic energy into translational motion.

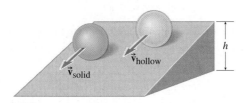

Figure 8.34
Rolling balls.

Discussion We can make this conceptual question into a quantitative one: what is the ratio of the speeds of the two balls at the bottom of the hill?

Let the height of the hill be h. Then for a ball of mass M, the loss of gravitational potential energy is Mgh. This amount of gravitational potential energy is converted into translational and rotational kinetic energy:

$$Mgh = K_{tr} + K_{rot} = (1 + \beta)K_{tr} = (1 + \beta)\frac{Mv_{CM}^2}{2}$$

Mass cancels out, as expected. We can solve for the final speed in terms of g, h, and β. The final speed is independent of the ball's mass and radius.

$$v_{CM} = \sqrt{\frac{2gh}{1 + \beta}}$$

The ratio of the final speeds for two balls rolling down the same hill is, therefore,

$$\frac{v_1}{v_2} = \sqrt{\frac{1 + \beta_2}{1 + \beta_1}}$$

To evaluate the ratio, we look up the rotational inertias in Table 8.1. The solid sphere has $\beta = \frac{2}{5}$ and the spherical shell has $\beta = \frac{2}{3}$. Then

$$\frac{v_{solid}}{v_{hollow}} = \sqrt{\frac{1 + \frac{2}{3}}{1 + \frac{2}{5}}} \approx 1.091$$

The solid ball's final speed is, therefore, 9.1% faster than that of the hollow ball. This ratio depends neither on the masses of the balls, the radii of the balls, the height of the hill, nor the slope of the hill.

Practice Problem 8.12 Fraction of Kinetic Energy That Is Rotational Energy

What fraction of a rolling ball's kinetic energy is rotational kinetic energy? Answer both for a solid ball and a hollow one.

✓ CHECKPOINT 8.7

Give an example of how a marble can move so that $K_{tr} > 0$ and $K_{rot} = 0$; (b) $K_{tr} = 0$ and $K_{rot} > 0$; (c) $K_{rot} = \frac{2}{5} K_{tr}$.

Acceleration of Rolling Objects What is the acceleration of a ball rolling down an incline? Figure 8.35 shows the forces acting on the ball. Static friction is the force that makes the ball rotate; if there were no friction, instead of rolling, the ball would just *slide* down the incline. This is true because friction is the only force acting that yields a

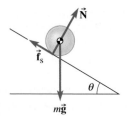

Figure 8.35 Forces acting on a ball rolling downhill.

nonzero torque about the rotation axis through the ball's center of mass. Gravity gives zero torque because it acts at the axis, so the lever arm is zero. The normal force points directly at the axis, so its lever arm is also zero.

The frictional force $\vec{\mathbf{f}}$ provides a torque

$$\tau = rf$$

where r is the ball's radius. An analysis of the forces and torques combined with Newton's second law in both forms enables us to calculate the acceleration of the ball in Example 8.13.

Example 8.13

Acceleration of a Rolling Ball

Calculate the acceleration of a solid ball rolling down a slope inclined at an angle θ to the horizontal (Fig. 8.36a).

Strategy The net torque is related to the angular acceleration by $\sum \tau = I\alpha$, Newton's second law for rotation. Similarly, the net force acting on the ball gives the acceleration of the center of mass: $\sum \vec{\mathbf{F}} = m\vec{\mathbf{a}}_{CM}$. The axis of rotation is through the ball's CM. As already discussed, neither gravity nor the normal force produce a torque about this axis; the net torque is $\sum \tau = rf$, where f is the magnitude of the frictional force. One problem is that the force of friction is unknown. We must resist the temptation to assume that $f = \mu_s N$; there is no reason to assume that static friction has its maximum possible magnitude. We do know that the two accelerations, translational and rotational, are related. We know that v_{CM} and ω are proportional since r is constant. To stay proportional they must change in lock step; their rates of change, a_{CM} and α, are proportional to each other by the same factor of r. Thus, $a_{CM} = \alpha r$. This connection should enable us to eliminate f and solve for the acceleration. Since the speed of a ball after rolling a certain distance was found to be independent of the mass and radius of the ball in Example 8.12, we expect the same to be true of the acceleration.

Solution Since the net torque is

$$\sum \tau = rf$$

the angular acceleration is

$$\alpha = \frac{\sum \tau}{I} = \frac{rf}{I} \qquad (1)$$

where I is the ball's rotational inertia about its CM.

Figure 8.36b shows the forces along the incline acting on the ball. The acceleration of the CM is found from Newton's second law. The component of the net force acting along the incline (in the direction of the acceleration) is

$$\sum F_x = mg \sin \theta - f = ma_{CM} \qquad (2)$$

Because the ball is rolling without slipping, the acceleration of the CM and the angular acceleration are related by

$$a_{CM} = \alpha r$$

Now we try to eliminate the unknown frictional force f from the previous equations. Solving Eq. (1) for f gives

$$f = \frac{I\alpha}{r}$$

Substituting this into Eq. (2), we get

$$mg \sin \theta - \frac{I\alpha}{r} = ma_{CM}$$

Now to eliminate α, we can substitute $\alpha = a_{CM}/r$:

$$mg \sin \theta - \frac{Ia_{CM}}{r^2} = ma_{CM}$$

Solving for a_{CM},

$$a_{CM} = \frac{g \sin \theta}{1 + I/(mr^2)}$$

For a solid sphere, $I = \frac{2}{5}mr^2$, so

$$a_{CM} = \frac{g \sin \theta}{1 + \frac{2}{5}} = \frac{5}{7}g \sin \theta$$

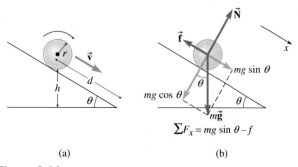

(a)   (b)

Figure 8.36

(a) A ball rolling downhill. (b) FBD for the ball, with the gravitational force resolved into components perpendicular and parallel to the incline.

$$\sum F_x = mg \sin \theta - f$$

continued on next page

Example 8.13 continued

Discussion The acceleration of an object *sliding* down an incline without friction is $a = g \sin \theta$. The acceleration of the rolling ball is smaller than $g \sin \theta$ due to the frictional force directed up the incline.

We can check the answer using the result of Example 8.12. The ball's acceleration is constant. If the ball starts from rest as in Fig. 8.36a, after it has rolled a distance d, its speed v is

$$v = \sqrt{2ad} = \sqrt{2\left(\frac{g \sin \theta}{1 + \beta}\right)d}$$

where $\beta = \frac{2}{5}$. The vertical drop during this time is $h = d \sin \theta$, so

$$v = \sqrt{\frac{2gh}{1 + \beta}}$$

Practice Problem 8.13 Acceleration of a Hollow Cylinder

Calculate the acceleration of a thin hollow cylindrical shell rolling down a slope inclined at an angle θ to the horizontal.

8.8 ANGULAR MOMENTUM

Newton's second law for translational motion can be written in two ways:

$$\sum \vec{F} = \lim_{\Delta t \to 0} \frac{\Delta \vec{p}}{\Delta t} \text{ (general form)} \quad \text{or} \quad \sum \vec{F} = m\vec{a} \text{ (constant mass)}$$

In Eq. (8-9) we wrote Newton's second law for rotation as $\sum \tau = I\alpha$, which applies only when I is constant—that is, for a rigid body rotating about a fixed axis. A more general form of Newton's second law for rotation uses the concept of **angular momentum** (symbol L).

The net external torque acting on a system is equal to the rate of change of the angular momentum of the system.

$$\sum \tau = \lim_{\Delta t \to 0} \frac{\Delta L}{\Delta t} \quad (8\text{-}13)$$

CONNECTION:

Note the analogy with

$$\sum \vec{F} = \lim_{\Delta t \to 0} \frac{\Delta \vec{p}}{\Delta t}$$

The angular momentum of a rigid body rotating about a fixed axis is the rotational inertia times the angular velocity, which is analogous to the definition of linear momentum (mass times velocity):

Angular momentum:

$$L = I\omega \quad (8\text{-}14)$$

(rigid body, fixed axis)

CONNECTION:

Note the analogy with $\vec{p} = m\vec{v}$. See the Master the Concepts section for a complete table of these analogies.

Either Eq. (8-13) or Eq. (8-14) can be used to show that the SI units of angular momentum are $kg \cdot m^2/s$.

For a rigid body rotating around a fixed axis, angular momentum doesn't tell us anything new. The rotational inertia is constant for such a body since the distance r_i between every point on the object and the axis stays the same. Then any change in angular momentum must be due to a change in angular velocity ω:

$$\sum \tau = \lim_{\Delta t \to 0} \frac{\Delta L}{\Delta t} = \lim_{\Delta t \to 0} \frac{I\Delta \omega}{\Delta t} = I \lim_{\Delta t \to 0} \frac{\Delta \omega}{\Delta t} = I\alpha$$

Conservation of Angular Momentum However, Eq. (8-13) is *not* restricted to rigid objects or to fixed rotation axes. In particular, if the net external torque acting on a

Conservation of angular momentum can be applied to any system if the net external torque on the system is zero (or negligibly small).

CONNECTION:

Another conservation law

⚠️

Application of angular momentum: figure skater

system is zero, then the angular momentum of the system cannot change. This is the **law of conservation of angular momentum**:

> **Conservation of angular momentum:**
>
> $$\text{If } \sum \tau = 0, \ L_\text{i} = L_\text{f} \tag{8-15}$$

Here L_i and L_f represent the angular momentum of the system at two different times. Conservation of angular momentum is one of the most basic and fundamental laws of physics, along with the two other conservation laws we have studied so far (energy and linear momentum). For an isolated system, the total energy, total linear momentum, and total angular momentum of the system are each conserved. None of these quantities can change unless some external agent causes the change.

With conservation of energy, we add up the amounts of the different forms of energy (such as kinetic energy and gravitational potential energy) to find the *total* energy. The conservation law refers to the total energy. By contrast, linear momentum and angular momentum *cannot* be added to find the "total momentum." They are entirely different quantities, not two forms of the same quantity. They even have different dimensions, so it would be impossible to add them. Conservation of linear momentum and conservation of angular momentum are *separate* laws of physics.

Changing Rotational Inertia In this section, we restrict our consideration to cases where the axis of rotation is fixed but where the rotational inertia is not necessarily constant. One familiar example of a changing rotational inertia occurs when a figure skater spins (Fig. 8.37). To start the spin, the skater glides along with her arms outstretched and then begins to rotate her body about a vertical axis by pushing against the ice with a skate. The push of the ice against the skate provides the external torque that gives the skater her initial angular momentum. Initially the skater's arms and the leg not in contact with the ice are extended away from her body. The mass of the arms and leg when extended contribute more to her rotational inertia than they do when held close to the body. As the skater spins, she pulls her arms and leg close and straightens her body to decrease her rotational inertia. As she does, her angular velocity increases dramatically in such a way that her angular momentum stays the same.

Figure 8.37 Figure skater Lucinda Ruh at the (a) beginning and (b) end of a spin. Her angular velocity is much higher in (b) than in (a).

(a)   (b)

✓ CHECKPOINT 8.8

If the skater then extends her arms and leg back to their initial configuration, does her angular velocity decrease back to its initial value, ignoring friction?

Many natural phenomena can be understood in terms of angular momentum. In a hurricane, circulating air is sucked inward by a low-pressure region at the center of the storm (the *eye*). As the air moves closer and closer to the axis of rotation, it circulates faster and faster. An even more dramatic example is the formation of a pulsar. Under certain conditions, a star can implode under its own gravity, forming a neutron star (a collection of tightly packed neutrons). If the Sun were to collapse into a neutron star, its radius would be only about 13 km. If a star is rotating before its collapse, then as its rotational inertia decreases dramatically, its angular velocity must increase to keep its angular momentum constant. Such rapidly rotating neutron stars are called pulsars because they emit regular pulses of x-rays, at the same frequency as their rotation, that can be detected when they reach Earth. Some pulsars rotate in only a few thousandths of a second per revolution.

Applications of angular momentum: hurricanes and pulsars

Example 8.14

Mouse on a Wheel

A 0.10-kg mouse is perched at point B on the rim of a 2.00-kg wagon wheel that rotates freely in a horizontal plane at 1.00 rev/s (Fig. 8.38). The mouse crawls to point A at the center. Assume the mass of the wheel is concentrated at the rim. What is the frequency of rotation in rev/s when the mouse arrives at point A?

Strategy Assuming that frictional torques are negligibly small, there is no external torque acting on the mouse/wheel system. Then the angular momentum of the mouse/wheel system must be conserved; it takes an external torque to change angular momentum. The mouse and wheel exert torques on one another, but these *internal* torques only transfer some angular momentum between the wheel and the mouse without changing the total angular momentum. We can think of the system as initially being a rigid body with rotational inertia I_i. When the mouse reaches the center, we think of the system as a rigid body with a different rotational

inertia I_f. The mouse changes the rotational inertia of the mouse/wheel system by moving from the outer rim, where its mass makes the maximum possible contribution to the rotational inertia, to the rotation axis, where its mass makes no contribution to the rotational inertia.

Solution Initially, all of the mass of the system is at a distance R from the rotation axis, where R is the radius of the wheel. Therefore,

$$I_i = (M + m)R^2$$

where M is the mass of the wheel and m is the mass of the mouse. After the mouse moves to the center of the wheel, its mass contributes nothing to the rotational inertia of the system:

$$I_f = MR^2$$

From conservation of angular momentum,

$$I_i\,\omega_i = I_f\,\omega_f$$

Substituting the rotational inertias and $\omega = 2\pi f$,

$$(M + m)R^2 \times 2\pi f_i = MR^2 \times 2\pi f_f$$

Factors of $2\pi R^2$ cancel from each side, leaving

$$(M + m)f_i = Mf_f$$

Solving for f_f,

$$f_f = \frac{M + m}{M}f_i = \frac{2.10\ \text{kg}}{2.00\ \text{kg}}\,(1.00\ \text{rev/s}) = 1.05\ \text{rev/s}$$

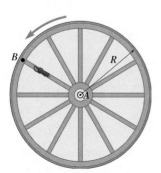

Figure 8.38
Mouse on a rotating wheel.

continued on next page

Example 8.14 continued

Discussion Conservation laws are powerful tools. We do not need to know the details of what happens as the mouse crawls along the spoke from the outer edge of the wheel; we need only look at the initial and final conditions.

A common mistake in this sort of problem is to assume that the initial rotational kinetic energy is equal to the final rotational kinetic energy. This is not true because the mouse crawling in toward the center expends energy to do so. In other words, the mouse converts some internal energy into rotational kinetic energy.

Practice Problem 8.14   Change in Rotational Kinetic Energy

What is the percentage change in the rotational kinetic energy of the mouse/wheel system?

Angular Momentum in Planetary Orbits

Application of angular momentum: planetary orbits

Conservation of angular momentum applies to planets orbiting the Sun in elliptical orbits. Kepler's second law says that the orbital speed varies in such a way that the planet sweeps out area at a constant rate (Fig. 8.39a). In Problem 104, you can show that Kepler's second law is a direct result of conservation of angular momentum, where the angular momentum of the planet is calculated using an axis of rotation perpendicular to the plane of the orbit and passing through the Sun. When the planet is closer to the Sun, it moves faster; when it is farther away, it moves more slowly. Conservation of angular momentum can be used to relate the orbital speeds and radii at two different points in the orbit. The same applies to satellites and moons orbiting planets.

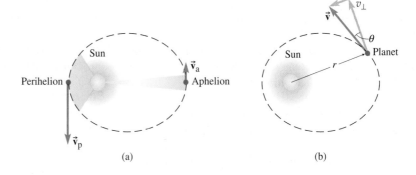

Figure 8.39 The planet's speed varies such that it sweeps out equal areas in equal time intervals. The eccentricity of the planetary orbit is exaggerated for clarity.

Example 8.15

Earth's Orbital Speed

At perihelion (closest approach to the Sun), Earth is 1.47×10^8 km from the Sun and its orbital speed is 30.3 km/s. What is Earth's orbital speed at aphelion (greatest distance from the Sun), when it is 1.52×10^8 km from the Sun? Note that at these two points Earth's velocity is perpendicular to a radial line from the Sun (see Fig. 8.39a).

Strategy We take the axis of rotation through the Sun. Then the gravitational force on Earth points directly toward the axis; with zero lever arm, the torque is zero. With no other external forces acting on the Earth, the net external torque is zero. Earth's angular momentum about the rotation axis through the Sun must therefore be conserved. To find

Earth's rotational inertia, we treat it as a point particle since its radius is much less than its distance from the axis of rotation.

Solution The rotational inertia of the Earth is

$$I = mr^2$$

where m is Earth's mass and r is its distance from the Sun. The angular velocity is

$$\omega = \frac{v_\perp}{r}$$

where $v_\perp$ is the component of the velocity perpendicular to a radial line from the Sun. At the two points under consideration,

continued on next page

Example 8.15 continued

$v_\perp = v$. As the distance from the Sun r varies, its speed v must vary to conserve angular momentum:

$$I_i \omega_i = I_f \omega_f$$

By substitution,

$$mr_i^2 \times \frac{v_i}{r_i} = mr_f^2 \times \frac{v_f}{r_f}$$

or

$$r_i v_i = r_f v_f \qquad (1)$$

Solving for v_f,

$$v_f = r_i/r_f \, v_i = \frac{1.47 \times 10^8 \text{ km}}{1.52 \times 10^8 \text{ km}} \times 30.3 \text{ km/s} = 29.3 \text{ km/s}$$

Discussion   Earth moves slower at a point farther from the Sun. This is what we expect from energy conservation. The potential energy is greater at aphelion than at perihelion.

Since the mechanical energy of the orbit is constant, the kinetic energy must be smaller at aphelion.

Equation (1) implies that the orbital speed and orbital radius are inversely proportional, but strictly speaking this equation only applies to the perihelion and aphelion. At a general point in the orbit, the *perpendicular component $v_\perp$* is inversely proportional to r (see Fig. 8.39b). The orbits of Earth and most of the other planets are nearly circular so that $\theta \approx 0°$ and $v_\perp \approx v$.

Practice Problem 8.15   Puck on a String

A puck on a frictionless, horizontal air table is attached to a string that passes down through a hole in the table. Initially the puck moves at 12 cm/s in a circle of radius 24 cm. If the string is pulled through the hole, reducing the radius of the puck's circular motion to 18 cm, what is the new speed of the puck?

8.9   THE VECTOR NATURE OF ANGULAR MOMENTUM

Until now we have treated torque and angular momentum as scalar quantities. Such a treatment is adequate in the cases we have considered so far. However, the law of conservation of angular momentum applies to *all* systems, including rotating objects whose axis of rotation changes direction. Torque and angular momentum are actually vector quantities. Angular momentum is conserved in *both magnitude and direction* in the absence of external torques.

An important special case is that of a symmetrical object rotating about an axis of symmetry, such as the spinning disk in Fig. 8.40. The magnitude of the angular momentum of such an object is $L = I\omega$. The direction of the angular momentum vector points along the axis of rotation. To choose between the two directions along the axis, a **right-hand rule** is used. Align your right hand so that, as you curl your fingers in toward your palm, your fingertips follow the object's rotation; then your thumb points in the direction of $\vec{L}$.

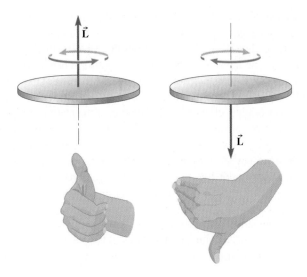

Figure 8.40 Right-hand rule for finding the direction of the angular momentum of a spinning disk.

Figure 8.41 Spinning like a top, the Earth maintains the direction of its angular momentum due to rotation as it revolves around the Sun (not to scale).

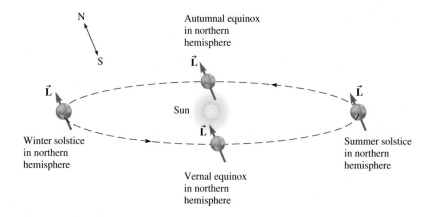

Application of angular momentum: the gyroscope

For rotation about a *fixed* axis, the net torque is also along the axis of rotation, in the direction of the *change* in angular momentum it causes. The sign convention we have used up to now for angular momentum and torque gives the sign of the *z-component of the vector quantity*, where the *z*-axis points toward the viewer (out of the page).

A disk with a large rotational inertia can be used as a *gyroscope.* When the gyroscope spins at a large angular velocity, it has a large angular momentum. It is then difficult to change the orientation of the gyroscope's rotation axis, because to do so requires changing its angular momentum. To change the direction of a large angular momentum requires a correspondingly large torque. Thus, a gyroscope can be used to maintain stability. Gyroscopes are used in guidance systems in airplanes, submarines, and space vehicles to maintain a constant direction in space.

The same principle explains the great stability of rifle bullets and spinning tops. A rifle bullet is made to spin as it passes through the rifle's barrel. The spinning bullet then keeps its correct orientation—nose first—as it travels through the air. Otherwise, a small torque due to air resistance could make the bullet turn around randomly, greatly increasing air resistance and undermining accuracy. A properly thrown football is made to spin for the same reasons. A spinning top can stay balanced for a long time, while the same top soon falls over if it is not spinning.

The Earth's rotation gives it a large angular momentum. As the Earth orbits the Sun, the axis of rotation stays in a fixed direction in space. The axis points nearly at Polaris (the North Star), so even as the Earth rotates around its axis, Polaris maintains its position in the northern sky. The fixed direction of the rotation axis gives us the regular progression of the seasons (Fig. 8.41).

A Classic Demonstration

A demonstration often done in physics classes is for a student to hold a spinning bicycle wheel while standing on a platform that is free to rotate. The wheel's rotation axis is initially horizontal (Fig. 8.42a). Then the student repositions the wheel so that its axis of rotation is vertical (Fig. 8.42b). As he repositions the wheel, the platform begins to rotate opposite to the wheel's rotation. If we assume *no* friction acts to resist rotation of the platform, then the platform continues to rotate as long as the wheel is held with its axis vertical. If the student returns the wheel to its original orientation, the rotation of the platform stops.

The platform is free to rotate about a vertical axis. As a result, once the student steps onto the platform, *the vertical component L_y of the angular momentum of the system* (student + platform + wheel) is conserved. The horizontal components of $\vec{L}$ are *not* conserved. The platform is not free to rotate about any horizontal axis since the floor can exert external torques to keep it from doing so. In vector language, we would say that only the vertical component of the external torque is zero, so only the vertical component of angular momentum is conserved.

Initially $L_y = 0$ since the student and the platform have zero angular momentum and the wheel's angular momentum is horizontal. When the wheel is repositioned so that it

Figure 8.42 A demonstration of angular momentum conservation.

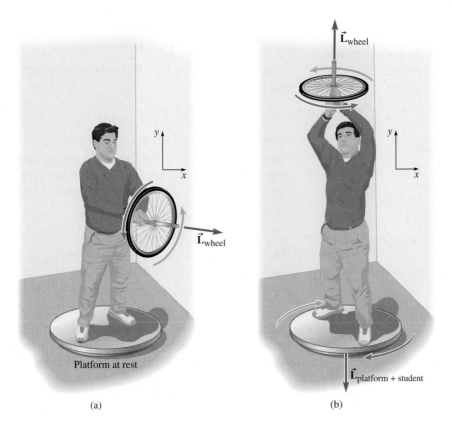

$\vec{L}_{\text{wheel}}$

$\vec{L}_{\text{wheel}}$

$\vec{L}_{\text{platform + student}}$

Platform at rest

(a)   (b)

spins with an upward angular momentum ($L_y > 0$), the rest of the system (the student and the platform) must acquire an equal magnitude of downward angular momentum ($L_y < 0$) so that the vertical component of the total angular momentum is still zero. Thus, the platform and student rotate in the opposite sense from the rotation of the wheel. Since the platform and student have more rotational inertia than the wheel, they do not spin as fast as the wheel, but their vertical angular momentum is just as large.

The student and the wheel apply torques to each other to transfer angular momentum from one part of the system to the other. These torques are equal and opposite and they have both vertical and horizontal components. As the student lifts the wheel, he feels a strange twisting force that tends to rotate him about a horizontal axis. The platform prevents the horizontal rotation by exerting unequal normal forces on the student's feet. The horizontal component of the torque is so counterintuitive that, if the student is not expecting it, he can easily be thrown from the platform!

Master the Concepts

- The rotational kinetic energy of a rigid object with rotational inertia I and angular velocity ω is

$$K_{\text{rot}} = \tfrac{1}{2}I\omega^2 \qquad (8\text{-}1)$$

  In this expression, ω must be measured in *radians* per unit time.

- Rotational inertia is a measure of how difficult it is to change an object's angular velocity. It is defined as:

$$I = \sum_{i=1}^{N} m_i r_i^2 \qquad (8\text{-}2)$$

  where r_i is the perpendicular distance between a particle of mass m_i and the rotation axis. The rotational inertia depends on the location of the rotation axis.

- Torque measures the effectiveness of a force for twisting or turning an object. It can be calculated in two equivalent ways: either as the product of the perpendicular component of the force with the shortest distance between the rotation axis and the point of application of the force

$$\tau = \pm r F_{\perp} \qquad (8\text{-}3)$$

continued on next page

Master the Concepts continued

or as the product of the magnitude of the force with its lever arm (the perpendicular distance between the line of action of the force and the axis of rotation)

$$\tau = \pm r_{\perp} F \qquad (8\text{-}4)$$

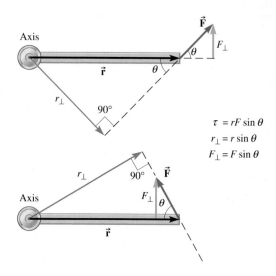

$$\tau = rF \sin\theta$$
$$r_{\perp} = r \sin\theta$$
$$F_{\perp} = F \sin\theta$$

- A force whose perpendicular component tends to cause rotation in the CCW direction gives rise to a positive torque; a force whose perpendicular component tends to cause rotation in the CW direction gives rise to a negative torque.
- The work done by a constant torque is the product of the torque and the angular displacement:

$$W = \tau \Delta\theta \quad (\Delta\theta \text{ in radians}) \qquad (8\text{-}6)$$

- The conditions for translational and rotational equilibrium are

$$\sum \vec{F} = 0 \text{ and } \sum \tau = 0 \qquad (8\text{-}8)$$

The rotation axis can be chosen *arbitrarily* when calculating torques in equilibrium problems. Generally, the best place to choose the axis is at the point of application of an unknown force so that the unknown force does not appear in the torque equation.

- Newton's second law for rotation is

$$\sum \tau = I\alpha \qquad (8\text{-}9)$$

where radian measure must be used for α. A more general form is

$$\sum \tau = \lim_{\Delta t \to 0} \frac{\Delta L}{\Delta t} \qquad (8\text{-}13)$$

where L is the angular momentum of the system.

- The total kinetic energy of a body that is rolling without slipping is the sum of the rotational kinetic energy about an axis through the CM and the translational kinetic energy:

$$K = \tfrac{1}{2}Mv_{\text{CM}}^2 + \tfrac{1}{2}I_{\text{CM}}\omega^2 \qquad (8\text{-}11)$$

- The angular momentum of a rigid body rotating about a fixed axis is the rotational inertia times the angular velocity:

$$L = I\omega \qquad (8\text{-}14)$$

- The law of conservation of angular momentum: if the net external torque acting on a system is zero, then the angular momentum of the system cannot change.

$$\text{If } \sum \tau = 0, \; L_i = L_f \qquad (8\text{-}15)$$

- This table summarizes the analogous quantities and equations in translational and rotational motion.

Translation	Rotation
m	I
$\vec{F}$	τ
$\vec{a}$	α
$\sum \vec{F} = m\vec{a}$	$\sum \tau = I\alpha$
Δx	$\Delta\theta$
$W = F_x \Delta x$	$W = \tau \Delta\theta$
$\vec{v}$	ω
$K = \tfrac{1}{2}mv^2$	$K = \tfrac{1}{2}I\omega^2$
$\vec{p} = m\vec{v}$	$L = I\omega$
$\sum \vec{F} = \lim\limits_{\Delta t \to 0} \dfrac{\Delta \vec{p}}{\Delta t}$	$\sum \tau = \lim\limits_{\Delta t \to 0} \dfrac{\Delta L}{\Delta t}$
If $\sum \vec{F} = 0$, $\vec{p}$ is conserved	If $\sum \tau = 0$, L is conserved

Conceptual Questions

1. In Fig. 8.2b, where should the doorknob be located to make the door easier to open?

2. Explain why it is easier to drive a wood screw using a screwdriver with a large diameter handle rather than one with a thin handle.

3. Why is it easier to push open a swinging door from near the edge away from the hinges rather than in the middle of the door?

4. A book measures 3 cm by 16 cm by 24 cm. About which of the axes shown in the figure is its rotational inertia smallest?

5. A body in equilibrium has only two forces acting on it. We found in Section 4.2 that the

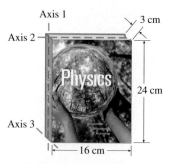

Conceptual Question 4

forces must be equal in magnitude and opposite in direction in order to give a translational net force of zero. What else must be true of the two forces for the body to be in equilibrium? [*Hint:* Consider the lines of action of the forces.]

6. Why do many helicopters have a small propeller attached to the tail that rotates in a vertical plane? Why is this attached at the tail rather than somewhere else? [*Hint:* Most of the helicopter's mass is forward, in the cab.]

7. In the "Pinewood Derby," Cub Scouts construct cars and then race them down an incline. Some say that, everything else being equal (friction, drag coefficient, same wheels, etc.), a heavier car will win; others maintain that the weight of the car does not matter. Who is right? Explain. [*Hint:* Think about the fraction of the car's kinetic energy that is rotational.]

8. A large barrel lies on its side. In order to roll it across the floor, you apply a horizontal force, as shown in the figure. If the applied force points toward the axis of rotation, which runs down the center of the barrel through the center of mass, it produces zero torque about that axis. How then can this applied force make the barrel start to roll?

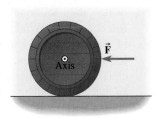

9. Animals that can run fast always have thin legs. Their leg muscles are concentrated close to the hip joint; only tendons extend into the lower leg. Using the concept of rotational inertia, explain how this helps them run fast.

10. Part (a) of the figure shows a simplified model of how the triceps muscle connects to the forearm. As the angle θ is changed, the tendon wraps around a nearly circular arc. Explain how this is much more effective than if the tendon is connected as in part (b) of the figure. [*Hint:* Look at the lever arm as θ changes.]

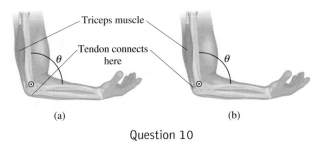

Question 10

11. Part (a) of the figure shows a simplified model of how the biceps muscle enables the forearm to support a load. What are the advantages of this arrangement as opposed to the alternative shown in part (b), where the flexor muscle is in the forearm instead of in the upper arm? Are the two equally effective when the forearm is

horizontal? What about for other angles between the upper arm and the forearm? Consider also the rotational inertia of the forearm about the elbow and of the entire arm about the shoulder.

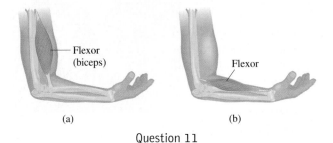

Question 11

12. In Section 8.6, it was asserted that the sum of all the internal torques (that is, the torques due to internal forces) acting on a rigid object is zero. The figure shows two particles in a rigid object. The particles exert forces $\vec{F}_{12}$ and $\vec{F}_{21}$ on each other. These forces are directed along a line that joins the two particles. Explain why the torques due to these two forces must be equal and opposite even though the forces are applied at different points (and, therefore, possibly different distances from the axis).

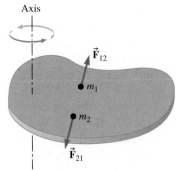

13. A playground merry-go-round (Fig. 8.5) spins with negligible friction. A child moves from the center out to the rim of the merry-go-round platform. Let the system be the merry-go-round plus the child. Which of these quantities change: angular velocity of the system, rotational kinetic energy of the system, angular momentum of the system? Explain your answer.

14. The figure shows a balancing toy with weights extending on either side. The toy is extremely stable. It can be pushed quite far off center one way or the other but it does not fall over. Explain why it is so stable.

15. Explain why the posture taken by defensive football linemen makes them more difficult to push out of the way. Consider both the height of the center of gravity and the size of the support base (the area on the ground bounded

by the hands and feet touching the ground). In order to knock a person over, what has to happen to the center of gravity? Which do you think needs a more complex neurological system for maintaining balance: four legged animals or humans?

16. The center of gravity of the upper body of a bird is located below the hips; in a human, the center of gravity of the upper body is located well above the hips. Since the upper body is supported by the hips, are birds or humans more stable? Consider what happens if the upper body is displaced a little so that its center of gravity is not directly above or below the hips. In what direction does the torque due to gravity tend to make the upper body rotate about an axis through the hips?

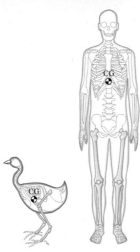

17. An astronaut wants to remove a bolt from a satellite in orbit. He positions himself so that he is at rest with respect to the satellite, then pulls out a wrench and attempts to remove the bolt. What is wrong with his method? How can he remove the bolt?

18. Your door is hinged to close automatically after being opened. Where is the best place to put a wedge-shaped door stopper on a slippery floor in order to hold the door open? Should it be placed close to the hinge or far from it?

19. You are riding your bicycle and approaching a rather steep hill. Which gear should you use to go uphill, a low gear or a high gear? With a low gear the wheel rotates less than with a high gear for one rotation of the pedals.

20. One way to find the center of gravity of an irregular flat object is to suspend it from various points so that it is free to rotate. When the object hangs in

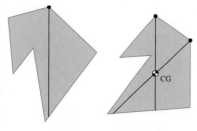

equilibrium, a vertical line is drawn downward from the support point. After drawing lines from several different support points, the center of gravity is the point where the lines all intersect. Explain how this works.

21. One of the effects of significant global warming would be the melting of part or all of the polar ice caps. This, in turn, would change the length of the day (the period of the Earth's rotation). Explain why. Would the day get longer or shorter?

Multiple-Choice Questions

1. A heavy box is resting on the floor. You would like to push the box to tip it over on its side, using the minimum force possible. Which of the force vectors in the diagram shows the correct location and direction of the force? The forces have equal horizontal components. Assume enough friction so that the box does not slide; instead it rotates about point P.

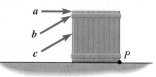

2. When both are expressed in terms of SI *base* units, torque has the same units as

   (a) angular acceleration   (b) angular momentum
   (c) force   (d) energy
   (e) rotational inertia   (f) angular velocity

Questions 3–4: A uniform solid cylinder rolls without slipping down an incline. At the bottom of the incline, the speed, v, of the cylinder is measured and the translational and rotational kinetic energies (K_{tr}, K_{rot}) are calculated. A hole is drilled through the cylinder along its axis and the experiment is repeated; at the bottom of the incline the cylinder now has speed v' and translational and rotational kinetic energies K'_{tr} and K'_{rot}.

3. How does the speed of the cylinder compare with its original value?

   (a) $v' < v$ (b) $v' = v$ (c) $v' > v$
   (d) Answer depends on the radius of the hole drilled.

4. How does the ratio of rotational to translational kinetic energy of the cylinder compare to its original value?

   (a) $\dfrac{K'_{rot}}{K'_{tr}} < \dfrac{K_{rot}}{K_{tr}}$ (b) $\dfrac{K'_{rot}}{K'_{tr}} = \dfrac{K_{rot}}{K_{tr}}$ (c) $\dfrac{K'_{rot}}{K'_{tr}} > \dfrac{K_{rot}}{K_{tr}}$

   (d) Answer depends on the radius of the hole drilled.

5. The SI units of angular momentum are

   (a) $\dfrac{rad}{s}$ (b) $\dfrac{rad}{s^2}$ (c) $\dfrac{kg \cdot m}{s^2}$

   (d) $\dfrac{kg \cdot m^2}{s^2}$ (e) $\dfrac{kg \cdot m^2}{s}$ (f) $\dfrac{kg \cdot m}{s}$

6. Which of the forces in the figure produces the largest magnitude torque about the rotation axis indicated?

   (a) 1   (b) 2   (c) 3   (d) 4

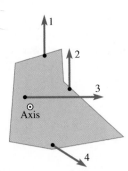

Multiple-Choice Questions 6–8

7. Which of the forces in the figure produces a CW torque about the rotation axis indicated?

   (a) 3 only   (b) 4 only   (c) 1 and 2
   (d) 1, 2, and 3   (e) 1, 2, and 4

8. Which pair of forces in the figure might produce equal magnitude torques with opposite signs?

   (a) 2 and 3   (b) 2 and 4   (c) 1 and 2
   (d) 1 and 3   (e) 1 and 4   (f) 3 and 4

9. A high diver in midair pulls her legs inward toward her chest in order to rotate faster. Doing so changes which of these quantities: her angular momentum L, her rotational inertia I, and her rotational kinetic energy K_{rot}?

   (a) L only   (b) I only   (c) K_{rot} only
   (d) L and I only   (e) I and K_{rot} only (f) all three

10. A uniform bar of mass m is supported by a pivot at its top, about which the bar can swing like a pendulum. If a force F is applied perpendicularly to the lower end of the bar as in the diagram, how big must F be in order to hold the bar in equilibrium at an angle θ from the vertical?

   (a) $2mg$ (b) $2mg \sin \theta$
   (c) $(mg/2) \sin \theta$ (d) $2mg \cos \theta$
   (e) $(mg/2) \cos \theta$ (f) $mg \sin \theta$

Problems

    ⚕   Combination conceptual/quantitative problem
    ⚕   Biological or medical application
    ✦   Challenging problem
Blue #   Detailed solution in the Student Solutions Manual
⒈ ⒉   Problems paired by concept
       Text website interactive or tutorial

8.1 Rotational Kinetic Energy and Rotational Inertia

1. Verify that $\frac{1}{2}I\omega^2$ has dimensions of energy.

2. What is the rotational inertia of a solid iron disk of mass 49 kg, with a thickness of 5.00 cm and radius of 20.0 cm, about an axis through its center and perpendicular to it?

3. A bowling ball made for a child has half the radius of an adult bowling ball. They are made of the same material (and therefore have the same mass *per unit volume*). By what factor is the (a) mass and (b) rotational inertia of the child's ball reduced compared with the adult ball?

4. Find the rotational inertia of the system of point particles shown in the figure assuming the system rotates about the (a) x-axis, (b) y-axis, (c) z-axis. The z-axis is perpendicular to the xy-plane and points out of the page. Point particle A has a mass of 200 g and is located at $(x, y, z) = (-3.0 \text{ cm}, 5.0 \text{ cm}, 0)$, point particle B has a mass of 300 g and is at (6.0 cm, 0, 0), and point particle C has a mass of 500 g and is at $(-5.0 \text{ cm}, -4.0 \text{ cm}, 0)$. (d) What are the x- and y-coordinates of the center of mass of the system?

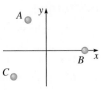

5. Four point masses of 3.0 kg each are arranged in a square on massless rods. The length of a side of the square is 0.50 m. What is the rotational inertia for rotation about an axis (a) passing through masses B and C? (b) passing through masses A and C? (c) passing through the center of the square and perpendicular to the plane of the square?

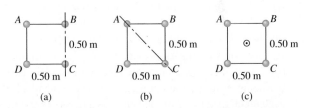

(a)   (b)   (c)

6. How much work is done by the motor in a CD player to make a CD spin, starting from rest? The CD has a diameter of 12.0 cm and a mass of 15.8 g. The laser scans at a constant tangential velocity of 1.20 m/s. Assume that the music is first detected at a radius of 20.0 mm from the center of the disk. Ignore the small circular hole at the CD's center.

7. Find the ratio of the rotational inertia of the Earth for rotation about its own axis to its rotational inertia for rotation about the Sun.

8. A bicycle has wheels of radius 0.32 m. Each wheel has a rotational inertia of 0.080 kg·m² about its axle. The total mass of the bicycle including the wheels and the rider is 79 kg. When coasting at constant speed, what fraction of the total kinetic energy of the bicycle (including rider) is the rotational kinetic energy of the wheels?

9. In many problems in previous chapters, cars and other objects that roll on wheels were considered to act as if they were sliding without friction. (a) Can the same assumption be made for a wheel rolling *by itself*? Explain your answer. (b) If a moving car of total mass 1300 kg has four wheels, each with rotational inertia of 0.705 kg·m² and radius of 35 cm, what fraction of the total kinetic energy is rotational?

10. A centrifuge has a rotational inertia of 6.5×10^{-3} kg·m². How much energy must be supplied to bring it from rest to 420 rad/s (4000 rpm)?

8.2 Torque

11. A mechanic turns a wrench using a force of 25 N at a distance of 16 cm from the rotation axis. The force is perpendicular to the wrench handle. What magnitude torque does she apply to the wrench?

12. The pull cord of a lawnmower engine is wound around a drum of radius 6.00 cm. While the cord is pulled with a force of 75 N to start the engine, what magnitude torque does the cord apply to the drum?

13. A child of mass 40.0 kg is sitting on a horizontal seesaw at a distance of 2.0 m from the supporting axis. What is the magnitude of the torque about the axis due to the weight of the child?

14. A 124-g mass is placed on one pan of a balance, at a point 25 cm from the support of the balance. What is the magnitude of the torque about the support exerted by the mass?

15. A uniform door weighs 50.0 N and is 1.0 m wide and 2.6 m high. What is the magnitude of the torque due to the door's own weight about a horizontal axis perpendicular to the door and passing through a corner?

16. A tower outside the Houses of Parliament in London has a famous clock commonly referred to as Big Ben, the name of its 13-ton chiming bell. The hour hand of each clock face is 2.7 m long and has a mass of 60.0 kg. Assume the hour hand to be a uniform rod attached at one end. (a) What is the torque on the clock mechanism due to the weight of one of the four hour hands when the clock strikes noon? The axis of rotation is perpendicular to a clock face and through the center of the clock. (b) What is the torque due to the weight of one hour hand about the same axis when the clock tolls 9:00 A.M.?

17. Any pair of equal and opposite forces acting on the same object is called a *couple*. Consider the couple in part (a) of the figure. The rotation axis is perpendicular to the page and passes through point *P*. (a) Show that the net torque due to this couple is equal to *Fd*, where *d* is the distance between the lines of action of the two forces. Because the distance *d* is independent of the location of the rotation axis, this shows that the torque is the same for any rotation axis. (b) Repeat for the couple in part (b) of the figure. Show that the torque is still

Fd if *d* is the *perpendicular* distance between the lines of action of the forces.

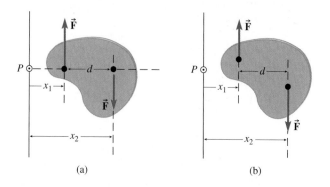

(a)   (b)

18. A 46.4-N force is applied to the outer edge of a door of width 1.26 m in such a way that it acts (a) perpendicular to the door, (b) at an angle of 43.0° with respect to the door surface, (c) so that the line of action of the force passes through the axis of the door hinges. Find the torque for these three cases.

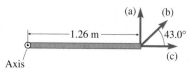

19. A trap door, of length and width 1.65 m, is held open at an angle of 65.0° with respect to the floor. A rope is attached to the raised edge of the door and fastened to the wall behind the door in such a position that the rope pulls perpendicularly to the trap door. If the mass of the trap door is 16.8 kg, what is the torque exerted on the trap door by the rope? (🕸 tutorial: deck hatch)

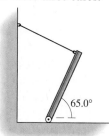

20. A weightless rod, 10.0 m long, supports three weights as shown. Where is its center of gravity?

5.0 kg   15.0 kg   10.0 kg

0.0   5.0 m   10.0 m

21. A door weighing 300.0 N measures 2.00 m × 3.00 m and is of uniform density; that is, the mass is uniformly distributed throughout the volume. A doorknob is attached to the door as shown. Where is the center of gravity if the doorknob weighs 5.0 N and is located 0.25 m from the edge?

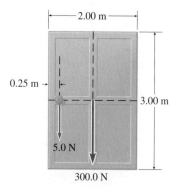

22. A plate of uniform thickness is shaped as shown. Where is the center of gravity? Assume the origin (0, 0) is

located at the lower left corner of the plate; the upper left corner is at $(0, s)$ and upper right corner is at (s, s).

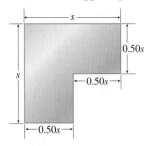

8.3 Calculating Work Done from the Torque

23. A stone used to grind wheat into flour is turned through 12 revolutions by a constant force of 20.0 N applied to the rim of a 10.0-cm-radius shaft connected to the wheel. How much work is done on the stone during the 12 revolutions?

24. The radius of a wheel is 0.500 m. A rope is wound around the outer rim of the wheel. The rope is pulled with a force of magnitude 5.00 N, unwinding the rope and making the wheel spin CCW about its central axis. Ignore the mass of the rope. (a) How much rope unwinds while the wheel makes 1.00 revolution? (b) How much work is done by the rope on the wheel during this time? (c) What is the torque on the wheel due to the rope? (d) What is the angular displacement $\Delta\theta$, in radians, of the wheel during 1.00 revolution? (e) Show that the numerical value of the work done is equal to the product $\tau\Delta\theta$.

✦25. A flywheel of mass 182 kg has an effective radius of 0.62 m (assume the mass is concentrated along a circumference located at the effective radius of the flywheel). (a) How much work is done to bring this wheel from rest to a speed of 120 rpm in a time interval of 30.0 s? (b) What is the applied torque on the flywheel (assumed constant)?

✦26. A Ferris wheel rotates because a motor exerts a torque on the wheel. The radius of the London Eye, a huge observation wheel on the banks of the Thames, is 67.5 m and its mass is 1.90×10^6 kg. The cruising angular speed of the wheel is 3.50×10^{-3} rad/s. (a) How much work does the motor need to do to bring the stationary wheel up to cruising speed? [*Hint:* Treat the wheel as a hoop.] (b) What is the torque (assumed constant) the motor needs to provide to the wheel if it takes 20.0 s to reach the cruising angular speed?

8.4 Rotational Equilibrium

27. A rod is being used as a lever as shown. The fulcrum is 1.2 m from the load and 2.4 m from the applied force. If the load has a mass of 20.0 kg, what force must be applied to lift the load?

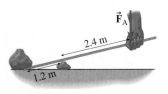

28. A weight of 1200 N rests on a lever at a point 0.50 m from a support. On the same side of the support, at a distance of 3.0 m from it, an upward force with magnitude F is applied. Ignore the weight of the board itself. If the system is in equilibrium, what is F?

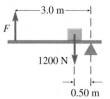

29. A sculpture is 4.00 m tall and has its center of gravity located 1.80 m above the center of its base. The base is a square with a side of 1.10 m. To what angle θ can the sculpture be tipped before it falls over? (tutorial: filing cabinet)

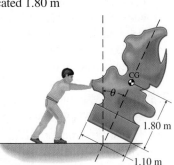

30. A house painter is standing on a uniform, horizontal platform that is held in equilibrium by two cables attached to supports on the roof. The painter has a mass of 75 kg and the mass of the platform is 20.0 kg. The distance from the left end of the platform to where the painter is standing is $d = 2.0$ m and the total length of the platform is 5.0 m. (a) How large is the force exerted by the left-hand cable on the platform? (b) How large is the force exerted by the right-hand cable?

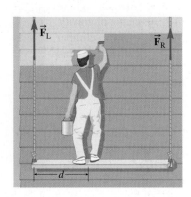

◎31. Four identical uniform metersticks are stacked on a table as shown. Where is the x-coordinate of the CM of the metersticks if the origin is chosen at the left end of the lowest stick? Why does the system balance?

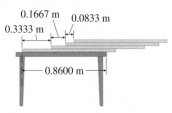

✦32. A uniform diving board, of length 5.0 m and mass 55 kg, is supported at two points; one support is located 3.4 m from the end of the board and the second is at 4.6 m from the end (see Fig. 8.19). What are the forces acting on the board due to the two supports when a diver of mass 65 kg stands at the end of the board over the water? Assume that these forces are vertical. (tutorial: plank) [*Hint:* In this problem, consider using two different torque equations about different rotation axes. This may help you determine the directions of the two forces.]

✦33. A house painter stands 3.0 m above the ground on a 5.0-m-long ladder that leans against the wall at a point

4.7 m above the ground. The painter weighs 680 N and the ladder weighs 120 N. Assuming no friction between the house and the upper end of the ladder, find the force of friction that the driveway exerts on the bottom of the ladder. (interactive: ladder; tutorial: ladder)

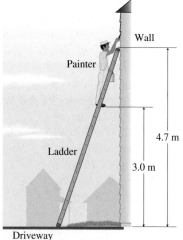

◆34. A mountain climber is rappelling down a vertical wall. The rope attaches to a buckle strapped to the climber's waist 15 cm to the right of his center of gravity. If the climber weighs 770 N, find (a) the tension in the rope and (b) the magnitude and direction of the contact force exerted by the wall on the climber's feet.

ⓒ35. A sign is supported by a uniform horizontal boom of length 3.00 m and weight 80.0 N. A cable, inclined at an angle of 35° with the boom, is attached at a distance of 2.38 m from the hinge at the wall. The weight of the sign is 120.0 N. What is the tension in the cable and what are the horizontal and vertical forces F_x and F_y exerted on the boom by the hinge? Comment on the magnitude of F_y.

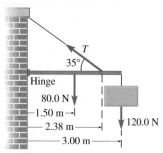

ⓒ36. A boom of mass m supports a steel girder of weight W hanging from its end. One end of the boom is hinged at the floor; a cable attaches to the other end of the boom and pulls horizontally on it. The boom makes an angle θ with the horizontal. Find the tension in the cable as a function of m, W, θ, and g. Comment on the tension at $\theta = 0$ and $\theta = 90°$.

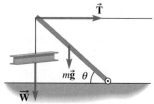

37. You are asked to hang a uniform beam and sign using a cable that has a breaking strength of 417 N. The store owner desires that it hang out over the sidewalk as shown. The sign has a weight of 200.0 N and the beam's weight is 50.0 N. The beam's length is 1.50 m and the sign's dimensions are 1.00 m horizontally × 0.80 m vertically. What is the minimum angle θ that you can have between the beam and cable?

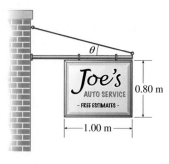

38. Refer to Problem 37. You chose an angle θ of 33.8°. An 8.7-kg cat has climbed onto the beam and is walking from the wall toward the point where the cable meets the beam. How far can the cat walk before the cable breaks?

39. A man is doing push-ups. He has a mass of 68 kg and his center of gravity is located at a horizontal distance of 0.70 m from his palms and 1.00 m from his feet. Find the forces exerted by the floor on his palms and feet.

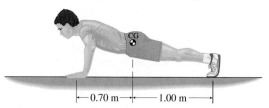

8.5 Equilibrium in the Human Body

▼40. Your friend balances a package with mass $m = 10$ kg on top of his head while standing. The mass of his upper body is $M = 55$ kg (about 65% of his total mass). Because the spine is vertical rather than horizontal, the force exerted by the sacrum on the spine ($\vec{\mathbf{F}}_s$ in Fig 8.32) is directed approximately straight up and the force exerted by the back muscles ($\vec{\mathbf{F}}_b$) is negligibly small. Find the magnitude of $\vec{\mathbf{F}}_s$.

▼41. Find the tension in the Achilles tendon and the force that the tibia exerts on the ankle joint when a person who weighs 750 N supports himself on the ball of one foot. The normal force $N = 750$ N pushes up on the ball of the foot on one side of the ankle joint, while the Achilles tendon pulls up on the foot on the other side of the joint.

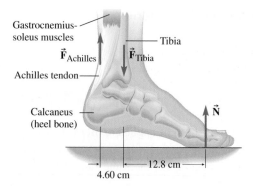

42. In the movie *Terminator*, Arnold Schwarzenegger lifts someone up by the neck and, with both arms fully extended and horizontal, holds the person off the ground. If the person being held weighs 700 N, is 60 cm from the shoulder joint, and Arnold has an anatomy analogous to that in Fig. 8.30, what force must *each* of the deltoid muscles exert to perform this task?

43. Find the force exerted by the biceps muscle in holding a 1-L milk carton (weight 9.9 N) with the forearm parallel to the floor. Assume that the hand is 35.0 cm from the elbow and that the upper arm is 30.0 cm long. The elbow is bent at a right angle and one tendon of the biceps is attached to the forearm at a position 5.00 cm from the elbow, while the other tendon is attached at 30.0 cm from the elbow. The weight of the forearm and empty hand is 18.0 N and the center of gravity of the forearm is at a distance of 16.5 cm from the elbow.

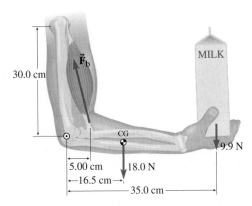

44. A person is doing leg lifts with 3.0-kg ankle weights. She is sitting in a chair with her legs bent at a right angle initially. The quadriceps muscles are attached to the patella via a tendon; the patella is connected to the tibia by the patellar tendon, which attaches to bone 10.0 cm below the knee joint. Assume that the tendon pulls at an angle of 20.0° *with respect to the lower leg*, regardless of the position of the lower leg. The lower leg has a mass of 5.0 kg and its center of gravity is 22 cm below the knee. The ankle weight is 41 cm from the knee. If the person lifts one leg, find the force exerted by the patellar tendon to hold the leg at an angle of (a) 30.0° and (b) 90.0° with respect to the vertical.

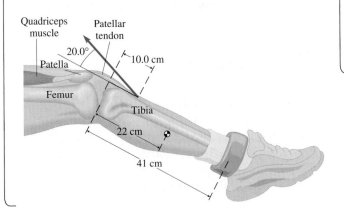

45. One day when your friend from Problem 40 is picking up a package, you notice that he bends at the waist to pick it up rather than keeping his back straight and bending his knees. You suspect that the lower back pain he complains about is caused by the large force on his lower vertebrae ($\vec{\mathbf{F}}_s$ in Fig. 8.32) when he lifts objects in this way. Suppose that when the spine is horizontal, the back muscles exert a force $\vec{\mathbf{F}}_b$ as in Fig. 8.32 (44 cm from the sacrum and at an angle of 12° to the horizontal). Assume that the CM of his upper body (including the arms) is at its geometric center, 38 cm from the sacrum. Find the horizontal component of $\vec{\mathbf{F}}_s$ when your friend is holding a 10-kg package at a distance of 76 cm from his sacrum. Compare this with the magnitude of $\vec{\mathbf{F}}_s$ found in Problem 40.

46. A man is trying to lift 60.0 kg off the floor by bending at the waist (see Fig. 8.32). Assume that the man's upper body weighs 455 N and the upper body's center of gravity is 38 cm from the sacrum (tailbone). (a) If, when bent over, the hands are a horizontal distance of 76 cm from the sacrum, what torque must be exerted by the erector spinae muscles to lift 60.0 kg off the floor? (The axis of rotation passes through the sacrum, as shown in Fig. 8.32.) (b) When bent over, the erector spinae muscles are a horizontal distance of 44 cm from the sacrum and act at a 12° angle above the horizontal. What force ($\vec{\mathbf{F}}_b$ in Fig. 8.32) do the erector spinae muscles need to exert to lift the weight? (c) What is the component of this force that compresses the spinal column?

8.6 Rotational Form of Newton's Second Law

47. Verify that the units of the rotational form of Newton's second law [Eq. (8-9)] are consistent. In other words, show that the product of a rotational inertia expressed in kg·m² and an angular acceleration expressed in rad/s² is a torque expressed in N·m.

48. A spinning flywheel has rotational inertia $I = 400.0$ kg·m². Its angular velocity decreases from 20.0 rad/s to zero in 300.0 s due to friction. What is the frictional torque acting?

49. A turntable must spin at 33.3 rpm (3.49 rad/s) to play an old-fashioned vinyl record. How much torque must the motor deliver if the turntable is to reach its final angular speed in 2.0 revolutions, starting from rest? The turntable is a uniform disk of diameter 30.5 cm and mass 0.22 kg.

50. A lawn sprinkler has three spouts that spray water, each 15.0 cm long. As the water is sprayed, the sprinkler turns around in a circle. The sprinkler has a total rotational inertia of

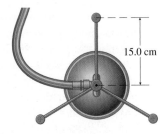

9.20×10^{-2} kg·m^2. If the sprinkler starts from rest and takes 3.20 s to reach its final speed of 2.2 rev/s, what force does each spout exert on the sprinkler?

51. A chain pulls tangentially on a 40.6-kg uniform cylindrical gear with a tension of 72.5 N. The chain is attached along the outside radius of the gear at 0.650 m from the axis of rotation. Starting from rest, the gear takes 1.70 s to reach its rotational speed of 1.35 rev/s. What is the total frictional torque opposing the rotation of the gear?

52. Four masses are arranged as shown. They are connected by rigid, massless rods of lengths 0.75 m and 0.50 m. What torque must be applied to cause an angular acceleration of 0.75 rad/s^2 about the axis shown?

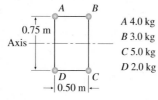

A 4.0 kg
B 3.0 kg
C 5.0 kg
D 2.0 kg

53. A bicycle wheel, of radius 0.30 m and mass 2 kg (concentrated on the rim), is rotating at 4.00 rev/s. After 50 s the wheel comes to a stop because of friction. What is the magnitude of the average torque due to frictional forces?

54. A playground merry-go-round (see Fig. 8.5), made in the shape of a solid disk, has a diameter of 2.50 m and a mass of 350.0 kg. Two children, each of mass 30.0 kg, sit on opposite sides at the edge of the platform. Approximate the children as point masses. (a) What torque is required to bring the merry-go-round from rest to 25 rpm in 20.0 s? (b) If two other bigger children are going to push on the merry-go-round rim to produce this acceleration, with what force magnitude must each child push? (tutorial: roundabout)

55. Two children standing on opposite sides of a merry-go-round (see Fig. 8.5) are trying to rotate it. They each push in opposite directions with forces of magnitude 10.0 N. (a) If the merry-go-round has a mass of 180 kg and a radius of 2.0 m, what is the angular acceleration of the merry-go-round? (Assume the merry-go-round is a uniform disk.) (b) How fast is the merry-go-round rotating after 4.0 s?

56. Refer to Atwood's machine (Example 8.2). (a) Assuming that the cord does not slip as it passes around the pulley, what is the relationship between the angular acceleration of the pulley (α) and the magnitude of the linear acceleration of the blocks (a)? (b) What is the net torque on the pulley about its axis of rotation in terms of the tensions T_1 and T_2 in the left and right sides of the cord? (c) Explain why the tensions cannot be equal if $m_1 \neq m_2$. (d) Apply Newton's second law to each of the blocks and Newton's second law for rotation to the pulley. Use these three equations to solve for a, T_1, and T_2. (e) Since the blocks move with constant acceleration, use the result of Example 8.2 along with the constant acceleration equation $v_{fy}^2 - v_{iy}^2 = 2a_y \Delta y$ to check your answer for a.

57. Derive the rotational form of Newton's second law as follows. Consider a rigid object that consists of a large number N of particles. Let F_i, m_i, and r_i represent the tangential component of the net force acting on the ith particle, the mass of that particle, and the particle's distance from the axis of rotation, respectively. (a) Use Newton's second law to find a_i, the particle's tangential acceleration. (b) Find the torque acting on this particle. (c) Replace a_i with an equivalent expression in terms of the angular acceleration α. (d) Sum the torques due to all the particles and show that

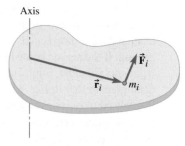

Axis

$$\sum_{i=1}^{N} \tau_i = I\alpha$$

8.7 The Motion of Rolling Objects

58. A solid sphere is rolling without slipping or sliding down a board that is tilted at an angle of 35° with respect to the horizontal. What is its acceleration?

59. A solid sphere is released from rest and allowed to roll down a board that has one end resting on the floor and is tilted at 30° with respect to the horizontal. If the sphere is released from a height of 60 cm above the floor, what is the sphere's speed when it reaches the lowest end of the board?

60. A hollow cylinder, a uniform solid sphere, and a uniform solid cylinder all have the same mass m. The three objects are rolling on a horizontal surface with identical translational speeds v. Find their total kinetic energies in terms of m and v and order them from smallest to largest.

61. A solid sphere of mass 0.600 kg rolls without slipping along a horizontal surface with a translational speed of 5.00 m/s. It comes to an incline that makes an angle of 30° with the horizontal surface. Ignoring energy losses due to friction, to what vertical height above the horizontal surface does the sphere rise on the incline?

62. A bucket of water with a mass of 2.0 kg is attached to a rope that is wound around a cylinder. The cylinder has a mass of 3.0 kg and is mounted horizontally on frictionless bearings. The bucket is released from rest. (a) Find its speed after it has fallen through a distance of 0.80 m. What are (b) the tension in the rope and (c) the acceleration of the bucket?

63. A 1.10-kg bucket is tied to a rope that is wrapped around a pole mounted horizontally on frictionless bearings. The cylindrical pole has a diameter of 0.340 m and a mass of 2.60 kg. When the

Problems 62 and 63